To Professor K. Venkatesan
with deep respects
N. Srinivasan

Biomolecular Forms and Functions

A Celebration of 50 years of the Ramachandran Map

IISc Press-WSPC Publication

BIOMOLECULAR FORMS AND FUNCTIONS

A Celebration of 50 years of the Ramachandran Map

Edited by

Manju Bansal

N Srinivasan

Molecular Biophysics Unit
Indian Institute of Science
Bangalore, India

IISc Press

World Scientific

NEW JERSEY • LONDON • SINGAPORE • BEIJING • SHANGHAI • HONG KONG • TAIPEI • CHENNAI

Published by

World Scientific Publishing Co. Pte. Ltd.
5 Toh Tuck Link, Singapore 596224
USA office: 27 Warren Street, Suite 401-402, Hackensack, NJ 07601
UK office: 57 Shelton Street, Covent Garden, London WC2H 9HE

British Library Cataloguing-in-Publication Data
A catalogue record for this book is available from the British Library.

BIOMOLECULAR FORMS AND FUNCTIONS
A Celebration of 50 years of the Ramachandran Map

ISBN 978-981-4449-13-7

Printed by FuIsland Offset Printing (S) Pte Ltd Singapore

PREFACE

The dream of J.D. Bernal and Dorothy Hodgkin in the year 1934, to study and visualize globular proteins at the atomic level using X-ray diffraction of protein crystals, had to wait for over 25 years, when Max Perutz and John Kendrew determined the three-dimensional structures of hemoglobin and myoglobin. In the mean time, the characterization of Pauling's single (α-) helix in fibrous proteins in 1951, Watson and Crick's double helical structure for DNA in 1953 and Ramachandran and Kartha's collagen triple helix in 1954, 1955, generated an expectation of highly regular structures for all biological macromolecules. Lack of overall symmetry noted for the first time when the myoglobin structure was determined was not only a surprise, but, also probably baffling at that time in understanding the structural diversity of proteins. Interestingly, the concerns regarding lack of information about possible conformations of polypeptide backbone was addressed quite effectively, almost at the same time as the determination of first structures of globular proteins, by the work of Ramachandran and coworkers in Madras (now, Chennai) on what became known subsequently as the "Ramachandran map". It was evident from this work that only about 30% of the total (φ-ψ) space is accessible ("allowed"). Although a molecular structure is stabilized by many different forces, consideration of the sole and simple concept of shortest possible distances between various non-bonded atoms, as found from an analysis of structures of small molecules, was shown to be sufficient to understand the extent of allowed conformational space. An even more interesting concept of "partially allowed" conformation was also introduced, indicating conformations that may be allowed under special conditions. This map, proposed in early 1960's by Ramachandran and co-workers, on the basis of fundamental principles of atomic sizes and around the same time as the first 3-D structure of any globular protein was determined, has proved to be a remarkably accurate representation of observed conformations for all the protein structures determined using X-ray diffraction and NMR analysis, with the number of such structures today in protein data bank reaching almost 85,000. This work serves as a source of inspiration, especially when one considers the limited resources available in Madras University. In fact, all the complex calculations necessary to create the Ramachandran map were manually performed using Clarks Tables and a Marchant calculator. This mechanical calculator was capable of performing only the 4 basic operations of addition, subtraction, multiplication and division, and had no memory storage! In addition one can only imagine the scientific seclusion of Ramachandran's school, from the Western world where much of the science was happening, during those days. Clearly, remarkable intuition and sheer determination to solve the problem of protein conformation, irrespective of the circumstances, fuelled the success.

Five decades later the branch of "Structural Biology" is witnessing a close integration with studies on various complex biological processes such as transcription, translation, signal transduction, immunology and metabolism. Its role in understanding the mechanistic basis of complex disease processes, as in diabetes, cancer, Alzheimers, TB etc and in the drug design process is undisputable. Determination of crystal structures of large macromolecular assemblies such as RNA polymerase and ribosome represent a huge leap forward in structural biology. Computational analyses of integrated datasets of three-dimensional structures and sequences add enormous value to genomic data which are accumulating at an unprecedented pace. Molecular dynamics and other simulation analyses continue to provide insights on time-dependent variability in biomolecular structures and interactions. Of course, modern structural biology has broadened in scope by inclusion of techniques such as atomic force microscopy, SAXS, FRET, SPR, ITC and mass spectrometry. Enormous technical advances made in crystallography, NMR, microscopy and computational resources, as well as development of new biophysical techniques have enabled structural biology to make both broad and deep impact in our understanding of biological phenomenon at the molecular and cellular level.

This volume is primarily a compilation of contributed articles representing the impact and application of Ramachandran's work on conformation analysis of biopolymers, as well as recent advances in macromolecular structural biology, involving diverse experimental and computational approaches. This volume is being published to coincide with a conference on "*Biomolecular Forms and Functions: A Celebration of 50 years of the Ramachandran Map*" being held at the Indian Institute of Science, Bangalore, in January 2013. Our endeavour in bringing out this volume will have served its purpose if young researchers are benefitted from the information put forth by the experts in the field of structural biology and its implications for biological function.

The culture of viewing teachers as "Gurus" (evolved guide held in highest esteem) is firmly embedded in the minds of most Indian children and they retain this mindset even during their time as a PhD student and beyond! GNR (as Professor G. N. Ramachandran is referred) was not only the Guru of one of us

(MB) but he is viewed as a very special Guru by the entire structural biology community, even in Today's India. This volume is our tribute to this outstanding Indian scientist, who did world class science based in a remote location, in the pre-computer and pre-internet era!

We are grateful to all the contributors for their enthusiastic response and excellent articles, as well as patience in undertaking multiple revisions. Critical assistance in formatting the manuscripts was essential and we thank Mr. Prasun Kumar, Mr. Naveen Kumar, Mr. Rakesh Ramachandran, Mr. Ashish Shelar, Mr. Gnanavel and Mr. G. Rajasekaran, for their timely and painstaking efforts. However any shortcomings that remain are entirely our responsibility.

September 2012

MANJU BANSAL
N. SRINIVASAN
Molecular Biophysics Unit
Indian Institute of Science
Bangalore 560012

CONTENTS

The Legacy of G. N. Ramachandran and the Development of Structural Biology in India 1
M. Vijayan

An Overview of Structural Studies of the Collagen Triple Helix 17
Rachel Kramer Green and Helen M. Berman

The Open-Ended Intellectual Legacy of GNR 38
George Rose

"The Plot" Thickens: More Data, More Dimensions, More Uses 46
Jane S. Richardson, Daniel A. Keedy and David C. Richardson

The Ramachandran Plot and Protein Structure Validation 62
Roman A. Laskowski, Nicholas Furnham and Janet M. Thornton

Non-Parametric Statistical Analysis of the Ramachandran Map 76
Maxim V. Shapovalov and Roland L. Dunbrack, Jr.

Non-Glycine Residues in the Disallowed Regions of the Ramachandran Map and their Neighborhood Preferences 95
Chanda R. Bhat, Moitrayee Bhattacharyya and Saraswathi Vishveshwara

Analysis of Dihedral Angle Variability in Related Protein Structures 107
Shekhar C. Mande, Ashwani Kumar and Payel Ghosh

Defining α-Helix Geometry by C^{α} Atom Trace *vs* (φ-ψ) Torsion Angles: A Comparative Analysis 116
Ashish Shelar, Prasun Kumar and Manju Bansal

Propensities of Aminoacid Residues in Proteins for Different Regions of the Ramachandran Map 128
B. Lakshmi, G. Archunan and N. Srinivasan

Stereochemistry of Polypeptide Conformation in Coarse Grained Analysis 136
Anil Korkut and Wayne A. Hendrickson

Multiprotein Assemblies: Modulating Cell Activity Through Targeting Protein-Protein Interfaces 148
Alicia P. Higueruelo, Harry Jubb and Tom L. Blundell

Prediction of Protein-Protein Complex Structures 160
Eiji Kanamori, Yoichi Murakami, Joy Sarmiento, Shide Liang, Daron M. Standley, Matsuyuki Shirota, Kengo Kinoshita, Yuko Tsuchiya, Junichi Higo and Haruki Nakamura

New Paradigms in Antibody Specificity: Structural Biology of Antigen Recognition by Germline Antibodies 173
Dinakar M. Salunke, Harmeet Kaur and Jasmita Gill

Proteomics, Protein Folding and Self-Association: New Insights from Recent Advances in Protein NMR Methodologies 184
Ramakrishna V. Hosur, Dinesh Kumar, Swagata Chakraborty, Manoj Kumar Rout, Jithender G. Reddy and Pushpa Mishra

Computational Approaches for Understanding the Recognition Mechanism of Protein Complexes 198
M. Michael Gromiha

Universalities in Protein Tertiary Structures: Some New Concepts 210
B. Jayaram, Aditya Mittal, Avinash Mishra, Chanchal Acharya and Garima Khandelwal

Cold-Shock Domains – Versatile Molecular Modules for Single-Stranded RNA Binding and Remodeling 220
Florian Mayr and Udo Heinemann

Insights into DNA Helical Transitions found in Protein-DNA Complexes 230
Andrew V. Colasanti, Yun Li, Gautam Singh, Xiang-Jun Lu and Wilma K. Olson

D-Amino Acids: Occurrence, Stereochemistry and Exclusion from Translational Machinery 247
Satya Brata Routh, Sadeem Ahmad and Rajan Sankaranarayanan

Induction of Folded Structures in Designed Peptides Using Conformationally Constrained Residues 264
S. Aravinda, N. Shamala and P. Balaram

Cyclic Beta-Amino Acids as Conformational Constraints 282
Gurunath Ramanathan and Anju Duley

Protein Disulfide Analysis and Design 296
Siddharth Patel, S. Indu, C. Ramakrishnan and Raghavan Varadarajan

Strongly Twisted and Coiled β-Hairpins and their Role in Protein Folding 312
Alexander V. Efimov

Identification and Conformational Analysis of Hinge Regions in Proteins that Undergo Domain Swapping 320
Prashant Shingate, Atul K. Upadhyay, U. S. Raghavender and R. Sowdhamini

Intrinsically Disordered Proteins: Revisiting the Structure-Function Paradigm 333
Meenakshi Anurag, Gajinder Pal Singh and Debasis Dash

Intrinsically Disordered Proteins: Regulation and Disease 346
M. Madan Babu, Robin van der Lee, Natalia Sanchez de Groot and Jörg Gsponer

Structural Basis for the Regulation of the T-Cell Tyrosine Kinase ZAP-70 362
Jonathan A. Winger, Patrick R. Visperas and John Kuriyan

Controlling Conformational Dynamics to Modulate Protein Function 378
Shahir S. Rizk, Marcin Paduch and Anthony A. Kossiakoff

Conformational Features of Sigma Factor/Anti-Sigma Complexes: 385
Anuja Joshi, Jinal K. Shukla, Ravi K. Jaiswal, Kapil Goutam and B. Gopal

Fold Type II Pyridoxal 5′-Phosphate Dependent Enzymes: Structure, Substrate Recognition and Catalysis 398
S. Bisht, S. R. Bharath, H. S. Savithri and M. R. N. Murthy

Intramolecular Isopeptide Bonds: Novel Post-Translational Modifications in Bacterial Pili and Cell-Surface Adhesins 417
Neil G. Paterson, Hae Joo Kang and Edward N. Baker

Structural Bioinformatics Approaches for Deciphering Biosynthetic Code of Secondary Metabolites 428
Shradha Khater and Debasisa Mohanty

Accelerated Molecular Dynamics: An Efficient Enhanced Sampling Method for Biomolecular Simulations 443
Levi Pierce, William Sinko and J. Andrew McCammon

Structural Dynamics of *E.coli* Porphobilinogen Deaminase During the Tetrapolymerisation of Porphobilinogen 455
Meenakshi Pradhan, Harini Srinivasan, Navneet Bung and Gopalakrishnan Bulusu

Plasticity of Hydrophobic Grooves in the Anti-Apoptotic MCL-1 and A1 Proteins 468
Vivek Modi, Dilraj Lama and Ramasubbu Sankararamakrishnan

Effectiveness of Fractal Dimension Based Measures to Investigate Protein Structures 482
Anirban Banerji and Indira Ghosh

The Inherent Structure Landscape of Met-Enkephalin Determined by the MOLS Technique 492
N. Balaji, L. Ramya and N. Gautham

THE LEGACY OF G.N. RAMACHANDRAN AND THE DEVELOPMENT OF STRUCTURAL BIOLOGY IN INDIA

M. VIJAYAN

Molecular Biophysics Unit, Indian Institute of Science, Bangalore-560012, India

G.N. Ramachandran is among the founding fathers of structural molecular biology. He made pioneering contributions in computational biology, modelling and what we now call bioinformatics. The triple helical coiled coil structure of collagen proposed by him forms the basis of much of collagen research at the molecular level. The Ramachandran map remains the simplest descriptor and tool for validation of protein structures. He has left his imprint on almost all aspects of biomolecular conformation. His contributions in the area of theoretical crystallography have been outstanding. His legacy has provided inspiration for the further development of structural biology in India. After a pause, computational biology and bioinformatics are in a resurgent phase. One of the two schools established by Ramachandran pioneered the development of macromolecular crystallography, which has now grown into an important component of modern biological research in India. Macromolecular NMR studies in the country are presently gathering momentum. Structural biology in India is now poised to again approach heights of the kind that Ramachandran conquered more than a generation ago.

1. Introduction

The nineteen fifties marked the beginning of the heroic age of structural molecular biology. Although perceptive leaders of science like J.D. Bernal and W.T. Astbury and their younger colleagues Dorothy (Crowfoot) Hodgkin and Max Perutz initiated biological macromolecular structural studies earlier than that, substantive results began to appear only in the fifties. The discovery of the α-helix and the β-sheet through simple modelling, published in 1951, constituted a major advance in structural studies on proteins. This was soon followed by the proposal of the double-helical model of DNA by Watson and Crick in 1953, a proposal that revolutionized modern biology. During the same period, Perutz and his colleagues demonstrated the feasibility of using the multiple isomorphous replacement for solving protein structures by X-ray crystallography. This eventually led to spectacular advances in macromolecular crystallography, which is central to structural biology and, indeed, modern biology as a whole. It was during this period when path-breaking results were emanating from leading centres like the MRC Laboratory at Cambridge, UK and CalTech, USA that G.N. Ramachandran, often known as GNR, burst onto the scene from a remote, comparatively obscure laboratory at Madras (now Chennai). GNR rose to become a pioneer and global leader in molecular modelling, computational biology and what we now call bioinformatics. He also made outstanding contributions to theoretical crystallography. Furthermore, his efforts provided much of the inspiration for the further development of structural biology in India, which now encompasses experimental approaches like crystallography and NMR as well.

2. The prelude: Bangalore to Madras

Educated at Ernakulam in Kerala and Trichy in Tamil Nadu, GNR joined the Physics Department of the Indian Institute of Science, Bangalore in 1942 as a research student of C.V. Raman. He obtained his M.Sc. and D.Sc. degrees working on the optics of heterogeneous media and X-ray topography and diffraction. He then proceeded to England in 1947 on an 1851 Exhibition Scholarship to work with W.A. Wooster in the Cavendish Laboratory at Cambridge, then headed by W.L. Bragg. The laboratory was then at the threshold of major advances in structural biology involving Max Perutz, John Kendrew and others with inspiration from Bragg. Ironically, there is no evidence of GNR, who himself subsequently became a pioneer in the area, having had any significant scientific interaction with them. He worked primarily on the determination of elastic constants using diffuse scattering, for which he was awarded the Ph.D. degree of the Cambridge University in 1949. He returned to the Physics department of the Indian Institute of Science during the same year and established an X-ray crystallography laboratory there with the help of Gopinath Kartha and others.

The legendary Lakshmanaswamy Mudaliar was then the Vice-Chancellor of the Madras University, one of the three oldest modern Universities of India. On his invitation, based on a recommendation of C.V. Raman, GNR joined the University in 1952 as the founder-head of the Department of Physics. That marked the beginning of an extraordinary productive period in Ramachandran's career. Mudaliar and Ramachandran had a unique relationship. Mudaliar provided all the administrative and financial support. Ramachandran's mandate was to produce world class science, unencumbered by excessive bureaucratic or financial constraints.

3. Collagen, Ramachandran map, other conformational studies

While Ramachandran was looking for important problems to work on in his newly established laboratory, it was J.D. Bernal, then on a visit to Madras, who suggested that he might look at the structure of collagen. The structures of the other two classes of fibrous proteins and indeed that of DNA, had been essentially solved. The structure of collagen remained unsolved, despite the efforts of many leading scientists of the time. Following the suggestion of Bernal, Ramachandran and his associate Kartha initiated fibre diffraction and modelling studies on collagen. Using all the relevant information available at

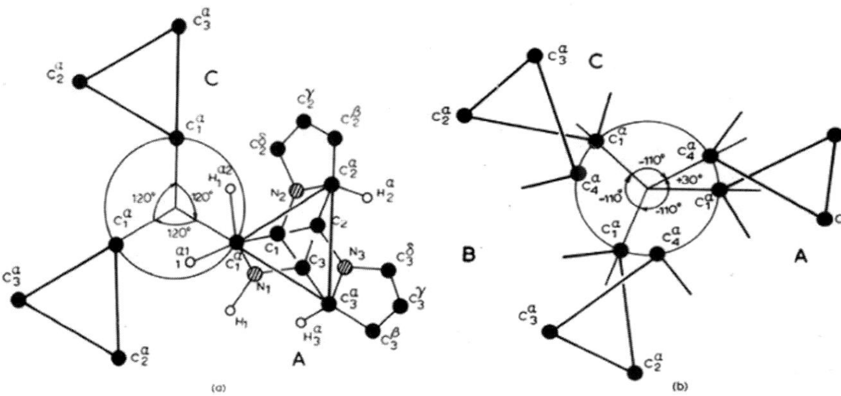

Figure 1(a): The first approximation to the triple helical model of collagen proposed by Ramachandran and Kartha. (b) The modified coiled-coil model. This figure and Figure 6 have been reproduced with permission from the doctoral thesis of Manju Bansal.[60]

that time, they proposed in 1954 a first approximation to their model consisting of three left-handed threefold helical polypeptide chains arranged at the apices of an equilateral triangle[1*] (Figure 1a). It was known that a third of collagen sequence is made up of glycines, the smallest of the 20 naturally occurring amino acids. From this information, Ramachandran and Kartha made the assumption that every third residue in the polypeptide chain was a glycine. In their model, the glycines occurred at the interface of the three helices. The model did not have any intra-chain hydrogen bonds. The hydrogen bonds were between chains. The proposed model, however, was not entirely compatible with the fiber pattern of collagen and was therefore modified[2]. The modified model, which was wholly compatible with the pattern, involved the coiling of the three helices, now with 3.3 residues per turn, around a common axis (Figure 1b). This coiled-coil model of collagen (Figure 2) has been widely accepted.

---

*Only selected representative publications during the period when Ramachandran was active and the immediate aftermath are cited in the text. Partial information on subsequent publications can be obtained from http://mbu.iisc.ernet.in/~mvlab/history.html

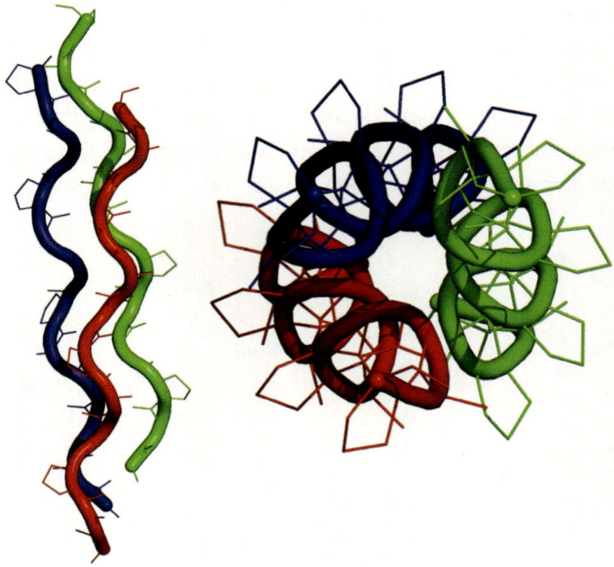

Figure 2: Perpendicular views of four turns of the collagen structure in cartoon representation. This figure and Figure 5 were generated using Pymol.[61]

The coiled-coil model of collagen, proposed by Ramachandran and Kartha, involved two hydrogen bonds per turn between adjacent chains. Francis Crick and Alex Rich suggested that only one such hydrogen bond is possible as two would lead to unacceptable steric contacts. The somewhat acrimonious debate on the two hydrogen bonds and the one hydrogen bond models led Ramachandran and his colleagues to critically examine the minimum interatomic distances between non-bonded atoms using all the available relevant crystal structures as the data base. It turned out that two non-bonded atoms can come substantially closer than the sum of their van der Waals radii. Based on the data, they arrived at two minimum limiting distances for each type of non- bonded interatomic distance, the normal limit that usually occurs and the extreme limit that is sometimes possible. As the main chain of a polypeptide can be considered as a series of planar peptide units hinged at C^{α} positions, they also realized that the conformation of the chain can be described by two rotation angles each, which we now call ϕ and ψ, about the single bonds that hinge at C^{α} positions (Figure 3). Using the limiting distances mentioned above, Ramachandran, Ramakrishnan and Sasisekharan worked out the allowed ranges of ϕ and ψ angles for an alanyl

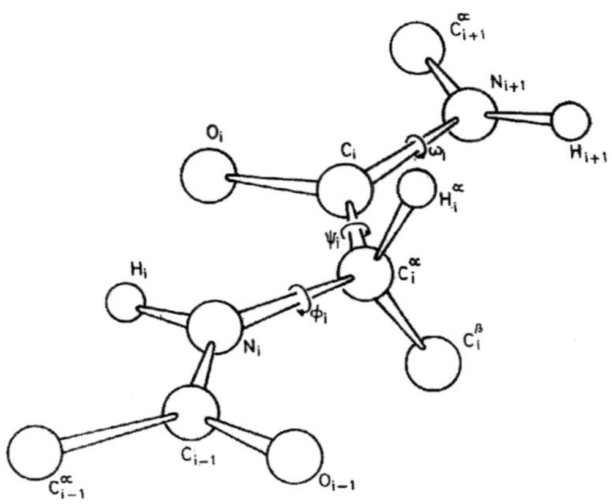

Figure 3: Two planar *trans* peptide units hinged at the ith α-carbon atom. ϕi and ψi are indicated. Reproduced with permission from ref. 62.

dipeptide. The two dimensional representation of the allowed ranges for the two torsion angles came to be known as the Ramachadran map[3]. The allowed regions are obviously much broader for the glycyl residue. The Ramachandran map, devised from first principles at the infancy of protein crystallography, remains the simplest descriptor and tool for validation of protein structures (Figure 4).

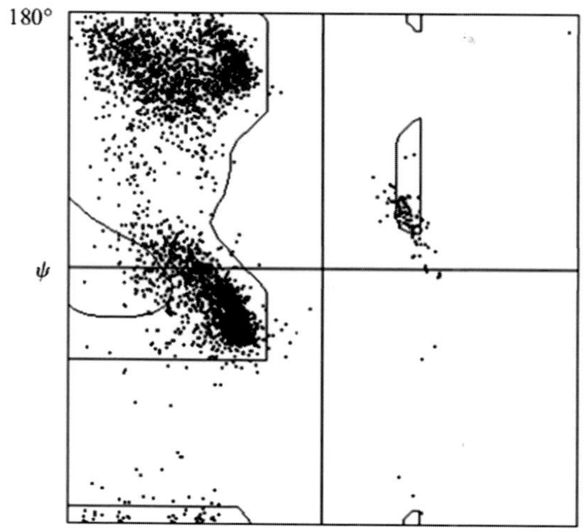

Figure 4: Ramachandran map generated from 21 selected high resolution structures. Open circles correspond to glycine residues. Reproduced with permission from ref. 63.

The computational efforts of Ramachandran at the Madras University encompassed many aspects of biomolecular conformation[4], although those on collagen and the Ramachandran map are the best known among them. He initiated computational studies on polysaccharide conformation[5]. His work on homopolypeptides is well known. It is interesting to note that he invoked C-H---O hydrogen bonds as early as in 1966 in a model of polyglycine[6] (Figure 5). His contributions to the understanding of hydrogen bonds, made in collaboration with R. Chidambaram are well-recognised.[7]

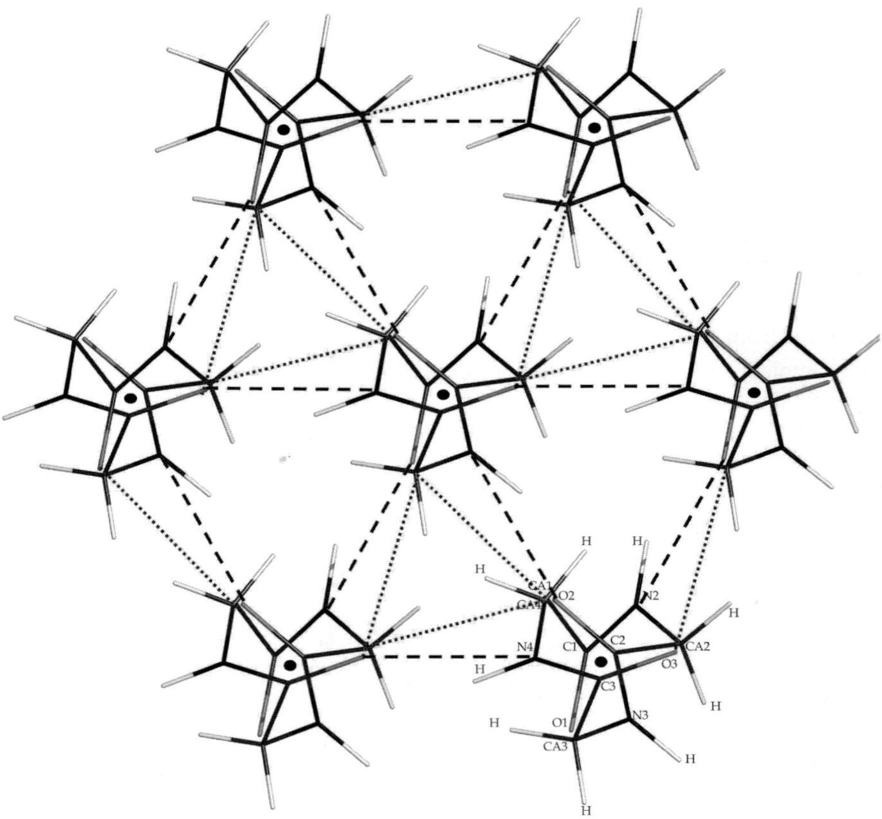

Figure 5: Model of the structure of polyglycine. N-H...O and C-H...O hydrogen bonds are represented by dashed and dotted lines respectively.

4. Crystallographic methodology

G.N. Ramachandran has been recognized as an outstanding crystallographer. He has many achievements to his credit in experimental crystallography, but his most important contributions have been in theoretical crystallography. Along with his colleagues, S. Raman and others, Ramachandran worked extensively on the use of anomalous dispersion in phase determination[8,9]. The methods developed by them have been used widely. He initiated and participated in, notably with R. Srinivasan and S. Parthasarathy, studies on crystallographic statistics[10,11]. The contributions of Ramachandran, Srinivasan and their colleagues concerned with Fourier transforms in crystallography have been outstanding[12,13]. In addition to applications, they illuminated the theoretical foundation of crystallography.

5. Back to Bangalore

Ramachandran found it difficult to adjust to the ambience that developed at the Madras University after the retirement of Lakshmanaswami Mudaliar. He left Madras in 1970 and joined the Indian institute of Science, Bangalore in 1971, after a year-long sojourn at Chicago. He established a vibrant school in the form of the Molecular Biophysics Unit at the Institute. His own personal research at Bangalore was by and large a continuation of the Madras efforts. During the transition period from Madras to Bangalore via Chicago, he completed some path breaking work on image reconstruction using convolutions[14]. Another issue he addressed was the non-planarity of the peptide group[15]. His work on the conformation of peptides containing L- and D- residues turned out to be of considerable importance in relation to peptide antibiotics[16]. As a fitting finale to his illustrious career in structural biology, which started with the proposal of the coiled coil structure of collagen, the last pieces of his original research in the area were primarily concerned with this fibrous protein. In particular, he and his last graduate student Manju Bansal worked on the role of hydroxyproline, the non-standard amino acid that occurs in collagen[17]. Its main role appeared to be the formation of a water bridge between adjacent chains (Figure 6).

This observation was interesting in relation to the earlier controversy on the number of hydrogen bonds between adjacent chains of the triple helix. It appeared that there could be one direct hydrogen bond and one water-mediated interaction between adjacent chains. Subsequently, many crystal structures incorporating collagen-like and natural collagen sequences became available. These structures lent further confirmation to the Ramachandran model of collagen, including water bridges.

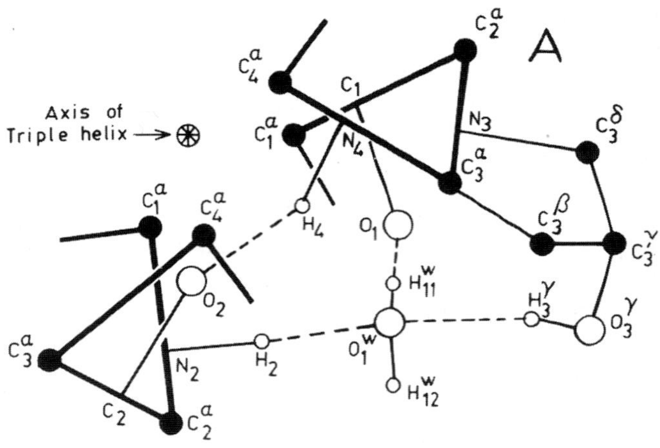

Figure 6: The proposed role of hydroxyproline in collagen involving a water bridge between adjacent helices.

6. The immediate aftermath

Many former students and colleagues of Ramachandran had already been established as leaders in the area even during the time he was active. They carried forward the efforts after the withdrawal of Ramachandran from the scene. A notable result to emerge during the late seventies and the early eighties was concerned with the variability in the structure of DNA. The concept of DNA as a monotonic helix was challenged by V. Sasisekharan and his colleagues[18-20]. They were the first to open up the debate on the structure of DNA, despite considerable resistance from many quarters. We now routinely discuss secondary structural features of DNA and their role in transcription. This aspect was also addressed as a follow up of theoretical investigations[21,22]. The computational work on DNA was carried forward by others[23-26]. The systematics of saccharide conformation were developed by V.S.R. Rao[27-29]. The work of C. Ramakrishnan on cyclic peptides[30] and hydrogen bonds[31] also deserve special mention. Indians were among the first to make use of the three-dimensional structures of proteins for bioinformatics analysis after the Protein Data Bank was established in the early seventies. Those who then effectively used the data included R. Srinivasan of the Madras University[32,33], to start with, among others[34-37]. Substantial progress was made during this phase in the crystallography of amino acids and peptides,[38-43] including the relevance of their supramolecular association to chemical evolution and origin of life[44], nucleic acids and their constituents[45], ionophores and related compounds[46,47] and medicinally important molecules[48-50].

The same is true about NMR studies on medium sized biomolecules[51-53]. The period also witnessed the beginnings of macromolecular crystallographic studies in India[54-59].

7. Computational efforts: a pause and the resurgence

By the mid eighties, the strength of computational biology in India apparently began to wane. By then the area became highly compute-intensive. India did not produce computers. Most of the time the country was under different kinds of sanctions and there were severe restrictions on supply of computers to India by developed countries. Inadequate computational facilities thus became a major handicap. Computational biology also became highly data based. Data-based operations thrive in an ambience where experimental data are produced in the same neighbourhood. India did not then produce much experimental data and this again proved to be a handicap. Happily, both these problems no longer exist. Computers are now universally available and a strong macromolecular crystallographic activity, spread over dozens of centres, has emerged in India. Macromolecular NMR studies are in a phase of expansion. Though late in starting, genome sequencing is also in progress at several centres of India. These developments have helped in a resurgence of computational biology and bioinformatics in the country.

The resurgence referred to above encompasses several groups and leaders. The work includes software development, construction of data bases and research pertaining to different biomolecules and their assemblies. The trend has been to carry out experimental and computational work in tandem, often by the same group. Much of the effort has been based on individual initiative, but conscious organized efforts also played a major role in the further development of the area. As an act of great foresight, a highly networked bioinformatics activity was initiated in 1987 by the Department of Biotechnology soon after the Department was established in 1986. This network continues to be the backbone of the efforts in the area in the country. Among the more recent initiatives, the coordinated activities at the Institute of Genomics and Integrative Biology (IGIB), particularly those related to the Open Source Drug Discovery (OSDD) programme of the Council of Scientific and Industrial Research (CSIR), merit special mention.

8. Macromolecular crystallography

India has a distinguished tradition in X-ray crystallography going back to the thirties of the last century. The theoretical contributions of G.N. Ramachandran from the fifties onwards have already been referred to. The Indian effort in small molecule crystallography encompassed a wide spectrum involving amino acids and peptides, nucleic acid components, medicinally important molecules, ionophores and related compounds, supramolecular assembly and peptide design. A few Indians have been involved in macromolecular crystallography projects abroad, notably Gopinath Kartha who led the structure solution of ribonuclease A. However, I was the first trained protein crystallographer to return to India in 1971, after participating in the structure solution of insulin in Dorothy Hodgkin's laboratory at Oxford. The general ambience, including the funding situation, was not propitious then for initiating macromolecular crystallography in the country. Therefore, the thrust of my early independent efforts at the Indian Institute of Science, Bangalore was on crystalline complexes involving amino acids and peptides, and their relevance to supramolecular association in general and chemical evolution and origin of life in particular. In the meantime, K.K. Kannan, who was involved in the structure solution of carbonic anhydrase at Uppsala, joined the Bhabha Atomic Research Centre, Bombay in 1978.

Macromolecular crystallography in India received the first major impetus when the Department of Science & Technology (DST) handsomely funded our group at Bangalore under their newly initiated Thrust Area Programme. The Bangalore centre was also mandated to function as a national nucleus for the development of the area in the country. Since then, work in the area has spread across the country and is now being pursued in about 30 institutions in India (Figure 7). In addition to DST, agencies like the Department of Biotechnology (DBT) and the Council of Scientific and Industrial Research (CSIR) have also been generously supporting the effort.

Much of the early efforts centered around lectins, protein hydration, plant viruses and carbonic anhydrase. In particular, the work on lectins, started at a time when the role of protein-sugar interactions in biological recognition was beginning to move into focus, has had considerable impact on the early development of macromolecular crystallography in the country, including manpower training. In addition to the areas mentioned above, the current activities in macromolecular crystallography in the country encompass several widely different systems. Particularly noteworthy is the work on proteins present in different body fluids. Proteins involved in inflammation and design of their inhibitors constitute an important area of study. Another effort involves exploration of molecular mimicry and its extension to fundamental immunological processes like antibody maturation. Proteases and their inhibitors

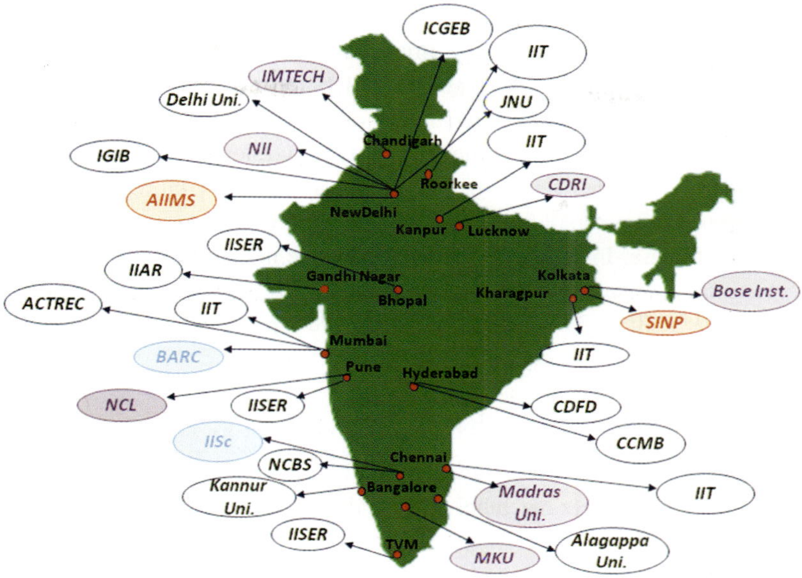

Figure 7: Institutions where macromolecular crystallographic studies were initiated before 1981 (sky blue), during 1981-91 (red), during 1991-2001 (purple) and from 2001 till date (black).

are studied in several laboratories. Another exciting systematic effort involves proteins engaged in editing during amino acid incorporation in peptide synthesis. Other systems being investigated include membrane proteins, ureases and transcription factors. Quite apart from the work on proteins, efforts are also underway to explore the sequence dependence of DNA structure.

Infectious diseases are of particular interest to a developing country like India. Therefore, a programme of structural biology of microbial pathogens was orchestrated around the turn of the country. The most successful component of the programme has been that on mycobacterial proteins, with special emphasis on TB proteins. As of now, ten institutions participate in the concerted structural effort on these proteins. The proteins studied include those involved in DNA replication, recombination, repair and modification; transcription and translation; amino acid synthesis, degradation and modification; fatty acid, mycolic acid and peptidoglycan synthesis; biosynthesis of cofactors, prosthetic groups and carriers; and signaling. Work is underway on chaperones/heat shock proteins and toxin-antitoxin systems as well. In fact, more than 10% of the *Mycobacterium tuberculosis* proteins of known structure have been X-ray analysed in Indian laboratories. Another bacterium, the proteins of which are being analysed in more than one laboratory is *Salmonella typhimurium*. Work on proteins from malarial parasites, particularly those involved in sexual development and fatty acid synthesis, is in progress in a couple of laboratories. Structural studies are underway on proteins form *Leishmania donovani* and *Entamoeba histolytica* as well. The work on plant viruses, though not directly relevant to human diseases, has provided a wealth of information of general interest on virus assembly. Another interesting piece of work currently underway involves a tethered mutant of the HIV protease. In addition to their importance in elucidating the biology of the concerned organisms, structural studies on proteins from microbial pathogens, particularly TB proteins, now provide a platform for structure-based inhibitor design in the country.

9. Macromolecular NMR studies

Ramachandran himself was involved in setting up a major Fourier transform NMR facility, a 270 MHz machine, at the Indian Institute of Science, Bangalore under the leadership of C.L. Khetrapal and Anil Kumar. Almost in parallel, serious NMR studies developed through the efforts of G. Govil and his colleagues at the Tata Institute of Fundamental Research, Mumbai as well. Both the groups have been deeply involved in the development of NMR methodology. Much of the early efforts centered around the two institutions have been on oligopeptides and oligonucleotides.

NMR facilities are now available in several institutions in the country. However, macromolecular structural studies are largely confined to the two centres mentioned above and the Central Drug Research Institute (CDRI), Lucknow. Such studies are now in the offing in a few other institutions. The proteins studied so far in India using NMR include calcium binding proteins, barstar, ubiquitin, toxins and proteins from mycobacteria, *Leishmania donovani* and the malarial parasite. An effort concerning the self-assembly process of GTPase effector domain of dynamin is also noteworthy. Although not directly relevant to structural biology, MRI and whole body NMR form a vibrant component of the NMR efforts in the country.

10. The legacy and the promise

Ramachandran has been among the global pioneers of structural molecular biology. The coiled-coil structure of collagen proposed by him, has stood the test of time and has formed the basis of much of subsequent collagen research. The Ramachandran map, widely used in protein structural research, has almost immortalized him. He has made seminal contributions to all aspects of biomolecular conformation. Ramachandran illuminated the theoretical foundations of crystallography through his work on Fourier transforms. His approach to the use of anomalous dispersion for phase determination combined theoretical insights and practical utility. It is now three and a half decades since Ramachandran stopped active research. Yet he remains an active presence in current scientific literature.

Ramachandran demonstrated that, with imagination and ingenuity, world class research can be done even when located in less endowed neighbourhoods. He is considered by many as the most distinguished scientist to have worked in independent India. Generations of scientists in India and, indeed, some abroad, have been inspired by his example. This inspiration, in no small measure, contributed to the further development of structural biology in the country. After a pause, research in computational biology and bioinformatics is now in a resurgent phase. Ramachandran was keen on initiating macromolecular crystallography in India, but available resources were not adequate when he was active. Subsequently, macromolecular crystallography grew into a major research area in the country, fittingly pioneered by a research school put together by him. Ramachandran contributed to the initiation of biological NMR studies in India. Macromolecular NMR investigations are now gathering momentum in the country. While Ramachandran and his colleagues concentrated on computational and theoretical approaches, structural biology in India today encompasses a wide spectrum involving experimental as well as computational investigations. It would not be perhaps too long before contributions with impact similar to those made by Ramachandran, begin to emanate from India again.

Acknowledgments

I thank K.V. Abhinav and Pankaja C. for help in preparing the figures and the text of the manuscript respectively and Manju Bansal for discussions.

References

1. G.N. Ramachandran and G. Kartha, *Nature* **174**, 269 (1954).
2. G.N. Ramachandran and G. Katha, *Nature* **176**, 593 (1955).
3. G.N. Ramachandran, V. Sasisekharan and C. Ramakrishnan, *J. Mol. Biol.* **7**, 95 (1963).
4. G.N. Ramachandran and V. Sasisekharan, *Adv. Prot. Chem.* **23**, 283 (1968).
5. G.N. Ramachandran and C. Ramakrishnan, *Biochim. Biophys. Acta* **63**, 307 (1962).
6. G.N. Ramachandran, V. Sasisekharan and C. Ramakrishnan, *Biochim. Biophys. Acta* **112**, 168 (1966).

7. R. Chidmabaram, R. Balasubramaniam and G.N. Ramachandran, *Biochim. Biophys. Acta* **221**, 196 (1970).
8. G.N. Ramachandran and S. Raman, Current *Science* **25**, 348 (1956).
9. S. Parthasarathy and G.N. Ramachandran, *Science* **150**, 212 (1965).
10. R. Srinivasan and G.N. Ramachandran, *Nature* **200**, 1090 (1963).
11. S. Parthasarathy and G.N. Ramachandran, *Acta Cryst.* **21**, 163 (1966).
12. G.N. Ramachandran and R. Srinivasan, *Nature* **190**, 159 (1961).
13. R. Srinivasan and G.N. Ramachandran, Fourier methods in Crystallography in Wiley Monographs in Crystallography, Wiley Interscience, (Ed. MJ Bueger) New York (1971).
14. G.N. Ramachandran and A.V. Lakshminarayanan, *Proc. Natl. Acad. of Sci. USA* **68**, 2236 (1971).
15. G.N. Ramachandran and A.S. Kolaskar, *Biochim. Biophys. Acta* **303**, 385 (1973).
16. R. Chandrasekharan, A.V. Lakshminarayanan, U.V. Pandya and G.N. Ramachandran, *Biochim. Biophys. Acta* **303**, 14 (1973).
17. G.N. Ramachandran, M. Bansal and R.S. Bhatnagar, *Biochim. Biophys. Acta* **322**, 166 (1973).
18. V. Sasisekharan, N. Pattabiraman, *Current Science* **45**, 779 (1976).
19. V. Sasisekharan, N. Pattabiraman, *Nature* **275**, 159 (1978).
20. G. Gupta, M. Bansal and V. Sasisekharan, *Proc. Natl. Acad. of Sci. USA* **77**, 6486 (1980).
21. M. Conrad, S.K. Brahmachari and V. Sasisekharan, *Biosystems* **19**, 123 (1986).
22. N. Ramesh, Y.S. Shouche and S.K. Brahmachari, *J. Mol. Biol.* **190**, 635 (1986).
23. S. Jayaraman and N. Yathindra, Biochem. Biophys. *Res. Commun* **97**, 1407 (1980).
24. R. Malathi and N. Yathindra, *J. Biomol. Stru. Dynamics* **3**, 127 (1985).
25. D. Bhattacharya and M. Bansal, *J. Biomol. Stru. Dynamics* **4**, 1027 (1987).
26. D. Bhattacharya and M. Bansal, *J. Biomol. Stru. Dynamics* **6**, 93 (1988).
27. N. Yathindra and V.S.R. Rao, *J. Polymer Science Part A-2: Polymer Physics* **10**, 1369 (1972).
28. M. Nagarajan and V.S.R. Rao, *Biopolymers* **18**, 1407 (1979).
29. M. Biswas, Y. Chandra Sekharudu, V.S.R. Rao, *Carbohydrate Research* **160**, 151 (1987).
30. C. Ramakrishnan, K. Ramnarayanan and G. Manjula, *Int. J. Peptide Protein Res.* **24**, 433 (1987).
31. J. Mitra and C. Ramakrishnan, *Int. J. Peptide Protein Res.* **4**, 401 (1981).
32. R. Srinivasan, R. Balasubramanian and S.S. Rajan, *J. Mol. Biol.***98**, 739 (1975).
33. R. Srinivasan, R. Balasubramanian and S.S. Rajan, *Science* **194**, 720 (1976).

34. T.N. Bhat, V. Sasisekharan and M. Vijayan, *Int. J. Peptide Protein Res.* **13**, 170 (1979).
35. K.V. Soman and C. Ramakrishnan, *J. Mol. Biol.* **170**, 1045 (1983).
36. A.S. Kolaskar and V. Ramabrahman, *Int. J. Peptide Protein Res.* **22**, 83-91 (1983).
37. R. Bhaskaran and P.K. Ponnuswamy, *Int. J. Peptide Protein Res.* **24**, 168 (1984).
38. D. Bhaduri and N.N. Saha, *J. Cryst. Mol. Struct.* **9**, 311 (1979).
39. R. Chidambaram, *Int. Jour. Qunatum Chem.* **20**, 449 (1981).
40. A. Sequiera, M. Ramanadham, H. Rajagopal and V.M. Padmababhan, *Acta Cryst.* **337**, 1839 (1981).
41. V. Lalitha, R. Murali, K. Subramanian, *Int. J. Peptide Protein Res.* **27**, 472 (1986).
42. I.L. Karle, Flippen-Anderson, Judith, K. Uma, P. Balaram, *Proc. Natl. Acad. of Sci. USA* **85**, 299 (1988).
43. P. Narula, H.C. Patel, T.P. Singh, V.S. Chauhan, A.K. Sharma, *Biopymers* **27**, 1595 (1988).
44. M. Vijayan, *Prog. Biophys. Mol. Biol.* **52**, 71 (1988).
45. M. A. Viswamitra, O. Kennard, W.B. Cruse, E. Egert, G.M. Sheldrick, P.G. Jones, M.J. Waring, L.P. Wakelin and R.K. Olsen, *Nature* **289**, 817 (1981).
46. S. Devarajan, C.M.K. Nair, K.R.K. Easwaran and M. Vijayan, *Nature* **286**, 640 (1980).
47. A.K. Francis, M. Iqbal, P. Balaram and M. Vijayan, *Biopolymers* **22**, 1499 (1983).
48. T.P. Singh and M. Vijayan, *J. Chem. Soc. Perkin II*, 693 (1977).
49. C. Chatterjee, J.K. Dattagupta and N.N. Saha, *Acta Cryst. B* **37**, 1835 (1981).
50. H.C. Patel and T.P. Singh, *Acta Cryst. C* **43**, 844 (1987).
51. M. Kumar, R.V. Hosur, K.B. Roy, H.T. Mules and G. Govil, *Biochemistry* **24**, 7703 (1985).
52. B.P. Shastri, M.B. Sankaran and K.R.K. Eswaran, *Biochemistry* **26**, 4930 (1987).
53. K. Chandrasekharan, M.K. Das, A. Kumar and P. Balaram, *Int. J. Peptide Protein Res.* **32**, 167 (1988).
54. D.M. Salunke, M. Islam Khan, A. Surolia and M. Vijayan, *J. Mol. Biol.* **154**, 177 (1982).
55. D.M. Salunke, B. Veerapandian, R. Kodandapani and M. Vijayan, *Acta Cryst. B* **41**, 431 (1985).
56. D.M. Salunke, M.J. Swamy, M. Islam Khan, S.C. Mande, A. Surolia and M. Vijayan, *J. Biol. Chem.* **260**, 13576 (1985).
57. V. Kumar, K. Sankaran and K.K. Kannan, *J. Mol. Biol.* **190**, 129 (1986).

58. S.K. Munshi, C.N. Hiremath, H.S. Savithri and M.R.N. Murthy, *Acta Cryst. B* **43**, 376 (1987).
59. V. Dhanaraj, S.R. Patanjali, A. Surolia and M. Vijayan, *J. Mol. Biol.* **203**, 1135 (1988).
60. M. Bansal, Ph.D. Thesis, Indian Institute of Science, Bangalore (1977).
61. W.L. DeLano, http://www.pymol.org (2002).
62. C. Ramakrishnan, *Resonance*, pp 48-56, October 2001.
63. M Vijayan and L.N. Johnson, *Biogr. Mem. Fell. R. Soc.* **51**, 367 (2005).

AN OVERVIEW OF STRUCTURAL STUDIES OF THE COLLAGEN TRIPLE HELIX

RACHEL KRAMER GREEN AND HELEN M. BERMAN
Department of Chemistry and Chemical Biology
Rutgers, The State University of New Jersey
174 Frelinghuysen Road, Piscataway, NJ 08854 USA
E-mail: berman@rcsb.rutgers.edu

Collagen, a protein found ubiquitously in vertebrate and invertebrate species, has a triple-helical structure that has been historically elucidated through fiber studies. More recently, single crystal X-ray experiments have corroborated, expanded upon, and clarified what was known from the earlier studies. Here we offer a survey of the triple-helical entries present in the Protein Data Bank (PDB) and a discussion of how these studies have provided new insight into the hydrogen bonding patterns, atomic structure, hydration and helical symmetry of the collagen molecule.

1. Introduction

Collagen is a triple-helical protein found in a large variety of vertebrate and invertebrate species and is one of the key components of the extracellular matrix. It is present in bone, skin, cartilage, tendon, cornea, and basement membranes; as much as half of an organism's protein may be collagenous. Collagen serves primarily as a structural protein, but its importance in development and physiology are becoming increasingly evident. Currently, there are known to be many genetically distinct types of collagen[1,2] with a tissue-specific distribution among these types. Collagen has the ability to self-assemble to form higher order structures that function only through their interaction with other molecules. The different types of collagen can be divided into two groups. One group, the fibril-forming family of collagen includes types I, II, III, V, and XI. These collagens form characteristic fibrils with an axial period of 67 nm (or D-period). The fibril-forming collagens have a high sequence homology and may be either homo- or heterotrimers. The second category of collagens is the nonfibrillar family of collagens whose members form a variety of different supermolecular structures. Of all the collagen types, type I is the most ubiquitous.

The triple-helical motif is also found as parts of other proteins such as the serum complement protein (C1q), the macrophage scavenger receptor, various collectins, the squirrel "hibernation protein," and the asymmetric form of acetylcholinesterase.[3]

The collagen molecule is known to be a triple-helical coiled-coil, in which each of three strands has a left-handed, extended polyproline II helical conformation. Each of these three strands then wraps around a common helical axis in a right-handed fashion. The three strands are held together with inter-chain hydrogen bonds in the Rich and Crick collagen II pattern.[4] Each strand of the triple-helix is composed of repeating Gly-X-Y triplets. The requirement for glycine in every third position results from a close packing of the helix that places the glycine Cα directed toward the center of the helix in such a way that any other residue would be too stereochemically bulky to fit. The X and Y positions are frequently occupied by the imino acids proline and hydroxyproline, respectively. Hydroxyproline is formed by the post-translational modification of proline by prolyl hydroxylase, which places a hydroxyl group at the 4 position. In general, only prolines on the amino side of a glycine are hydroxylated. In the collagen literature, the letter "O" is used as the one letter code for hydroxyproline.

Here we offer a survey of the triple-helical entries present in the Protein Data Bank (PDB)[5,6] and a discussion of how these studies have confirmed, clarified, and expanded on what was proposed in the early years.

2. Fiber Studies

Much of what is currently known about the structure of the collagen molecule is the result of fiber diffraction studies of actual collagen and collagen-like peptides, as the crystallization of collagen itself has generally proven to be difficult. Table 1 provides an overview of the important milestones in the study of collagen and collagen-like peptide structure.

Early X-ray patterns of collagen showed that the backbone was involved in some sort of coiling or folding, but many years passed and many attempts were made to define the structure before a comprehensive model was elucidated in the mid 1950's which would entirely fit the X-ray pattern.

By the mid to late 1940's it was recognized that glycine accounted for one third of the residues in collagen and that a high percentage of the rest were the imino acids proline and hydroxyproline. In 1954, Ramachandran and Kartha proposed a triple-helical structure for collagen based on their X-ray studies of various animal tendons. Their model accounted for the high glycine content by placing one at every third position. In this model, three equivalent left-handed

Table 1: Historical timeline of collagen and collagen like peptide structural studies.

Year	Discovery
1954, 1955	Triple-helical structure for collagen proposed with glycine at every third position and containing two N–H•••O=C hydrogen bonds for every three residues.[7,8]
1955, 1961	Collagen I and II model proposed with one hydrogen bond per tripeptide; collagen II reported to be the better of the two.[4,9]
1966	Unit cell parameters determined for $(Gly\text{-}Pro\text{-}Pro)_n$; collagen II found to be the only model with reasonable intermolecular contacts that could also accommodate an imino acid in the Y position.[10,11]
1968	Collagen-like oligopeptides created with defined and uniform length.[12]
1968	Proposal of additional water-mediated interchain hydrogen bond in imino acid-poor regions.[13]
1972	First single crystals of a collagen-like synthetic peptide, $(Pro\text{-}Pro\text{-}Gly)_{10}$.[14]
1976	Collagen II model confirmed through theoretical studies.[15]
1979	Collagen II model refined against X-ray fiber diffraction intensities.[16]
1981	Crystal structure of $(Pro\text{-}Pro\text{-}Gly)_{10}$ determined.[17]
1994	First atomic resolution crystal structure of a fully crystalline collagen-like peptide.[18]
1998-2012	Large number of structural determinations of model peptides.[19-37]
2000-2012	Several structures of collagen-like triple helices bound to globular molecules published.[38-45]

helical chains were placed in parallel and were connected by N–H•••O=C hydrogen bonds.[7] In 1955, Ramachandran and Kartha revised their model to better account for the X-ray data by including a non-integral screw axis.[8] This condition was satisfied by winding the three left-handed helices around a central axis in a right-handed manner and staggering the three chains with respect to each other by one residue to form a triple-helical coiled-coil conformation. This structure had two N–H•••O=C hydrogen bonds for every three residues.

Soon afterwards however, Rich and Crick suggested that the model should contain only one hydrogen bond per tripeptide and proposed two models that, while similar to that of Ramachandran and Kartha, were purported to be stereochemically improved.[9] In 1961, Rich and Crick reported that their second model "collagen II" was the better of the two.[4] Many years of debate over the relative merits of the two group's structures ensued, and it was only after various studies of the structures of collagen-like polypeptides that Rich and Crick's collagen II was generally accepted to be correct (Figure 1). Rich and Crick's model was later refined against X-ray fiber diffraction intensities by Fraser, MacRae, and Suzuki.[16] In this, as with other fiber models, each tripeptide in a helix can be generated from a single tripeptide by applying a specific rotation and a specific translation.

During this same period, investigations were also made into the fiber diffraction structures of synthetic polypeptides that contained large proportions of glycine, proline and hydroxyproline. The similarity of the results to those from fiber studies of actual collagen confirmed the utility of peptides as good models for collagen.

In their efforts to understand the structure of collagen, in 1955, Cowan and McGavin proposed a structure for what they believed to be the closely related poly-L-proline.[46] In this structure, each chain folded as a left-handed helix with a three-fold screw axis. This conformation has been termed polyproline II. Investigation of this structure led Cowan *et al.* to suggest that while the overall configuration of the backbone in the model proposed by Ramachandran and Kartha for collagen in 1955[8] was correct, various details were not. Also in 1955, Crick and Rich proposed a structure of polyglycine,[47] and it was from this structure that they were able to derive their two models, collagen I and collagen II.

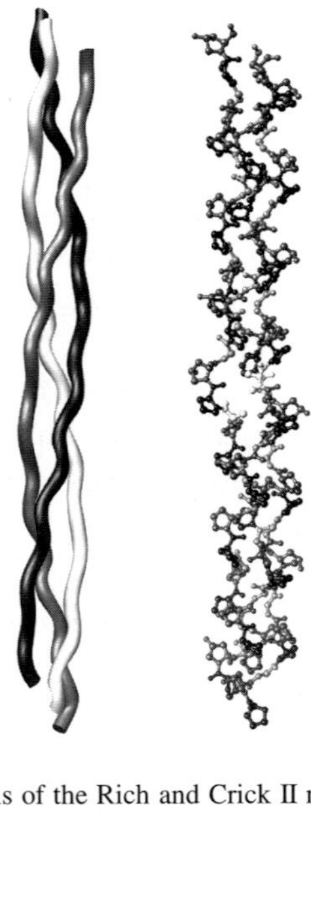

Figure 1: View along the helical axis of the Rich and Crick II model[4] (left) and Gly->Ala structure 1CAG (right).[18]

During the next decade, the fiber patterns of various polypeptides were investigated; $(Pro\text{-}Pro\text{-}Gly)_n$ was one of the most systematically and extensively analyzed. In 1965, Shibnev *et al.* obtained an X-ray pattern of poly-(Gly-Pro-Pro) and noted that the pattern was similar to that of collagen.[48] The structure was further elucidated by Yonath and Traub in 1966[10] and 1969.[11] Because they were working with a specific sequence and not a consensus sequence, they were able to test many possible conformations against X-ray intensities rather than having to compare the intensities against an approximate, average composition of actual collagen. In addition, they noted that the molecules of $(Pro\text{-}Pro\text{-}Gly)_n$ molecules packed in a more regular fashion than those of collagen. As a result they were able to determine unit-cell parameters for the first time, and investigate conformations that would be plausible in terms of packing. They found that Rich and Crick's collagen II was the only model that had reasonable intermolecular contacts and could accommodate an imino acid in the Y position, and at the same time could agree with the observed X-ray data. The collagen II model was further confirmed in 1976 by Scheraga and coworkers[15] through theoretical studies on poly(Pro-Pro-Gly) in which they demonstrated that this was the conformation lowest in potential energy.

3. Model Peptide Studies

In 1968, using solid-phase synthesis, Sakakibara was able to create collagen-like oligopeptides that were of defined and uniform length, that were therefore suitable for single crystal studies.[12] Then in 1972, Sakakibara obtained the first single crystals of a collagen-like synthetic peptide, $(Pro\text{-}Pro\text{-}Gly)_{10}$.[14] In 1981, Okuyama and coworkers determined the crystal structure and refined it to 2.2 Å resolution.[17]

In 1994, Bella *et al.* determined the first atomic resolution structure of a collagen-like peptide (Figure 1). This structure, termed Gly->Ala, is a designed peptide with sequence $(Pro\text{-}Hyp\text{-}Gly)_4$-Pro-Hyp-Ala-$(Pro\text{-}Hyp\text{-}Gly)_5$ (PDB IDs 1CAG and 1CGD).[18,49] This structure displays regular triple-helical conformation at the Pro-Hyp-Gly repeats at both ends and a bulging in the center of the molecule where one alanine in each chain is substituted for a glycine. This results in the untwisting of one end of the overall 7-fold triple helix with respect to the other.

In subsequent years, additional crystal structures of imino acid-rich collagen-like peptides (Table 2a) have corroborated, expanded upon and clarified what was known from fiber models, and provided new insight into hydrogen bonding patterns, atomic structure, hydration, and helical symmetry.

Host-guest models of collagen-like peptides (Table 2b), where a guest triplet is inserted in the center region of each chain of a stable imino acid-rich host peptide, have provided information on the conformation of specific residues in imino acid-rich environments.

Table 2: Collagen Structures in PDB.

a) Imino acid-rich collagen-like peptide entries

PDB ID	Sequence	Resolution	Helical Properties
1A3I, 1A3J[19]	$(PPG)_{10}$	1.97 Å 1.60 Å	7-fold infinite triple helix[b]
1G9W[21]	$(PPG)_{10}$	1.30 Å	7-fold infinite triple helix[b]
1ITT[22]	$(PPG)_9$	1.00 Å	7-fold infinite triple helix[b]
1K6F[23]	$(PPG)_{10}$	1.30 Å	7-fold triple helix
1V4F, 1V6Q, 1V7H[24]	$(PPG)_{11}$ $(PPG)_{11}$ $(PPG)_{10}$	1.26 Å 1.25 Å 1.25 Å	7-fold infinite triple helix[b]
1WZB[25]	$(OOG)_{10}$	1.50 Å	7-fold triple helix
1YM8[27]	$(GOO)_{10}$	1.55 Å	7-fold triple helix
2CUO[28]	$(PPG)_9$	1.33 Å	7-fold triple helix
1EI8[20]	$(POG)_4PG(POG)_5$	2.00 Å	7-fold triple helix with local disruption
3IPN[50]	$(mepFlpgly)_7$ mep = (2S, 4R)-4-methylproline Flp = (2S, 4R)-4-fluoroproline	1.21 Å	7-fold triple helix
3AH9	$(PPG)_9$	1.08 Å	7-fold triple helix
3AI6	$(PPG)_9$	1.10 Å	7-fold triple helix Prolines are D-enantiomer
3B0S[51]	$(GPO)_9$	1.45 Å	7-fold triple helix

[a]O is (2S,4R)-4-hydroxyproline (HYP); *indicates that experimental terminal groups (NH_2, ACE) have been left off of the sequence description; 3 identical chains are listed once in the sequence; non-identical chains are separated with a semi-colon; [b]"Infinite triple helix" indicates the asymmetric unit is shorter along the helical axis than the entire peptide.

b) Host-guest collagen-like peptide entries

PDB ID	Sequence	Resolution	Helical Properties
1CAG, 1CGD[18, 49]	$(POG)_4POA(POG)_5$	1.85 Å 1.85 Å	7-fold triple helix
1BKV[53]	$(POG)_3ITGARGLAG(POG)_4$	2.00 Å	7-fold symmetry at ends; 10-fold in the center
1QSU[54]	$(POG)_4EKG(POG)_5$	1.75 Å	7-fold triple helix
1X1K[26]	$(PPG)_4$(Pro-4(*S*)Hyp-Gly)$(PPG)_4$	1.10 Å	7-fold triple helix
2G66[30]	$(GPO)_3$(Gly-3(*S*)Hyp-Hyp)$_2(GPO)_4G$	1.80 Å	7-fold triple helix
2DRT, 2DRX[55]	$(POG)_4LOG(POG)_5$ $(POG)_4\ (LOG)_2(POG)_4$	1.60 Å 1.40 Å	7-fold triple helix Relaxed helical parameters in central region of 2DRX
2D3F, 2D3H[29]	$(PPG)_4POG(PPG)_4$ $(PPG)_4OOG(PPG)_4$	1.26 Å 1.22 Å	7-fold triple helix
3A08, 3A19[56]	$(PPG)_4OOG(PPG)_4$ $(PPG)_4OOG(PPG)_4$	1.22 Å 1.55 Å	7-fold triple helix
3A0A	$(PPG)_4OPG(PPG)_4$	1.36 Å	7-fold triple helix
3ABN	$(PPG)_4ODG(PPG)_4$	1.02 Å	7-fold triple helix
3B2C[57]	$(POG)_4$(4(*S*)Hyp-Pro-Gly)$_2(POG)_4$	1.36 Å	7-fold triple helix
3A0M, 3A1H, 3ADM[31]	$(PPG)_4OVG(PPG)_4$ $(PPG)_4OTG(PPG)_4$ $(PPG)_4OSG(PPG)_4$	1.02 Å 1.08 Å 1.18 Å	7-fold triple helix
3P46[32]	*$(GPO)_2GLOGEA(GPO)_2$*	1.70 Å	[c]
3T4F, 3U29[33]	$(POG)_3PKGEOG(POG)_3$ $(POG)_3PKGDOG(POG)_3$	1.68 Å 2.00 Å	7-fold triple helix

[c]Structural analysis of helical properties is not available in reference cited.

c) Other collagen-like peptide entries

PDB ID	Sequence	Category	Resolution	Helical Properties
1NAY[58]	$(GPP)_9$GSGYIPEAP RDGQAYVRKDGE WVLLSTFL	chimera	2.60 Å	7-fold triple helix
3HQV, 3HR2[59]	Extended	fiber	5.16 Å	
3DMW[60]	GPIGPOGPRGNR GERGSEGSOGHO G#MOGPOGPOGA OGPCCGG #M = selenomethione	natural sequence	2.30 Å	10-fold helix in central non-imino acid containing region; 7-fold helix in C-terminal and N-terminal imino acid-rich regions
2KLW[61]	$(PKG)_{10}$; $(DOG)_{10}$; $(POG)_{10}$*	heterotrimer	NMR	Heterotrimer Helical twist closer to 7-fold than 10-fold
2LLP[62]	PPGPQGIAGQRGV VGLPG	natural sequence	NMR	c

d) Collagen-like peptide entries involved in binding

PDB ID	Sequence	Bound Protein
1DZI[38]	$(GPO)_2$GFPGER$(GPO)_3$*	Integrin
1Q7D[39]	$(GPO)_2$GFOGER$(GPO)_3$PO*	Un-bound collagen companion of 1DZI
2V53[41]	$(GPO)_3$GPSGPRGQOGVMGFOGPKGPOGAO	Sparc
2WUH[42]	*$(GPO)_3$GPRGQOGVL^GFO$(GPO)_2$G* L^ = NLE = norleucine	Discoidin Domain Receptor 2
4DMU[45]	*$(POG)_3$GPRGQOGVMGFOG$(POG)_2$PO*	von Willebrand factor
4DMT[45]	*$(POG)_3$GPRGQOGVMGFOG$(POG)_2$PO*	Un-bound collagen companion of 4DMU
2Y5T[43]	$(GPO)_5$GARGLTGROGDAG$(POG)_2$; $(GPO)_5$GARGLTGROGDAG$(POG)_2$; $(GPO)_5$GARGLTGROGDAG$(POG)_2$KKYG	Pathogenic Autoantibody CIIC1
2F6A[40]	$(GPO)_4$GPRGRT$(GPO)_4$	*Staphylococcus aureus* CNA
3POB[44]	*$(GPO)_4$GKL$(GPO)_4$*	Mannan-binding lectin serine protease 1
3PON[44]	*$(GPO)_4$GKL$(GPO)_4$*	Un-bound collagen companion of 3POB
3POD[44]	*$(GPP)_3$GKL$(GPO)_4$*	Un-bound collagen companion of 3POB

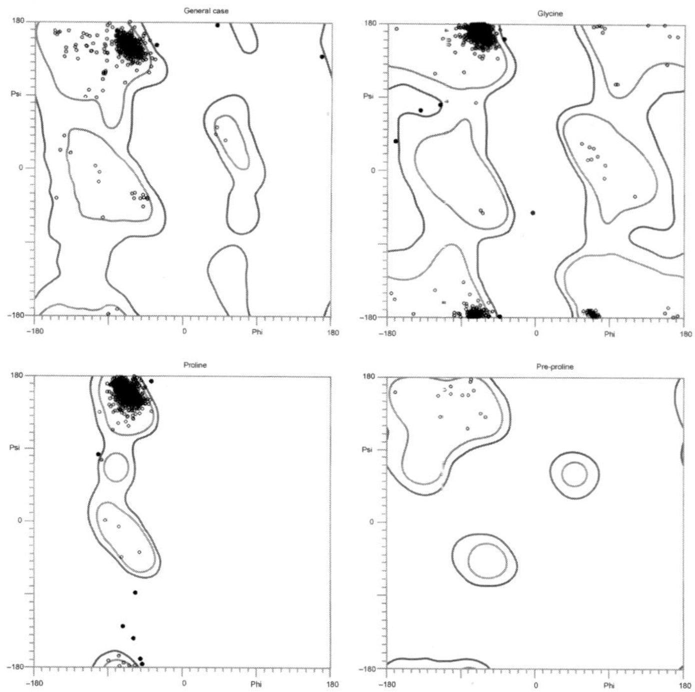

Figure 2: Superimposed Ramachandran plots[52] of the 52 triple-helical collagen-like entries found in the PDB (two additional entries from table 1 contain only a CA trace). 41 of these entries contain hydroxyproline. While this figure does not include results for these 1413 hydroxyprolines, we have checked these residues against the reference geometry of proline (plot not shown), and found only 0.5% of these to be outside of the allowed region. For structures from NMR ensemble entries, the plot includes results for the first model in the entry only. These plots were created using Molprobity.[63]

In general, these studies have confirmed the basic structure of the collagen triple-helix wherein three chains are held together in parallel with hydrogen bonds in the Rich and Crick II pattern[4] between glycyl N-H groups and C=O groups of the prolines in the X position of the neighboring chain.

An inspection of the Ramachandran plots[52] of the collagen-like peptides across the PDB archive indicates that these structures occupy a consistent conformational space (Figure 2).

3.1. *Helical twist*

Given the extended nature of the rod-like collagen molecule, and the need for collagen molecules to self-assemble with each other in microfibril formation, much consideration has been given to helical pitch, in that even small changes would alter the amino acid environment on the surface. Rich and Crick's model of collagen, obtained from studies of native collagen, proposes 10/3 (or 10-fold) triple helical symmetry, while Okuyama's single crystal analysis of the imino acid-rich collagen model peptide $(Pro\text{-}Pro\text{-}Gly)_{10}$, displays 7/2 triple helical symmetry (Figure 3).

Currently most of the collagen-like peptides found in the PDB archive are primarily imino acid-rich. Uniformly, these entries display 7-fold symmetry (1ITT,[22] 1EI8,[20] 2CUO,[28] 1NAY,[58] 1V7H,[24] 1V6Q,[24] 1V4F,[24] 2DRT,[55] 2DRX,[55] 2D3F,[29] 2D3H[29]).

In contrast, entries that have regions where this imino acid content is disrupted display relaxation of the helical symmetry and show that pitch can vary by region depending on imino acid content. The T3-785 peptide (1BKV),[53] $(Pro\text{-}Hyp\text{-}Gly)_3$-Ile-Thr-Gly-Ala-Arg-Gly-Leu-Ala-Gly$(Pro\text{-}Hyp\text{-}Gly)_4$, was designed to model the 785-797 region of type III collagen on the C-terminal side of the collagenase cleavage site[64,65] following Gly 781. While the steric restrictions imposed on the chain by imino acids provide considerable stability,[66] it has been proposed that inherent instability in this imino acid-poor region could induce conformational changes that would be important for enzymatic action.[64] The T3-785 molecule can be divided into three regions: the N-terminal $(Pro\text{-}Hyp\text{-}Gly)_3$ region, the central imino acid-poor region (Ile-Thr-Gly-Ala-Arg-Gly-Leu-Ala-Gly), and the C-terminal $(Pro\text{-}Hyp\text{-}Gly)_4$ region. In each of the terminal regions, approximate 7-fold symmetry is maintained, whereas the central region more closely matches a 10-fold symmetric model (Figure 3).

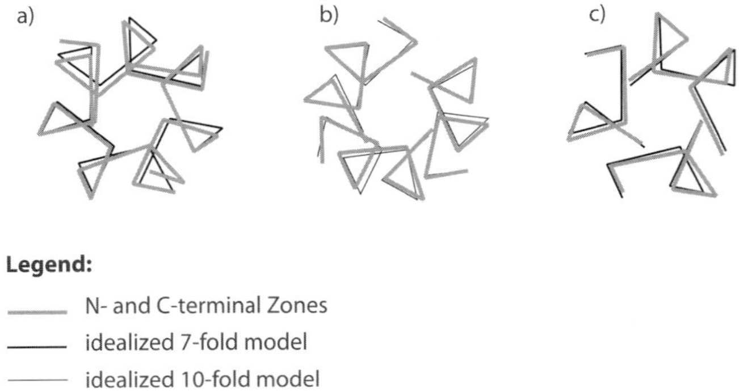

Figure 3: Views down the helical axis of the T3-785 structure 1BKV:[53] a) the N-terminal region superimposed on a 7-fold idealized model; b) the central region superimposed on an idealized 10-fold model; c) the C-terminal region superimposed on an idealized 7-fold model.

Similar results are seen in entries 1DZI[38] and 3DMW.[60] Entry 1DZI contains a collagen like peptide with sequence $(GPO)_2GFPGER(GPO)_3$. Entry 3DMW contains a collagen-like peptide with imino acid-rich ends and a non-imino acid containing center. In both of these cases, 7-fold symmetry was observed in the imino acid-rich ends, and 10-fold symmetry in the central region.

Other entries that contain shorter imino acid-poor regions, or subsitutions for an imino acid in the central region, show a relaxing or perturbation of helical symmetry in those areas.[20,55]

3.2. *Interchain-hydrogen bonding*

In addition to the interchain hydrogen bonds in the Rich and Crick II pattern between glycine N-H groups and X-position proline carbonyl groups, many entries (1ITT,[22] 1QSU,[54] 1BKV,[53] 1CAG,[18] 1A3I,[19] 1A3J,[19] 1EI8,[20] 2CUO[28]) report the observation of additional Cα-H...O interchain hydrogen bond interactions between the glycine Cα of one chain and the glycine carbonyl of an adjacent chain as well as between the Cα of the Y position residue and the carbonyl of the X position carbonyl. One of these bonds was originally suggested by Ramachandran *et al.*[67] to be important to the triple-helical structure.

3.3. *Water-mediated inter-chain hydrogen bonds*

In 1968, Ramachandran and Chandrasekharan proposed that in regions in which neither the X nor the Y position are occupied by imino acids, an additional N-H group is available for hydrogen bonding and does so *via* a water molecule to the C=O group of a neighboring chain.[13] This interaction would occur with the chain that is not already involved in Rich and Crick Collagen II bonds.

This interaction, later observed in entries containing non-imino acid-rich region (1BKV,[53] 1QSU,[54] 2DRT,[55] 2DRX[55]) appears to be a general feature of collagen structure (Figure 4a,b) It has also been observed in these entries that this water molecule can make an additional contact to a hydroxyproline group in the same chain as the Gly carbonyl. In entry 1BKV, this type of interaction was additionally observed with threonine sidechains. The involvement of hydroxyproline in water-mediated hydrogen bonds between amide nitrogens and carbonyl groups was suggested by Ramachandran in 1973.[68]

3.4. *Structural waters and hydration*

Residues in the X and Y positions are considerably exposed to solvent because of the extended nature of the triple helix. With only one interchain hydrogen bond per tripeptide, many polar groups are available for hydrogen binding to water molecules. From the beginning it was noticed that native collagen required hydration to maintain its structure; without it collagen became denatured. However, the structure and the positions of bound water molecules have been the subject of much debate.[11,16,68-71]

The Gly->Ala structure showed a repetitive distribution of water molecules around the carbonyl groups of Hyp and Gly (the carbonyl of Pro is involved in interchain hydrogen bonds). Here the glycine carbonyl has one water binding position and the proline carbonyl has two. In addition, three water sites were observed surrounding the Hyp OD. These ordered waters make up a first hydration shell (Figure 4c). Additional ordered waters created a second hydration shell and then form a repetitive series of water bridges connecting carbonyl and hydroxyproline hydroxyl groups both between and within chains (Figure 4d). Here Hyp residues participate extensively in the building of the water network providing extra anchoring points for hydrogen bonding to or from the peptide surface.

The observation of an ordered and repetitive water structure has been made in a variety of other high-resolution collagen-like peptide structures (1ITT,[22] 1K6F,[23] 1V7H,[24] 1V6Q,[24] 1V4F,[24] 2CUO,[28] 2DRT,[55] 2DRX[55]). Although it has been suggested that these water networks may be important for triple helical stability and macromolecular assembly, more work remains to determine the exact role they play.

3.5. *Stability issues related to hydroxyproline and ring puckering*

Hydroxyproline is common in collagens and in proteins containing collagen-like regions although it is a rare amino acid in proteins overall. It was known from as early as 1924 that a deficiency of vitamin C would interfere with proper collagen formation in healing wounds. Later, it became apparent that vitamin C was necessary for the proper hydroxylation of proline and that hydroxyproline was critical for stability of the collagen molecule. Melting temperature studies have showed a distinct disparity in the stability of the two triple-helical peptides $(\text{Pro-Pro-Gly})_{10}$ and $(\text{Pro-Hyp-Gly})_{10}$ wherein $(\text{Pro-Pro-Gly})_{10}$ is significantly less stable than its hydroxylated counterpart.[76] Analogous experiments using collagen yielded similar results,[77] demonstrating that collagen that lacked hydroxyproline would be unstable at biological temperatures. Furthermore, there is evidence that animals with higher body temperatures have collagens with higher percentages of hydroxyproline content.[78]

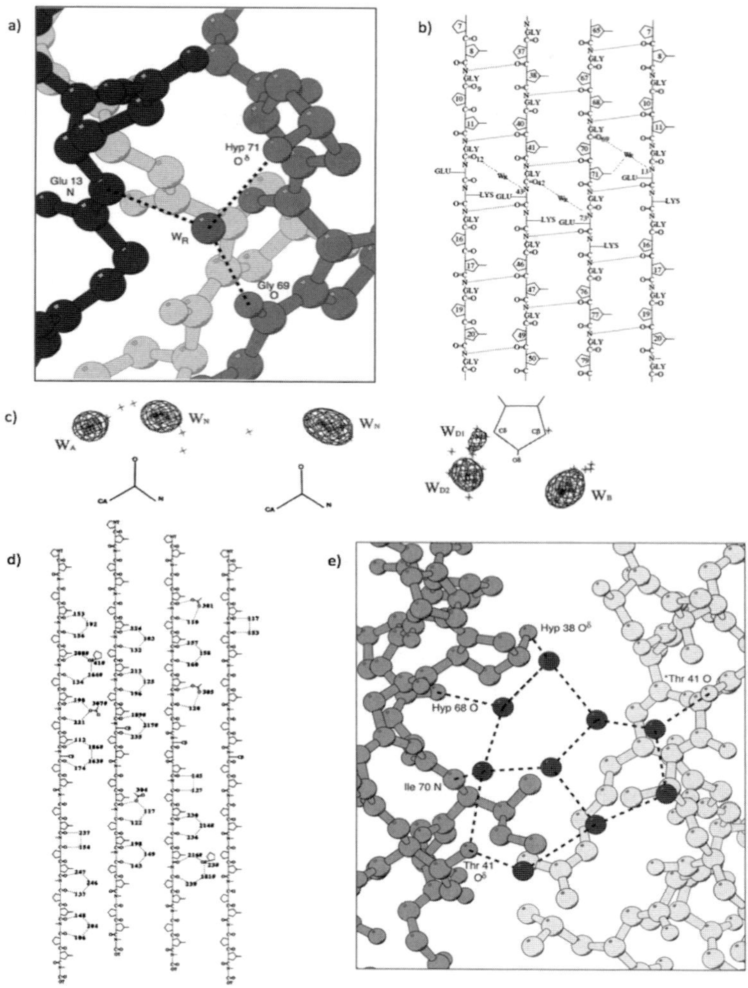

Figure 4: Hydration patterns in collagen-like peptides: a) The water-mediated interaction between Glu 13 N, Gly 69 O, and Hyp 71 Oδ in the EKG structure 1QSU.[54] The three chains are colored in light, medium, and dark grey, respectively. Reprinted from Berman and Kramer[72] with permission from Universities Press. b) Hydrogen bonding topology in the EKG structure. Water molecules connecting Gly C=O groups from one chain with N-H groups from the X position of an adjacent chain are shown with double dashed lines. Interchain hydrogen bonds are shown with single dashed lines. Additional contacts are made between the waters involved in water-mediated interchain hydrogen bonds and either threonine side chains in the same chain as the glycine carbonyl or hydroxyproline hydroxyl groups in the analogous position. Reprinted from Berman and Kramer[72] with permission from Universities Press. c) Orthogonal views of the distribution of water molecules

around the carbonyl groups of Hyp and glycine (or alanine) (shown center) residues. Waters have been selected by their distance to the carbonyl oxygen atoms, applying a cutoff of 3.25 Å. A few oxygen atoms from acetic acid molecules have been included because they also fulfill this criterion. 3D contours have been calculated using the method of Schneider *et al.*[73] Water sites are labeled according to their proximity to atoms in the N-C(=O)-Cα group: WA site for the Cα atom and WN site for the N atom. Figures reprinted from Bella *et al.*[49] with permission from Elsevier. d) Carbonyl-carbonyl water bridges in the collagen zones of the Gly->Ala peptide.[18] Chains 1, 2 and, 3 of the triple helix are shown from left to right in open cylindrical projection with chain 1 repeated on the right to provide a clearer description of the chain 3 to chain 1 interchain water bridges. Figure reprinted from Bella *et al.*[49] with permission from Elsevier. d) Examples of pentagonal clathrate-like water bridges in the T3-785 structure (grey). A symmetry-related helix is shown in light grey. These bridges occur in the region between the layers that interact through their centers. Water molecules are shown in dark grey. The image, originally from Kramer *et al.*,[74] was generated with MOLSCRIPT.[75] Figure reprinted with permission from Elsevier.

While there have been various explanations for the position-dependent stability hydroxyproline proffers (for example POG_{10} forms a stable triple helix, whereas $(OPG)_{10}$ does not), unanswered questions remain and much work has been done to try to clarify this issue. Early hypotheses suggested that hydroxyproline stabilized collagen through interchain hydrogen bonds and likened the effects on stability to that of cystine in keratin.[79-81] Later, this idea was deemed to be impossible given the proposed model of helical structure.

To get at the nature of the stability that Hyp offers, many aspects of triple helical structure have been studied using a variety of collagen-like peptides (Table 2a and b). Investigations of the stereochemistry of hydroxyproline (e.g. 4R vs. 4S), the positional preference of Hyp for the Y position, the imino acid ring puckering propensity at the X and Y positions, the interactions of X and Y position imino acids with the water network, conformational constraints of imino acids within the triple helix, and the inductive effects of the hydroxyl OH group[50,82] have all brought much insight to the issue, but the precise influence of this imino acid is still unclear.

4. Binding and Native Collagen Studies

A few structures have now been determined in which a protein binds to collagen as a ligand (Table 2d, Figure 5). Many of the conformational and interaction features seen in the model peptides, are seen in these complexes. Integrin's I domain is part of a family of cell surface proteins that bind to the extracellular matrix. The availability of the unbound collagen peptide, the unbound domain, and the complex makes it possible to compare the changes that occur to both the ligand and the protein upon complexation. The α2β1 domain (1DZI)[38] binds to a triple helical collagen peptide with the sequence GFOGER. Three loops in the domain coordinate to a metal ion that also coordinates to a glutamate residue in the collagen peptide. The phenylanine in the peptide fits into a hydrophobic region of the integrin domain. The apo protein has a conserved beta sheet motif; there is conformational rearrangement of the other parts of the structure that occurs upon binding to the collagen. The collagen in the complex shows features very similar those seen in the 785 peptide;[53] it has 7-fold helical symmetry at the imino acid-rich C and N termini and 10-fold symmetry in the center. In addition, an arginine in one strand hydrogen bonds with the carbonyl oxygen of a glycine in another strand, also similar to what was observed in T3-785. The isolated peptide maintains the 7-fold symmetry at the imino acid-rich termini, and a conformation intermediate between 10-fold and 7-fold in the center. It is suggested that the flexibility of the central regions may be important for binding. SPARC (2V53)[41] is involved in cell-matrix interactions and collagen assembly and recognizes the GVMGFO sequence. In the structure of the SPARC-collagen complex, the N- and C-termini have 7-fold helical symmetry and the center has an intermediate conformation. As in the integrin domain, a phenylalanine is in a deep pocket at the C terminal end of the triple helix; polar residues surrounding that pocket form hydrogen bonds with the collagen. The structures of discoidin (2WUH)[42] and also of the von Willebrand Factor (4DMU)[45] bound to this same sequence have been determined. Both have the deep phenylalanine pocket. Unlike the collagen in the SPARC complex, the collagens in these complexes are straight.

The structure of CII collagen peptide found in cartilage bound to an antibody (2Y5T)[43] provides structural descriptions relevant to arthritis. In this complex, there are many direct and water-mediated hydrogen bonds with arginine residues from all three chains playing important roles in binding. A novel binding motif called CAN is found in the collagen binding adhesins of the microbial surface component recognizing adhesive matrix molecules (MSCRAMM) of Gram positive bacteria. (2F6A).[40] In the complex the collagen interacts with two CAN molecules spaced apart by a linker, forming a "collagen hug"; in which most of the interactions are hydrophobic in nature. The collagen itself has a similar conformation to that seen in other collagen-like peptides.

In addition to these complexes, a structure of a protein with a triple-helical domain, mannan-binding lectin (3POB),[44] has been reported. In this case, calcium plays an important role in binding.

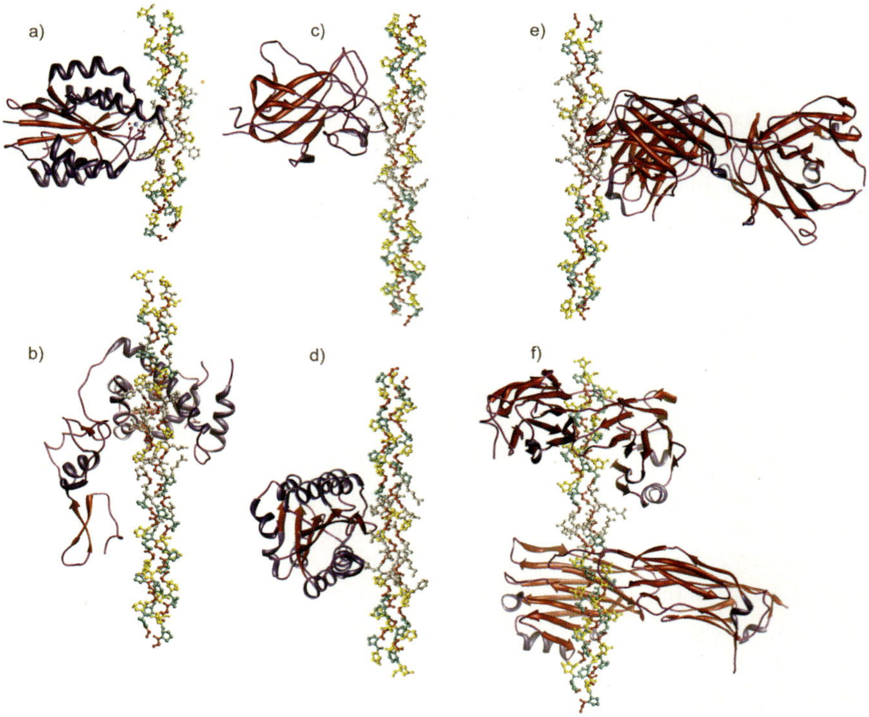

Figure 5: Triple-helical collagen-like peptide bound to globular proteins: a) 1DZI;[38] b) 2V53;[41] c) 2WUH;[42] d) 4DMU;[45] e) 2Y5T;[43] f) 2F6A.[40] In the triple helices, glycine is colored red, proline green, and hydroxyproline yellow.

5. Packing of Collagen

Early fiber work established that collagen triple helices packed in a quasi-hexagonal manner. In the Gly->Ala structure,[18] the collagen peptide is hexagonally close packed with center-to-center distances of approximately 14Å similar to what was observed by Fraser *et al.* in their study of collagen fibers.[83] In Gly->Ala there are no direct contacts between helices; instead there were many water-mediated contacts. In the PPG model peptide,[19] the pattern is pseudo hexagonal with square and triangular arrangements very similar to the pattern proposed earlier by Sasisekharan and Bansal.[84] While the details of the packing motifs have been attributed to variations in sequence, in one very interesting study the same peptide crystallized with two different packing arrangements: pseudohexagonal and pseudotetragonal (3A08).[56] In some triple helical structures there are direct contacts between the hydroxyprolines (1QSU,[54]

1V7H, 1V6Q, 1V4F)[24] and in one very recent study there are charge pair interactions between helices (3U29).[33] A notable exception to the pseudo hexagonal packing seen in most of these structures is the layered packing seen in T3-785 in which the helices are at a 50-degree angle with respect to one another. This type of packing appears to be similar to the fibrillar packing seen in cornea tissue.[85]

In a fiber study of type I collagen *in situ* (3HR2),[59] the collagen helices pack in a quasi-hexagonal pattern with neighboring microfibrils interdigitated with one another. The overall structure has an appearance of being pleated. It has been suggested that the resultant interruptions make serve as recognition points for binding of other extracellular matrix proteins.

6. Conclusion

The early work on collagen by Ramachandran and by Rich and Crick in which it was proposed that collagen was triple helical has been confirmed by subsequent analyses of fibers and of single crystal model peptides. Although Ramachandran's initial model was not correct in detail, his proposal of C-H...O hydrogen bonds as well as interchain water mediated hydrogen bonds has been validated in the single crystal studies.

The single crystal work on model peptides has demonstrated that the helical geometry is sequence dependent, with imino acid-rich sequences forming 7-fold helices and more heterogeneous sequences forming 10-fold helices. This subtle plasticity enables collagen to form specific interactions with proteins. The involvement of glycine and hydroxyproline in forming water networks within and between the triple helices has been seen in the structures of the many model peptides that have been analyzed since the first structure, in which these intricate networks were first seen, was determined.

The properties of collagen, which has a rigid triple helical framework and yet is flexible enough to form specific interactions with other macromolecules, may help explain its many roles in biology.

Acknowledgments

We are grateful to Barbara Brodsky, who first introduced us to the collagen world, and to Jordi Bella, whose meticulous analyses of collagen structure led to the enduring understanding we now have of its structure.

References

1. J. C. Brown and R. Timpl, *Int Arch Allergy Immunol* **107**, 484 (1995).
2. D. J. Prockop and K. I. Kivirriko, *Annu Rev Biochem* **64**, 403 (1995).
3. H.-J. Hoppe and K. B. M. Reid, *Protein Sci* **3**, 1143 (1994).
4. A. Rich and F. H. C. Crick, *J Mol Biol* **3**, 483 (1961).
5. H. M. Berman, K. Henrick and H. Nakamura, *Nat Struct Biol* **10**, 980 (2003).
6. F. C. Bernstein, T. F. Koetzle, G. J. B. Williams, E. F. Meyer Jr., M. D. Brice, J. R. Rodgers, O. Kennard, T. Shimanouchi and M. Tasumi, *J Mol Biol* **112**, 535 (1977).
7. G. N. Ramachandran and G. Kartha, *Nature* **174**, 269 (1954).
8. G. N. Ramachandran and G. Kartha, *Nature* **176**, 593 (1955).
9. A. Rich and F. H. C. Crick, *Nature* **176**, 915 (1955).
10. W. Traub and A. Yonath, *J Mol Biol* **16**, 404 (1966).
11. A. Yonath and W. Traub, *J Mol Biol* **43**, 461 (1969).
12. S. Sakakibara, Y. Kishida, Y. Kikuchi, R. Sakai and K. Kakiuchi, *Bull Chem Soc Jpn* **41**, 1273 (1968).
13. G. N. Ramachandran and R. Chandrasekharan, *Biopolymers* **6**, 1649 (1968).
14. S. Sakakibara, Y. Kishida, K. Okuyama, N. Tanaka, T. Ashida and M. Kakudo, *J Mol Biol* **65**, 371 (1972).
15. M. H. Miller and H. A. Scheraga, *J Polym Sci, Symp* **54**, 171 (1976).
16. R. D. B. Fraser, T. P. MacRae and E. Suzuki, *J Mol Biol* **129**, 463 (1979).
17. K. Okuyama, K. Okuyama, S. Arnott, M. Takayanagi and M. Kakudo, *J Mol Biol* **152**, 427 (1981).
18. J. Bella, M. Eaton, B. Brodsky and H. M. Berman, *Science* **266**, 75 (1994).
19. R. Z. Kramer, L. Vitagliano, J. Bella, R. Berisio, L. Mazzarella, B. Brodsky, A. Zagari and H. M. Berman, *J Mol Biol* **280**, 623 (1998).
20. J. Bella, J. Liu, R. Kramer, B. Brodsky and H. M. Berman, *J Mol Biol* **362**, 298 (2006).
21. L. Vitagliano, R. Berisio, L. Mazzarella and A. Zagari, *Biopolymers* **58**, 459 (2001).
22. C. Hongo, V. Nagarajan, K. Noguchi, S. Kamitori, K. Okuyama, Y. Tanaka and N. Nishino, *PlymJ* **33**, 812 (2001).
23. R. Berisio, L. Vitagliano, L. Mazzarella and A. Zagari, *Protein Sci* **11**, 262 (2002).
24. K. Okuyama, C. Hongo, R. Fukushima, G. Wu, H. Narita, K. Noguchi, Y. Tanaka and N. Nishino, *Biopolymers* **76**, 367 (2004).
25. K. Kawahara, Y. Nishi, S. Nakamura, S. Uchiyama, Y. Nishiuchi, T. Nakazawa, T. Ohkubo and Y. Kobayashi, *Biochemistry* **44**, 15812 (2005).

26. N. Jiravanichanun, C. Hongo, G. Wu, K. Noguchi, K. Okuyama, N. Nishino and T. Silva, *Chembiochem* **6**, 1184 (2005).
27. M. Schumacher, K. Mizuno and H. P. Bachinger, *J Biol Chem* **280**, 20397 (2005).
28. C. Hongo, K. Noguchi, K. Okuyama, Y. Tanaka and N. Nishino, *J Biochem* **138**, 135 (2005).
29. K. Okuyama, C. Hongo, G. Wu, K. Mizuno, K. Noguchi, S. Ebisuzaki, Y. Tanaka, N. Nishino and H. P. Bachinger, *Biopolymers* **91**, 361 (2009).
30. M. A. Schumacher, K. Mizuno and H. P. Bachinger, *J Biol Chem* **281**, 27566 (2006).
31. K. Okuyama, K. Miyama, T. Morimoto, K. Masakiyo, K. Mizuno and H. P. Bachinger, *Biopolymers* **95**, 628 (2011).
32. C. Byrne, P. A. McEwan, J. Emsley, P. M. Fischer and W. C. Chan, *Chem Commun (Camb)* **47**, 2589 (2011).
33. J. A. Fallas, J. Dong, Y. J. Tao and J. D. Hartgerink, *J Biol Chem* **287**, 8039 (2012).
34. K. Okuyama, K. Miyama, T. Kawaguchi and N. Nishino, *DOI: 102210/pdb3ai6/pdb*).
35. K. Okuyama, M . Shimura, T. Kawaguchi, K. Noguchi, K. Mizuno and H. P. Bachinger, *DOI: 102210/pdb3abn/pdb*).
36. K. Okuyama, T. Morimoto,C. Hongo,N. Hosaka and N. Nishino, *DOI: 102210/pdb3ah9/pdb*).
37. K. Okuyama, T. Morimoto,C. Hongo, A. Katagiri, Y. Tanaka and N. Nishino, *DOI: 102210/pdb3a0a/pdb*).
38. J. Emsley, C. G. Knight, R. W. Farndale, M. J. Barnes and R. C. Liddington, *Cell* **100**, 47 (2000).
39. J. Emsley, C. G. Knight, R. W. Farndale and M. J. Barnes, *J Mol Biol* **335**, 1019 (2004).
40. Y. Zong, Y. Xu, X. Liang, D. R. Keene, A. Hook, S. Gurusiddappa, M. Hook and S. V. Narayana, *EMBO J* **24**, 4224 (2005).
41. E. Hohenester, T. Sasaki, C. Giudici, R. W. Farndale and H. P. Bachinger, *Proc Natl Acad Sci U S A* **105**, 18273 (2008).
42. F. Carafoli, N. J. Clout and E. Hohenester, *J Biol Chem* **284**, 22786 (2009).
43. D. Dobritzsch, I. Lindh, H. Uysal, K. S. Nandakumar, H. Burkhardt, G. Schneider and R. Holmdahl, *Arthritis Rheum* **63**, 3740 (2011).
44. A. R. Gingras, U. V. Girija, A. H. Keeble, R. Panchal, D. A. Mitchell, P. C. Moody and R. Wallis, *Structure* **19**, 1635 (2011).
45. T. H. Brondijk, D. Bihan, R. W. Farndale and E. G. Huizinga, *Proc Natl Acad Sci U S A* **109**, 5253 (2012).
46. P. M. Cowan and S. McGavin, *Nature* **176**, 501 (1955).
47. F. H. C. Crick and A. Rich, *Nature* **176**, 780 (1955).

48. G. N. Ramachandran. in *Treatise on Collagen*; G. N. Ramachandran, Ed.; (Academic Press, New York), 103 (1967).
49. J. Bella, B. Brodsky and H. M. Berman, *Structure* **3**, 893 (1995).
50. M. D. Shoulders, K. A. Satyshur, K. T. Forest and R. T. Raines, *Proc Natl Acad Sci U S A* **107**, 559 (2010).
51. K. Okuyama, K. Miyama, K. Mizuno and H. P. Bachinger, *Biopolymers* **97**, 607 (2012).
52. G. N. Ramachandran, C. Ramakrishnan and V. Sasisekharan, *J Mol Biol* **7**, 95 (1963).
53. R. Kramer, J. Bella, P. Mayville, B. Brodsky and H. M. Berman, *Nat Struct Biol* **6**, 454 (1999).
54. R. Z. Kramer, M. Venugopal, J. Bella, P. Mayville, B. Brodsky and H. M. Berman, *J Mol Biol* **301**, 1191 (2000).
55. K. Okuyama, H. Narita, T. Kawaguchi, K. Noguchi, Y. Tanaka and N. Nishino, *Biopolymers* **86**, 212 (2007).
56. K. Okuyama, T. Morimoto, H. Narita, T. Kawaguchi, K. Mizuno, H. P. Bachinger, G. Wu and K. Noguchi, *Acta Crystallogr D Biol Crystallogr* **66**, 88 (2010).
57. D. Motooka, K. Kawahara, S. Nakamura, M. Doi, Y. Nishi, Y. Nishiuchi, Y. Kee Kang, T. Nakazawa, S. Uchiyama, T. Yoshida, T. Ohkubo and Y. Kobayashi, *Biopolymers* **98**, 111 (2011).
58. J. Stetefeld, S. Frank, M. Jenny, T. Schulthess, R. A. Kammerer, S. Boudko, R. Landwehr, K. Okuyama and J. Engel, *Structure* **11**, 339 (2003).
59. J. P. Orgel, T. C. Irving, A. Miller and T. J. Wess, *Proc Natl Acad Sci U S A* **103**, 9001 (2006).
60. S. P. Boudko, J. Engel, K. Okuyama, K. Mizuno, H. P. Bachinger and M. A. Schumacher, *J Biol Chem* **283**, 32580 (2008).
61. J. A. Fallas, V. Gauba and J. D. Hartgerink, *J Biol Chem* **284**, 26851 (2009).
62. I. Bertini, M. Fragai, C. Luchinat, M. Melikian, M. Toccafondi, J. L. Lauer and G. B. Fields, *J Am Chem Soc* **134**, 2100 (2012).
63. I. W. Davis, L. W. Murray, J. S. Richardson and D. C. Richardson, *Nucleic Acids Res* **32**, W615 (2004).
64. G. B. Fields, *J Theor Biol* **153**, 585 (1991).
65. J. M. Seyer, C. Mainardi and A. H. Kang, *Biochemistry* **19**, 1583 (1980).
66. G. Némethy and H. A. Scheraga, *Biopolymers* **23**, 2781 (1984).
67. G. N. Ramachandran and V. Sasisekharan, *Biochim Biophys Acta* **109**, 314 (1965).
68. G. N. Ramachandran, M. Bansal and R. S. Bhatnagar, *Biochim Biophys Acta* **322**, 166 (1973).
69. J. Engel, H.-T. Chen, D. J. Prockop and H. Klump, *Biopolymers* **16**, 601 (1977).

70. G. N. Ramachandran and V. Sasisekharan, *Adv Protein Chem* **23**, 283 (1968).
71. E. Suzuki and R. D. B. Fraser. in *Peptides, Polypeptides and Proteins*; E. R. Blout; F. A. Bovey; M. Goodman; N. Lotan, Eds.; (John Wiley & Sons, New York) , 449. (1974)
72. R. Kramer and H. M. Berman, *Perspectives in Structural Biology*, 169 (1999).
73. B. Schneider, D. M. Cohen, L. Schleifer, A. R. Srinivasan, W. K. Olson and H. M. Berman, *Biophys J* **65**, 2291 (1993).
74. R. Z. Kramer, J. Bella, B. Brodsky and H. M. Berman, *J Mol Biol* **311**, 131 (2001).
75. P. Kraulis, *J Appl Cryst* **24**, 946 (1991).
76. S. Sakakibara, K. Inouye, K. Shudo, Y. Kishida, Y. Kobayashi and D. J. Prockop, *Biochim Biophys Acta* **303**, 198 (1973).
77. J. Rosenbloom, M. Harsch and S. Jimenez, *Arc Biochem Biophys* **158**, 478 (1973).
78. K. H. Gustavson. The chemistry and reactivity of collagen; Academic Press Inc.: New York, (1956).
79. K. H. Gustavson, *Acta Chem Scand* **8**, 1298 (1954).
80. K. H. Gustavson, *Acta Chem Scand* **8**, 1299 (1954).
81. K. H. Gustavson, *Nature* **175**, 70 (1955).
82. S. K. Holmgren, K. M. Taylor, L. E. Bretcher and R. T. Raines, *Nature* **392**, 666 (1998).
83. R. D. B. Fraser, T. P. MacRae, A. Miller and E. Suzuki, *J Mol Biol* **167**, 497 (1983).
84. V. Sasisekharan and M. Bansal, *Curr Sci* **1990**, 863 (1990).
85. A. Craig and D. Parry, *J Ultrastruct Res* **74**, 232 (1981).

THE OPEN-ENDED INTELLECTUAL LEGACY OF GNR

GEORGE D. ROSE

Jenkins Department of Biophysics
Johns Hopkins University
3400 N. Charles Street
Baltimore, MD 21218 USA
E-mail: grose@jhu.edu

The Ramachandran Plot, first introduced half a century ago, maps the backbone torsion angles – ϕ and ψ – for a dipeptide, showing that steric restrictions eliminate approximately ¾ of conceivable conformational space. This chapter revisits and then generalizes this early result, a follow-on to ideas implicit in the Plot. Ramachandran's dramatic result notwithstanding, subsequent work seemed to indicate that dipeptide restrictions have a negligible impact on overall protein conformation. Consequently, later decades witnessed an ever-increasing emphasis on the significance of attractive forces in folding simulations, with concomitant neglect of the organizing power of excluding forces. *Attractive forces* act by stabilizing native interactions, while *excluding forces* such as steric clashes act by destabilizing non-native interactions. This chapter summarizes some recent advances that pertain to both types of excluding forces: (a) systematic local steric clashes beyond the dipeptide and (b) the strongly destabilizing influence of buried polar groups that lack hydrogen bond partners.

1. Introduction

I met Professor G. N. Ramachandran (GNR, as he was known throughout India) just once, in 1977 when he was a Fogarty Scholar at the National Institutes of Health in Bethesda, MD. After quickly dispensing with the customary preliminaries, he produced a large dipeptide model attached to a ring stand and, with a few deft movements, locked in a conformation of interest and assigned some unsolicited homework. It was a memorable afternoon, but only later did I come to realize the deeper implications of my homework assignment.[1]

No biochemistry textbook is complete without a Ramachandran Plot,[2] the ϕ,ψ-map of allowed conformations for a dipeptide[*]. Steric restrictions alone eliminate approximately ¾ of conceivable conformational space for the dipeptide, a conclusion that has since been validated in thousands of high resolution X-ray structures.[3,4] Like Kepler's laws, the theory accounts for the

---

[*]More precisely, the compound Cα-CO-NH–CαHR–CO-NH-Cα, which has two degrees of backbone freedom like a dipeptide.

data well enough that it has become synonymous with reality. The plot is a compact and accessible representation of a profound idea, one that has thoroughly conditioned our thinking about the structure of proteins.

The Plot was first introduced into the literature half a century ago,[2] and its impact became widespread after John Edsall invited Ramachandran to contribute a review to Advances in Protein Chemistry.[5] Similar thinking during this era[6] led Pauling to develop space-filling models.[7] Such concepts stimulated considerable work on generalizing from hard sphere sterics to distance-dependent potentials,[8] resulting in versions of the plot with refined margins but unchanged overall conclusions. This situation was described by Fred Richards, who commented:

> *"For chemically bonded atoms the distribution is not spherically symmetric nor are the properties of such atoms isotropic. In spite of all this, the use of the hard sphere model has a venerable history and an enviable record in explaining a variety of different observable properties. As applied specifically to proteins, the work of G.N. Ramachandran and his colleagues has provided much of our present thinking about permissible peptide chain conformations."*[9]

Implicitin the eponymous Plot[2,5] is a larger question. Do sterically disfavored conformations impose substantial organizing restrictions on overall protein structure? Dipeptide restrictions are strictly local. In this chapter, I summarize evidence that related restrictions extend well beyond adjacent chain neighbors. This accumulating body of evidence is the basis for a generalized Ramachandran-type analysis across a longer length scale than the dipeptide. Generalized-Ramachandran concepts are inherent in the dipeptide map although results of the analysis can no longer be represented in a convenient two-dimensional plot. At this golden anniversary celebration, the Ramachandran Plot is still providing new insights with far-reaching consequences.

2. Attractive forces vs. excluding forces

Today, predictive work in protein folding is based largely on *attractive forces*, which are represented by force fields, knowledge-based potentials, Gō models, contact maps, and the like. In contrast, steric clashes are *excluding forces* that operate by destabilizing non-native interactions rather than by stabilizing native interactions. Conformers with even a single steric clash are of such high energy that they make only a negligible contribution to the thermodynamic population, and their organizing effect is exerted by reducing the magnitude of available

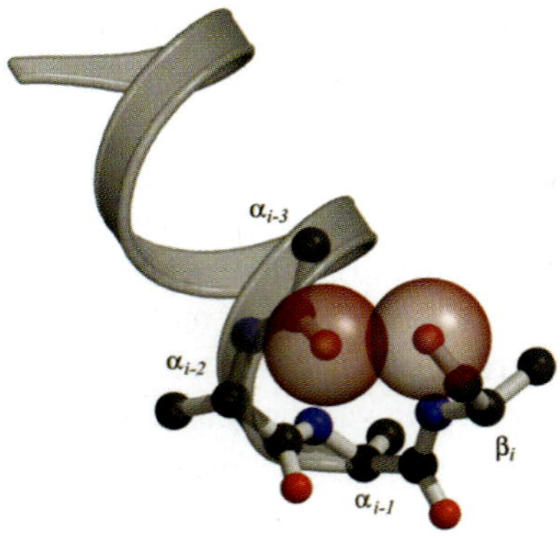

Figure 1. An α-helix cannot be followed immediately by a β-strand without encountering a steric clash. Here, ball-and-stick backbone atoms for three residues of an α-helix (α_{i-3} - α_{i-1}) are shown superimposed on a longer helical ribbon, followed by a single β-strand residue (β_i). This conformation forces an inescapable steric overlap between atoms O_i and O_{i-3}, shown here as transparent van der Waals spheres. The clash is caused by a steric conflict between backbone atoms and is therefore independent of specific side chain identities. Atoms are rendered using conventional CPK colors, i.e. carbon:black, nitrogen: blue and oxygen:red. (Reproduced from Fitzkee and Rose [20] with permission.)

conformational space. Consequently, any approach that is limited to the analysis of native folds is necessarily blind to excluding forces.[10] Generalized-Ramachandran analysis, which is based on excluding forces, is unaffected by this limitation, but apart from dipeptide interactions, the relevance of this approach has been in doubt.

The conviction that local steric constraints exert a negligible impact on global protein conformation dates back to Flory's isolated pair hypothesis, the simplifying assumption that each ϕ,ψ-pair is sterically independent of all but its adjacent chain neighbors.[11] This hypothesis has been a conceptual linchpinin subsequent protein folding studies, such as Tanford's extensive work on protein denaturation.[12,13] It would seem that longer-range organization originates elsewhere.[14]

Eventually however, this well-accepted assumption was re-assessed.[15-19] From exhaustive enumeration of the conformational space accessible to a polyalanyl 7-mer, Pappu et al. found that the isolated pair hypothesis holds for fully extended conformers but fails for more collapsed conformers.[18] In other words, systematic local steric effects can extend beyond adjacent-chain neighbors. Seeking specific examples of such effects, Fizkee and Rose described multiple instances of longer range systematic steric clashes.[20,21] For example, an

α-helical segment cannot be followed immediately by a β-strand segment without encountering a steric clash between the backbone carbonyl oxygens of the antepenultimate residue of the helix and the first residue of the strand (Figure. 1). Consequently, the segments must be connected by an intervening linker. Intuitively, it may seem that slight relaxation of the ϕ,ψ–angles could overcome the clash, but rigorous analysis shows that this is not the case. Consistent with this analysis, no apparent violations of this conclusion were found within an extensive database of high-resolution X-ray structures.[20] Nature's exclusion of α–β hybrid segments is built into proteins at the covalent level, with far-reaching consequences for protein organization.

The crucial role of excluding forces was recognized long ago, but in recent years, it has fallen into neglect. Fred Richards had it exactly right:

> *"In summary it seems reasonable that there is a generalized driving force with no particular specificity produced by the tendency of non-polar groups to "escape" from aqueous contact. Permissible folded structures are then the result of the relative absence of two energetically unfavorable factors: (a) the lack of pairing of buried polar groups, and (b) poor geometrical packing of internal groups both polar and non-polar. The combination of these latter two restrictions is perhaps more severe than might be imagined and should markedly reduce the number of possible native structures even without further knowledge of the interaction parameters."*[22]

Systematic steric clashes at longer length scales represent one type of excluding force – Richards' factor (b). The other type of excluding force involves conformers with unsatisfied hydrogen bonds – his factor (a).

3. Hydrogen bond satisfaction — the other excluding force

As described above, the organizational impact of local steric constraints has been under-appreciated in recent decades. Instead, algorithms to predict protein folding have focused on attractive forces, with ever greater emphasis on improved force fields. Despite impressive advances (see for example Ref. 23), strategies grounded in pattern recognition still outperform purely *ab initio* approaches,[24] perhaps because these latter methods fail to capture essential aspects of the problem.

Indeed, hydrogen-bond satisfaction, the other excluding force, is often underweighted in predictive algorithms. Yet, its organizing power exceeds even that of steric constraints, as was already suggested when the Plot was first introduced. Quoting Charles Tanford:

> *"It is thus evident that not just any globular structure with non-polar groups on the inside, and charged groups on the outside, will be stable. It is also necessary that most of the polar groups which find themselves on the inside of the globular structure form suitable hydrogen bonds. Since formation of hydrogen bonds is possible only for highly specific relative orientations of the participating groups, whereas formation of hydrophobic regions carries no such requirement, it is likely that the necessity for forming such hydrogen bonds is the crucial factor which limits the number of possible structures which will satisfy all of the conditions needed for stability of a compact conformation."*[25]

Even a single backbone polar group (-NH or -C=O) that is satisfied by a hydrogen bond to solvent in the unfolded state but buried and left unsatisfied in the folded state would come at an energetic cost of ~5 kcal/mol, rivaling the entire free energy of stabilization, $\Delta G_{U \to N} \approx$ [-5 to -15] kcal/ mol, for a typical globular protein.[26] As a consequence, protein folds are – of necessity – built on scaffolds of α-helices and strands of β-sheet.[10] These two secondary structure elements, and only these two, can be extended indefinitely without encountering a steric clash while at the same time providing readymade hydrogen bond partners for every backbone polar group removed from solvent access upon folding.

It then follows that only a limited number of distinct folds is possible, regardless of other energetic considerations.[27-29] These folds can be partitioned into a remarkably small set of frequently observed backbone motifs. Just seven recurrent motifs account for ~90% of protein structure on average,[30] and they can be rationalized by simple models that include only sterics and hydrogen-bond satisfaction.[31] The predominant motifs are α-helices, β-strands, polyproline II, and β-turns together with three additional minor motifs. All seven motifs are familiar structures that range in length from 3-5 residues. Consequently, mapping ϕ,ψ-angles onto these iconic structures reveals a dramatic and previously unsuspected backbone-based structural degeneracy.[32]

Recently, we extended the classical Ramachandran plot by including hydrogen bond satisfaction (Figure 2). Sterics alone eliminate approximately ¾ of ϕ,ψ-space for the alanyl dipeptide. Adding hydrogen-bonding constraints to these steric criteria eliminates another substantial region of ϕ,ψ-space; for conformers within this region, an amide hydrogen is solvent-inaccessible, depriving it of a hydrogen-bonding partner.[33]

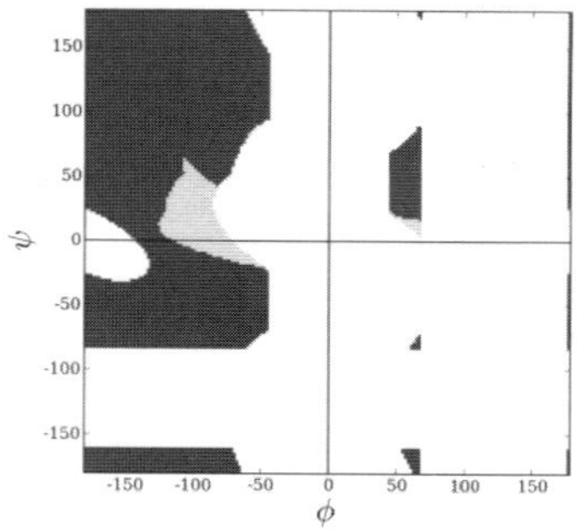

Figure 2. Ramachandran plot modified to include hydrogen bond satisfaction. Adding a hydrogen bond satisfaction constraint to Ramachandran's calculation of sterically allowed backbone dihedral space for a blocked alanine monomer winnows the bridge and the α_L basin.[33] Sterically allowed regions disfavored by hydrogen bond satisfaction are shown in light gray.

Yet, this "forbidden" region is well populated in folded proteins, which can provide longer-range intramolecular hydrogen-bond partners for these otherwise unsatisfied polar groups. Consequently, accessible conformational space expands under folding conditions, a paradigm-shifting realization with significant implications for understanding the mechanism of protein folding.

4. Summary and conclusions

The organizing power of excluding forces was already appreciated at the time the Plot was first introduced. However, assumptions about the limited impact of sterics led to a prevailing emphasis on attractive forces. During this time frame, the hydrophobic effect[34] came to be accepted as the dominant factor in protein folding – see Baldwin for a quantitative assessment of the forces involved.[35,36] Concomitantly, early ideas about the central importance of hydrogen bonding[37-39] were overturned, as described at length in a recent review.[40] These historical trends overshadowed alternative ideas. Only recently has there been renewed interest in the generalized Ramachandran-type analysis discussed in this chapter.

What we see depends upon the perspective from which we look. Viewed from the perspective of non-specific excluding forces, it is readily apparent that protein folds are of necessity built upon scaffolds of α-helices and strands of β-sheet.[25,22,10] It follows that only a limited number of distinct folds are possible, regardless of other energetic considerations.[27-29] This organizational infrastructure underpins dry molten globule species, conformers in which the overall chain topology has been established but detailed side-chain contact energies are not yet fully realized.[41]

Additional insights into the mechanism of protein folding are poised to emerge as further work on excluding forces continues. The tools and methods for these inquiries were launched half a century ago, thanks in large measure to the pioneering work of Ramachandran, Ramakrishnan and Sasisekharan.[2]

Acknowledgments

I thank Buzz Baldwin for calling my attention to Fred Richards' Cold Spring Harbor Symposium paper.[22] Support from the NSF and the Humboldt Foundation is gratefully acknowledged.

References

1. G. D. Rose, *Protein Sci* **10**, 1691 (2001).
2. G. N. Ramachandran, C. Ramakrishnan and V. Sasisekharan, *J Mol Biol* **7,** 95 (1963).
3. R. A. Laskowski, M. W. MacArthur, D. S. Moss and J. M. Thornton, *J Appl Cryst* **26,** 283 (1993).
4. H. M. Berman, J. Westbrook, Z. Feng, G. Gilliland, T. N. Bhat, H Weissig, I. N. Shindyalov and P. E. Bourne , *Nucleic Acids Res* **28**, 235 (2000).
5. G. N. Ramachandran and V. Sasisekharan, *Adv Prot Chem* **23**, 283 (1968).
6. D. Eisenberg, *Proc Natl Acad Sci USA* **100**, 11207 (2003).
7. W. L. Koltun, *Biopolymers* **3**, 665 (1965).
8. D. A . Brant, W. G. Miller and P. J. Flory, *J Mol Biol* **23**, 47 (1967).
9. F. M. Richards, *Ann Rev Biophys Bioeng* **6**, 151 (1977).
10. G. D. Rose, P. J. Fleming, J. R. Banavar and A. Maritan, *Proc Natl Acad Sci USA* **103**, 16623 (2006).
11. P. J. Flory, Wiley, New York. **8**, 699 (1969).
12. C. Tanford, *Adv Prot Chem* **23**,121 (1968).
13. Tanford C, *Adv Prot Chem* **24**,1 (1970).
14. K. A. Dill, *Protein Sci* **8,**1166 (1999).
15. H. Qian and J. A. Schellman, *J Phys Chem* **96**,3987(1992).
16. T. P. Creamer, *Proteins* **33**, 218 (1998).
17. R. L. Baldwin and B. H. Zimm, *Proc Natl Acad Sci USA* **97,** 12391 (2000).
18. R. V. Pappu, R. Srinivasan and G. D. Rose, *Proc Natl Acad Sci USA* **97**, 12565 (2000).
19. M. H. Zaman, M .Y. Shen, Berry, K. F. Freed and T. R. Sosnick, *J Mol Biol* **331**, 693(2003).
20. N. C. Fitzkee and G. D. Rose, *Protein Sci* **13**, 633 (2004).
21. N. C. Fitzkee and G. D. Rose, *J Mol Biol* **353**, 873(2005).

22. F. M. Richards, H. W. Wyckoff, W. D. Carlson, N. M. Allewell, B. Lee and Y. Mitsui, *Cold Spring Harbor symposia on quantitative biology* **36**, 35 (1972).
23. J. W. Ponder, C. Wu, P. Ren, V. S. Pande, J.D. Chodera, M. J. Schnieders, I. Haque, D. L. Mobley, D. S. Lambrecht, R. A. DiStasio, Jr., M. Head-Gordon, G. N. Clark, M. E. Johnson and T. Head-Gordon, *J Phys Chem B* **114**, 2549 (2010).
24. J. Moult, K. Fidelis, A. Kryshtafovych and A. Tramontano, *Proteins*, **79**, 1 (2011).
25. C. Tanford, *J Am Chem Soc* **84**, 4240 (1962).
26. P. J. Fleming and G. D. Rose, *Protein Sci* **14**,1911 (2005).
27. T. Przytycka, R. Aurora and G. D. Rose, *Nat Struct Biol* **6**, 672 (1999).
28. T. X. Hoang, A. Trovato, F. Seno, J. R. Banavar and A. Maritan, *Proc Natl Acad Sci USA* **101**,7960 (2004).
29. A. P. Kamat and A. M. Lesk, *Proteins* **66**,869 (2007).
30. L. L. Perskie, T. O. Street and G. D. Rose, *Protein Sci* **17**,1151 (2008).
31. L. L. Perskie and G. D. Rose, *Prot Science* **19**,1127 (2010).
32. G. D. Chellapa and G. D. Rose, *Protein Sci* **21**,1231(2012).
33. L. L. Porter and G. D. Rose, *Proc Natl Acad Sci USA* **108**, 109 (2011).
34. W. Kauzmann, *Adv Protein Chem* **14**, 1 (1959).
35. R. L. Baldwin, *J Mol Biol* **371**, 283 (2007).
36. R. L. Baldwin, *Proc Natl Acad Sci USA* **109**, 7310 (2012).
37. L. Pauling, R. B.Corey and *Proc Natl Acad Sci USA* **37**, 251 (1951).
38. L. Pauling, R. B. Corey and H. R. Branson, *Proc Natl Acad Sci USA* **37**, 205 (1951).
39. J. A. Schellman, *Compt Rend Trav Lab Carlsburg, Ser Chim* **29**, 230 (1955).
40. D. W. Bolen and G. D. Rose, *Annu Rev Biochem* **77**, 339 (2008).
41. R. L. Baldwin, C Frieden and G. D. Rose, *Proteins* **78**, 2725 (2010).

"THE PLOT" THICKENS: MORE DATA, MORE DIMENSIONS, MORE USES

JANE S. RICHARDSON, DANIEL A. KEEDY AND DAVID C. RICHARDSON
Department of Biochemistry, Duke University
Durham, North Carolina 27710 USA
E-mail: jsr@kinemage.biochem.duke.edu

The scientific history of the Ramachandran plot is reviewed, emphasizing relationships to the title theme and to trends in current research. The growth and quality of macromolecular structure data have enabled the understanding of relationships with further variables and the application to an ever-widening set of uses such as prediction, simulation, design, motif identification, and structure validation and improvement. Then a current research example is explored, using a new dataset of 8000 selected protein chains and >1.5 million quality-filtered residues. The new Ramachandran plots show an unprecedented level of detail, allow the valid addition of more subcategories and more dimensions, and point the way toward the feasibility of an essentially complete and robust treatment of torsional conformation in the near future.

1. Historical Introduction

This volume celebrates the 50th anniversary of the 1963 paper by Ramachandran, Ramakrishnan, and Sasisekharan[1] that introduced the "Ramachandran plot" of the ϕ vs ψ backbone torsion angles for protein structures. The concept of that plot was recognized as groundbreaking very early on, and has continued to be of central importance for the understanding of protein structure, energetics, and folding. The two dimensions of the Ramachandran plot are necessary and sufficient to encapsulate the primary relationships: necessary because ϕ and ψ do not show meaningful preferences individually but only work as a joint distribution; sufficient because the approximation of ideal geometry and trans planar peptides is close enough that further effects are second order. The residue-centric ϕ,ψ plot is much more informative than a peptide-centric ψ,ϕ plot because the dominant steric interactions occur between adjacent peptides. And, given that they are the right two dimensions, such a plot is suited to be easily shown, intuitively understandable, and even iconic.

1.1 Early Developments: Conventions, Calculations and Turns

The basic Ramachandran plot illuminates protein conformation by outlining the allowed conformations of the Ala dipeptide. In 1963 there were no established standards for the conventions and nomenclature of the conformational variables, and the original paper chose to call the backbone torsion angles ϕ and ϕ', with definitions suited to the dipeptide rather than generic: 0° in the trans position for ϕ but in cis for ϕ', giving a flat and near-symmetric reference conformation, with values defined from 0° to 360° and the plot origin at lower left. On the original Ramachandran plot, those conventions put extended conformation in the lower left quadrant and helix in the upper left. In 1966, standards were established that uniformly used trans as the 0° reference state, which produced a Ramachandran plot that looked the way it does now but kept the origin at lower left (e.g. see Watson 1969[2] or Dickerson and Geis 1st edition[3]). In 1970, the IUPAC-IUB Commission report[4] redefined the reference for all torsions as cis with values from -180° to +180° to match chemical nomenclature, which gave the same Ramachandran plot appearance but with the origin at the center, so that mirror-image conformations have since then been symmetric around the origin. Figure 1 shows the modified hard-sphere Ramachandran contours[5] on the modern axes, with regions such as α, β, 3_{10}, Lα, bridge, etc. labeled.

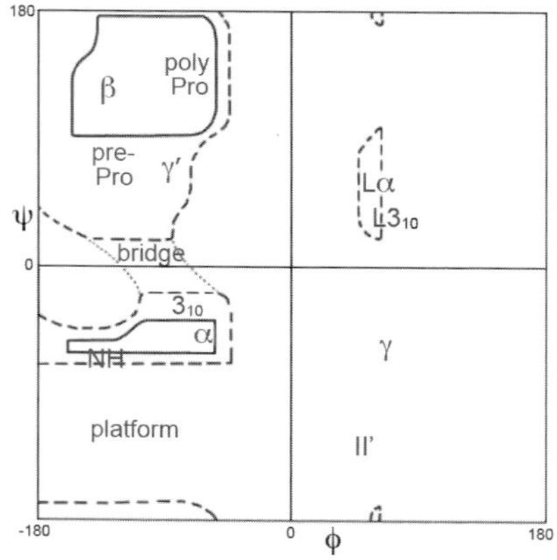

Figure 1 - Modified original hard-sphere Ramachandran contours plotted on the modern axis system, with labels for regions discussed here.

Note that the original Ramachandran plot contours were calculated from hard-sphere sterics for the inner "fully allowed" contours and with slightly smaller atomic radii for the "outer limits". Conclusions were drawn about the acceptability of various proposed backbone conformations, such as showing that left-handed α-helix is much less stable than right-handed. It was mentioned that changes in τ (the N-Cα-C bond angle) could make a difference, and a modified φ,ψ plot was published later by Ramachandran and Sasisekharan[5] showing the bridge of added conformations between α and β that are allowed if τ is opened up (foreshadowing an extra dimension now gaining use). In 1963 there was empirical confirmation of specific allowed conformations from a few small peptide crystal structures. The 1960 myoglobin structure[6] was not used or even mentioned, because the coordinates were not available and so φ and ψ could not be calculated. That fact underlines the crucial function of public databases and open availability for the results of publicly supported research. It was not until the Protein Data Bank[7] (PDB) and the Feldmann microfiche atlas[8], in the late 1970s, that Ramachandran plots could show datapoints from protein crystal structures. The 1976 PDB had 27 deposited structures, about half of which reached 2Å resolution (1b5c cytochrome b5, 1cpa carboxypeptidase A, 1cyt ferricytochrome C, 1fxn flavodoxin, 1lhb lamprey hemoglobin, 1mbn met myoglobin, 2pab prealbumin, 1rei immunoglobulin light chain, 1rns ribonuclease S, 1sns Staphylococcal nuclease, 2cha chymotrypsin, 2cna concanavalin A, 2pti pancreatic trypsin inhibitor, 2rxn rubredoxin).

The advent of the Ramachandran plot resulted in adding φ,ψ criteria to hydrogen bonding patterns as defining characteristics of the α-helix and β-sheet secondary structures. For example, both Chothia[9] and Salemme[10] studied the relationships between shape parameters for β-sheet (twist, pleat, curl, etc.) and direction of change or alternation on the φ,ψ plot. That same process also happened for smaller, non-repetitive structural motifs. Tight turns, or β-turns, were first defined by Venkatachalam in 1968 as the conformations of a tripeptide unit that form a CO(i) - NH(i+3) H-bond[11]. When turn examples were compiled[12,13], the H-bond requirement was found to be an artificial distinction, and the definition shifted to requiring a Cα(i) - Cα(i+3) distance <7Å with turn types defined by φ,ψ values for the two central residues, often plotted explicitly[14].

The much rarer γ conformation (Figure 1) has a history of wild swings between over-use and complete obscurity. First postulated in 1943 by Huggins[15] and then forgotten, it was proposed again in 1972[16] as forming the γ turn (with a CO(i) - NH(i+2) H-bond) and an example was identified in thermolysin[17]. The γ conformation is overemphasized in some calculations (see below), and was for a long period invoked too readily to explain peptide spectra, while it was again forgotten by the protein community and classed as an outlier by all the early validation programs (e.g., ProCheck[18]). Recently it has been reinstated as allowed but unfavorable, and validly over-represented at active sites[19].

The discovery of helix N-caps and C-caps was motivated by a search for ways that the overall polypeptide chain direction could change dramatically at the ends of secondary structures. It was carried out by plotting contours for the angle between successive CO groups onto the Ramachandran diagram, which showed that the most promising conformations would be poly-Pro, 3-10, and L3-10, with nearly perpendicular successive COs[20]. A similar CO angle now shows promise as a backbone validation tool at low resolution.

Once enough protein crystallographic data became available for comparison, reproducing the Ramachandran plot became a popular, and demanding, test of molecular energy calculations. For either classical or quantum calculations on the Ala dipeptide in vaccuum, the two global minima are not α-helix and extended, but the γ and γ' conformations[21-23], which are the only possible ways of forming a reasonable H-bond between two adjacent peptides. The angle is not good, but the lack of water drives formation of that H-bond in the dipeptide. Occurrence statistics show that this is not the right answer, because the γ' conformation is allowed but not optimal, and γ occurs but has a small clash between O(i-1) and Cβ and is very rare[19]. Increasingly sophisticated simulations with dielectric terms, explicit waters, all hydrogens, free energies, and mixed quantum/classical strategies do an increasingly good job[24-26], but many details remain elusive, such as the shape of the steep diagonal cliff above right of α, the relatively unoccupied region left of α, and especially the distribution for Gly. Some energy functions now include an empirical Ramachandran-plot term[27,28], although double-counting of the same physical effect is a caution, offset by means such as eliminating repetitive, H-bonded secondary structures from the reference data[29]. However, both the structural data analyses and the computational methods are actively improving, and it seems promising that they may fairly soon converge on agreement supported by a more complete physical understanding.

1.2 Use for Validation

The idea of "validating" macromolecular crystal structures started around 1990, with Ponder and Richards defining the first sidechain rotamer library in 1987[30], Branden and Jones pointing out some serious errors in deposited structures in 1990[31], Engh and Huber surveying small-molecule crystal structures to better determine mean and σ values for bond lengths and angles in 1991[32], Brunger defining Rfree in 1992[33], and the Thornton lab developing

ProCheck in 1992-3[18,34]. ProCheck used geometry and rotamer measures, but its centerpiece was Ramachandran-plot criteria as a solidly based quality check independent of the target functions in crystallographic refinement. They recognized that the database of solved macromolecular structures had achieved adequacy for empirically defining such criteria, but it still required a *tour-de-force* of interpretation to extract reliable measures from the very noisy plot of all 100,000 residues of data in the 1991 PDB - all resolutions, all copies, all B-factors. For the last 20 years of protein crystallography, ProCheck's "core", "allowed", and "generously allowed" Ramachandran regions have been the most reproducibly reported quality measures after resolution and Rfree, are part of the PDB deposition process[35], and are widely recognized to have reduced the frequency of errors[36].

After several years' growth in PDB depositions, increase in high-resolution structures enabled by cryocooling and synchrotron sources, and community experience in applying ϕ,ψ validation, those Ramachandran criteria were re-examined. Examples (such as γ turns) of residues in regions disallowed by the modified hard-sphere Ramachandran calculations[5] or by the ProCheck empirical criteria were carefully investigated and judged valid[37-39], while some parts of ProCheck's "generously allowed" regions now seemed too generous. In response, both Hooft et al. for What-Check[40] and Kleywegt and Jones for Oops[41] defined very similar and well-behaved ϕ,ψ outlines of a "favored" region that encloses 98% of the reference data, but declined to make the more difficult distinction between somewhat disfavored conformations and truly disallowed outliers. The 98% level was a good choice, because its outline has stayed virtually constant up through the most recent updates. Percent Ramachandran favored residues (or % core and % allowed by ProCheck) for a given protein crystal structure provides good validation of its overall quality, and has guided the development of model-building and refinement methodologies, but such single numbers do not help very much in identifying and fixing individual errors.

A reliable distinction of disfavored from outlier ϕ,ψ conformations is only a decade old, because it required B-factor filtering at the residue level, smoothed contours rather than 10° bins, and enough high-resolution, high-quality data to define a robust contour at the 3.5σ level (99.95%, or 1 in 2000) for general-case residues. That was achieved for MolProbity in 2003[19] from a 500-structure dataset, using the by-then classic 98% contour as favored, and the 99.95% contour to divide allowed from outliers. As well as omitting Gly and Pro from the Ramachandran general case (done since ProCheck), MolProbity explicitly scores them from individual distributions. Following the lead of Karplus[42], pre-

Pro residues are also scored separately. For those three individual cases, a well-behaved outer contour could be defined only at the level of 99.8% of the filtered reference data. These scores are reported for individual residues in Coot[43] and Phenix[44] as well as MolProbity[45], to help identify and correct local fitting errors when used in combination with hydrogen addition and all-atom contact analysis[46].

1.3 Use in Prediction and Design

The relationships shown by the Ramachandran plot were one factor in encouraging development of the early tries at predicting protein secondary structure and then 3D structure[47-49]. Backbone conformation will always be the central goal of such predictions, and although their success is still judged primarily by Cα rmsd measures[50], the full Ramachandran relationships in some form are essential to achieving such success and have increasingly been incorporated explicitly into most prediction software[28,51].

For structure validation, the outermost contour between allowed and outlier conformations is the most crucial; it is determined by backbone and Cβ clashes within the dipeptide, and is therefore essentially identical for 16 of the 20 amino acids (the "general case"). Testing further up than the 98% contour is actually detrimental, because it makes global validation scores dependent on composition of secondary structures. The MolProbity web service is tailored for structure validation, aiming to identify outliers for correction with a minimum of false alarms.

In contrast, for purposes of prediction or design one is most interested in identifying the especially favorable combinations of sequence and conformation. Those depend on shape and H-bonding groups all the way out the sidechain, and thus are individual for each amino acid. Individual Ramachandran plots for each amino acid have often been made[21,52] and interesting specific points learned from comparing them, but until recently there was not enough data to allow their systematic use. Two main strategies are used to improve reliability and smooth probabilities in the favored Ramachandran regions: fit of analytic, differentiable functions[53]and use of Bayesian statistics[54]. The Dunbrack lab has recently applied Bayesian methodology from their Scwrl sidechain rotamer tools[55,56], widely used in the prediction/design community, to the calculation of Ramachandran probabilities[57]. Now that the protein design field is moving past redesign on rigid backbones[58-60], some form of Ramachandran term has become important either for perturbation or for full *de novo* design of backbone. Such use primarily depends on the favorable ϕ,ψ regions and their correlation with sequence.

1.4 Adding Dimensions

Although the long-lasting success of the Ramachandran plot depended strongly on the initial decision to drop all but the two most important dimensions of ϕ and ψ, we now have enormous datasets and powerful visualization tools that give us the luxury of investigating the second-order gains from adding other dimensions. Reflecting Ramachandran's use of a wider τ angle to allow population of the "bridge" region between α and β, the accepted mean value of τ is 111.8°,[32] well above tetrahedral 109.6°. In 1996 Karplus showed that the variation of τ as a function of ϕ,ψ is quite large[43] (as shown in Figure 2). That τ variation is the centerpiece of a "conformation-dependent library"[61,62](CDL) from Karplus and Dunbrack which defines all near-backbone bonds and angles as a function of ϕ, ψ.

The most obvious parameter to add is of course ω, the torsion around the peptide bond. That is tricky, however, because the 180° difference between cis and trans forms overwhelms the significant variability within each form. Dunbrack and Karplus have studied the variability of trans ω seen in the well-ordered parts of atomic-resolution structures and found that ϕ,ψ dependence can account for about 25% (+/-8°) of the variance[63]. That effect is now part of the CDL. The cis/trans correlation with ϕ,ψ is now handled by separate Pro ϕ,ψ plots in MolProbity.

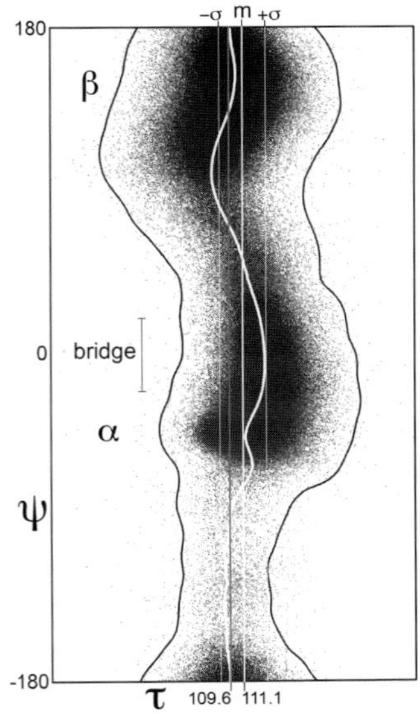

Figure 2 - Plot of τ angle variation vs ψ. Wavy white centerline plots local mean τ value.

A dramatic expansion in dimensionality has been the neighbor-dependent Ramachandran distributions[57], which add 840 left-neighbor and right-neighbor plots to the basic 21 neighbor-independent plots. Pre-Pro makes the largest difference, but aromatics and branched Cβ residues also have significant effects. This expansion of categories is enabled by a hierarchical Direchlet treatment, which mixes information from plots that share a dimension.

The most powerful extra dimension to the Ramachandran plot found so far is $\chi 1$, the first sidechain torsion angle. From the opposite direction, the Scwrl rotamer libraries[55] include "backbone dependent" terms that give rotamer frequencies as a function of ϕ,ψ and the "penultimate" rotamer library[64] gives frequencies in terms of secondary structure. In the next section, it will be shown that at current data densities it should be highly productive to make use of individual 3D amino-acid Ramachandran plots versus $\chi 1$.

2. Millions of Quality-filtered Residues for the Ramachandran Plot

Our ability to understand and use the information in Ramachandran plots has benefited greatly over the years by growth and improvement in the empirical database of protein crystal structures. This section will explore some highlights of what can now be seen from an order-of-magnitude more high-quality data than produced the original MolProbity system.

2.1 Selecting and Filtering the Top8000 Dataset

The Top8000 database was selected from protein crystal structures at <2Å resolution in the PDB release as of March 29, 2011. Reduce[65] was run to add and optimize hydrogens, including "flips" of Asn/Gln/His sidechains where needed, and the MolProbity score[66] (an estimate of the resolution at which the structure's all-atom clashscore[67], rotamer quality, and Ramachandran quality would be average) was calculated for each non-identical protein chain of >37 residues. Only chains with <2.0MolProbity scores, <5% residues with bond length or angle outliers, and <5% Cβ deviation outliers were kept. The PDB's homology clusters as of that same date were used to construct a non-redundant dataset at the 70% level by choosing the example with the best average of resolution and MolProbity score within each cluster. Datasets were also selected at other homology levels, and with a requirement for availability of maps from the Electron Density Server[68], but the simple 70% dataset is the one used here and called the Top8000. It contains 7,957 protein chains, compared to the 500 chains used in 2003[19].

After the above filtering at the file and chain level, the residue level is also quality-filtered. For purposes of Ramachandran-plot analysis, we omit any residue that has a backbone atom with B-factor ≥30 or an alternate conformation. That leaves 1,573,349 total residues for analysis, compared with 97,368 in 2003. Individual-residue validation scores, all torsion angles and other relevant parameters are stored and queried from our local MySQL database.

2.2 Ramachandran Plots for the New PDB Validation

Recently the wwPDB[69] has required the deposition of experimental data as well as coordinates, broadening the possibilities for structure validation, and has convened validation task forces (VTFs) for various structural biology methods. The VTF recommendations will be implemented in new information for depositors, referees, and end-users of macromolecular structures. The X-ray VTF report[70] recommends the adoption of many standardly used criteria, but often with some updating such as percentile scores for each structure both relative to its resolution cohort and also relative to all x-ray crystal structures in the PDB. By sometime in 2013, then, both global and per-residue Ramachandran scores will be easily available for every deposited protein structure.

For Ramachandran validation the existing MolProbity system[45] of general-case, Gly, Pro, and pre-Pro needs expansion to 6 categories, with cis and trans Pro separated and an additional category for Ile/Val. Those changes were decided by comparing contours for all amino-acid types (see supplement to VTF report). The outer contours that separate allowed from outlier were essentially identical for 16 general-case residues, and differed strongly among the 6 recommended categories whose Top8000 B<30 distributions are shown in Figure 3.

Some residues change from allowed to outlier or *vice versa* with the new criteria, either because of small shifts in the contour position or because of substantial changes for the new categories. 1A88 cis Pro31 (Figure 5a) with ϕ=-106°, ψ=+56° was previously scored as outlier, but is within the allowed region of the now-separate Top8000 cis Pro distribution. Clear electron density and flanking mainchain H-bonds validate its conformation. Furthermore, this residue's location adjacent to the enzyme's catalytic triad, with a putative H-bond (dotted line) thought to be important at a later step in the reaction, gives a plausible reason for its unusual but geniune conformation. 1BU8 Val246 (Figure 5b) with ϕ,ψ just below α-helix, was previously scored as allowed, but is now scored as an outlier by the separate Ile/Val distribution. Other validation criteria confirm that it is badly misfit: many all-atom clashes, neighboring Ramachandran outliers, bad Cβ geometry, and difference density peaks.

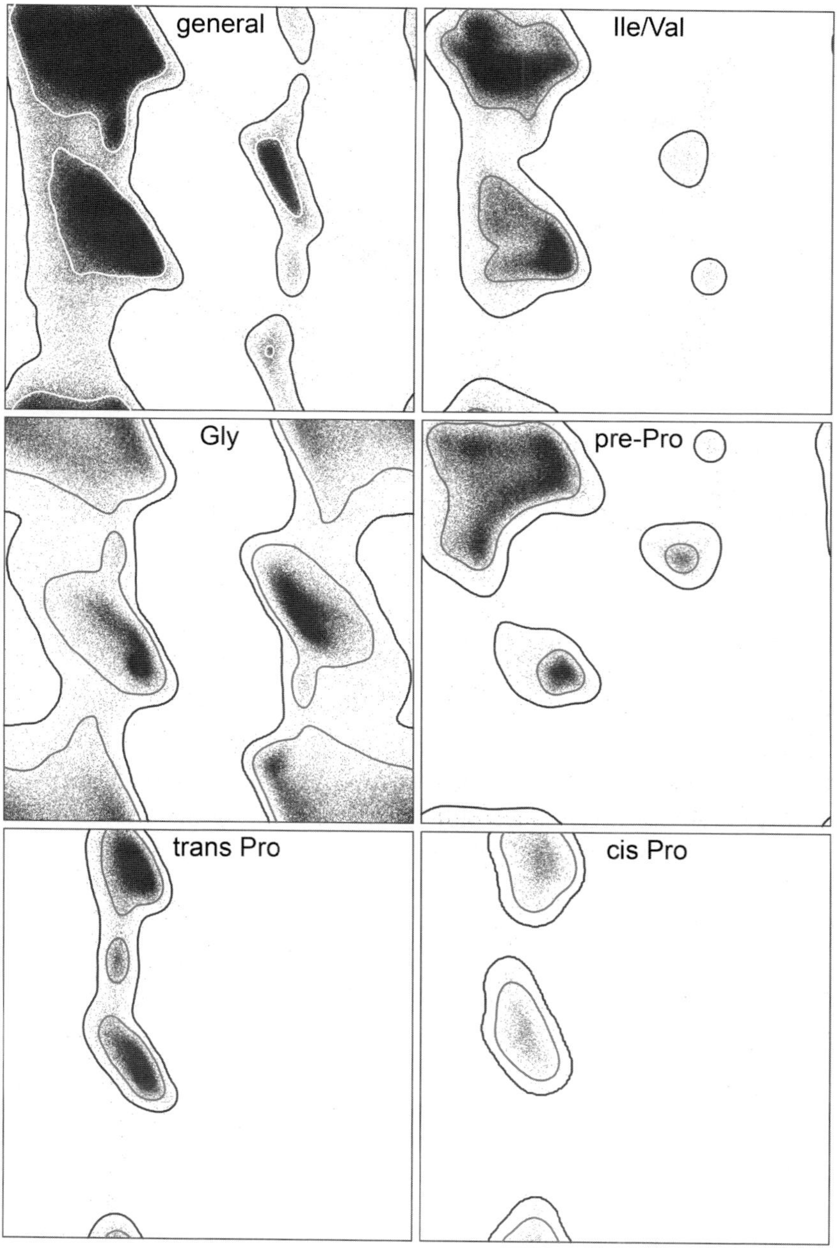

Figure 3 - Top8000 Ramachandran plots for the 6 categories recommended by the wwPDB VTF. Outer contour divides allowed from outliers; inner is at 98% level.

2.3 *Exploiting More Details and More Dimensions*

The 1.6 million quality-filtered datapoints in the Top8000 distributions illuminate new details such as small clusters indicating local structural motifs. The Gly distribution gains in understandability from repositioning its axes to make the populated areas continuous, as done in Figure 4, with color-coded datapoint density. The prominent, round cluster at top left contains glycines that enable type II' turns (example in Figure 5c). Even more evident here than for

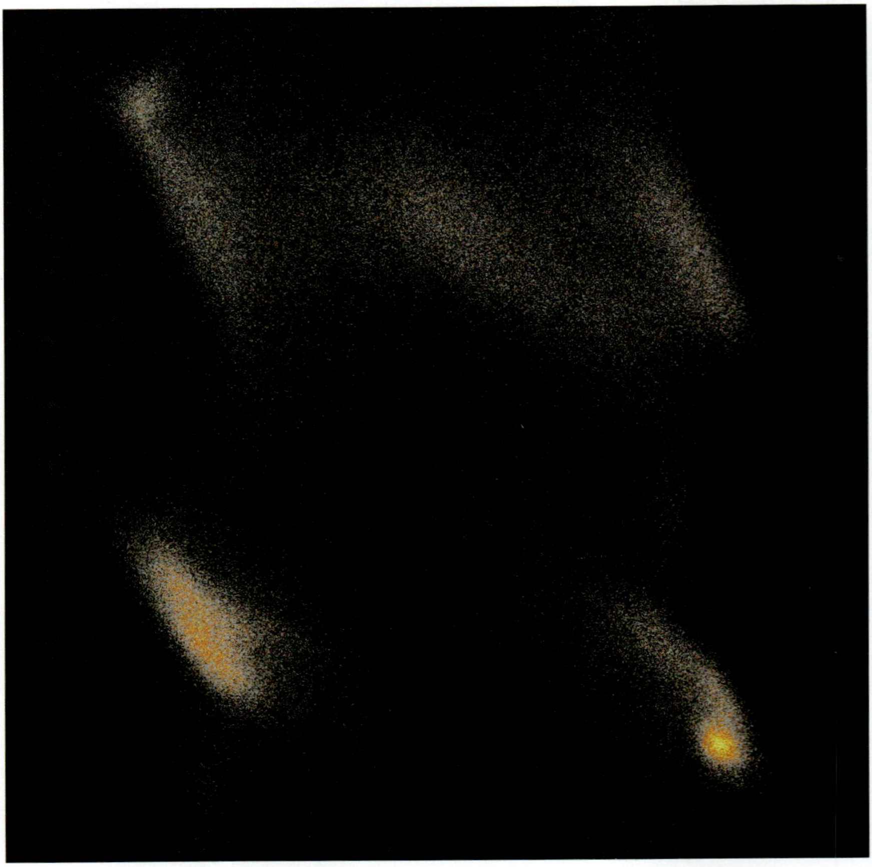

Figure 4 - Top8000 Ramachandran plot for Gly, with axes repositioned to make the populated areas continuous: ϕ is shifted by 180°, ψ by -90°. α-helix is at lower right and extended conformations across the top. This figure (and related images) is available at 3600x3600px full resolution on http://commons.wikimedia.org/wiki/User:Dcrjsr.

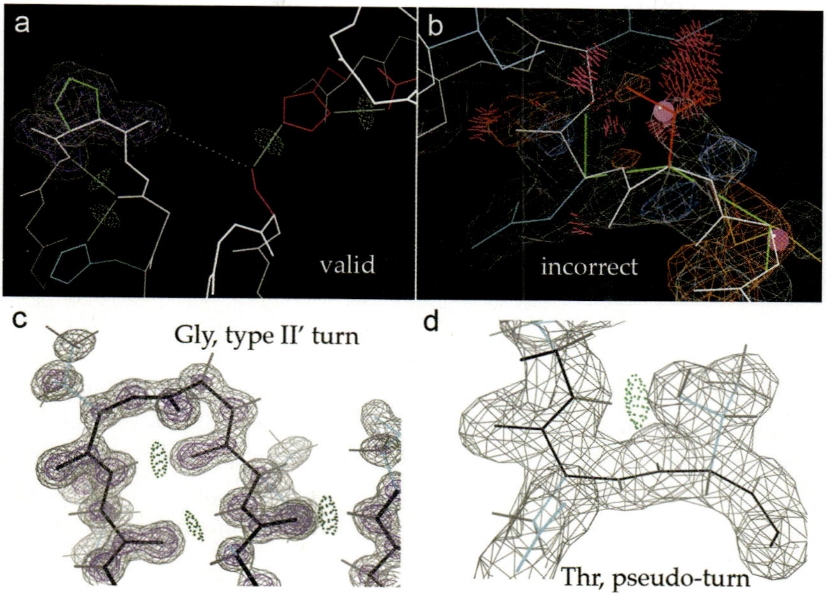

Figure 5 –a) newly allowed case; b) newly recognized outlier; c) II' turn; d) pseudo-turn.

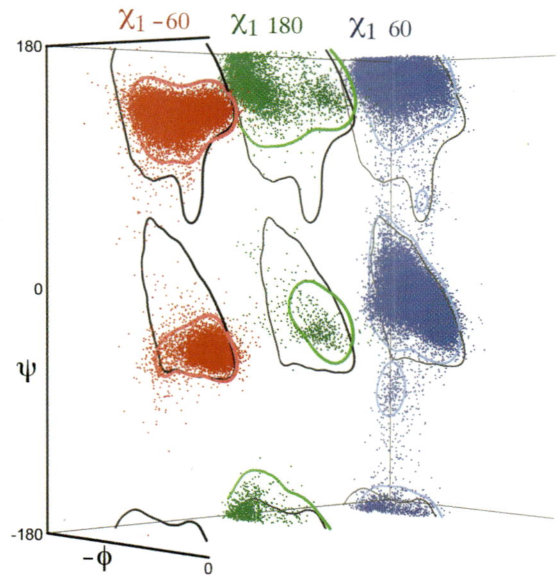

Figure 6 - Top8000 plot of $\phi,\psi,\chi 1$ for Thr, showing large differences between $\chi 1$ classes.

the general case are the diagonal sweeps produced by correlated changes of ϕ and ψ along the edges of regions forbidden by backbone clashes.

Motif clusters and other features often are seen only when backbone and sidechain dimensions are combined. A prominent example is the Thr Ramachandran distribution versus the $\chi1$ dimension, shown in Figure 6. Although the outer contour for Thr matches the general case very well, the areas populated by each $\chi1$ group have almost no overlap. The 3_{10} region above left of α is populated entirely by $\chi1$ p threonines. $\chi1$ m threonines lie low in the extended region and $\chi1$ p sit high, while $\chi1$ t strongly prefers the far left side -- most of those form pseudo-turns[71] (Figure 5d) where Oγ1 imitates a tight turn by H-bonding with NH(i+2).

The most important conclusion from multi-dimensional distributions like Figure 6 is that we had been missing an important aspect of ϕ,ψ validation. As the next step for MolProbity we will doubtless be using three $\chi1$ classes of Ramachandran distribution for each amino-acid group, and will also explore what these re-imagined, high-dimensional Ramachandran plots can teach us about local structural motifs.

Acknowledgments

We acknowledge support from National Institutes of Health grants GM073930 and GM073919, and we thank Bryan Arendall for his all-PDB MySQL database and his help with the Top8000 selection.

References

1. G. N. Ramachandran, C. Ramakrishnan and V. Sasisekharan, *J. Mol. Biol.* **7**, 95 (1963).
2. H. C. Watson, *Prog. Stereochem.* **4**, 299 (1969).
3. R. E. Dickerson and I. Geis, Structure and Action of Proteins, Harper, New York (1969).
4. IUPAC-IUB Commission on Biochemical Nomenclature, *J. Biol. Chem.* **245**, 6489 (1970).
5. G. N. Ramachandran and V. Sasisekharan, *Adv. Protein Chem.* **23**, 284 (1968).
6. J. C. Kendrew, R. E. Dickerson, B. E. Strandberg, R. G. Hart, D. R. Davies, D. C. Phillips and V. C. Shore, *Nature* **185**, 422 (1960)
7. F. C. Bernstein, T. F. Koetzle, G. J. B. Williams, E. F. Meyer Jr., M. D. Brice, J. R. Rodgers, O. Kennard, T. Shimanouchi and M. Tasumi, *J. Mol. Biol.* **112**, 535 (1977).

8. R. J. Feldmann, *Atlas of Macromolecular Structure on Microfiche,* Tracor-Jitco Inc., Rockville MD (1976).
9. F. R. Salemme and D. Weatherford, *J. Mol. Biol.***146**, 119 (1981).
10. C. Chothia, *J. Mol. Biol.***163**, 107 (1983).
11. C. M. Venkatachalam, *Biopolymers* **6**, 1425 (1968).
12. P. N. Lewis, F. A. Momany and H. A. Scheraga, *Biochem. Biophys. Acta* **303**, 211 (1973).
13. P. Y. Chou and G. D. Fasman, *J. Mol. Biol.* **115**, 135 (1977).
14. J. S. Richardson, *Adv. Protein Chem.* **34**, 167 (1981).
15. M. L. Huggins, *Chem. Rev.* **32**, 195 (1943).
16. G. Nemethy and M. P. Prinz, *Macromolecules* **5**, 755 (1972).
17. B. W. Matthews, *Macromolecules* **5**, 818 (1972).
18. R. A. Laskowski, M. W. MacArthur, D. S. Moss and J. M. Thornton, *J. Appl. Crystallogr.* **26**, 283 (1993).
19. S. C. Lovell, I. W. Davis, W. B. Arendall III, P. I. W. de Bakker, J. M. Word, M. G. Prisant, J. S. Richardson and D. C. Richardson, *Proteins: Struc. Func. Genet.* **50**, 437 (2003).
20. J. S. Richardson and D. C. Richardson, chapter 1 in *Prediction of Protein Structure and the Principles of Protein Conformation*, Ed. G. D. Fasman Plenum Press (1988).
21. T. C. Cheam and S. Krimm, *J. Mol. Struct. Theochem.* **206**, 173 (1990).
22. T. Head-Gordon, M. Head-Gordon, M. J. Frisch, C. L. Brooks III and J. A. Pople, *J. Amer. Chem. Soc.* **113**, 5989 (1991).
23. I. R. Gould and P. A. Kollman, *J. Phys. Chem.* **96**, 9255, (1992).
24. I. K. Roterman, M. H. Lambert, K. D. Gibson and H. A. Scheraga, *J. Biomol. Struct. Dyn.* **7**, 421 (1989).
25. J. A. D'Aquino, J. Gomez, V. J. Hilser, K. H. Lee, L. M. Amzel and E. Freire, *Proteins* **25**, 143 (1996).
26. H. Hu, M. Elstner and J. Hermans, *Proteins* **50**, 451 (2003).
27. M. Buck, S. Bouguet-Bonnet, R. W. Pastor and A. D. MacKerell, *Biophys. J.* **90**, L36 (2006).
28. C. A. Rohl, C. E. Strauss, K. M. Misura and D. Baker, *Methods Enzymol.* **383**, 66 (2004).
29. M. B. Swindells, M. W. MacArthur and J. M. Thornton, *Nature Struct. Biol.* **2**, 596 (1995).
30. J. W. Ponder and F. M. Richards, *J. Mol. Biol.* **193**, 775 (1987).
31. C.-I. Branden and T. A. Jones, *Nature* **343**, 687 (1990).
32. R. A. Engh and R. Huber, *Acta Crystallogr. A***47**, 392 (1991).
33. A. T. Brunger, *Nature* **355**, 472 (1992).
34. A. L. Morris, M. W. MacArthur, E. G. Hutchinson and J. M. Thornton, *Proteins* **12**, 345 (1992).

35. H. M. Berman, J. Westbrook, Z. Feng, G. Gilliland, T. N. Bhat, H. Weissig, I. N. Shindyalov and P. E. Bourne, *Nucleic Acids Res.* **28**, 235 (2000).
36. G. J. Kleywegt and T. A. Jones, *Structure* **10**, 465 (2002).
37. O. Herzberg and J. Moult, *Proteins* **11**, 223 (1991).
38. K. Gunasekaran, C. Ramakrishnan and P. Balaram, *J. Mol. Biol.* **264**, 191 (1996).
39. D. Pal and P. Chakrabarti, *Biopolymers* **63**, 195 (2002).
40. R. W. W. Hooft, G. Vriend, C. Sander and E. E. Abola, *Nature* **381**, 272 (1996).
41. G. J. Kleywegt and T. A. Jones, *Structure* **5**, 1557 (1996).
42. P. A. Karplus, *Protein Sci.* **5**, 1406 (1996).
43. P. Emsley, B. Lohkamp, W. G. Scott and K. Cowtan, *Acta Crystallogr. D***66**, 486 (2010).
44. P. D. Adams, P. V. Afonine, G. Bunkoczi, V. B. Chen, I. W. Davis, N. Echols, J. J. Headd, L.-W. Hung, G. J. Kapral, R. W. Grosse-Kunstleve, A. J. McCoy, N. W. Moriarty, R. Oeffner, R. J. Read, D. C. Richardson, J. S. Richardson, T. C. Terwilliger and P. H. Zwart, *Acta Crystallogr. D***66**, 213 (2010).
45. V. B. Chen, W. B. Arendall III, J. J. Headd, D. A. Keedy, R. M. Immormino, G. J. Kapral, L. W. Murray, J. S. Richardson and D. C. Richardson, *Acta Crystallogr. D***66**, 12 (2010).
46. W. B. Arendall III, W. Tempel, J. S. Richardson, W. Zhou, S. Wang, I. W. Davis, Z.-J. Liu, J. P. Rose, W. M. Carson, M. Luo, D. C. Richardson and B.-C. Wang, *J. Struct. Funct. Genomics* **6**, 1 (2005).
47. O. Ptitsyn, *J. Mol. Biol.* **42**, 501 (1969).
48. P. Y. Chou and G. D. Fasman, *Biochemistry* **13**, 222 (1974).
49. J. Garnier, D. J. Osguthorpe and B. Robson, *J. Mol. Biol.* **120**, 97 (1978).
50. V. Mariani, F. Kiefer, T. Schmidt, J. Haas and T. Schwede, *Proteins* **79** Suppl **S10**, 37 (2011).
51. D. Xu, J. Zhang, A. Roy and Y. Zhang, *Proteins* **79** Suppl **S10**, 147 (2011).
52. P. Chakrabarti and D. Pal, *Prog. Biophys. Mol. Biol.* **76**, 1 (2001).
53. E. D. Amir, N. Kalisman and C. Keasar, *Proteins* **72**, 62 (2008).
54. A. Pertsemlidis, J. Zelinka, J. W. Fondon III, R. K. Henderson and Z. Otwinowski, *Stat. Appl. Genet. Mol. Biol.* **4**, Article35 (2005).
55. R. L. Dunbrack and F. E. Cohen, *Protein Sci.* **6**, 1661 (1997).
56. G. G. Krivov, M. V. Shapalov and R. L. Dunbrack, *Proteins* **77**, 778 (2009).
57. D. Ting, G. Wang, M. V. Shapalov, R. Mitra, M. I. Jordan and R. L. Dunbrack Jr., *PLoS Comp. Biol.* **6**, e1000763 (2010) .
58. B. Kuhlman, G. Dantas, G. C. Ireton, G. Varani, B. L. Stoddard and D. Baker, *Science* **302**, 1364(2003).
59. C. A. Smith and T. Kortemme, *J. Mol. Biol.* **380**, 742 (2008).

60. M. A. Hallen, D. A. Keedy and B. R. Donald, *Proteins* epub, doi: 10.1002/prot.24150 (2012)
61. D. S. Berkholz, M. V. Shapalov, R. L. Dunbrack and P. A. Karplus, *Structure* **17**, 1316 (2009).
62. D. E. Tronrud and P. A. Karplus, *Acta Crystallogr. D***67**, 699 (2011).
63. D. S. Berkholz, C. M. Driggers, M. V. Shapalov, R. L. Dunbrack and P. A. Karplus, *Proc. Nat. Acad. Sci. USA* **109**, 449 (2012).
64. S. C. Lovell, J. M. Word, J. S. Richardson and D. C. Richardson, *Proteins* **40**, 389 (2000).
65. . M. Word, S. C. Lovell, J. S. Richardson and D. C. Richardson, *J. Mol. Biol.* **285**, 1735 (1999).
66. I. W. Davis, A. Leaver-Fay, V. B. Chen, J. N. Block, G. J. Kapral, X. Wang, L. W. Murray, W. B. Arendall III, J. Snoeyink, J. S. Richardson and D. C. Richardson, *Nucl. Acids Res.* **35**, W375 (2007).
67. J. M. Word, S. C. Lovell, T. H. LaBean, H. C. Taylor, M. E. Zalis, B. K. Presley, J. S. Richardson and D. C. Richardson, *J. Mol. Biol.* **285**, 1711 (1999).
68. G. J. Kleywegt, M. R. Harris, J. Y. Zou, T. C. Taylor, A. Wahlby and T. A. Jones, *Acta Crystallogr. D***60**, 2240 (2004).
69. H. M. Berman, K. Henrick and H. Nakamura, *Nature Struct. Biol.***10**, 980 (2003).
70. R. J. Read, P. D. Adams, W. B. Arendall III, A. T. Brunger, P. Emsley, R. P. Joosten, G. J. Kleywegt, E. B. Krissinel, T. Lutteke, Z. Otwinowski, A. Perrakis, J. S. Richardson, W. H. Scheffler, J. L. Smith, I. J. Tickle, G. Vriend and P. Zwart, *Structure* **19**, 1395 (2011).
71. D. Rees, M. Lewis, J. Lipscomb, *J. Mol. Biol.* **168**, 367 (1983).

THE RAMACHANDRAN PLOT AND PROTEIN STRUCTURE VALIDATION

ROMAN A LASKOWSKI, NICHOLAS FURNHAM AND JANET M THORNTON

European Bioinformatics Institute
Wellcome Trust Genome Campus
Hinxton, Cambridge
CB10 1SD, UK

The Ramachandran plot has been the mainstay of protein structure validation for many years. Its detailed structure has been continually analysed and refined as more and more experimentally determined models of protein 3D structures have become available, particularly at high and ultra-high resolution. As more data has accumulated, so the differences in the plot for different amino acids, and with different residue neighbours, or in different structural contexts, have become apparent. Validation software has had to adapt to these subtle changes and a single Ramachandran plot can no longer accurately depict 'favourable' and 'disallowed' regions for all amino acid types.

1. Introduction

The classic 1963 paper of Ramachandran *et al.*[1] (and subsequent work[2,3]) showed how certain combinations of residue φ-ψ torsion angles in a protein's backbone should be unattainable as a result of steric clashes. These clashes – between the residues' backbone atoms, or between backbone and side chain C^{β} atoms – were characterized by 'disallowed' regions on the 2D plot of φ against ψ (Figure 1). The remaining, 'allowed' regions were classified according to the local 3D conformation of the protein's backbone – for example, residues in α-helices have φ-ψ values close to -57° and -45°, respectively. The plot – termed the Ramachandran plot or the Ramachandran diagram – was devised before the first protein structures had been solved and its general form was confirmed as soon as structural models of proteins started to be determined.

Very soon the plot became a validation tool used by X-ray crystallographers to check the reasonableness of their protein structural models. Models having residues with φ-ψ combinations lying in the disallowed regions of the plot needed to be checked or refined further. Its use as a validation tool was strengthened in 1990 when the details of a number of seriously wrong structural models were reported[4]. In that paper the authors spelt out various sources of error that can occur during structure determination, and identified the

Ramachandran plot as one of several crucial validation tools to help avoid such serious errors.

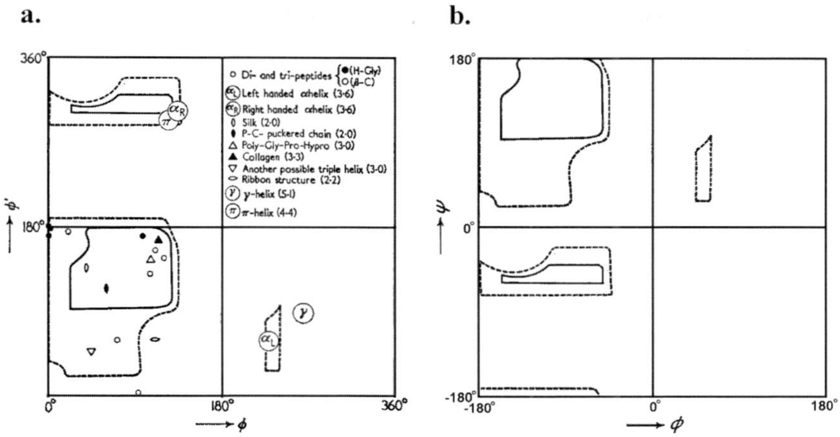

Figure 1. Original Ramachandran plot (a) as reported in the 1963 paper, and (b) using current-day definitions of the φ-ψ torsion angles. The 'fully allowed' regions are those enclosed by solid lines, while the 'outer limit' regions are bounded by dashed lines.

Protein structure validation became a hot topic throughout the early 1990s. The standard measures of 'quality' – the resolution at which the structure had been solved and the *R*-factor achieved at the end of refinement – provided only a rough guide to 'quality'. In particular, the *R*-factor was vulnerable to manipulation, whether deliberate or inadvertent. Furthermore, these measures applied to the structure as a whole, and hence provided no information on local structural quality – ie which regions were more reliable than others. Admittedly, the individual atomic *B*-factors could provide a qualitative indication of where the structure was less well defined by the experimental data, but even these were a poor guide to whether the structure was wholly or partially correct.

As more protein structures were solved, so the number of experimentally observed residue φ-ψ values increased. Differences between the theoretical φ-ψ distribution of Ramachandran and the experimental data became apparent, as will be mentioned later. So, the experimentally observed distributions took over as the basis of the 'allowed' and 'forbidden' regions of the plot.

In 1992, Morris *et al.*[5] looked at the Ramachandran plot as part of their analysis of various geometrical properties of proteins. The aim of their study was to identify geometrical properties that correlated with resolution. The idea was that, at higher and higher resolution, certain properties would tend towards their 'ideal' values – and hence give an indication of what these ideal values might be. For the Ramachandran plot, they first obtained the φ-ψ distribution for all 121,870 residues, other than Pro and Gly, in the 462 protein structural models available from the Protein Data Bank[6] (PDB) at the time (Oct 1990). They then counted the number of data points in each 10°×10° cell of the plot and assigned the cells to one of four classes according to the number of points: CORE regions (containing >100 data points), ALLOWED (>7 residues), GENEROUS (extending two cells all round the ALLOWED region), and OUTSIDE. These regions can be seen as the coloured background areas in Figure 2*a*. The authors found that models solved at higher resolution tended to have more residues in the CORE region and fewer in the disallowed OTHER regions. However, some models bucked this trend and these tended to be ones that had high *R*-factors or had been incompletely refined or were known to contain errors.

2. Use of the Ramachandran plot in validation

2.1. *PROCHECK*

The work of Morris *et al.* led directly to the development of PROCHECK[7] – a suite of protein structure validation programs that analysed the geometry of any given protein model and assessed it on the basis of its geometrical properties. The Ramachandran plot analysis employed the same Morris *et al.* regions shown in Figure 2*a*. Pro and Gly were excluded from this plot as they have very different distributions of φ-ψ values. The number of residues in the most favoured 'CORE' region provided a rough estimate of a model's quality. A rule of thumb was that, for atomic resolution models, the percentage of CORE residues should be >90%, with this percentage dropping as a function of resolution to around 70% for 3.0-4.0Å structures.

PROCHECK also provided separate Ramachandran plots for Gly and Pro residues: glycine has a wider and freer allowed region due to its lack of a C^{β} atom, and proline has its φ restricted to values around -65°. Indeed, PROCHECK also provided a separate plot for each of the 20 amino acid residues. Rather than using distinct regions, as in the main plot, these amino acid-based plots compared the φ-ψ values against the density of point in the given 10°×10° cell, as obtained from a high quality data set.

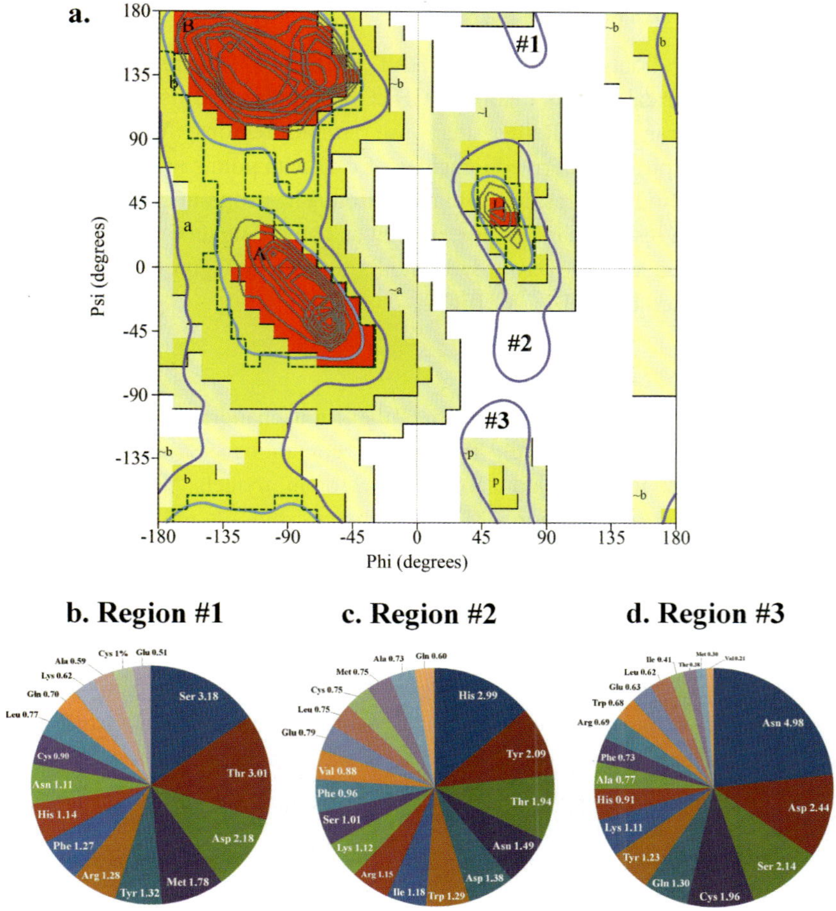

Figure 2. A comparison of some of the Ramachandran plot regions, for all residues except Gly and Pro, as implemented in some commonly-used validation programs. The background coloured regions are those of PROCHECK: red = most favoured, yellow = additional allowed, cream = generously allowed, and white disallowed. The curved blue contours define the MolProbity regions (which also exclude pre-Pro residues): light blue = favoured, dark blue = allowed. The dashed green lines define the boundary of the Kleywegt allowed regions, while the grey contour lines are WHATCHECK's contours. The regions labelled #1, #2 and #3 are regions considered disallowed by PROCHECK, but allowed by MolProbity. These tend to be occupied by specific amino acids, as illustrated by the three pie charts at the bottom. The pie charts show the propensities of each amino acid type to be found in each of these three regions. The propensities are calculated by normalizing the count of each amino acid type in the region by the total number of that amino acid type in the data set.

2.2. *WHAT_CHECK*

Another early validation tool was WHAT_CHECK[8]. This provided a wider range of geometrical checks than PROCHECK and also had its own version of the Ramachandran plot. Rather than use simple regions, WHAT_CHECK converted its empirically-derived densities of φ-ψ data points at each location to a probability of finding a residue at that location. The program could then assign a *z*-score to each residue's φ-ψ value, as well as computing an overall *z*-score for the structure as a whole. The grey contours shown in Figure 2*a* indicate the densities computed by WHAT_CHECK.

2.3. *OOPS*

In 1996, Kleywegt & Jones simplified the Ramachandran plot to just two regions: 'core' and 'non-core', the former containing 98% of non-glycine residues[9]. They used a data set of 403 protein models, solved by X-ray crystallography at a resolution of 2.0Å or better, and filtered so that no two proteins in the data set had more than 95% sequence identity. The borders of their 'core' regions are shown as the dashed green contours in Figure 2*a*. The percentage of residues lying in the 'non-core' regions provided a quality measure for the OOPS validation program. The authors showed that, although it was possible to apply φ-ψ restraints during refinement and obtain 'better-looking' structures – in terms of the φ-ψ values – this was at the cost of worse *R*-factor and R_{free} measures, so use of such restraints was not recommended. They also suggested another validation use of the Ramachandran plot in cases where the structural model exhibited non-crystallographic symmetry (NCS): the plot could provide a helpful way of visualizing any differences in the corresponding φ-ψ values of equivalent residues.

2.4. *MolProbity*

Perhaps the best-used protein structure validation program nowdays is MolProbity[10]. Its Ramachandran plot employs contours obtained from a careful analysis of 100,000 residues in 500 high resolution structures by Lovell *et al.*[11] In the original version of the program, four cases were treated separately: Pro, Gly, pre-Pro, and all other residues. The contours for each plot were derived such that the inner ones encompassed 98% of the φ-ψ values, while the outer ones 99.95%. Figure 2*a* shows the MolProbity contours for the general case as the light-blue and dark-blue curved contour lines.

The most recent analysis for MolProbity has separated out the Ile/Val residues into their own, separate, distribution, as well as having separate plots for trans- and cis-Pro[10], making six Ramachandran plots in all. The data set on which the data are based is now also considerably larger, comprising 8,000 structures and 1.5 million filtered residues[12].

3. The power of the Ramachandran plot

The Ramachandran plot has provided a potent validation check on protein structure models, whether experimentally or computationally determined (eg by homology modelling or *ab initio* prediction). Its principal strength is that it is a difficult measure to artificially 'fix' if the structure has been refined without φ-ψ restraints. A demonstration of this was provided by the deliberate tracing of a protein chain *backwards* through the electron density and then refining the backwards model until reasonable quality indicators were obtained[13]. While some quality measures could be manipulated to reasonable-looking values, the Ramachandran plot refused to yield, portraying many residues in the disallowed regions and exhibiting none of the usual clustering of residues in the most favourable regions. This fake structure, and its Ramachandran plot, can be found in the PDBsum database[14] under its dummy PDB code of 'fake': http://www.ebi.ac.uk/pdbsum/fake.

4. Differences in the Ramachandran plots used for validation

One can see from Figure 2*a* that there are some differences in the regions defined as favourable and unfavourable by the different validation programs. The main cause of these differences are the numbers of structural models used for defining the regions; as more models have been solved, so the extra data has revealed subtle new features of the plot. The PROCHECK plot, in particular, suffers from having been derived using very old data when there were not enough structures to allow selection of only the best.

The PROCHECK regions show a number of differences from the MolProbity regions, the most significant of which are labelled #1, #2 and #3 in Figure 2*a*. These regions are 'allowed' by MolProbity, but 'disallowed' by PROCHECK. A look at the φ-ψ distribution in the original Morris *et al.* paper[5] shows that there were indeed data points in these three regions, but too few to appear significant. One reason for this is that two of the regions – #1 and #3 – have a preference for specific amino acids. Figure 2*b* shows that region #1 has a preponderance of Ser, Thr and Asp residues, while region #3 is dominated by

Asn, Asp and Ser residues. This would account for the low density of points in the original distribution. Region #2 does not have any such marked preferences, yet nevertheless shows a low density in the Morris *et al.* distribution.

The sizes (ie areas) of the 'disallowed' regions also differ quite markedly across the validation methods. MolProbity has the smallest 'disallowed' area, covering approximately 735 of the 10°×10° cells. PROCHECK flags residues that fall in both its 'disallowed' areas of 386 cells and its 'generous' areas of 404 cells – making 790 'disallowed' cells in all – while OOPS has the largest 'disallowed' region at 1,041 cells.

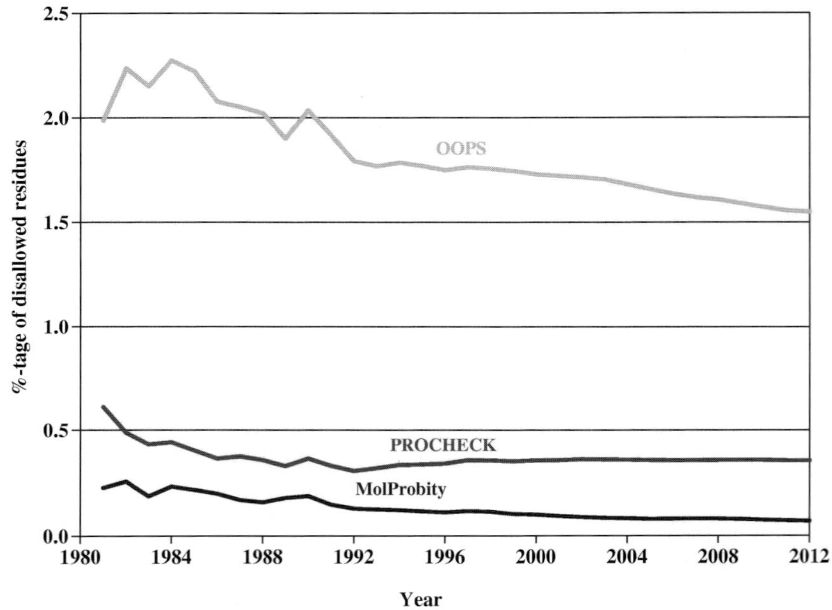

Figure 3. Percentages of residues in the 'disallowed' regions of the Ramachandran plot – as defined by three different programs – over time (1981 to 2012). The structures in each year were chosen to be X-ray structures solved to 2.0Å or better, and with an *R*-factor of no worse than 0.2. Further, they were filtered to exclude redundant entries at higher than 99% sequence identity. The three programs represented are: OOPS, PROCHECK and MolProbity. The PROCHECK 'disallowed' regions are defined as the 'OUTSIDE' and 'GENEROUS' regions, combined. The number of non-redundant files in each year increases from 10 in 1981 to 1,923 in 2011 (the data for 2012 were collected in August, so did not cover a full year).

5. Changes as a function of time

As mentioned above, the Morris *et al.* analyses[5] showed that, at higher resolutions, the residues tend to converge on the CORE regions in φ-ψ space. This has been confirmed by analyses of atomic resolution structures – ie structures solved to 1.4Å or better. These high-quality structures exhibit very tight clustering in the most favourable regions of the plot[15].

One might expect, therefore, that this tighter clustering would also be observed as a function of time, as techniques improve and the average resolution of protein structural models improves. Figure 3 shows the variation in the average percentage of residues in the 'disallowed' regions of the Ramachandran plot (as defined by three of the validation programs) for each year since 1981. The average percentage varies wildly over the first few years due to the small numbers of unique protein structures solved in those years. But then the percentage decreases according to both the OOPS and MolProbity definitions from about 1993 onwards. The OOPS allowed regions were defined to contain 98% of the data points, so it is interesting how the percentage of disallowed residues starts off near the 2% mark, but then drops to nearly 1.5%. The MolProbity allowed regions aim to encompass 99.95% of the data points, and even here the percentage disallowed has dropped below the expected value of 0.05 to 0.03. Unlike the other two plots, the PROCHECK percentage seems to level out, suggesting that its Ramachandran regions are not detecting significant changes in their disallowed regions.

Are these apparent improvements, as detected by OOPS and MolProbity, the result of improvements in resolution? In fact, it would seem not, as there is no discernible shift in average resolution over the years since 1993 for the data used for Figure 3 (which, it should be noted, excludes all structures solved at lower than 2.0Å). The improvements are probably a consequence of the higher emphasis that has been placed on structure validation since the start of the 1990s.

More interesting is the question of where on the plot the changes are occurring. To answer this question, we counted the percentage of residues in each 10°×10° cell of the plot for every year since 1995. For any cell this percentage may have risen, or dropped or stayed fairly constant over time. To identify which cells showed marked increases or decreases, we calculated the correlation coefficient between the year and the percentage observed for that year for every cell. Cells with a positive correlation are those whose residue density has increased over time, while the negatively correlated cells show a depletion of residues with time.

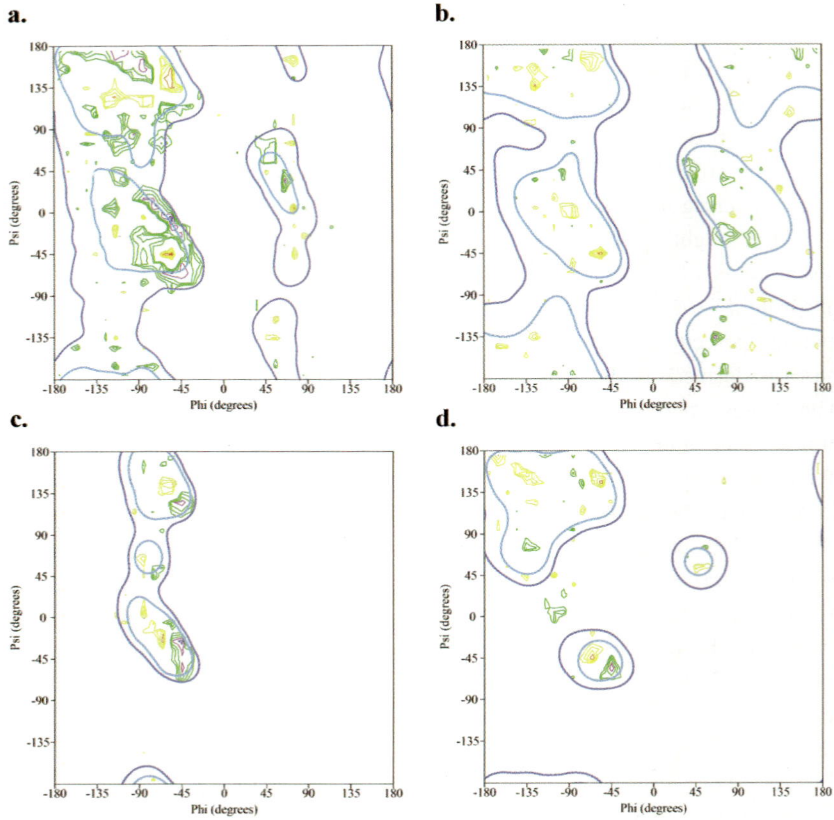

Figure 4. Contour plots showing the regions of the Ramachandran plot that have become more populated over time (yellow, orange and red contours) and those that have become less populated (green, violet and purple contours). The blue and light blue contours are those of MolProbity, shown for comparison. The plots show the results obtained for: *a*. all residues except Gly, Pro and pre-Pro; *b*. Gly residues only; *c*. Pro residues only; and *d*. pre-Pro residues only. The data were derived from the counts of residues in each 10°×10° cell, for each year from 1995 to 2012. The counts were converted to percentage values of the total number of points for that year. For each cell a correlation coefficient was computed for year *vs* percentage value. The yellow, orange and red contours show regions of high positive correlation – ie the percentage of points in the given cell has risen during the years 1995-2012. Conversely, the green, violet and purple contours signify negative correlation, and consequently a relative loss of data points in those cells.

Figure 4 shows the contour plots of these correlation coefficients for four cases: *a*. all residues except Gly, Pro and pre-Pro; *b*. Gly residues only; *c*. Pro residues only; and *d*. pre-Pro residues only. The smooth blue lines correspond to the MolProbity contours. The yellow, orange and red contours show regions of high positive correlation where the percentage of points in the given cell has risen during the years 1995-2012. Conversely, the green, violet and purple contours signify negative correlation, and consequently a relative loss of data points in those cells. As expected, the increases in all four plots occur in the highly favoured regions, such as those corresponding to α-helices and β-sheets. The decreases in Figs 4a and 4c, together with the decrease near the α–helical region of Fig 4*d,* appear to come from regions adjacent to the highly-favoured 'hot spots', as though the peaks there are getting sharper. The decreases in the glycine plot of Figs 4*b* appear more randomly distributed.

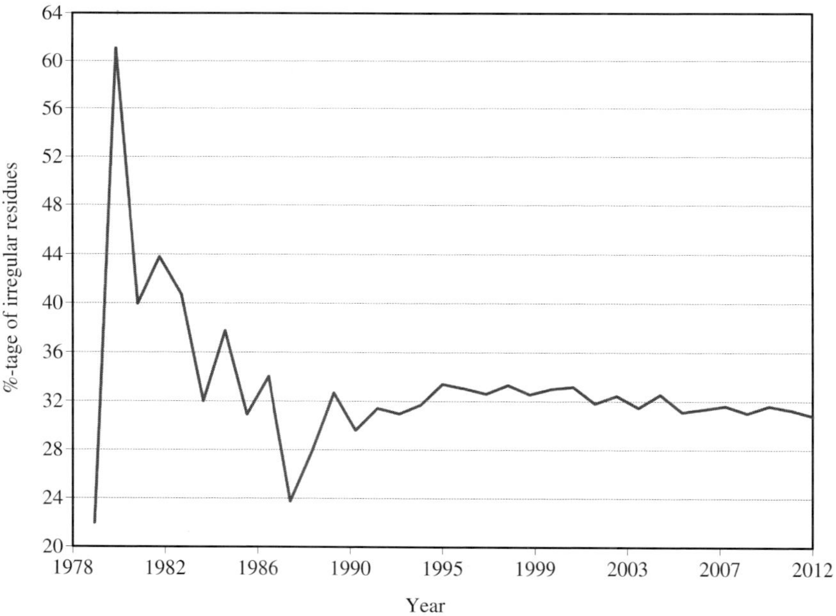

Figure 5. Average percentage of residues in irregular secondary structure, as a function of time. Secondary structure was calculated using sstruc, which uses a modified version of the DSSP algorithm, and assigns secondary structure on the basis of hydrogen bonding patterns, rather than on φ-ψ combination. Residues in irregular secondary structure were taken to be those not assigned to helix or strand by sstruc (ie codes H, h, G, E, and e).

Figure 5 shows one apparent consequence (or cause) of this tightening of the core Ramachandran regions. It is a plot of the average percentage of irregular secondary structure in proteins solved in each year. There appears to be a trend for this percentage to decrease from 1995 to the present day – and consequently the percentage of residues in regular secondary structure to increase. Earlier years are represented by fewer structures, and so have a greater degree of variability. The recent downward trend is not huge – 33.4% in 1995 to 30.8% in 2012 – but the implication is that, as more residues have moved into the more favourable regions of the Ramachandran plot over time (Figs. 3 and 4), this has led to more being assigned to the regular secondary structures as H-bonding criteria are met. Conversely, it could be that restraints on hydrogen-bonding potentials have driven the φ-ψ values towards the highly favoured regions of the regular secondary structures.

6. Closer analyses of the Ramachandran plot

Quite apart from its use in validation, the Ramachandran plot has found many other applications, including derivation of potentials of mean force for use in refinement of experimental structures and theoretical models[16], for *ab initio* modelling, as in Rosetta[17,18], and for defining secondary structure[19,20].

There have been numerous studies of the finer aspects of its structure as more and more data, particularly at higher resolution, have become available. The individual distributions of each of the 20 amino acid types, in different secondary structure conformations, have been analysed[21,21]. The effects that different neighbouring residues have on the φ-ψ distribution of each amino acid type have been studied[18,23,24], in loop regions[25], as influenced by a following proline[26], and as a function of sequence[27]. The similarities and differences in the φ-ψ plots of the 20 amino acids have formed the basis of a classification of the amino acids[28]. Ramachandran distributions have also found a use in refinement of protein structural models determined by NMR[29] or by X-ray crystallography[30-32]. Such use, however, rather nullifies the usefulness of the Ramachandran plot for validation of these models.

The original Ramachandran analysis was based on a simple model involving hard-sphere steric clashes. However, the theoretically determined forbidden regions were not reproduced exactly by the observed distributions from experimentally determined structures, such as in the Lovell *et al.* analysis[11]. Specifically the observed diagonal distributions of points in the α_R and α_L regions of the plot seemed at odds with the theoretical expectations of these regions to be horizontal and vertical, respectively. Ho *et al.*[33] were able to show, using simulations, that the diagonal regions were a result of various dipole-dipole interactions in the protein. They further showed that the disallowed regions of the plot are the result of a single steric clash, whereas forbidden regions were caused by more than one steric clash. They also showed that the differences in φ-ψ values along a helix are caused by backbone hydrogen bond restraints, with the C-terminus being more flexible than the N-terminus[34] and consequently residues at the carboxyl terminus of a helix showing a more diffuse set of φ-ψ values than those in the rest of the helix[35].

A recent analysis having a direct bearing on protein structure validation was carried out by the Validation Task Force, a consortium including members of the wwPDB[36] with a particular concern about the validation of structures deposited at the wwPDB and submitted to journals for review[37]. In their section on the Ramachandran plot, they showed that the percentage of outliers (defined as residues outside the MolProbity 99.95% contour lines in each of the 6 Ramachandran plots: Gly, trans-Pro cis-Pro, pre-Pro, Ile/Val and all other residues) remains highly correlated with resolution. The relative rank of a structure within its resolution range provides a 'good measure of comparative quality'. Consequently the task force recommended the following measure of quality, based on the six MolProbity category-specific Ramachandran plots: the number and percentage of outliers in the structure, together with the structure's percentile rank both globally and within its resolution class. They also suggested that any validation reports should include the six Ramachandran plots, with all outliers being flagged. Plus they emphasised the need for the Ramachandran density distributions to be regularly updated as the number of structures in the wwPDB continues to grow.

7. Summary

The Ramachandran plot has proved to be an invaluable tool in protein structure validation, principally because it provides a largely independent way of verifying the geometry of protein structural models. The early, experimentally determined, distributions of residues on the plot differed in a number of respects from the theoretical distribution. Hence, as more protein structures were solved, the experimental distributions came to define the 'expected' φ-ψ distributions. These were refined in the early 90s as more data became available, with minor adjustments being regularly made since then. The growth in structural data in the wwPDB has revealed subtle differences in the plots of different amino acids as well as differences in the plots for the same amino acid in different contexts – eg with different sequence neighbours or in different secondary structures. As the amount of data has grown, so the validation tools have adjusted their Ramachandran distributions and consequently improved their ability to spot possible errors or regions of strain in protein structural models. The Ramachandran plot remains a powerful tool to capture how well proteins reflect the basic physical chemistry of the peptide main chain whilst forming complex 3D structures.

Acknowledgments

We would like to acknowledge financial support from EMBL and the following grants: Wellcome Trust (Grant No. 081989/Z/07/A to N.F.), and in part by US Department of Energy Contract (DE-AC02-06CH11357 to R.A.L.) as part of the Midwest Center for Structural Genomics.

References

1. G. N. Ramachandran, C. Ramakrishnan, and V. Sasisekharan, *J. Mol. Biol.* **7**, 95 (1963).
2. C. Ramakrishnan and G. N. Ramachandran, *Biophys. J.* **5**, 909 (1965).
3. G. N. Ramachandran and V. Sasisekharan, *Adv. Protein Chem.* **23**, 283 (1968).
4. C-I. Brändén and T.A. Jones, *Nature* **343**, 687 (1990).
5. A. L. Morris, M. W. MacArthur, E. G. Hutchinson *et al.*, *Proteins* **12**, 345 (1992).
6. F. C. Bernstein, T. F. Koetzle, G. J. Williams *et al.*, *J. Mol. Biol.* **112**, 535 (1977).
7. R.A. Laskowski, M.W. MacArthur, D.S. Moss *et al.*, *J Appl Crystallogr* **26**, 283 (1993).
8. R. W. Hooft, G. Vriend, C. Sander *et al.*, *Nature* **381**, 272 (1996).
9. G. J. Kleywegt and T. A. Jones, *Structure* **4**, 1395 (1996).
10. V. B. Chen, W. B. Arendall, 3rd, J. J. Headd *et al.*, *Acta Crystallogr. D Biol. Crystallogr.* **66**, 12 (2010).
11. S. C. Lovell, I. W. Davis, W. B. Arendall, 3rd *et al.*, *Proteins* **50**, 437 (2003).
12. J.S. Richardson, *Biopolymers* (2013) in press.
13. G. J. Kleywegt and T. A. Jones, *Structure* **3**, 535 (1995).
14. R. A. Laskowski, *Nucleic Acids Res.* **37** (Database issue), D355 (2009).
15. EU 3D Validation Network, *J. Mol. Biol.* **276**, 417 (1998).
16. S. Koppole and M. Schaefer, *J Comput Chem* **33**, 791 (2012).
17. C. A. Rohl, C. E. Strauss, K. M. Misura *et al.*, *Methods Enzymol.* **383**, 66 (2004).

18. D. Ting, G. Wang, M. Shapovalov *et al.*, *PLoS Comput. Biol.* **6**, e1000763 (2010).
19. P. Y. Chou and G. D. Fasman, *Biochemistry (Mosc.)* **13**, 211 (1974).
20. V. Munoz and L. Serrano, *Proteins* **20**, 301 (1994).
21. S. Hovmöller, T. Zhou, and T. Ohlson, *Acta Crystallogr. D Biol. Crystallogr.* **58**, 768 (2002).
22. A. K. Jha, A. Colubri, M. H. Zaman *et al.*, *Biochemistry (Mosc.)* **44**, 9691 (2005).
23. P. Chakrabarti and D. Pal, *Prog. Biophys. Mol. Biol.* **76**, 1 (2001).
24. M. R. Betancourt and J. Skolnick, *J. Mol. Biol.* **342**, 635 (2004).
25. M. B. Swindells, M. W. MacArthur, and J. M. Thornton, *Nat. Struct. Biol.* **2**, 596 (1995).
26. M. W. MacArthur and J. M. Thornton, *J. Mol. Biol.* **218** (2), 397 (1991).
27. O. Keskin, D. Yuret, A. Gursoy *et al.*, *Proteins* **55**, 992 (2004).
28. D. B. Dahl, Z. Bohannan, Q. Mo *et al.*, *J. Mol. Biol.* **378**, 749 (2008).
29. J. Kuszewski, A. M. Gronenborn, and G. M. Clore, *Protein Sci.* **5**, 1067 (1996).
30. A. T. Brünger, *X-PLOR: version 3.1: a system for x-ray crystallography and NMR.* (Yale University Press, New Haven, 1992).
31. A. T. Brünger, P. D. Adams, G. M. Clore *et al.*, *Acta Crystallogr. D Biol. Crystallogr.* **54**, 905 (1998).
32. J. J. Headd, N. Echols, P. V. Afonine *et al.*, *Acta Crystallogr. D Biol. Crystallogr.* **68**, 381 (2012).
33. B. K. Ho, A. Thomas, and R. Brasseur, *Protein Sci.* **12**, 2508 (2003).
34. S. M. Miick, K. M. Casteel, and G. L. Millhauser, *Biochemistry (Mosc.)* **32**, 8014 (1993).
35. M. Petukhov, K. Uegaki, N. Yumoto *et al.*, *Protein Sci.* **11**, 766 (2002).
36. H. Berman, K. Henrick, H. Nakamura *et al.*, *Nucleic Acids Res.* **35** (Database issue), D301 (2007).
37. R. J. Read, P. D. Adams, W. B. Arendall, 3rd *et al.*, *Structure* **19**, 1395 (2011).

NON-PARAMETRIC STATISTICAL ANALYSIS OF THE RAMACHANDRAN MAP

MAXIM V. SHAPOVALOV AND ROLAND L. DUNBRACK, JR.
Institute for Cancer Research, Fox Chase Cancer Center
333 Cottman Avenue, Philadelphia PA 19111, USA

The Ramachandran map consisting of the dihedral angles ϕ and ψ has been an object of study for 50 years, since its presentation by Ramachandran, Ramakrishnan, and Sasisekharan in July 1963. As the number of structures has grown, it has become possible to apply modern non-parametric statistics to develop proper probability density estimates for the Ramachandran distribution for each amino acid type and different input data sets. It has also become possible to derive classification functions and regressions as a function of ϕ and ψ. For example our most recent backbone-dependent rotamer library consists of the classification probability of the side-chain rotamers and a regression of the mean dihedral angles and variances. The backbone and side-chain bond angles of amino acids also vary with ϕ, ψ, as observed in non-parametric regressions from sub-Ångstrom crystal structure data. This variation confirms Ramachandran's early work that some regions of the Ramachandran map are only accessible with larger values of the backbone bond angle N-Cα-C. Most non-parametric statistical methods can be appropriately modified for pairs of angles and applied to the Ramachandran variables, including kernel density estimates and kernel regressions with the bivariate von Mises distribution, hierarchical Dirichlet processes, and Gaussian processes.

1. Introduction

Ramachandran, Ramakrishnan, and Sasisekharan presented what later became called the Ramachandran map in a paper in the *Journal of Molecular Biology* in July 1963[1] (as it turns out, the month I was born). At that time, structural data and models were available for some di- and tri-peptides, the left and right-handed α helices, and Ramachandran's and Sasisekharan's successful model of the collagen triple helix.[2,3] What was lacking was a way to analyze these structures in a unified manner. Assuming a trans peptide bond, Ramachandran and colleagues created a plot of the two flexible dihedral angles of the peptide backbone, which they called ϕ and ϕ', corresponding to what later became our values of ϕ+180° and ψ. Using fixed bond lengths and bond angles from experimental structures then available and assumed van der Waals radii, they worked out what regions of the ϕ, ϕ' space would be physically accessible. The resulting plot showed elegantly that the known structures fit into the regions without significant steric conflicts. The first atomic-resolution protein structures also confirmed the analysis first presented in 1963.[4] A number of authors have continued Ramachandran's work and investigated the origins of the boundaries between regions of the map with observations in very high-resolution crystal structures and effectively disallowed regions with few or no such observations[5-8]

Over the ensuing decades as many protein structures have been determined, the Ramachandran plot, a scatterplot of ϕ vs ψ, has become a standard tool for analyzing new protein structures. Tools such as PROCHECK[9] and WHATCHECK[10] were developed in the 1990s to validate new structures based on the number of residues in disallowed regions. In these programs, accessible regions were determined from two-dimensional histograms of ϕ, ψ data on 10°x10° grids, and boundaries were drawn around "allowed" and "generously allowed" regions. For the WHATCHECK program, Hooft, Sander and Vriend produced two-dimensional histograms for each of the 20 amino acids separately for helix, sheet and coil residues.[10]

A histogram is a crude form of *probability density estimate*, which is a function that when integrated over some region of its dependent variables produces the probability that data points lie in that region. They are in fact forms of *non-parametric* density estimates. Parametric density distributions are expressed as functions with a small number of parameters, such as a Gaussian distribution or a short Fourier series. Non-parametric density distributions are not simple functions, but may be expressed as sums over functions where the number of parameters scales with the number of data points. A histogram is a non-parametric density estimate because it is a simple sum over the data points in each bin of the histogram. The number of parameters is equal to the number of data points. Non-parametric density estimates are appropriate when the data are widely distributed, multi-model, or otherwise sufficiently irregular that they cannot readily be fit to a simple parameterized distribution function.

Surprisingly, it has only been relatively recently that modern non-parametric statistical methods beyond simple histograms have been used to define probability density distributions on the Ramachandran map.[11-17] These include kernel density estimates, Dirichlet process and hierarchical Dirichlet process methods. These methods produce smooth and differentiable distributions and generally minimize the effects of outliers on the density estimate.

Further, modern non-parametric statistical methods can be used to produce *classification functions* and *regressions* on the Ramachandran map. A classification function reflects the probability of each member of a discrete set of objects at every combination of its dependent variables. The "backbone-dependent rotamer library" is an example of this, and contains the probability of each rotamer of a residue type given the local backbone conformation as expressed by ϕ and ψ.[11,13,18,19] A local regression of a conformational variable may also be expressed as a function on the Ramachandran map. The backbone bond angles, particularly N-Cα-C, have been shown to vary significantly as a function of ϕ and ψ, and have been determined using kernel regression methods.[20]

In this chapter, I review the application of modern non-parametric statistical methods to data described by the Ramachandran map variables ϕ and ψ, including probability density estimates, classification functions, and regression functions. I also discuss the application of these results to protein structure determination, validation, and prediction.

2. Probability density estimation on the Ramachandran map

After histograms, the most common method for estimating probability densities is *kernel density estimation*.[21] In a kernel density estimate, instead of counting data values in fixed intervals, each data point is spread out using a kernel. A kernel K is a function that is a probability density function (that is, $K \geq 0$ everywhere and integrates to 1.0), centered on the actual data point. For instance, to model a density, the kernel estimate is given by:

$$\hat{f}(x) = \frac{1}{N} \sum_{i=1}^{N} K_h(x - x_i) \qquad (1)$$

where the n data points are designated as x_i, and K_h is (usually) a symmetric function concentrated at 0 with effective width defined by a parameter h. A commonly used kernel is the Gaussian function:

$$K_h(x - x_i) = \frac{1}{\sqrt{2\pi}h} \exp\left(-\frac{(x - x_i)^2}{2h^2} \right) \qquad (2)$$

The kernel density estimate allows us to smooth the density beyond the standard histogram, which depends on arbitrary bin positions and bin widths.

A modification of Eq. 1 may be made such that the bin width varies with the data point and is inversely related to the local density. The result is an *adaptive kernel density estimate*:

$$\hat{f}(x) = \frac{1}{N} \sum_{i=1}^{N} K_{h_i}(x - x_i) \qquad (3)$$

The local density-dependent bandwidths can be achieved by means of a *pilot estimate*, $\hat{f}_0(x_i)$, such that

$$h_i = \left(\frac{g}{\hat{f}_0(x_i)} \right)^{\alpha} h \tag{4}$$

where

$$g = \left(\prod_{i=1,N} \hat{f}_0(x_i) \right)^{\frac{1}{N}} \tag{5}$$

h is a base bandwidth and g is the geometric mean of the pilot estimate at the data points. The pilot estimate does not need to be very accurate; for instance, it can be derived from a fixed bandwidth kernel density estimate. A value of $\alpha = 1/2$ is commonly used.[22,23]

Surprisingly, kernel density estimators, especially with adaptive kernels, have been used relatively rarely in protein structural bioinformatics. They have been applied primarily to analyzing dihedral angles in molecular dynamics trajectories using either fixed width Gaussian kernels[24] or with nearest neighbor adaptive bandwidths with Epanechnikov kernels.[25] Wall et al. used fixed-width Gaussian kernel density estimators to develop potentials of mean force for atom-atom interactions based on pairwise atom distances.[26] In terms of the Ramachandran map, we used a two-fold periodic kernel, proportional to $\exp\left(-k\sin^2(\phi-\phi_i) - k\sin^2(\psi-\psi_i)\right)$ to count residues of each rotamer type in order to produce a prior distribution in a Bayesian implementation of the backbone-dependent rotamer library.[11] Lovell et al. used an Epanechnikov-like kernel proportional to $\cos\left(k\sqrt{(\phi-\phi_i)^2 + (\psi-\psi_i)^2}\right)$ in a small region around (ϕ_i, ψ_i) and equal to zero elsewhere. In addition, Lovell et al. varied the width of their kernel by altering k and the size of the non-zero region according to a pilot estimate of the local density using the same kernel function. This is in effect an adaptive kernel density estimate,[23] although they did not describe it as such. By establishing contours around 98% and 99.5% of the density, these authors determined boundaries for the allowed and generously allowed regions of the Ramachandran map for validation of protein structures in their MolProbity software.[27]

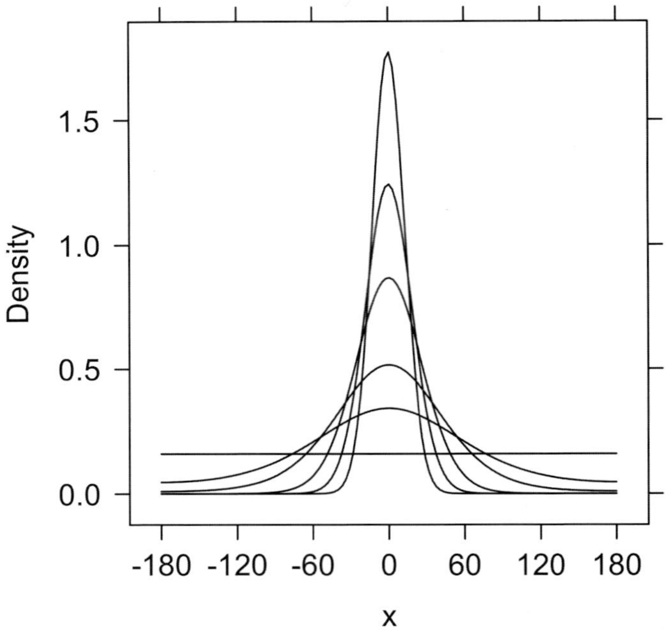

Figure 1: Von Mises probability density functions used for kernel density estimation for angular variables. The curves are shown for κ=0 (horizontal line) and κ=1, κ=2, κ=5, κ=10, and κ=20 (from broadest to narrowest distributions).

In the well-established field of directional statistics,[28] the proper kernel for an angular variable that ranges from 0 to 2π is the von Mises function[29]:

$$f(x) = \frac{1}{2\pi I_0(\kappa)} \exp\left(\kappa \cos\left(x - x_i\right)\right) \tag{6}$$

which is shown in Figure 1 for various values of κ which acts as the bandwidth variable for the von Mises kernel. $I_0(\kappa)$ is a normalization constant given by the Bessel function of the first kind of order 0.

We have used this kernel to develop an adaptive kernel density estimation procedure for the Ramachandran map:[13]

$$\hat{f}(\phi,\psi) = \frac{1}{4\pi^2 N} \sum_{i=1}^{N} \frac{1}{I_0^{\,2}(\kappa_i)} \exp\left(\kappa_i \left(\cos\left(\phi - \phi_i\right) + \cos\left(\psi - \psi_i\right)\right)\right) \tag{7}$$

where the adaptive kernel bandwidth is given by:

$$\kappa_i = \left(\frac{\hat{f}_0(\phi_i, \psi_i)}{\left(\prod_{j=1}^{N} \hat{f}(\phi_j, \psi_j) \right)^{\frac{1}{N}}} \right)^{\alpha} \kappa \tag{8}$$

where $\alpha > 0$. Eq. 8 is the inverse of Eq. 4 because smaller values of κ lead to larger bandwidths, in contrast to the Gaussian kernel, where larger values of h lead to larger bandwidths.

The results of the kernel density estimation procedure depend on the input data as well as the geometric mean of the kernel bandwidth, κ. The input data are usually calculated from a set of structures culled by resolution, R-factor, and maximum mutual sequence identity, which our PISCES server provides.[30,31] Individual residue criteria may also be applied, such as maximum B-factor values, elimination of residues with atomic overlaps,[32] and electron density cutoffs.[33]

Depending on the purpose for which the Ramachandran densities are intended, such as structure validation, structure determination, or structure prediction, the residue list may be limited to specific amino acid types and/or to specific structural elements. For instance, structure prediction methods typically employ Ramachandran distributions for residues in non-regular secondary structure regions to avoid the very large probability spike in the α-helix region.[34] Finally, larger values of κ result in sharper distributions, and the desired value may depend on the application. Typically, structure validation may utilize narrower bandwidths since a sharp delineation of the boundary between accessible and inaccessible regions is the main purpose of the calculation. Structure prediction methods may require broader bandwidths, so that scoring functions based on $-\log(p)$ may have broader basins of attraction when the scoring function is minimized. Several examples are given in Figure 2, exploring the probability distributions for different residue types, different secondary structure content, and different values of κ.

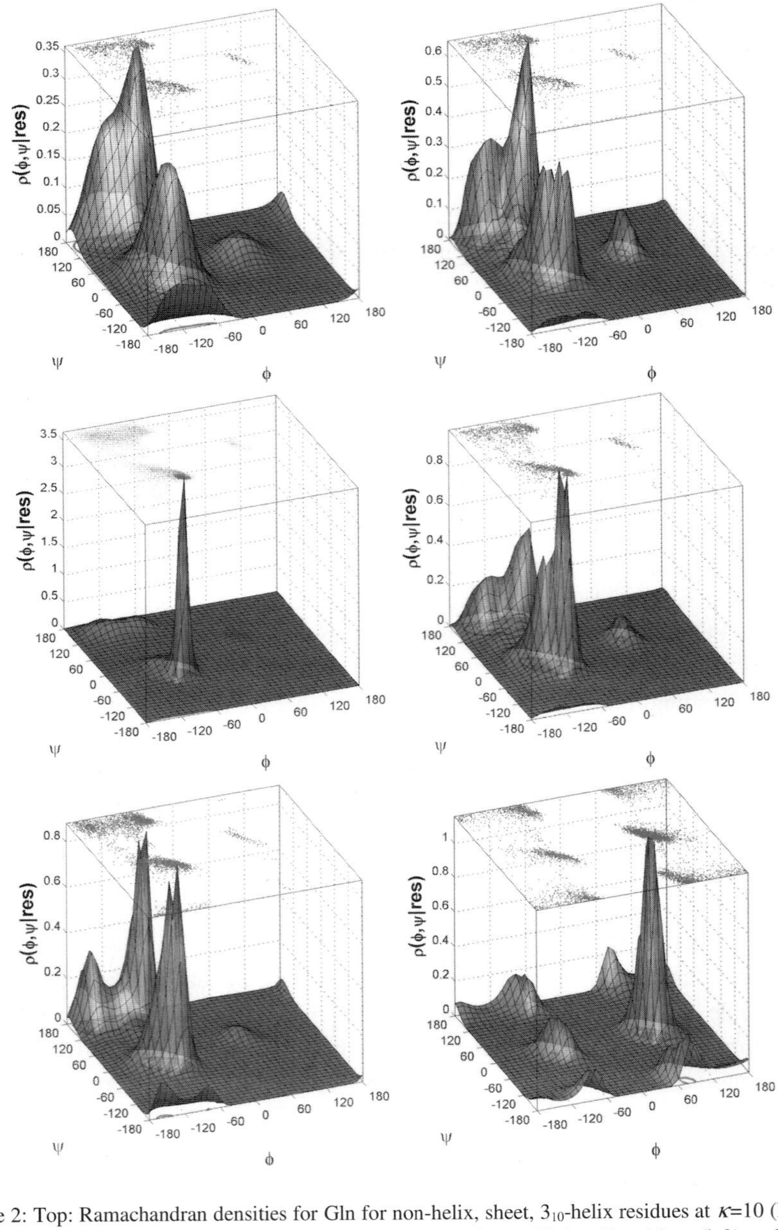

Figure 2: Top: Ramachandran densities for Gln for non-helix, sheet, 3_{10}-helix residues at κ=10 (left) and κ=50 (right). Middle: Ramachandran densities for Gln for κ=50 for all residues (left) and for non-helix, sheet but including 3_{10}-helix residues (right). Bottom: Ramachandran densities for Ser (left) and Gly (right) non-helix, sheet residues.

Ramachandran distributions may also be affected by the identity or conformation of neighboring amino acids. In particular, it has long been known that residues that precede proline have quite different Ramachandran distributions,[35] with significantly less density in the α and left-handed regions of the Ramachandran map. They also exhibit additional density in the so-called ζ region near ϕ,ψ=(-130°, +80°),[36] due to favorable van der Waals and electrostatic interactions.[5] A number of groups have discussed the effect of local sequence, usually plus or minus one amino acid, on backbone conformational distributions.[16,37-42] This has been examined as a violation of the Flory isolated pair hypothesis, which states that conformations of individual dihedral angle pairs in a polymer are approximately independent of the conformations and/or residue identity of their neighbors.[43]

Several estimations of neighbor-dependence have been calculated from crude histograms. Jha et al. examined ϕ,ψ distributions in coiled regions and found strong neighbor dependence on the populations in the α, β, and Polyproline II (PPII) regions of the Ramachandran map.[37] Erman et al. also examined neighbor residue-type dependence of ϕ,ψ regions (α, β, etc.) as a method for predicting these regions given local sequence context.[41,42] Betancourt and Skolnick determined the conformational properties of triplets of amino acids, in terms of occupied basins of the Ramachandran map, as well as other distributions such as the pair ψ_i,ψ_{i+1}.[39] They used the data to produce a low-resolution potential energy function for backbone conformations that depends on local sequence. By contrast, Lennox et al. used a bivariate von Mises distribution and a Bayesian non-parametric density estimation procedure to derive smooth density estimates for ϕ_{i+1},ψ_i pairs, which span a pair of residues and thus capture a limited form of neighbor dependence.[16]

To determine neighbor-dependent Ramachandran distributions, we employed a hierarchical non-parametric statistical analysis based on the Dirichlet process.[17] A hierarchical model treats the data as arising from several related but distinct distributions. Each distinct distribution might have a different value for a categorical variable. For Ramachandran distributions, we treated data for one amino acid type, say glutamine, as arising from a set of 20 related distributions, one for each amino acid type one residue to the right or left. One way to do this is to model the Ramachandran distribution of glutamines as a *mixture model* of a limited number of component distribution functions. These might be Gaussian or von Mises distributions with appropriate mean values and variances that represent the modes and peak widths of the observed distribution. The distribution of glutamine for any specific residue type to the right (or left) would be modeled with proportions of the mixture components specific for that residue type. Gln with Pro to the right would have a smaller fraction of α region components and a larger fraction of β region components

The hierarchical Dirichlet process is a Bayesian method for determining such mixture models.[44] It is implemented as a Monte Carlo simulation such that the mixture components are shared by all of the subgroups and such that the number, positions and variances of the components fluctuate during the simulation. It successfully ties together the different distributions while modeling any idiosyncratic differences that exist between them.

In Figure 3, I show the distribution of glutamine in non-regular secondary structure (i.e., not helices or sheets) with several different neighbors to the right. Certain neighbor residue types characteristically shift the Ramachandran distribution of most residue types in the same direction – either towards α or β region. Many of these are easily explained based on physical considerations such as the formation or blocking of hydrogen bonds or hydrophobic interactions. As noted earlier, Pro to the right has a profound effect, knocking out most of the density in the α region and in the left-handed helical region (ϕ>0°). We and others (T. Kortemme, personal communication) have found that the neighbor-dependent Ramachandran distributions improve structure prediction of protein loops with the program Rosetta.

3. Classification functions on the Ramachandran map

A second form of statistical function of the Ramachandran map variables ϕ and ψ is the *classification function* of the form $p\left(C_k \middle| \phi, \psi\right)$, where C_k is one of K discrete class variables, such that:

$$\sum_{k=1,K} p\left(C_k \middle| \phi, \psi\right) = 1 \tag{9}$$

From the kernel density estimates described in the last section (or any density estimation method), we can calculate $p\left(\phi, \psi \middle| C_k\right)$, the probability density of C_k as a function of ϕ,ψ, for each value of k. Using Bayes' rule, we can calculate the classification function for C_k:

$$p\left(C_k \middle| \phi, \psi\right) = \frac{p\left(\phi, \psi \middle| C_k\right) p\left(C_k\right)}{\sum_{k'=1,K} p\left(\phi, \psi \middle| C_{k'}\right) p\left(C_{k'}\right)} \tag{10}$$

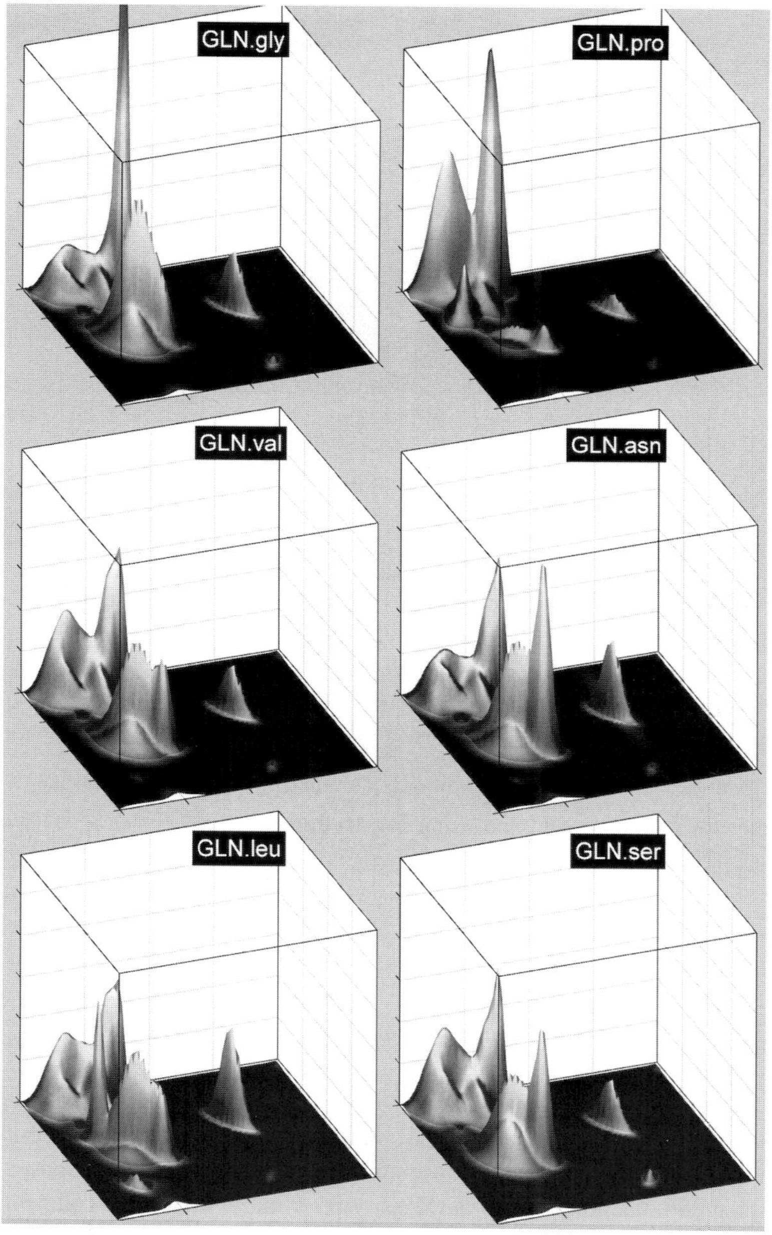

Figure 3: Neighbor-dependent Ramachandran distributions for glutamine with different neighbors (listed in lower case) as calculated by a hierarchical Dirichlet process model.

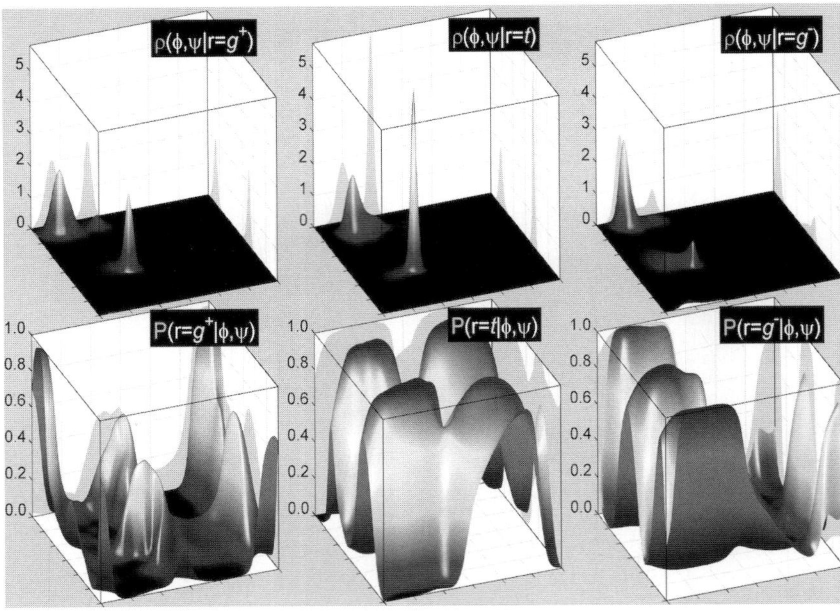

Figure 4: Kernel estimation of the backbone-dependent rotamer library for valine. Top row: probability density estimates of the Ramachandran distribution of the three χ_1 rotamers of valine. Bottom row: classification functions for the three χ_1 rotamers of valine.

We have used such a method for developing the latest version of the backbone-dependent rotamer library by calculating the ϕ, ψ probability of each rotamer for an amino acid and using Eq. 10 to determine $p\left(r\middle|\phi,\psi,R\right)$ where r is a rotamer of residue type R. The level of smoothing can be controlled by the base value of κ in Eq. 8. This value may be optimized by maximizing the log likelihood of data points in a cross-validation procedure. That is, we maximize the value of:

$$\kappa_{Optim} = \arg\max_{\kappa} \mathcal{L}\left(\kappa\right) = \arg\max_{\kappa} \sum_{i=1}^{N} \log\left(P\left(r_i\middle|\phi_i,\psi_i\right)\right) \tag{11}$$

Examples of these functions are given in Figure 4. These functions are significant improvements in terms of smoothness and continuity from simple histogram estimates our previous version of the library from 2002 (based on a Bayesian combination of a kernel density prior and a histogram count in each 10°x10° bin).

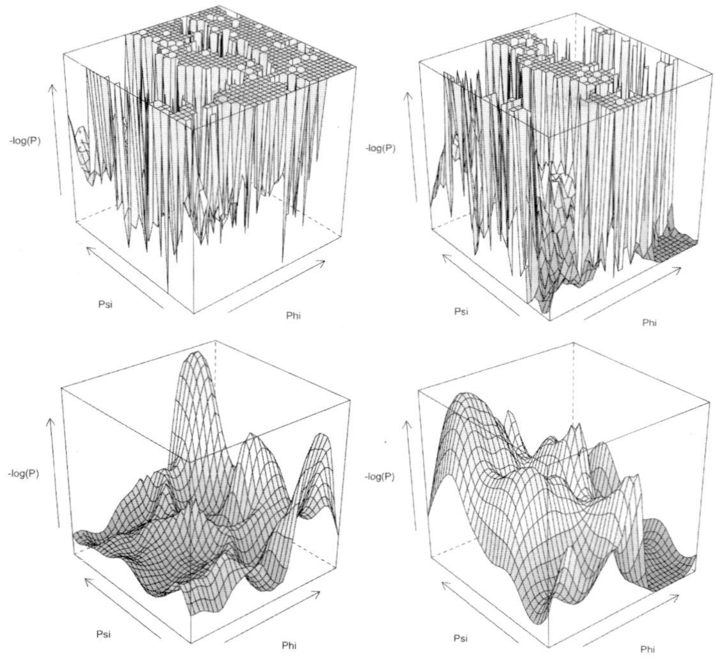

Figure 5: Classification functions for different residues types as a function of the Ramachandran variables ϕ and ψ. Top row: $-\log p(R|\phi,\psi)$ for Lys (left) and Gly (right) from the Rosetta scoring function. Bottom row: $-\log p(R|\phi,\psi)$ for Lys and Gly from inverting kernel density estimates of the Ramachandran distribution using Eq. 12.

Another classification function on the Ramachandran variables, one that is often used in protein design, is a set of residue propensities. These functions are given by

$$p(R|\phi,\psi)=\frac{p(\phi,\psi|R)p(R)}{\sum_{R'}p(\phi,\psi|R')p(R')} \tag{12}$$

$p(\phi,\psi|R)$ is simply the Ramachandran probability distribution of amino acid type R and $p(R)$ is the frequency of amino acid type R. Such a function is used in the program Rosetta.[45] A comparison of the kernel density estimate version of this function and the histogram-based Rosetta function is shown in Figure 5.

In terms of design, a useful feature of Eq. 12 is that the values of $p(R)$ need not be the same as the values in the data set used to produce the various

$p(\phi,\psi|R)$. For instance, if we chose to remove certain residue types from consideration, such as cysteine, for physical-chemical reasons, the classification function can easily be recalculated.

4. Regression analysis on the Ramachandran map

Finally, non-parametric statistical methods have been developed for the purpose of local regression functions that may take on any form (as opposed to fitting to a single parametric function such as a cubic polynomial or a short Fourier series) and which can be tuned to any desired level of smoothness. The simplest of these methods is the kernel regression function, where the data, x_i, are modeled as:

$$x_i = m(\phi_i, \psi_i) + \nu^{\frac{1}{2}}(\phi_i, \psi_i)\varepsilon_i \tag{13}$$

where $m(\phi_i, \psi_i)$ is the unknown regression function, $\nu(\phi_i, \psi_i)$ is the variance, and ε_i are random observation errors normally distributed with a mean of zero and variance 1. Written this way, the model is *heteroscedastic* in that the variance may vary as a function of the independent variables. The zeroth order or Nadaraya-Watson estimators may be written:

$$m(\phi,\psi) = \frac{\sum_{i=1,N} K(\phi-\phi_i, \psi-\psi_i)x_i}{\sum_{i=1,N} K(\phi-\phi_i, \psi-\psi_i)} \tag{14}$$

$$\sigma^2 = \frac{\sum_{i=1,N} K(\phi-\phi_i, \psi-\psi_i)\left(m(\phi,\psi)-x_i\right)^2}{\sum_{i=1,N} K(\phi-\phi_i, \psi-\psi_i)} \tag{15}$$

The appropriate kernel for regression onto the angles ϕ and ψ is again a symmetric two-dimensional von Mises kernel:

$$K(\phi-\phi_i, \psi-\psi_i) = \frac{1}{4\pi^2\left(I_0(\kappa)\right)^2}\exp\left(\kappa\left(\cos(\phi-\phi_i)+\cos(\psi-\psi_i)\right)\right) \tag{16}$$

This kernel can also be made adaptive. We have found that we get better regression functions if the kernel is not adaptive to the position of the data point (ϕ_i, ψ_i) but rather to the local density of the query point, ϕ, ψ.[13] We have used this method to determine the backbone-dependent variation in the dihedral angles of each rotamer of each residue type as part of the 2010 version of the backbone-dependent rotamer library.[13] Examples of these functions are shown in Figure 6.

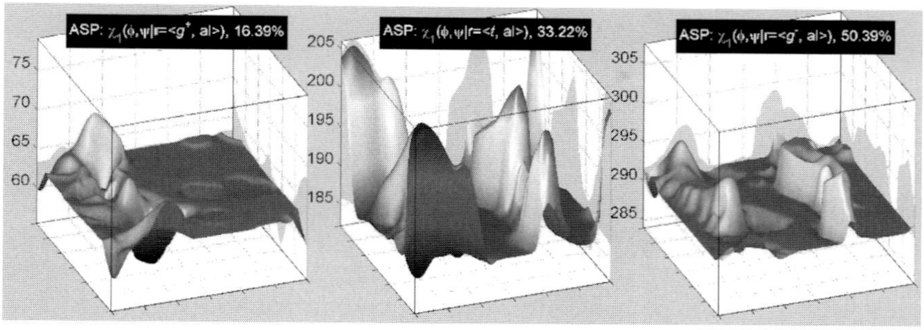

Figure 6. Regression functions for the χ_1 angle of Asp for the g^+, *trans* and g^- rotamers.

More recently, we have also used this method to perform ϕ,ψ-dependent regressions of the backbone bond angles C_{i-1}-N-Cα, N-Cα-C, and Cα-C-N_{i+1} and the backbone dihedral angle ω. We found that the angle τ, N-Cα-C, varies by up to 7° as a function of ϕ and ψ, and most noticeably there is an expansion of this angle in the so-called bridge region between the α-helix and β-sheet region. In the earliest papers in which Ramachandran and colleagues determined the allowed regions of ϕ and ψ using a hard-sphere model, they were careful to perform these calculations for different values of τ (105°, 110°, and 115°) and found that the largest value of τ was required to make the bridge region accessible. This anticipated by decades a sufficient number of high-resolution structures helped us to determine that this was indeed the case. The functions for τ and ω are shown in Figure 7.

5. Discussion

For fifty years, the Ramachandran map has been a unifying principle in the analysis of protein structure. With the rapid increase in high-resolution structures, it is now possible to apply modern, sophisticated forms of non-parametric statistical analysis to functions of the Ramachandran map variables, ϕ and ψ. Each of these functional forms may play a role in structure determination, structure validation, structure prediction, and other forms of modeling and simulation. I have outlined three classes of functions of the

Ramachandran variables: probability densities, classification functions, and regression functions, and some of the modern methods used to determine them.

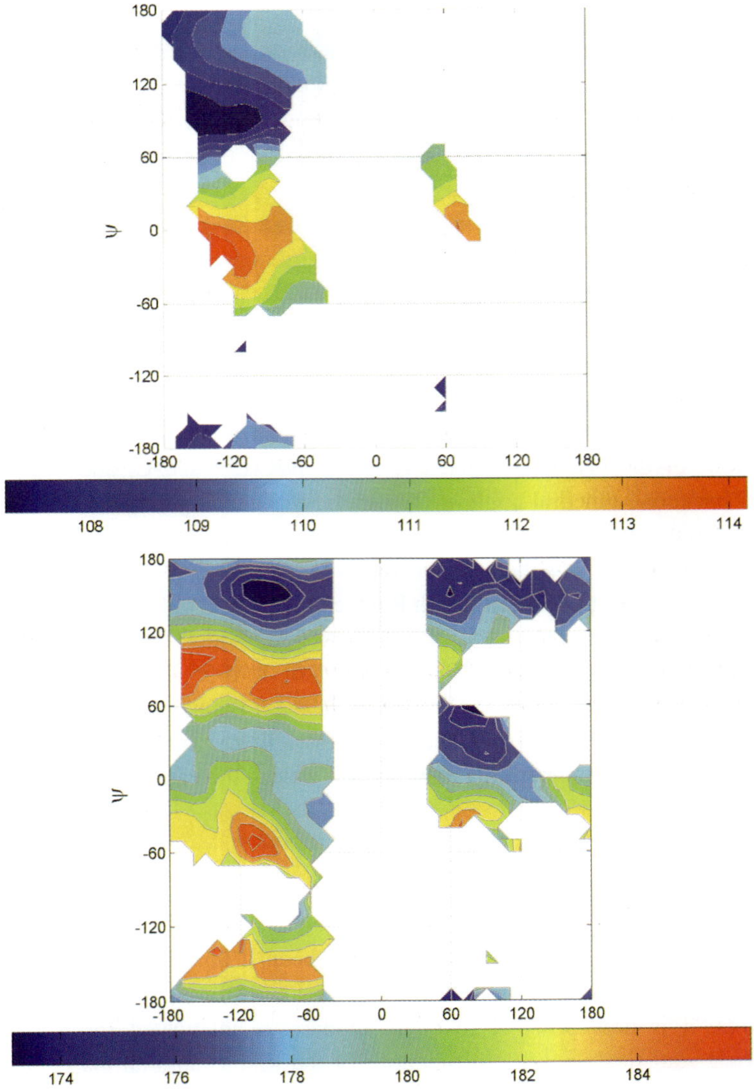

Figure 7: Top: Regression of the N-Cα-C angle on the Ramachandran variables. Bottom: Regression of the ω dihedral angle on ψ of the previous amino acid on the y-axis and ϕ of the same amino acid (i.e., immediately following the ω dihedral) on the x-axis.

Probability density distributions on the Ramachandran map are used in structure validation, most recently as part of the MolProbity program.[12,27] This program is now part of the recommended validation procedure by the PDB,[46] and the smoothed non-parametric distributions represent an important advance in structure validation over the previous generation of histograms in programs such as WhatCheck and ProCheck.

Ramachandran probability distributions are also used extensively in protein structure prediction. The Rosetta program, for instance, has a Ramachandran probability term in its scoring function. The current default version is a partially smoothed histogram analysis produced in 2001.[34,45] We have been working with the Rosetta development groups to implement our kernel density estimates as a replacement for these older distributions. One important advantage of the kernel density estimates or the Dirichlet process density estimates is their smoothness. During minimization of its scoring function, Rosetta takes first derivatives of $-\log\left(p\left(\phi,\psi\middle|R\right)\right)$ with respect to ϕ and ψ. Until recently, this was performed as a bilinear interpolation of p, $dp/d\phi$, and $dp/d\psi$ of the Ramachandran distributions given on a 10°x10° grid. This leads to very poor values for the derivatives since the probability density distributions change by orders of magnitude over short ranges of ϕ and ψ. The code has been modified so that now the derivatives are calculated based on bicubic splines of log(p) with greatly improved agreement between distributions of dihedral angles in decoys and experimental structures.[47] Our neighbor-dependent Ramachandran distributions are also implemented in Rosetta for loop modeling, and have been found to significantly increase the accuracy of loops structure prediction.

Classification functions of the Ramachandran variables are used in both structure prediction and protein design. The backbone-dependent rotamer library is a classification function that provides the probability of each of the rotamers of each amino acid as a function of ϕ and ψ. It is used heavily in side-chain conformation prediction methods[48-54] and in protein design.[55-59] Similarly, the probability of each residue type as a function of ϕ and ψ is a classification function that is used primarily in protein design.[45]

Regression functions may be applied to any variable that can be measured for amino acids in proteins given their values of ϕ and ψ. We have applied this kind of analysis to the backbone bond angles, the backbone dihedral angle ω, and the side-chain dihedral angles for each distinct rotamer of each amino acid. These functions have not yet been applied to structure validation or structure prediction, but the backbone-dependent bond angles are being incorporated into structure determination programs (P.A. Karplus, personal communication). We are currently implementing a more robust form of local regression, Gaussian processes, to the problem of placing the Cβ atom with the angles N-Cα-Cβ and the improper dihedral angle C-N-Cα-Cβ. These angles vary significantly both with the χ_1 rotamer and with the ϕ and ψ.

In summary, modern non-parametric statistical method provide much better estimates of densities, classifications, and regressions than simple histograms, sums, or averages of data. With the large number of crystal structures now available and these sophisticated methods, many interesting features of protein structure can be addressed in the future. Ramachandran's legacy is alive and well.

Acknowledgments

This work was supported by National Institutes of Health grants R01-GM84453 and R01-GM76222. I want to thank Elizabeth Davenport for reading and commenting on the manuscript.

References

1. G. N. Ramachandran, C. Ramakrishnan and V. Sasisekharan, *J. Mol. Biol.* **7**, 95 (1963).
2. G. N. Ramachandran and V. Sasisekharan, *Nature* **190**, 1004 (1961).
3. G. N. Ramachandran and V. Sasisekharan, *Biochim. Biophys. Acta* **109**, 314 (1965).
4. G. N. Ramachandran and V. Sasisekharan, *Adv. Protein Chem.* **23**, 283 (1968).
5. B. K. Ho and R. Brasseur, *BMC Struct. Biol* **5**, 14 (2005).
6. L. L. Porter and G. D. Rose, *Proc. Natl. Acad. Sci. U.S.A.* **108**, 109 (2011).
7. B. K. Ho, A. Thomas and R. Brasseur, *Protein Sci.* **12**, 2508 (2003).
8. A. Q. Zhou, C. S. O'Hern and L. Regan, *Protein Sci.* **20**, 1166 (2011).
9. R. A. Laskowski, M. W. MacArthur, D. S. Moss and J. M. Thornton, *J. Appl. Cryst.* **26**, 283 (1993).
10. R. W. Hooft, C. Sander and G. Vriend, *Comput. Appl. Biosci.* **13**, 425 (1997).
11. R. L. Dunbrack, Jr. and F. E. Cohen, *Protein Sci.* **6**, 1661 (1997).
12. S. C. Lovell, I. W. Davis, W. B. Arendall, 3rd, P. I. de Bakker, J. M. Word, M. G. Prisant, J. S. Richardson and D. C. Richardson, *Proteins* **50**, 437 (2003).
13. M. V. Shapovalov and R. L. Dunbrack, Jr., *Structure* **19**, 844 (2011).

14. W. Boomsma, K. V. Mardia, C. C. Taylor, J. Ferkinghoff-Borg, A. Krogh and T. Hamelryck, *Proc. Natl. Acad. Sci. U.S.A.* **105**, 8932 (2008).
15. D. B. Dahl, Z. Bohannan, Q. Mo, M. Vannucci and J. Tsai, *J. Mol. Biol.* **378**, 749 (2008).
16. K. P. Lennox, D. B. Dahl, M. Vannucci and J. W. Tsai, *J. Am. Stat. Assoc.* **104**, 586 (2009).
17. D. Ting, G. Wang, M. Shapovalov, R. Mitra, M. I. Jordan and R. L. Dunbrack, Jr., *PLoS Comput. Biol.* **6**, e1000763 (2010).
18. R. L. Dunbrack, Jr. and M. Karplus, *J Mol Biol* **230**, 543 (1993).
19. R. L. Dunbrack, Jr., *Curr. Opin. Struct. Biol.* **12**, 431 (2002).
20. D. S. Berkholz, M. V. Shapovalov, R. L. Dunbrack, Jr. and P. A. Karplus, *Structure* **17**, 1316 (2009).
21. B. W. Silverman. Density Estimation for Statistics and Data Analysis, New York: Chapman & Hall; 1986. 175 pp.
22. L. Breiman, J. H. Friedman and E. Purcell, *Technometrics* **19**, 135 (1977).
23. I. S. Abramson, *Ann. Statist.* **10**, 1217 (1982).
24. M. Machuqueiro and A. M. Baptista, *Biophys. J.* **92**, 1836 (2007).
25. F. A. Hamprecht, C. Peter, X. Daura, W. Thiel and W. F. van Gunsteren, *J. Chem. Phys.* **114**, 2079 (2001).
26. M. E. Wall, S. Subramaniam and G. N. Phillips, Jr., *Protein Sci.* **8**, 2720 (1999).
27. I. W. Davis, L. W. Murray, J. S. Richardson and D. C. Richardson, *Nucleic Acids Res.* **32**, W615 (2004).
28. K. V. Mardia and P. E. Jupp. Directional Statistics, London: Wiley; 2000.
29. R. von Mises, *Phys. Z.* **19**, 490 (1918).
30. G. Wang and R. L. Dunbrack, Jr., *Bioinformatics* **19**, 1589 (2003).
31. G. Wang and R. L. Dunbrack, Jr., *Nucleic Acids Res.* **33**, W94 (2005).
32. J. M. Word, S. C. Lovell, T. H. LaBean, H. C. Taylor, M. E. Zalis, B. K. Presley, J. S. Richardson and D. C. Richardson, *J. Mol. Biol.* **285**, 1711 (1999).
33. M. V. Shapovalov and R. L. Dunbrack, Jr., *Proteins* **66**, 279 (2007).
34. C. A. Rohl, C. E. Strauss, K. M. Misura and D. Baker, *Methods Enzymol.* **383**, 66 (2004).
35. M. W. MacArthur and J. M. Thornton, *J. Mol. Biol.* **218**, 397 (1991).
36. P. A. Karplus, *Protein Sci.* **5**, 1406 (1996).
37. A. K. Jha, A. Colubri, M. H. Zaman, S. Koide, T. R. Sosnick and K. F. Freed, *Biochemistry* **44**, 9691 (2005).
38. M. H. Zaman, M. Y. Shen, R. S. Berry, K. F. Freed and T. R. Sosnick, *J. Mol. Biol.* **331**, 693 (2003).
39. M. R. Betancourt and J. Skolnick, *J Mol Biol* **342**, 635 (2004).
40. R. V. Pappu, R. Srinivasan and G. D. Rose, *Proc. Natl. Acad. Sci. U.S.A.* **97**, 12565 (2000).

41. L. Ormeci, A. Gursoy, G. Tunca and B. Erman, *Proteins* **66**, 29 (2007).
42. O. Keskin, D. Yuret, A. Gursoy, M. Turkay and B. Erman, *Proteins* **55**, 992 (2004).
43. P. J. Flory. Statistical Mechanics of Chain Molecules, New York: Wiley; 1969.
44. Y. W. Teh, M. I. Jordan, M. J. Beal and D. M. Blei, *J. Am. Stat. Assoc.* **101**, 1566 (2006).
45. B. Kuhlman and D. Baker, *Proc. Natl. Acad. Sci. U.S.A.* **97**, 10383 (2000).
46. R. J. Read, P. D. Adams, W. B. Arendall, 3rd, A. T. Brunger, P. Emsley, R. P. Joosten, G. J. Kleywegt, E. B. Krissinel, T. Lutteke, Z. Otwinowski, A. Perrakis, J. S. Richardson, W. H. Sheffler, J. L. Smith, I. J. Tickle, G. Vriend and P. H. Zwart, *Structure* **19**, 1395 (2011).
47. A. Leaver-Fay, M. J. O'Meara, M.Tyka, R. Jacak, Y. Song, E. H. Kellogg, J. Thompson, I. W. Davis, R. A. Pache, S. Lyskov, J. J. Gray, T. Kortemme, J. S. Richardson, J. J. Havranek, J. Snoeyink, D. Baker and B. Kuhlman, *Methods Enzymol.,* in press (2012).
48. M. J. Bower, F. E. Cohen and R. L. Dunbrack, Jr, *J. Mol. Biol.* **267**, 1268 (1997).
49. J. Mendes, H. A. Nagarajaram, C. M. Soares, T. L. Blundell and M. A. Carrondo, *Biopolymers* **59**, 72 (2001).
50. S. Liang and N. V. Grishin, *Protein Sci.* **11**, 322 (2002).
51. A. A. Canutescu, A. A. Shelenkov and R. L. Dunbrack, Jr., *Protein Sci.* **12**, 2001 (2003).
52. K. M. Misura, A. V. Morozov and D. Baker, *J. Mol. Biol.* **342**, 651 (2004).
53. J. Xu and B. Berger, *J. Assoc. Comp. Machinery* **53**, 553 (2006).
54. G. G. Krivov, M. V. Shapovalov and R. L. Dunbrack, Jr., *Proteins* **77**, 778 (2009).
55. G. Dantas, B. Kuhlman, D. Callender, M. Wong and D. Baker, *J. Mol. Biol.* **332**, 449 (2003).
56. H. Kono and J. G. Saven, *J. Mol. Biol.* **306**, 607 (2001).
57. J. R. Calhoun, H. Kono, S. Lahr, W. Wang, W. F. DeGrado and J. G. Saven, *J. Mol. Biol.* **334**, 1101 (2003).
58. D. N. Bolon and S. L. Mayo, *Proc. Natl. Acad. Sci. U.S.A.* **98**, 14274 (2001).
59. M. C. Saraf, G. L. Moore, N. M. Goodey, V. Y. Cao, S. J. Benkovic and C. D. Maranas, *Biophys J.* **90**, 4167 (2006).

NON-GLYCINE RESIDUES IN THE DISALLOWED REGIONS OF THE RAMACHANDRAN MAP AND THEIR NEIGHBORHOOD PREFERENCES

CHANDA R. BHAT, MOITRAYEE BHATTACHARYYA AND SARASWATHI VISHVESHWARA

Molecular Biophysics Unit, Indian Institute of Science Bangalore-560012, India

The allowed and the "disallowed" regions in the celebrated Ramachandran map ([ϕ-ψ] map) was elegantly deduced by Ramachandran, Ramakrishnan and Sasisekharan even before the protein crystal structures became available. This powerful map was derived based on rigid geometry of the peptide group and later several investigations on protein crystal structures reported the occurrence of a small fraction of the [ϕ-ψ] torsion angles in the disallowed region. The question is what factors make these residues adopt disallowed conformations? Is it driven by the necessity to maintain the overall topology or is it associated with function or is it just that the disallowed conformations are extreme limits of the allowed conformations? Today, with the availability of a large number of high resolution crystal structures, we have revisited this problem. Apart from validating some of the earlier findings such as residue propensities, preferred location in the secondary structure, we have explored their spatial neighborhood preferences using the protein structure network [PSN] approach developed in our lab. Finally, the structural and functional implications of the disallowed conformations are examined.

1. Introduction

The Ramachandran map[1] has stood the test of time by exhibiting high propensity of torsion angles [ϕ, ψ] in the allowed regions. A small population in the "disallowed" regions by non-glycine residues had led to several investigations[2,3], with the aim of identifying special features associated with them, if any. Enormous increase in the protein structural data over the last decade has inspired us to take up a systematic analysis to examine whether the increased dataset provides any new insights into the residues that take up sterically disallowed conformations. With the increase in the data , one can probe their location in the protein structure, the residue propensities, and question if this conformation is adopted at the cost of deviations in internal geometry, or by the need to maintain overall protein topology or is it required for function. Further, one can even ask the question whether the limits of the allowed regions have to be extended. We present our finding in the following sections.

2. General features of residues in the disallowed region

Several previous studies probed the amino acid residues with sterically disallowed conformations in the Ramachandran map, highlighting their special structural features. It was reported that only 0.3% of the residues adopt disallowed conformation in 110 protein structures[2]. Later, a similar analysis performed with a dataset of 353 protein structures showed that 0.4% of the residues are conformationally disallowed[3]. The present study on a dataset of 3021 monomeric protein structures (resolution < 2Å and R-factor < 0.3) shows about 0.47% of the residues to be in disallowed regions [although the conformations are technically "allowed", we refer to them as "disallowed conformations"]. 1700 protein structures are identified with at least one residue in disallowed conformation. While most of these proteins contain one/two disallowed residues, it is worth mentioning that some structures have > ten disallowed residues [Fig.1a] (PDB ids 1T1U, 2EVA, 1G7S, 2ZWS, 1MSC, 3EVN have 18, 17, 16, 15, 14 and 13 disallowed residues respectively). While 0.47% is rather a small percentage for disallowed non-glycine residues (in the dataset of 3021 proteins containing 818251 non-glycine residues), it is not negligible (3912). The values taken by these residues appear as if the extreme limits of the allowed regions have been extended in certain regions of the map [Fig.1b-c]. It was shown that certain regions of conformational space indeed expand under folding conditions due to hydrogen bonding constraints[4]. The map of glycine however has not significantly altered [Fig.1c], since it has an extended allowed space.

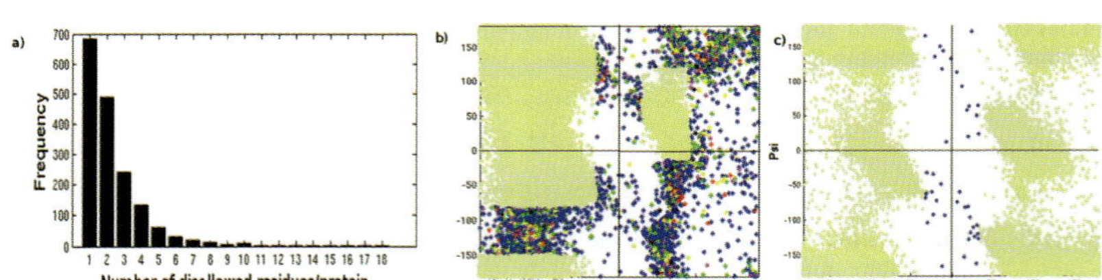

Figure 1. (a) Distribution of the number of disallowed residues/protein in the chosen dataset. Ramachandran plots for all the (b) non-glycine and (c) glycine residues from 3021 monomeric structures in the dataset. Pale yellow [in (b) and (c)] corresponds to the allowed regions, while the other colors correspond to [ϕ, ψ] values in the disallowed regions. Blue to red colors correspond to the residues in low-contact to high-contact environments.

Investigation of the residue propensities[2] shows that small polar/charged residues like Asn, Asp, Ser, Thr and His exhibit a higher tendency to adopt disallowed conformations [Fig.2a], in agreement with the earlier findings[2,3]. The disallowed values are preferred in irregular secondary structural elements (calculated using DSSP[5]) like loops (L), turns (T) and bends (B) [Fig.2b], consistent with earlier studies[3]. A significant number of disallowed Cys and an unusually high propensity for Trp are observed in the current study.

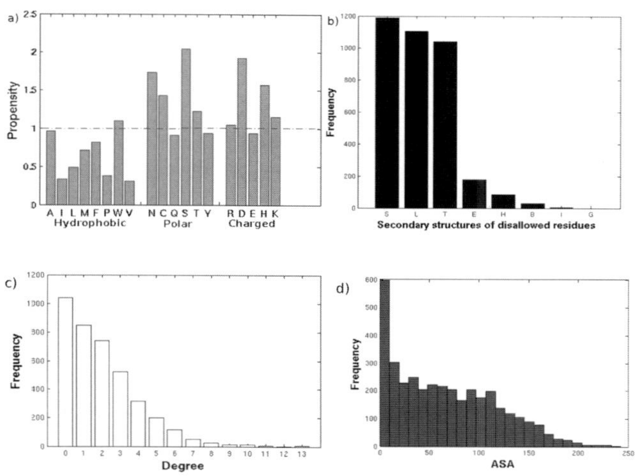

Figure 2. (a) Propensity of residues to be in the disallowed region of the Ramachandran plot. The polar/charged residues [above dotted line] have greater propensity. (b) Propensity of the location of disallowed residues in different secondary structures. (c) Degree distribution profile of disallowed residues reveals preference for low contact (exposed) environment. (d) ASA profile of disallowed residues.

3. Spatial environment of the disallowed residues

3.1. *Construction of Protein Structure Network (PSN)*

A brief description of the method adopted to identify the spatial neighbors is presented below. Protein structure networks [PSNs] capture the protein non-covalent interactions from a global perspective. PSNs are constructed by considering amino acid residues as nodes and edges are constructed between the nodes on the basis of non-covalent side-chain interactions, as evaluated from the normalized number of contacts between the interacting residues (with the exception of Gly where CA atom is considered). The interactions between two adjacent sequence neighbors are ignored. The interaction between two residues i and j has been quantified previously in our lab as[6]:

$$I_{ij} = \frac{n_{ij}}{\sqrt{(N_i \times N_j}} \times 100 \qquad (1)$$

Where n_{ij} is the number of distinct atom pairs between the side-chains of residues i and j, which come within a distance of 4.5Å and Ni and Nj are the normalization factors for residues i and j. The pairs of amino acid residues having interaction strength (I_{ij}) greater than a user-defined cut-off (I_{min}) are connected by edges in the PSN for a given interaction strength I_{min}. Generally, I_{min}s in the PSNs vary from 1% to 15%. The lower the I_{min}, the higher is the connectivity. We further construct an adjacency matrix (a mathematical representation of the PSN) at a given I_{min} of 3% [the rationale for the choice of I_{min}=3% is given in ref.7 and ref.8].

The degree of a residue is the number of connections it makes with its neighbors. The advantage of degree evaluation is that, it not only depicts the approximate location of the residue in protein topology [lower/higher degree correspond to exposed/buried residues respectively], but also provides information about its neighbors.

The degree distribution of the disallowed residues varies from zero-thirteen [Fig.2c]. It is obvious from the figure that the preference to be in lower contact (<3) is greater than to be in higher contact environments. The [ϕ-ψ] values of these residues are plotted in the Ramachandran map [Fig.1b]. It is evident that the residues of different degree have no particular [ϕ-ψ] value preference in the disallowed regions. The expected approximate inverse correlation between the degree of a residue and its accessible surface area (ASA) is found to be weak [correlation coefficient ~ -0.3663] and the ASA [evaluated by the program NACCESS[9]] profile is presented in Fig.2d for comparison.

3.2. *Contact based residue propensity*

We evaluate the distributions of different amino acids with disallowed conformations in bins of high (>4), mid (3, 4) and low degree (<3). The low and high degree profiles highlight the differences in their distribution [Fig.3a-b (left panel)]. These distributions are further compared [Fig.3a-b (right panel)] with a control dataset comprising of residues [199973 residues] with allowed conformations in loops, turns and bends. At low contact, it is seen that mostly polar/charged residues such as Asp, Glu, Ser, and Thr are preferred. Aromatic residues like Phe, Trp, and Tyr gain prominence in the high degree bin, suggesting their role in tight hydrophobic packing (due to stacking interactions).

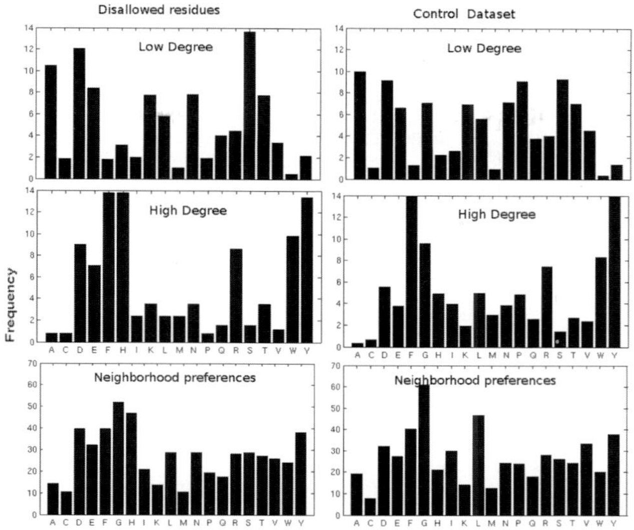

Figure 3. Histograms for contact based residue frequency (a) at low degree (<3) and (b) high degree regions (>4) for the disallowed (left panels) and the control residues (right panels) in normalized percentage. (c) The neighborhood propensities of disallowed residues (left panel) and allowed residues (right panel).

The control dataset show a similar trend, with polar/charged residues predominating at lower contact and aromatic residues showing up at higher contact. However, it is noteworthy that the frequency of disallowed His is prominently greater in high contact bin as compared to that in the control dataset. Thus, disallowed residue propensities are very similar to that of allowed residues in loops, turns and bends, with the exception of His.

3.3. *Neighborhood preferences*

The distribution of different amino acid residues that are in the neighborhood of disallowed residues is evaluated at an $I_{min}=3\%$. It is similar to the distribution of those in the neighborhood of the control dataset at low/mid contact with relatively high intensity of Gly and Leu in both cases (result not shown). However, in the high contact bin, apart from the high interaction propensity with Gly, the residues like Asp, Glu, Phe and Tyr also show increased participation in the neighborhood of both disallowed and control set of residues [Fig.3c]. His has a higher tendency to be in the neighborhood of disallowed residues as compared to the control residues [Fig.3c]. Similarly, Leu shows a diminished propensity in the neighborhood of disallowed residues in the high contact bin.

4. Hydrogen bonding interactions

Hydrogen bonding interactions of disallowed residues with their neighbors are calculated using the program HBPLUS, with default criteria for assigning H-bonding connectivity[10]. A large fraction of the disallowed residues is involved in hydrogen bonding interactions irrespective of their degrees [Fig.4a first left panel]. Similar degree distribution profile is also exhibited by the control dataset [Fig.4a first right panel], suggesting that the amino acid residues, both in the disallowed and the control dataset are similar in terms of their packing due to hydrogen bonding interactions.

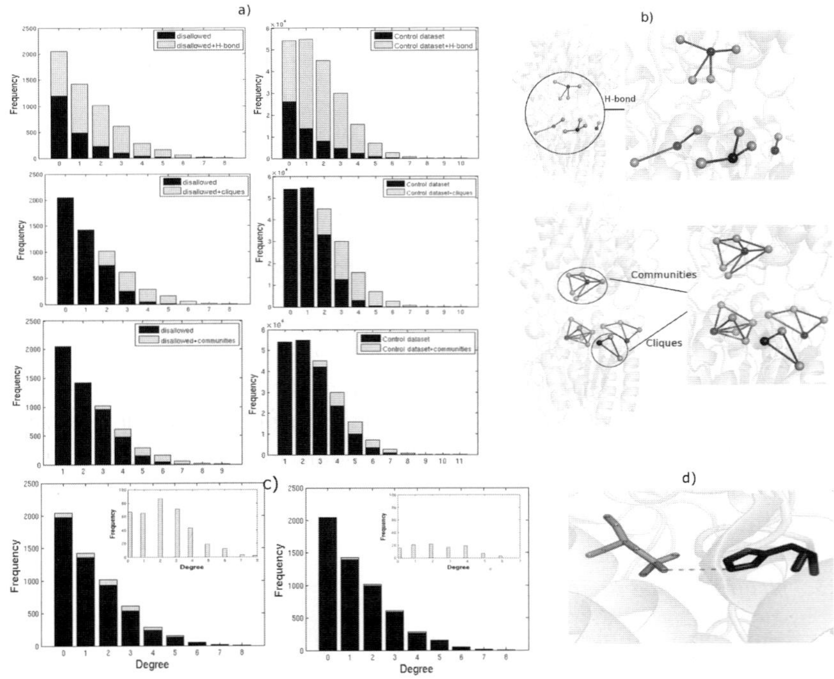

Figure 4. Fraction of disallowed residues (as a function of degree) with special features: (a) Fraction of disallowed (on the left) and control residues (on the right) involved in hydrogen bonding interactions (first panel), involved in the formation of cliques (second panel), and involved in the formation of communities (third panel). (b) Pictorial representations of hydrogen bonding interactions and cliques/communities formed by the disallowed residues (black van der Waals' sphere) with other neighboring residues (grey van der Waals' sphere). (c) Distribution of disallowed residues involved in ligand interaction through Van der Waals' contact (on the left) and through hydrogen bonding interactions (on the right) at various degrees. (d) Pictorial representation of the interaction of a disallowed residue (stick representation in black) with its neighboring ligand (stick representation in grey).

5. Packing of disallowed residues identified from network parameters

The spatial environment of disallowed residues is probed by a judicious use of several network parameters. The non-covalent interactions result in spatial contact with their neighbors in the 3D structural environments. While the number of connections defines the degree of a residue, the inter-connectivity of the interacting residues defines higher-order connectivity, representing the extent of packing around the chosen residue. This can be captured by graph theoretic parameters known as cliques and communities (a clique is a set of points in which all points are connected to all other points and communities are an assemblage of cliques).

The structural and functional importance of such tightly packed entities formed by the side-chain interactions in proteins has been discussed in earlier studies [11,12]. Here we have explored the cliques and communities containing the disallowed residues to investigate the side-chain packing features. The community search approach is based on the algorithm proposed in ref.13. Most of the amino acid residues in protein structures form cliques of size k=3 and less often of size k=4. The degree distribution of the disallowed residues, involved in the formation of cliques/communities, exhibited a similar profile to those of the control residues [Fig.4a (second and third panel)]. This suggests that both the allowed and the disallowed residues in loops, turns and bends contribute similarly to the overall packing of residues in proteins, even at the higher level connectivity. In order to capture the residue-wise detail of interaction preferences within cliques, an analysis of the pairwise residue interactions between the disallowed residues and their partner residues in cliques reveals that the disallowed residues show more preference for partner residues in cliques as compared to the allowed ones [Fig.5a-b]. High preference between a disallowed His residue and His, Asp and Glu as clique neighbors is evident from Fig.5a. Additionally, disallowed Arg prefers Asp and Glu and Phe prefers Trp. In contrast, the control residues exhibit much less preference. This may be due to the diminished steric constraints that they experience and thus have relatively flexible choosing criteria for their interacting partners in a highly packed and rigid union such as cliques. Thus, the disallowed residues do not exhibit any additional bias towards formation of cliques as compared to the control set of residues; however, they show an enhanced preference in choosing their neighbors in cliques.

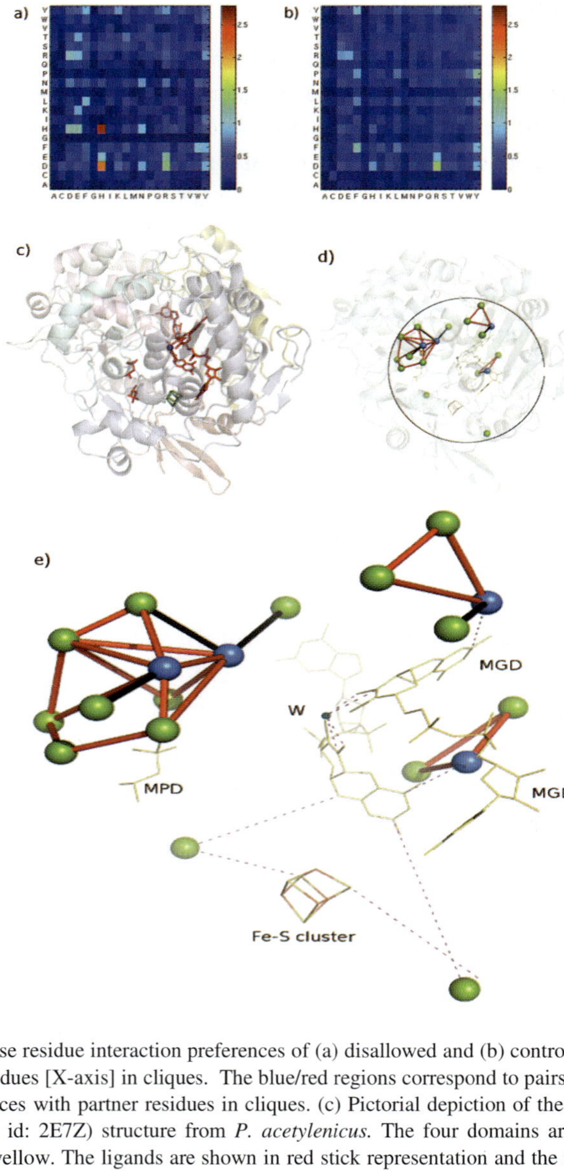

Figure 5. Pairwise residue interaction preferences of (a) disallowed and (b) control residues [Y-axis] with partner residues [X-axis] in cliques. The blue/red regions correspond to pairs with lower/higher contact preferences with partner residues in cliques. (c) Pictorial depiction of the enzyme acetylene hydratase (PDB id: 2E7Z) structure from *P. acetylenicus.* The four domains are colored in pink, cyan, blue and yellow. The ligands are shown in red stick representation and the tungsten metal ion in blue van der Waals sphere. (d) The complex network architecture mediated by the four disallowed residues (depicted as blue van der Waals' spheres). (e) Zoomed in view of the network as encircled in (d). Hydrogen bonding interactions are shown in black and other non covalent interactions in red. The ligand MGD and MPD and the FE-S cluster are shown in yellow stick representation.

6. Residues with disallowed conformation-ligand interactions

Binding of ligands (several ion/small organic molecules/peptides) or their presence near the active sites in proteins is often of biological relevance. The study of the disallowed residues interacting with ligands would throw light on their specific functionality and here it is investigated in two ways: based on the hydrogen bonding interaction and spatial contact (within 4.5Å of the ligand) between the ligand and the disallowed residues. This is again calculated using the program HBPLUS. About 110 and 370 disallowed residues interact with ligand (excluding solvent and $S0^{4-}$ and $P0^{4-}$ molecules) by hydrogen bonding interactions and spatial contact criteria respectively. The degree distributions of the disallowed residues involved in interactions with ligands through both hydrogen bonding and spatial contact criteria reveal similar profiles [Fig.4c], with disallowed residues of lower degree participating to a greater extent in ligand interactions as compared to those in the higher degree bin. This result is also intuitively rationalized as the residues on the surface of the protein (usually in the lower degree bin) are more prone to interactions with ligand in contrast to the buried ones (usually in the higher degree bin).

7. Biological relevance of the disallowed residues in proteins

It would be further interesting to probe whether the disallowed residues adopt non-standard backbone conformations only as a 3D-structural consequence or if they bear some biological importance. Interactions of a significant fraction of the disallowed residues with ligands (~ 10%) motivated us to examine them for functional implications. In this context, the non-redox-active tungsten/ [4Fe:4S] enzyme acetylene hydratase (PDB id: 2E7Z) from *P. acetylenicus,* which catalyses the hydration of acetylene to acetaldehyde[14], proves to be a convincing example. The enzyme is a monomeric, tetradomain (Domains I-IV) protein with 730 residues and is bound to an [4Fe:4S] cluster and two Molybdopterin guanine dinucleotide cofactors (MGD), with a tungsten metal centre. Four residues (142I, 387M, 423Y and 612Q) in this protein are disallowed, three of which belong to domains II (423Y), III (142I), and IV (612Q) respectively. The loop connecting domain II and III (residues 327-393) also contains a disallowed residue (387M). Domain II and III, including the disallowed residues 423Y and 142I, display multiple hydrogen bonding interactions with each of the MGD cofactor [Fig.5(c-e)]. Interestingly, the entire region connecting domains II and III in this enzyme undergoes rearrangement in comparison to its homologues in the DMSO reductase family. As a consequence of this rearrangement, acetylene gains an easy access of the tungsten ion through a channel that contains the

4Fe:4S cluster. This overall rearrangement, typical to acetylene reductase, may be a cause or result of the presence of the four disallowed residues in this protein. Additionally, the disallowed residue 142I participates in the hydrophobic ring (containing 6 residues from domain I, II and III) which forms a cavity to bind to the substrate acetylene, playing a major role in the determination of substrate specificity.

At the level of higher order connectivity, all the four disallowed residues participate in the formation of well packed regions, in terms of cliques, around the active site. Two of them also participate in an assemblage of cliques known as communities which are bigger units of rigid regions. Fig.5(c-e) clearly depicts that the four disallowed residues, through networks of varying rigidity, like cliques/communities, H-bonding and non-covalent interactions, form an architecture around the metal centre and the ligand MGD, which may eventually be of functional relevance. The direct or indirect interactions of these four residues with the ligands and metal centre indicate their importance towards the biological outcome.

8. Conclusion

The mapping of a large number of high resolution crystal structures has shown that about 0.47% of the non-glycine residues are in the "disallowed" region of the [ϕ-ψ] map. They occur mainly in irregular secondary structures and are predominantly taken up by small, polar/charged residues, in agreement with previous studies.

Graph theory based method is used to identify the location of disallowed residues in the context of protein topology. It provides not only the number of non-bonded connections (degree/contact of a residue) made by the residue, but also provides the details of the interacting residues. The propensities of disallowed Asp, Glu, Ser, and Thr are high in the low contact regions and that of Gly, Asp, Glu, Phe, Tyr and His are high in the high contact regions. Surprisingly, this is not unique to the disallowed residues (with the exception of His in high contact region), since a control dataset also shows the same profile. Further, the propensities of the interacting residue types in both the disallowed and the control sets are less discriminatory. A substantial fraction of the disallowed residues participate in hydrogen bonds, cliques/communities (represents the tight packing with neighbors). The same feature is also exhibited by the control set. Strikingly, the analysis of pairwise residue interactions in cliques reveals that disallowed residues show enhanced preference for their partners in contrast to the control dataset. A high preference between a disallowed His residue and His, Asp and Glu as clique neighbors is observed.

Finally, there is a hint of functional relevance for the adoption of disallowed conformation, from the interaction with ligands. As expected, most of the interactions are with the low contact residues. Surprisingly, some of the ligands bind to the disallowed residues in the highly connected regions. A specific case is presented in which four disallowed residues involved in hydrogen bonds, cliques/communities, work in unison with Fe-S cluster, tungsten to interact with the ligand.

In summary, the protein backbone conformations that are discouraged from the standard Ramachandran plot analysis appear more often in conformational space than expected and they occur predominantly in loops, turns and bends. Interestingly, the excess occurrences are in similar neighborhood environments as seen for the allowed residues in similar secondary structural locations. A subtle balance between the backbone conformational preference and the overall protein topology, in combination with possible functional relevance seem to be the principal reasons driving this excess.

Acknowledgement

SV and MB thank CSIR, Government of India for Emeritus Scientist Fellowship and Senior Research Fellowship respectively. Microsoft Research supported the fellowship of Chanda R. Bhat. This article is dedicated to the memory of Professor G.N. Ramachandran whose deep, penetrating work greatly influenced the development of the field of Structural Biology.

References

1. G.N.Ramachandran, C.Ramakrishnan and V.Sasisekharan, *J. Mol. Biol.* **7**, 95 (1963).
2. K.Gunasekaran, C.Ramakrishnan and P.Balaram, *J. Mol. Biol.* **264**, 191 (1996).
3. D. Pal and P.Chakrabarti, *Biopolymers.* **63**, 195 (2002).
4. L.L.Porter and G.D.Rose, *Proc Natl Acad Sci U S A.* **108**, 109 (2011).
5. W. Kabsch and C.Sander, *Biopolymers.* **22**, 2577 (1983).
6. N. Kannan and S.Vishveshwara, *J. Mol. Biol.* **292**, 441 (1999).
7. K.V. Brinda and S.Vishveshwara, *Biophys. J.* **89**, 4159 (2005).
8. A.Ghosh and S.Vishveshwara, *Biochemistry.* **47**, 11398 (2008).
9. S.J. Hubbard and J.M.Thornton, 'NACCESS', Computer Program, Department of Biochemistry and Molecular Biology, University College London (1993).

10. I.K. McDonald and J.M. Thornton, *J. Mol. Biol.* **238**, 777 (1994).
11. M. Bhattacharyya and S.Vishveshwara, *Biochemistry.* **50**, 6225 (2011).
12. D.Deb, S.Vishveshwara and S.Vishveshwara, *Biophys. J.* **97**, 1787 (2009).
13. G.Palla, I.Derenyi, I. Farkas and T. Vicsek, *Nature.* **435**, 814 (2005).
14. G.B.Seiffert, G.M.Ullmann, A.Messerschmidt, B.Schink, P.M.H Kroneck and O.Einsle, *Proc. Natl. Acad. Sci. U.S.A.* **104**, 3073 (2007).

ANALYSIS OF DIHEDRAL ANGLE VARIABILITY IN RELATED PROTEIN STRUCTURES

SHEKHAR C. MANDE[1], ASHWANI KUMAR[1,2] AND PAYEL GHOSH[1,2]

[1]*National Centre for Cell Science, Pune and*

[2]*Centre for DNA Fingerprinting and Diagnostics, Hyderabad*

Similarity or variability of protein structures is usually measured by superposition of structures and then assessed by root mean square deviation of the superposed main chain atoms. The relationship between equivalent main chain atoms (or Cα atoms) and the variability in their dihedral angles has been assessed in this work. Proteins with identical sequences (or point mutants), those with homologous structures, and those whose structures have been determined by NMR were used to assess such relationship.

1. Introduction

Protein structures are often analysed from the viewpoint of their main chain dihedral angles, limits of which were defined in the seminal paper by Ramachandran and colleagues[4]. The main chain dihedral angles thereby represent the overall characterization of polypeptide conformation. These also routinely serve as a useful tool in the validation of experimental structure determinations.

It is not surprising that conformational differences among closely related polypeptides are brought about by changes in the main chain dihedrals of only a handful of residues. Domain motions, for example in multi-domain proteins, are brought about by residues which act as conformational hinges. Such conformational changes at the hinge residues are a result of perturbations in main chain dihedral angles of only a few residues. Thus, changes in a few conformational parameters of the main chain can potentially result in dramatically different conformations of polypeptides.

Protein structures with similar tertiary folds are often compared to each other by superposition of their Cartesian coordinates. Sequential superposition of the coordinates of a pair of protein structures, primarily judged by root mean square deviation (rmsd) among equivalent main chain atoms, characterizes the extent of structural similarity between the two proteins. The rmsd values serve as global indicators in suggesting relatedness of their folds, and/or evolutionary divergence of their structures[2]. Values typically in the range of 1-3 Å indicate global similarity in the structures. The rmsd values, however, represent a single global parameter and therefore often fail to yield satisfactory comparative evidence in proteins where large conformational changes have occurred due to domain motions.

In most related protein structures, the core regions of proteins align with each other far more significantly compared to the loop regions. Although it is well accepted that loop regions are more variable than core regions of proteins, the underlying variability of dihedral angles that gives rise to conformational heterogeneity of the loops is less studied. Moreover, conformational variations even within structurally well aligned regions, or those in similar tertiary structures, are less well understood in terms of the variability of dihedral angles. It should now be possible to calculate such variability as there are a large number of examples where the structures of identical proteins (or their point mutants) have been determined. Moreover, structures of many families of proteins, which are closely related evolutionarily, have also been determined. It should thus be possible to calculate variations in main dihedrals and correlate such variations to conformational differences in protein structures.

We have undertaken analysis of variations of main chain dihedral angles in related protein structures in this work. We examine the variations in structures which are not only identical in their sequences (or the point mutants), but also examine dihedral angle variations in a few families of protein structures which are evolutionarily related. Some interesting insights are obtained by these comparisons which might lead to better understanding of relationship between conformational variability and main chain dihedral angle changes.

2. Materials and Methods

We chose to study 7 families of proteins whose structures have been determined by X-ray crystallography to high resolutions, and which are nearly identical in their sequences (Table 1). These structures include proteins with point mutations or those where a protein-ligand complex was available. Some of these structures were of the identical protein but the crystals were either grown in different environmental conditions, or determined by a different group of investigators.

Apart from the 7 families of nearly identical structures, we chose 5 families of proteins which are known to be evolutionarily related. These structures included enzymes whose functions have been well characterized (Table 1). The natural sources of these enzymes are different, although they might have been expressed in *E. coli* for the purpose of crystallographic studies. We also chose to

Table 1: Selected protein families. The RMSD values are with respect to the average structure in each family.

Protein name	Structures/ Models (Aligned residues)	Method	Range of maximum resolution (Å)	Range of RMSD (Å)
Hen Egg White Lysozyme	39 (128)	X-ray	1.60 – 2.50	0.60 – 0.88
T4 Lysozyme	69 (162)	X-ray	1.70 – 3.00	0.15 – 0.47
Serum Albumin	73 (576)	X-ray	1.90 - 3.30	0.26 – 2.50
Cytochrome c	29 (123)	X-ray	0.84 - 2.70	0.29 – 0.44
Trypsin	305 (221)	X-ray	0.97 – 3.0	0.53 – 0.99
Thioredoxin	24 (106)	X-ray	1.68 – 2.95	0.30 – 0.57
Triosephosphate Isomerase (*Trypanosoma brucei*)	39 (246)	X-ray	1.15 – 2.40	0.26 – 0.54
Azurin	91 (112)	X-ray	0.98-2.83	0.23-0.83
Cutinase	51 (174)	X-ray	0.85-2.70	0.18-0.75
Flavodoxin	69 (107)	X-ray	1.20-2.62	0.55-1.31
Malate Dehydrogenase	45 (171)	X-ray	1.45-3.20	0.66-1.20
Triosephosphate Isomerase (non-identical structures)	100 (175)	X-ray	0.82-2.90	0.44-1.13
Gag Polyprotein (1GWP)	20 (151)	NMR	-	0.94-2.68
Autophagy-related Protein 8 (2KWC)	30 (116)	NMR	-	0.58-1.52
Interleukin-21 (2OQP)	20 (134)	NMR	-	2.97-5.15
Voltage-gated Potassium Channel (2KYH)	20 (147)	NMR	-	2.56-5.45
Transcription elongation regulator 1 (2KIQ)	20 (62)	NMR	-	0.19-0.40

investigate 5 protein structures which were determined by NMR. The NMR structural investigations typically require 20-30 independent structural models to be deposited in the PDB, all of which satisfy the experimental constraints.

Structural superpositions were carried out using GROMACS and other freely available tools[6,7]. Having superposed all the related structures on each other, we calculated an average structure for each of the families. The phi and psi angles and other structural features were calculated using appropriate BioPython scripts[1] and the freely available MMTSB toolset[3]. Averages and standard deviations for all the main chain dihedral angles of structurally aligned residues were calculated using in-house perl/python scripts. Care was taken to ensure that values close to 180° were appropriately averaged.

3. Results and Discussion

Table 1 lists all the structures that were analysed for phi, psi deviations. The first 7 structures listed in the Table 1 were almost identical in their sequences and therefore might represent deviations arising out of experimental variations. The following 5 structures belong to the same protein family with variable sequences. Most of the protein structures considered in our analysis were determined at high resolutions. The trypsin family represents the largest number of structures in the same family with 305 structures and those determined at resolutions from 0.95 to 3.2 Å. Finally, the last five structures that we analysed were determined by NMR, and therefore represent dihedral angle deviations obtained from a different experimental technique. It was anticipated that the deviations for NMR structures would be higher.

The standard deviations of the dihedral angles, phi and psi, vary between 5° and 10° among the identical structures (Figure 1). In the two families, namely common cold virus and HIV protease, the standard deviations show significantly low or high values respectively. It is likely that restraints used in the refinement of common cold virus structures led to small deviations in dihedral angles. The basis of large deviations in HIV protease structures is not easily understood (data not shown here). Overall, deviations in dihedral angles are seen to be high in regions where RMSD from mean structure is high (e.g. Figure 1a). However, this is not seen in every polypeptide (e.g. Figure 1c).

In the five structures that are not identical in their sequences, the deviations in dihedrals of aligned residues are similar to those observed for identical structures (Figure 2). Large deviations are observed in loops, or places which have insertion or deletion of residues. It is interesting to note that RMSD values of the superposed structures are not always high at places where deviations in dihedrals are high. Thus, the regions of low RMSD from the average structure in the evolutionary distant proteins exhibit similar phi, psi deviations as those in the identical protein structures.

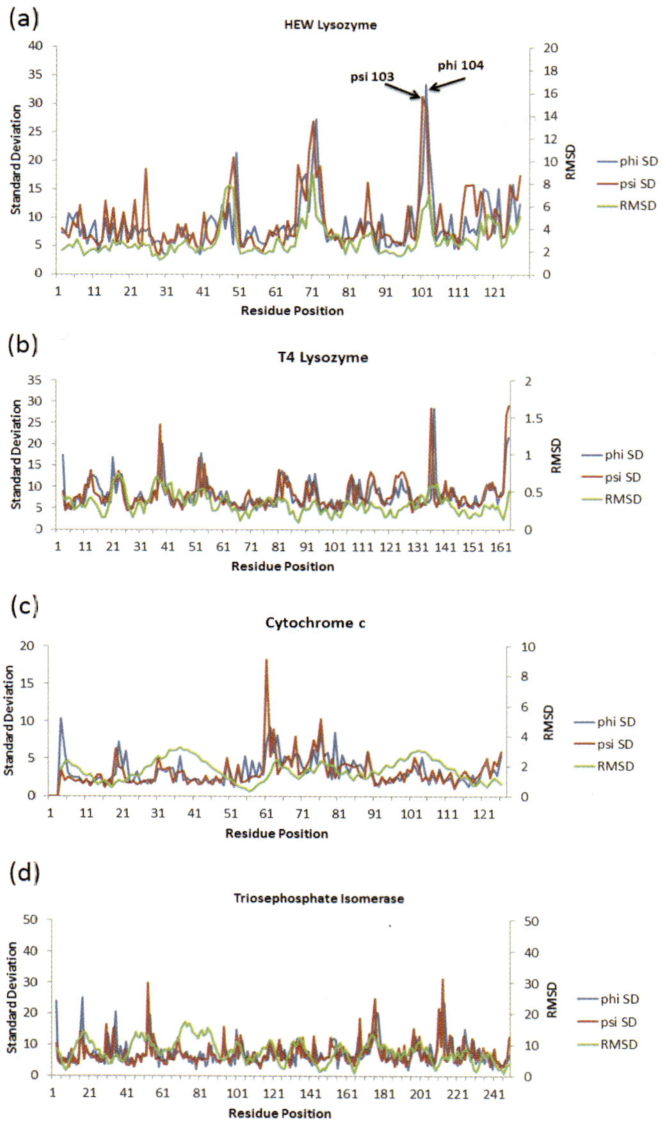

Figure 1: Standard deviations of phi and psi angles and root mean square deviation (RMSD) of amino acids for each aligned position. Deviations in psi are shown in red and those in phi in blue. RMSD of each aligned Cα position from the mean structures is plotted in green colour. The phi, psi deviations are shown in (°) and RMSD values in Å. (a) HEW Lysozyme, (b) T4 lysozyme, (c) Cytochrome c, (d) Triosephosphate Isomerase (*Trypanosoma brucei*).

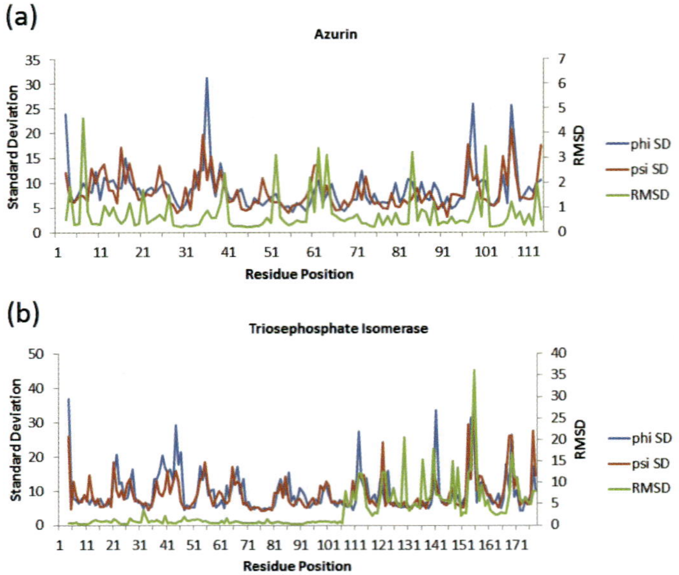

Figure 2: Standard deviations of phi and psi angles and RMSD at aligned positions in evolutionary distant protein structures. Deviations in phi and psi are shown in blue and red respectively. RMSD at each position are shown in green. The phi, psi deviations are shown in (°) and RMSD values in Å. (a) Azurin, and (b) Triosephosphate Isomerase.

Analysis of regions where phi, psi deviations were high, revealed remarkable plasticity of polypeptide conformations. For example, HEW lysozyme structures exhibit a peptide flip at residue Asn 103[8]. Consequently, two residues- Asn 103 and Gly 104 show high RMSD values in the aligned structures. Yet all the polypeptides align well in the flanking regions of these two residues. The dihedral angles at these two residues also show high variability in this region, with standard deviations in excess of 20° (Figure 1a). It is clearly seen that polypeptide regions on either side of the peptide bond align well, despite high structural variability near the peptide flip region (Figure 3a).

It is interesting to analyse the regions of polypeptide where conformational variability is observed. Special attention might be given to check if the polypeptide conformational variability is distributed over a few residues continually, or if there are distinct clusters of conformations of a few residues in such regions. If the former, conformational variability might arise out of continuous change in dihedral angles spread over a few residues. On the other hand, if the latter is true, polypeptides might be assumed to have restricted

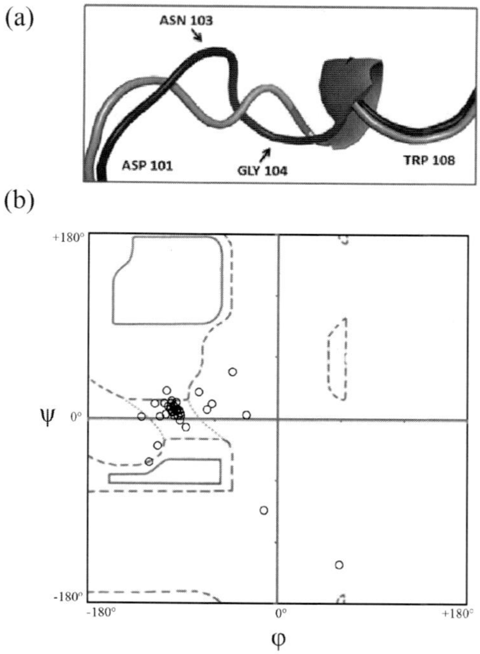

Figure 3: The region of HEW lysozyme structure that shows peptide flip. Asn 103 - Gly 104 peptide shows a flip in many HEW structures. Consequently, although all the flanking regions superpose well in the HEW lysozyme structures, there are structural deviations in the Asn 103-Gly 104 region. (a) Two structures in which the peptide bond is flipped, (b) Ramachandran map showing changes in phi, psi at Asn 103. Figure 3(a) was generated using PyMol[5].

conformational possibilities. We find that regions where large conformational variations are seen, these variations arise out of a few distinct conformations in a stretch of residues. For example, two distinct conformations are seen for Gly 37 of Azurin structures (Figure 4a). These two conformations lead to conformational variability of polypeptide immediately following this residue. Similarly, in all such regions analysed in different protein structures suggest that conformational variability arises out of restricted conformations of a few residues, rather than a continuous change in a stretch of residues.

An interesting comparison was between triosephosphate isomerase (TIM) structures where large number of structures are available for the same TIM (*Trypanosoma brucei*), as well as equally large number of structures from different sources. Analysis of dihedral deviations showed that regions of conformational variability are identical in both these comparisons (Figure 1d and 2b). As anticipated, evolutionary distant proteins exhibit higher deviation compared to the identical TIMs, as seen in residues 210-215. The dihedral variations in this region of polypeptide arise principally at residue Gly 209, which has two distinctly possible conformations. Ramachandran map shows two clusters of conformations for this residue in the evolutionary distant proteins (Figure 4b). Thus, it can be seen that variability within each cluster is of similar range as that observed at aligned positions. Consequently conformational changes in regions of polypeptide appear to arise out of two distinct conformations of a residue, rather than an overall change distributed over a length of a polypeptide.

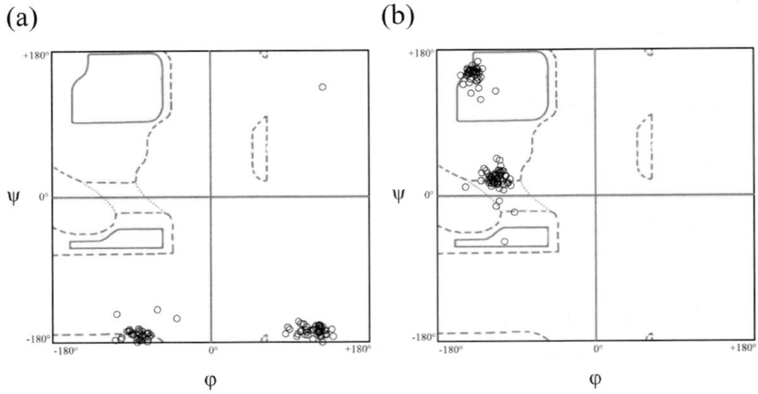

Figure 4: (a) Ramachandran map of Gly 37 in Azurin structures. It can be seen clearly that this residue has two possible conformations (b) Ramachandran map of Gly 209 in Triosephosphate Isomerase structures. It can be seen clearly that this residue has two possible conformations.

Our analysis reveals that there is no generalized relationship between conformational variations in a polypeptide and localized changes in dihedral angles. In regions of polypeptide, the two are correlated well with each other, whereas there are also regions in polypetides where the dihedral angle changes are spread over a stretch of residues in order to bring about conformational variations. More analysis will be required to comment on the relationships among other parameters, such as intrinsic flexibility of polypeptides localized on the surface residues, compensatory changes in variations between phi and psi angles, propensity of different amino acids to accept large variations among the dihedral angles etc. With the availability of a large number of protein structures, it should now be possible to address these fundamental questions on polypeptide conformation.

Acknowledgments

Authors thank the National Centre for Cell Science for providing computational facilities. The work was supported by Centre of Excellence grants: DBT Centre of Excellence for Microbial Biology BT/01/COE/07/02 and Virtual Centre of Excellence on multidisciplinary approaches aimed at interventions against *Mycobacterium tuberculosis* BT/01/COE/07/02.

References

1. P.J.A. Cock, T. Antao, J.T. Chang, B.A. Chapman, C.J. Cox, A. Dalke, I. Friedberg, T. Hamelryck, F. Kauff, B. Wilczynski, and M.J.L. de Hoon, *Bioinformatics*, **25** 1422 (2009).
2. C. Chothia, and A.M. Lesk, *EMBO J.*, **5** 823 (1986).
3. M. Feig, J. Karanicolas, and C.L. 3rd. Brooks, *J. Mol. Graph Mod.*, **22** 377 (2004).
4. G.N. Ramachandran,C. Ramakrishnan, and V. Sasisekharan, *J. Mol. Biol.*, **7** 95 (1963).
5. W.L. DeLano, The PyMOL Molecular Graphics System, DeLano Scientific, San Carlos, CA, USA (2002).
6. M. Shatsky, R. Nussinov, and H.J. Wolfson, *Proteins*, **56** 143 (2004).
7. D. van der Spoel, E. Lindahl, B. Hess, G. Groenhof, A.E. Mark, and H.J. Berendsen, *J. Comput. Chem.*, **26** 1701 (2005).
8. M.A. Walsh, T.R. Schneider, L.C. Sieker, Z. Dauter, V.S. Lamzin, and K.S. Wilson, *Acta Crystallogr D Biol Crystallogr.*, **54** 522 (1998).

DEFINING α-HELIX GEOMETRY BY C^{α} ATOM TRACE *vs* (φ-ψ) TORSION ANGLES: A COMPARATIVE ANALYSIS

ASHISH SHELAR[*], PRASUN KUMAR[*] AND MANJU BANSAL
Molecular Biophysics Unit, Indian Institute of Science
Bangalore, 560 012, India

A regular secondary structure is described by a well defined set of values for the backbone dihedral angles (φ, ψ and ω) in a polypeptide chain. However in real protein structures small local variations give rise to distortions from the ideal structures, which can lead to considerable variation in higher order organization. Protein structure analysis and accurate assignment of various structural elements, especially their terminii, are important first step in protein structure prediction and design. Various algorithms are available for assigning secondary structure elements in proteins but some lacunae still exist. In this study, results of a recently developed in-house program ASSP have been compared with those from STRIDE, in identification of α-helical regions in both globular and membrane proteins. It is found that, while a combination of hydrogen bond patterns and backbone torsional angles (φ-ψ) are generally used to define secondary structure elements, the geometry of the C^{α} atom trace by itself is sufficient to define the parameters of helical structures in proteins. It is also possible to differentiate the various helical structures by their C^{α} trace and identify the deviations occurring both at mid-positions as well as at the terminii of α-helices, which often lead to occurrence of 3_{10} and π-helical fragments in both globular and membrane proteins.

1. Introduction

The most general description of 'helix' is a smooth 3D curve that lies on a conical or cylindrical surface. Helices which constitute the major secondary structure elements in proteins take up this structure due to repeating values of the backbone torsion angles (φ, ψ and ω) accompanied by regular hydrogen bond patterns between the backbone NH and CO groups of amino acids. Numerous secondary structure assignment algorithms which use atomic coordinate data have been developed and can be broadly classified into three categories: (i) algorithms based on backbone torsion angles and hydrogen bond patterns (ii) algorithms based on 3D geometry (iii) hybrid methods, which use both (i) and (ii). Programs like DSSP[1], STRIDE[2] and PROSS[3] fall into the first category, DEFINE-S[4], P-CURVE[5] come under second category whereas KAKSI[6], PALSSE[7] etc. fall under the third category. All these algorithms

---

[*]Both authors contributed equally.

identify the main body of the α-helix, however they often differ in their definition of helix terminii. Even in the main body, sometimes differences occur in the assignment because of small local deviations from the uniform helical character, arising due to solvent induced distortions[8], peptide bond distortions[9,10] or presence of Proline[11-13], Serine and Threonine residues[14,15]. Blundell *et.al.*[8] carried out the first survey on the irregularities in helices and concluded that a majority of the helices are curved, which has been confirmed by subsequent studies[16]. α-helices also show distinct preferences for amino acids at their N and C terminal regions which define the helix initiation and termination motifs[17-19] as well as at mid-positions[20].

Assignment of β-strands is comparatively more complex, as the residues involved in the hydrogen bonds, can be far from each other in sequence space. Several experimental[21-23] and statistical[24,25] analyses have not conclusively indicated any specificity for the terminal residues in β-strands.

In this article, we describe the results of an analysis of helix terminii in α-helices only, since they are most numerous and of sufficient length to be well defined geometrically. An α-helix is defined as extending from residues N1 to C1 which make up the main body of the helix while Ncap and Ccap are the immediately preceding and succeeding non-helical residues. We have recently developed a program, designated ASSP (Assignment of Secondary Structure in Proteins) as an extension of our in-house program HELANAL-Plus[26], which identifies helical regions based on the geometry of the C^α trace. The most widely used program STRIDE (**Str**uctural **ide**ntification) uses hydrogen bond patterns along with the backbone torsional angles φ and-ψ to define helical regions. Results of analysis using both the algorithms, for helices in globular as well as membrane proteins, have been compared. Our analysis using ASSP indicates that in some cases, even if the main chain-main chain hydrogen bonds are missing at the terminii, the overall helical geometry is retained by the C^α atoms and hence the corresponding residues should be categorized as being part of a helix, while in other cases an α-helix is found to be interspersed or flanked by a small, but identifiable, fragment of 3_{10} or π-helix.

2. Methods

2.1 Dataset of globular and membrane proteins

A non-redundant dataset of globular proteins was created at the fold level, using the ASTRAL[27] compendium in the SCOP[28] database. The ASTRAL compendium groups proteins with the same major secondary structures, with the same arrangement and topological connections. Total of 1195 representative folds were downloaded from ASTRAL-1.75 release database. The dataset was further refined by the following steps: (i) Domains with SPACI score less than 0.4 (resolution worse than 2.5 Å) were excluded (total 266 folds removed) (ii) Proteins which had a missing ATOM record for any of the residues were excluded (total 1 fold removed) (iii) proteins with 'all beta' fold were excluded (total 174 folds removed) and (iv) membrane and cell surface proteins were excluded (total 58 folds removed). The final dataset consisted of 626 representative folds.

A separate non-homologous X-ray crystal structure dataset of membrane proteins was created by selecting non redundant protein structures from PDBTM[29]. Proteins with a resolution better than 2.5 Å, R factor less than 0.25 and sequence identity less than 25% were selected using the PISCES server[30]. The dataset contains 75 proteins comprising 181 chains and includes 48 out of the 58 folds defined by the SCOP database.

Coordinates for the culled datasets of globular and membrane proteins were then downloaded from PDB[31].

2.2 Helix assignment and identification of H-bonds

In-house program ASSP and commonly used STRIDE program were used for defining helices in both the datasets. Helical parameters and local bending angles for successive steps have been calculated using HELANAL-Plus[26]. Hydrogen bonds were identified using HBPLUS v3.06[32].

3. Results and Discussion

3.1 Length distribution of the helices in both datasets

The globular protein dataset contained a total of 3615 and 4028 α-helices as assigned by STRIDE and ASSP respectively. The length of helices defined by STRIDE for the globular and membrane protein dataset varies from 1 to 80 and 1 to 68 residues respectively. However, for ASSP it varies from 4 to 80 and 4 to 69 residues respectively, since ASSP defines helices based on the geometric trace of C^{α} atoms and requires a minimum of four residues. The mean length of helices assigned by ASSP in both datasets is hence found to be smaller, particularly in case of membrane proteins, where the mean length is 16 according to STRIDE and 13 according to ASSP. The length distribution of the helices with length >4 residues, in both the datasets, is shown in Figure 1. Interestingly a few of the long helices assigned by STRIDE are broken and defined as two or more helices by ASSP, thereby increasing the number of shorter helices (≤15 residues). In this study, we have analyzed in detail only α-helices with length >8 residues, since in these helices at least one residue has both N-H and C=O groups involved in hydrogen bond formation and hence these are expected to be structurally well defined. In globular protein dataset, 2680 and 2665 STRIDE and ASSP assigned helices were found to be of length >8 residues. In membrane proteins, out of 1164 and 1404 STRIDE and ASSP assigned helices, 865 and 883 helices respectively were found to be of length >8 residues.

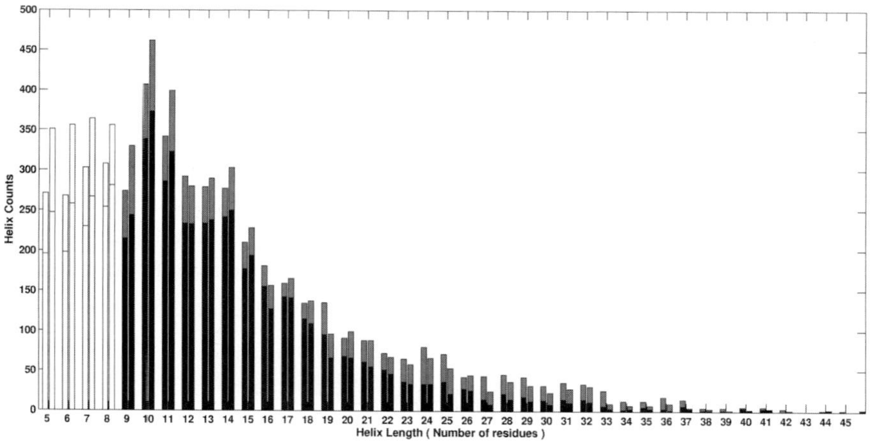

Figure 1: Stacked bar diagram showing length distribution for α-helices assigned by STRIDE and ASSP in globular (black bars) and membrane (grey bars) proteins. α-helices with ≤8 amino acid residues are shown as white bars and not considered in present analysis. In each pair of stacked bars, first bar represents helix assignment by STRIDE and second one corresponds to ASSP. A few α-helices with length >45 residues are also defined by both programs. Such helices are very few in number (<5) and the data are not displayed in the plot.

3.2 *Mismatch in the helix terminus assignment*

The variation in helix length found by STRIDE and ASSP analysis clearly indicates that there are differences in identifying the helix terminii by the two programs. Table 1 gives the statistics of the STRIDE and ASSP helix assignments with exact match at both terminii, match at either N or C terminii, as well as mismatch at both ends. 1307 and 320 helices in globular and membrane proteins match at both N and C terminii, as assigned by both STRIDE and ASSP. A significantly higher number of helices are observed to have only the N-terminii matching, as compared to those with only their C-termini matching.

Table 1: Statistics of the matching and non-matching helix terminal assignments by STRIDE and ASSP algorithms for α-helices in globular and membrane protein datasets.

Dataset	No. of Helices		No. of Helices (Length > 8)		Both ends matching	Only N-terminii matching	Only C-terminii matching	Both terminii not matching
	STRIDE	ASSP	STRIDE	ASSP				
Globular	3615	4028	2680	2665	1307	918	403	37
Membrane	1164	1404	865	883	320	339	190	16

The terminal regions of the helices show a wider spread in their (φ-ψ) values as compared to the MID regions. However, the commonly used structural parameters, viz. unit rise and unit twist, calculated for these α-helices, do not show large variations, as also reported in our earlier analysis of long α-helices[33].

The (φ-ψ) distribution for helices with identical assignments by STRIDE and ASSP for globular (1307) and membrane (320) protein datasets lies well within the allowed regions of the Ramachandran Map (data not shown). The (φ-ψ) distribution for N2 to C2 residues in helices with one residue mismatch at either terminus, according to STRIDE and ASSP assignments, shows a similar spread.

3.3.1 Helices with one residue mismatch at the N terminus

In helices where STRIDE assignment of N terminus starts one residue before ASSP, the STRIDE assigned N1 position becomes the Ncap position for ASSP. There are 154 and 46 helices showing such mismatches in globular and membrane proteins datasets respectively. The N1 residues in this case show two distinct clusters for backbone torsion angles (φ-ψ), as seen on the Ramachandran map (Figure 2a).

The largest cluster of torsion angles is observed with (φ-ψ) values shifted away from α-helical region, towards the 3_{10} helical and bridge regions. STRIDE identifies these residues as N1 of the helix whereas ASSP defines these positions as Ncap and starts the helix assignment from the next residue. The twist value for the Ncap-N1-N2-N3 step in these helices also deviates considerably from α-helical value, in both datasets, leading to the one residue later start assigned by ASSP. A small cluster is also seen in the bottom right quadrant (φ = 35 to 70° and ψ = -120 to -180°) a region disallowed for non-glycine residues. Interestingly all 24 residues in this cluster are Glycines that can take up these torsion angle values, but are not expected to be categorized as being helical. Since only the carbonyl group at N1 position is involved in H-bond formation, it is not affected by these non-helical torsion angles.

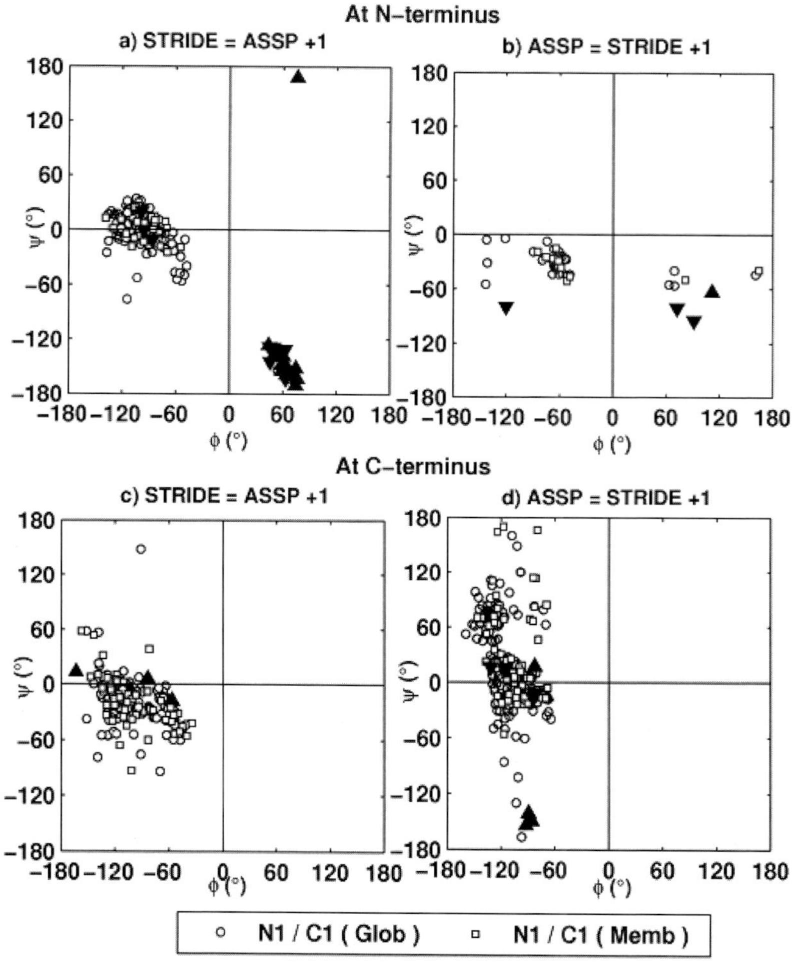

Dataset	N–terminus		C–terminus	
	a	b	c	d
Globular	154	27	170	203
Membrane	46	14	67	78

Figure 2: (φ-ψ) distribution for amino acid residues at N1 position (subplots a and b) and C1 position (subplots c and d) in helices with one residue mismatch in assignment by STRIDE and ASSP. Glycine residues in globular and membrane proteins are represented by '▲' and '▼' respectively. The residue position nomenclature in subplots a) and c) is according to STRIDE (ASSP identifies helices that are one residue shorter) whereas in b) and d) it is according to ASSP assignment.

In helices where ASSP assignment of the N terminus starts one residue earlier than STRIDE, the STRIDE assigned Ncap position is the N1 position for ASSP. Only 27 and 14 such cases are seen, with 9 and 5 residues respectively showing non α-helical backbone torsions, in globular and membrane proteins (Figure 2b) The (φ-ψ) of all other residues fall within the allowed α-helical regions of the Ramachandran Map, validating their assignment as N1 residues by ASSP, contrary to STRIDE which designates them as Ncap.

3.3.2 Helices with one residue mismatch at the C terminus

In helices where STRIDE assigns the C terminus one residue later than ASSP, the STRIDE assigned C1 position corresponds to the Ccap for ASSP. 170 and 67 helices with these mismatches are reported in globular and membrane proteins respectively. In both the datasets the (φ-ψ) for residues at C1 position shows a large variation, with several examples lying in the 3_{10} and π as well as the bridge regions (Figure 2c). This leads to the twist and rise values at the C3-C2-C1-Ccap step deviating considerably from the α-helical values in both globular and membrane proteins (as shown in Figure 3a) due to which ASSP truncates the helix one residue earlier. It is interesting to note that even though it is not essential for the backbone amide of Cap to form a main chain-main chain hydrogen-bond, 115 of 4→1 type, 14 of 5→1 and 1 of 6→1 type hydrogen bonds, corresponding to 3_{10}, α and π-helical hydrogen bond patterns, were found in globular proteins, while the corresponding figures for membrane proteins were 45, 4 and 11 respectively (Figure 3a).

In helices where ASSP definition at the C terminus ends one residue later than STRIDE, the ASSP assigned C1 position corresponds to the Ccap according to STRIDE. 203 and 78 helices with mismatches of this type are seen in globular and membrane proteins respectively. In spite of deviations in the (φ − ψ) values from α-helical values at C1 position according to ASSP (Figure 2d), the twist values for the C3-C2-C1-Ccap step lie between 78° to 115.6° and 80.3° to 112° for globular and membrane proteins respectively (Figure 3b). The corresponding ranges for rise values are between 0.8 Å to 2.3 Å and 0.9 Å to 2 Å for the two datasets. The Ccap residues in these helices were examined for the presence of a hydrogen bond involving their backbone amide group. 102 of 4→1 type and 37 of 5→1 type hydrogen bonds, corresponding to 3_{10} and α-helical hydrogen bond patterns, were found in globular proteins, while the corresponding figures for membrane proteins were 26 and 29 respectively (Figure 3b). A correlation was observed between the helical parameters

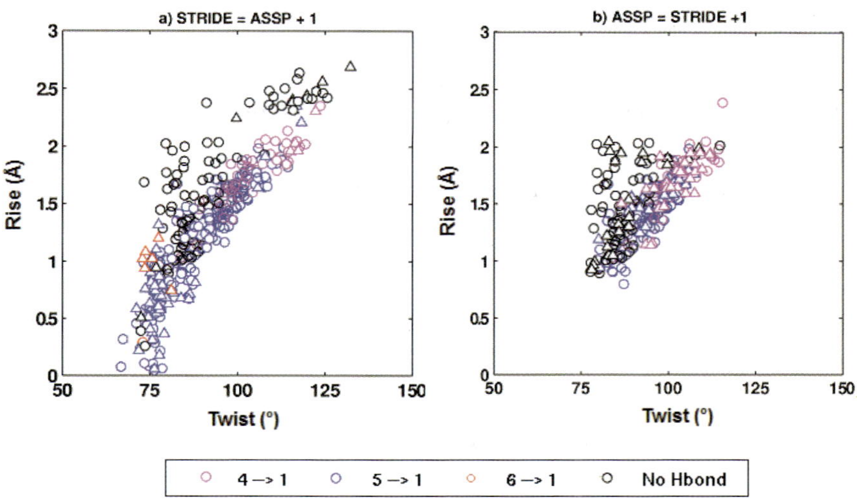

Figure 3: Rise *vs* Twist plots for the step defined by C^{α} atoms of the C3-C2-C1-Ccap residues in the helices with mismatch at C terminii. The color scheme is based on the hydrogen bond pattern observed for the backbone N-H of the Ccap residue. Globular protein data points are represented by 'o' while those of membrane proteins are represented by 'Δ' of the corresponding color.

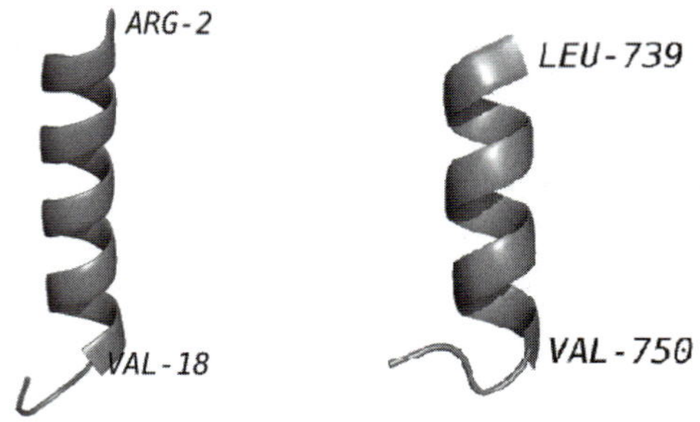

PDB ID	1L9L; A	1OFC; X
STRIDE	2-17	739-749
ASSP	2-18	739-750
C1 (ϕ, ψ)	-108.8. 160.6	-97.3, -166.4
Twist (°)	99.7*, 81.3#	99.2*, 84.6#
Rise (Å)	1.8*, 1.9#	1.6*, 1.8#

Figure 4: Examples of helices with (φ-ψ) of C1 residue (as defined by ASSP) showing deviations from helical values, as seen in Figure 2d. The twist and rise values indicated by '*' and '#' correspond to the C4-C3-C2-C1 and C3-C2-C1-Ccap steps respectively and lie within the accepted helical range.

calculated and the type of hydrogen bond observed. A proline is present at Ccap position in several cases where no hydrogen bond is observed.

Figure 2d shows that, in a few cases the ψ values for the residues at C1 position deviate considerably, taking up either large positive or negative values. An analysis of two representative examples from among these helices (shown in Figure 4) reveals that twist values at the C3-C2-C1-Ccap step lie within the α-helical range and the backbone amide group of C1 is involved in a 5→1 hydrogen bond. The local bending angles calculated by HELANAL-Plus[25] at this step, for the two example helices, are 11° and 1° respectively, indicating that residues assigned as C1 by ASSP are indeed a part of the helix and do not produce a discontinuity or even a kink.

3.3.3 Helices with more than one residue mismatch at the terminii

The length distribution of helices shown in Figure 1 indicated that several STRIDE assigned long α-helices were divided into two or more α-helices by ASSP. In some cases these are interspersed with 3_{10} or π-helical fragments.

For example, in periplasmic hydrogenase protein (PDB id: 1WUI), STRIDE assigns residues 252-284 of 'L' chain as a single α-helix (Figure 5a), while ASSP breaks this 33 residue helix into two α-helical fragments (252-268 and 274-284) with an intervening π-helix (269-273). HELANAL-Plus gives the average twist and rise per residue for the region 269-273 as 77.9° and 0.99 Å, respectively, close to the π-helix values and 6→1 hydrogen bonds are also observed between 273(NH)→268(CO) and 274(NH)→269(CO), confirming the correct assignment of this region as a π-helix by ASSP.

Similarly in Aquaglyceroporin, a membrane protein (PDB id: 3C02), STRIDE assigns residue numbers 210–248 of 'A' chain as a single helix (Figure 5b), while ASSP breaks this into three α-helical segments (211-214; 220-239 and 242-248) and surprisingly also identifies a left-handed α-helical fragment (between residues 216-218). HELANAL-Plus gives (Twist, Rise) values of (-96.59°, 1.53Å) and (-99.3°, 1.56Å) for the first and second step respectively. When the (φ-ψ) for the residues in this helix were checked, residue numbers 216-218 were found to have positive values for both φ and ψ, confirming the occurrence of a left-handed region, as assigned by ASSP.

There are also a few cases where ASSP defined helices are much longer than STRIDE. For example, in both chains 'A' and 'B' of protein yheA (PDB id: 2OEE), ASSP defines a single kinked α-helix between residues 81-111 with large bending angles of ~44° at position M105 and K106 (Figure 5c), while STRIDE breaks it into two α-helical parts (81-85 and 96-111) and a 3_{10} helix

(86-89) for both 'A' and 'B' chains. Interestingly PDB identifies two α-helices (80-105 and 105-112) in chain 'A' while in the 'B' chain, a single helix spanning residues 80-112 is identified. The (φ-ψ) values and helical parameters in the 86-89 region do not indicate the presence of a 3_{10} helix.

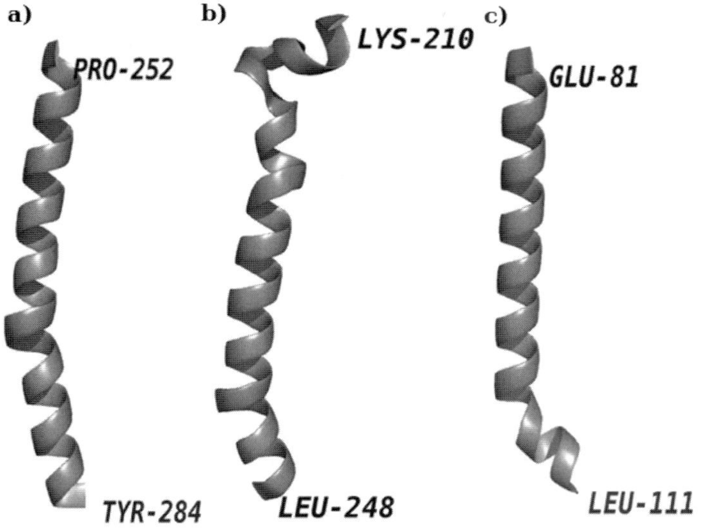

Figure 5: Helices with STRIDE and ASSP helix assignment differing significantly. Representative helices taken from a) Periplasmic hydrogenase (PDB id: 1WUI) b) Aquaglyceroporin (PDB id: 3C02) and c) Protein yheA (PDB id: 2OEE). In examples a) and b) the helix start and end residues indicated are as defined by STRIDE for a single long helix, while in c) the residue assignment is according to ASSP.

4. Conclusion

In this study we have compared α-helix assignments by two algorithms, one based on C^{α} trace (ASSP) and the other using a combination of hydrogen bond pattern along with backbone torsion angles φ and ψ (STRIDE). We find that local variations/deviations in the helical geometries are better detected by ASSP, leading to it breaking longer helices into shorter ones and even identifying some short 3_{10} or π helical regions. A systematic comparison shows higher number of mismatches at the C terminii as compared to the N terminii of α-helices in both globular and membrane protein datasets, indicating that the helix initiation site is better defined. ASSP often identifies a π-helical region at the C-terminii, leading to a large number of mismatches in assignment between the two algorithms.

A geometrically uniform C^α trace defines the helical nature of the protein chain, but the exact orientation of the peptide unit about the virtual C^α - C^α bond determines the backbone torsion angles (φ-ψ) and hydrogen bond pattern between the backbone amides and carbonyls. Specifically in helices where ASSP extends the helix at the C terminus (ASSP = STRIDE+1), the (φ-ψ) of residues at the C1 position assigned by ASSP deviate from α-helical values in some cases, which is accompanied by absence of hydrogen bonds involving the backbone amide of Ccap. Hence while STRIDE truncates the helix one residue earlier, these residues are included as C1 in the helix definition by ASSP, since the C3-C2-C1-Ccap step has acceptable twist and rise values. Thus, while the backbone torsion angles and hydrogen bonds are required to describe the detailed features of a helix and other secondary structures, geometry defined by the C^α atom trace seems to be a good alternative to define variations in helical parameters, which can better determine the extent of a helix and differentiate between various helical conformations. It can also identify helices in low resolution structures, where (φ-ψ) values may not be highly reliable. We believe that insights from our study will help in improving the understanding of residue preferences at helix initiation and termination sites, as well as in modeling tertiary structures of large protein complexes.

Acknowledgments

MB would like to express her gratitude to Professor Ramachandran for his unfailing encouragement and inspiring mentorship. This work was supported by DBT India. Prasun Kumar acknowledges DBT-BINC, India for a fellowship. MB is a recipient of J. C. Bose National Fellowship of Department of Science and Technology, Government of India.

References

1. W. Kabsch and C. Sander, *Biopolymers* **22**, 2577 (1983).
2. D. Frishman and P. Argos, *Proteins* **23**, 566 (1995).
3. R. Srinivasan and G.D. Rose, *Proc. Natl. Acad. Sci. USA,* **96**, 14258. (1999).
4. F.M. Richards and C.E. Kundrot, *Proteins*, **3**, 71(1988).
5. H. Sklenar, et al., *Proteins* **6**, 46 (1989).
6. Martin,J. et al., *BMC Struct. Biol.*, **5**, 17 (2005).
7. Majumdar,I. et al. (2005). *BMC Bioinformatics*, **6**, 202.8 15 (2005).
8. T. Blundell, D. Barlow, N. Borkakoti and J. Thornton, *Nature* **306**, 281 (1983).

9. W. E. Love, P.A. Klock, E. E. Lattman, E. A. Padlan, K. B. Ward, Jr. and W. A. Hendrickson, *Cold Spring Harb Symp Quant Biol* **36**, 349 (1972).
10. D. J. Barlow and J. M. Thornton, *J Mol Biol* **201**, 601 (1988).
11. P. Chakrabarti, M. Bernard and D. C. Rees. *Biopolymers* **25**, 1087 (1986).
12. M. W. MacArthur and J. M. Thornton, *J Mol Biol* **264**, 1180 (1996).
13. R. Sankararamakrishnan and S. Vishveshwara, *Biopolymers* **30**, 287 (1990).
14. J. A. Ballesteros, X. Deupi, M. Olivella, E. E. Haaksma and L. Pardo, *Biophys J* **79**, 2754 (2000).
15. X. Deupi, M. Olivella, C. Govaerts, J. A. Ballesteros, M. Campillo and L. Pardo, *Biophys J* **86**, 105 (2004).
16. S. Kumar and M. Bansal, *Biophys J* **75,** 1935 (1998).
17. R. Aurora, G. D. Rose, *Protein Sci.* **7**, 21 (1998).
18. S. Kumar, M. Bansal, *Proteins: Struct Funct Genet* **31**, 460 (1998).
19. K. Gunasekaran, H. A. Nagarajaram, C. Ramakrishnan, P. Balaram, *J Mol Biol* **275**, 917 (1998).
20. D. E. Engel and W.F. DeGrado, *J Mol Biol* **337**, 1195 (2004).
21. S. M. Zaremba, L. M. Gregoret, *J Mol Biol* **291**, 463 (1999).
22. J. S. Merkel, L. Regan, *Fold Des* **3,** 449 (1998).
23. J. S. Merkel, J. M. Sturtevant, L. Regan, *Fold Des* **7,** 1333 (1999).
24. M. A. Wouters, P. M. Curmi, *Proteins* **22**, 119 (1995).
25. E. G. Hutchinson, R. B. Sessions, J. M. Thornton, D. N. Woolfson, *Protein Sci* **7**, 2287 (1998).
26. P. Kumar, M. Bansal, *J Biomol Struct Dyn* **30**, 773 (2012).
27. S. E. Brenner, P. Koehl and M. Levitt, *Nucleic Acids Res* **28**, 254 (2000).
28. A. Andreeva, D. Howorth, J. M. Chandonia, S. E. Brenner, T. J. Hubbard, C. Chothia and A. G. Murzin, *Nucleic Acids Res* **36**, D419 (2008).
29. G. E. Tusnady, Z. Dosztanyi and I. Simon, *Nucleic Acids Res* **33**, D275 (2005).
30. G. Wang and R. L. Dunbrack, Jr, *Bioinformatics* **19**, 1589 (2003).
31. H. M. Berman, J. Westbrook, Z. Feng, G. Gilliland, T. N. Bhat, H. Weissig, I. N. Shindyalov and P. E. Bourne. *Nucleic Acids Res* **28**, 235 (2000).
32. I. K. McDonald and J. M. Thornton. *J Mol Biol* **238**, 777 (1994).
33. S. Kumar, M. Bansal, *Biophys J* **75**, 1935 (1998).

PROPENSITIES OF AMINO ACID RESIDUES IN PROTEINS FOR DIFFERENT REGIONS OF THE RAMACHANDRAN MAP

B. LAKSHMI[1,2], G. ARCHUNAN[2] AND N. SRINIVASAN[1]

[1]*Molecular Biophysics Unit, Indian Institute of Science, Bangalore 560012, India*

[2]Presently at *Centre for Pheromone Technology, Department of Animal Science Bharathidasan University, Tiruchirappalli 620024, India*

Using a dataset of 1164 crystal structures of largely non-homologous proteins defined at a resolution of 1.5Å or better, we have investigated the (ϕ,ψ) preferences of 20 residue types by considering the residues which occur in loops. Propensities of residue types to occur in the loops with (ϕ,ψ) values in the α_R region of the Ramachandran map has a poor correlation coefficient of 0.48 to the Chou-Fasman propensities of the residue types to occur in the α-helical segments. However the correlation coefficient between propensities of residues in loops to adopt β conformations and those in β-sheet is much higher (0.95). These observations suggest that α-helix formation is well influenced by the local amino acid sequence while intrinsic preference of residue types for β-sheet plays a major role in the formation of β-sheet. The main chain polar groups of residues in loops, that can affect the (ϕ,ψ) values, can be involved in intra-molecular hydrogen bonding. Therefore we investigated further by considering subset of residues in loops with low (0 to 2) number of intra-molecular hydrogen bonds per residue involving main chain polar atoms. For this subset, the correlation coefficients between propensities for α-helix and α_R region and between β-sheet and β-region are 0.26 and 0.64 respectively. This reiterates higher intrinsic tendency of β-region favouring residues to adopt β-sheet than α_R region favouring residues to adopt α-helical structure.

1. Introduction

In the year 1984 one of us (NS), like many other graduate students at that time in the Molecular Biophysics Unit, Indian Institute of Science, had the privilege of learning about the Ramachandran map straight from the inspiring lectures of Professor C. Ramakrishnan, one of the authors of the seminal paper in *Journal of Molecular Biology*[1]. Ramachandran maps for L-Ala[1] and Gly residues[2] showed that accessible conformational space for residues in peptides and proteins is highly restricted with only about 30% of the conformational space allowed for non-glycyl residues. The "fully allowed" regions in the Ramachandran map encompass right handed α-helical (α_R) and the β-sheet conformations which turned out to be the most frequent secondary structures in proteins. Although "partially allowed" (ϕ,ψ) combinations, such as left-handed α-helical (α_L)

conformation, are usually seen at the termini of helical and β-strand regions and in loops, residues with α_R and β (ϕ,ψ) values do occur in the loop regions.

The final folded structure of a protein is primarily determined by the amino acid sequence[3]. As α-helical and β-strand regions are characteristic features of folded proteins it is conceivable that the amino acid sequence plays a major role in the formation of regular secondary structure regions. The residues with high propensities to occur in α-helical and β-sheet can be perceived as favourable residues to adopt α_R and β-regions of the Ramachandran map. However it is not clear what is the role of spontaneous tendency of a local region to form a helix or β-strand in conferring a general apparent preference of a residue for a local region in the Ramachandran map.

In this chapter, using a large dataset of known 3-D structures of globular proteins, we investigate the tendencies of 20 residue types to adopt α_R, β and α_L regions of the Ramachandran map by considering only those residues in the loop regions. The observed tendencies of the residues in loops to adopt conformations in α_R and β regions are compared with Chou-Fasman propensities of the residues for α-helix and β-sheet respectively.

2. Dataset and analysis

A non-redundant dataset of 1348 monomeric protein structures with the resolution of 1.5Å or better obtained from the protein data bank[4] was considered first. The B-factors were normalized based on the method proposed by Parthasarathy and Murthy[5] and the cases with normalized B-factor ≤ 2, which resulted in 1164 protein structures, were considered for further analysis. The non-helical and non-β-sheet (loop) residues have been picked up using the Kabsch and Sander's DSSP program[6]. Further, the loop residues have been classified into three categories using Ramachandran (ϕ, ψ) values as α_R region with ϕ (-120 to 0), ψ (-60 to 0); α_L region with ϕ (0 to 120), ψ (0 to 120) and β-region with ϕ (-180 to -60), ψ (60 to 180). The hydrogen bonds were identified for the loop residues in α_R and β-region of the Ramachandran map using the HBOND program[7]. The propensity of any residue type (X) to occur in a given conformational state is calculated using the formula given below

$$Propensity\,(X) = \frac{\dfrac{Number\ of\ X\ in\ beta/\alpha R/\alpha L\ region}{Total\ number\ of\ X\ in\ loop\ regions}}{\dfrac{Number\ of\ residues\ in\ beta/\alpha R/\alpha L\ region}{Total\ number\ of\ residues\ in\ loop\ regions}}$$

3. Propensities of residue types for local regions of Ramachandran map

A dataset of 1164 largely non-homologous proteins of known 3-D structures determined at a resolution of 1.5Å or better has been used in the current analysis. As the analysis is confined to the residues in the loop regions we removed those residues with high normalized B-factor as it could add uncertainty to the conformation at those residues. Figure 1 shows the (ϕ,ψ) distribution in the α_R, α_L and β-regions of the Ramachandran map, used for the current analysis encompassing 84665 points. A dense cluster of points can be noticed outside the allowed region corresponding to α_L conformation. Many of these residues correspond to Gly and these residues prefer a (ϕ,ψ) value closer to (90,0) rather than α_L as the main chain polar groups maximally get exposed to water[8] that is a characteristic of residues in loops.

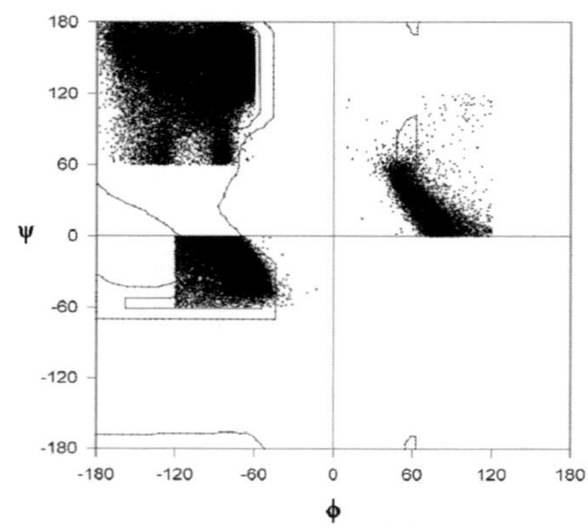

Figure 1: (ϕ,ψ) plot of residues in loops with α_R, β and α_L conformations superimposed on the Ramachandran map.

Propensity values of the 20 residue types present in loops to occur in the α_R, β and α_L regions of the Ramachandran map were calculated and these values are represented as a bar diagram in Figure 2.

It can be noted in Figure 2 that Ala, Glu, Lys, Pro, Gln, Arg, Ser and Thr in the loop regions show reasonably high propensity to occur in α_R region. Interestingly, it is well known that the Chou-Fasman propensities of some of these residues such as Ser, Thr and Pro have low overall propensity for α-helical regions in proteins. Therefore, it is clear that not all residues that has apparent intrinsic preference for α_R region are favoured in α-helical regions. In contrary, residues such as Leu and Val which occur in reasonable frequency in α-helical regions do not show high propensity to occur in α_R region.

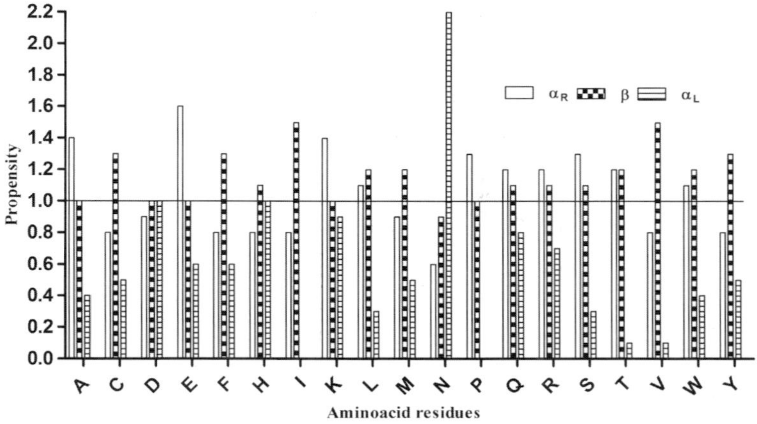

Figure 2: Propensities of non-glycyl residues in loops to adopt conformations in α_R, β and α_L regions of the Ramachandran map.

Residues in loops with high propensity for β-region include Cys, Phe, Ile, Leu, Met, Thr, Val, Trp and Tyr. Almost all of these residues are known to be preferred in β-sheets in proteins. Most of the residues with low propensity for β-region are also known to have low propensity for β-sheet regions.

Spectacular high propensity of Asn to adopt α_L conformation is well known since as early as 80's[9-11] though the reason is not clear yet. Interestingly, Asp and Gln, the two most similar residues to Asn do not show high propensity to adopt conformations in α_L region.

These observations prompted us to make a direct comparison of propensities of residues in loops to adopt α_R and β-conformations with propensities of residues to occur in α-helical and β-sheet regions.

3.1 Correlation between propensities of residues in loops for α_R region and those in α-helical secondary structure

We used the dataset of 1164 protein structures and calculated the Chou-Fasman propensities of 20 residue types to occur in α-helical regions in proteins. Figure 3a shows scatter plot depicting correlation between these Chou-Fasman propensities for α-helical regions and propensities of residues in loops with α_R conformation in the Ramachandran map. Similar plots were obtained by Swindells et al[12] for a smaller dataset of protein structures.

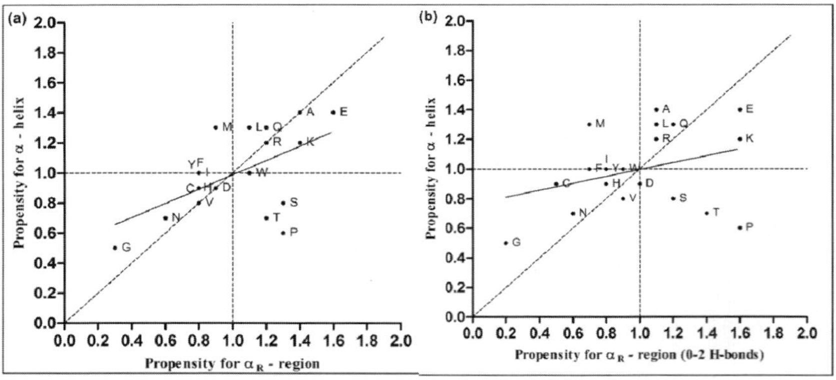

Figure 3: (a) Propensities of 20 residues in loops to adopt conformations in α_R region of the Ramachandran map plotted against Chou-Fasman propensities of residues to occur in α-helical regions in proteins. The three dashed lines represent propensity value of 1 along X and Y axes and 45° line. The solid line represents the best fit straight line for the points in the plot. (b) Same as in (a) but by choosing a sub-set of residues in loops with α_R conformation with 0 to 2 intra-molecule hydrogen bonds present involving amide and carbonyl groups in the main chain flanking the residue of interest.

From Figure 3a it can be observed that most of the residues are not in the 45° line suggesting that the propensity values of residues in loops with α_R conformation is quite different from the Chou-Fasman propensity values for α-helical regions. Indeed the correlation coefficient for this distribution is only 0.48. Some of the residues which contribute to weak correlation are Ser, Thr and Pro. These residues have high propensities for α_R region when they occur in loops, but, have low propensities for α-helical regions though Pro is well known to be the most frequent residue at the first position of the helices. Further, Met and Leu which have good propensity to occur in α-helical regions in proteins have not a high propensity for α_R conformation when they occur in loops. It appears that sequential neighborhood contributes to an extent in favouring Leu and Met to occur in helices.

The poor correlation in Figure 3a seems to suggest that for a few residue types their intrinsic high tendency to adopt α_R conformation is not translated into their preference to occur in α-helical segments. This point can be reconciled very well for prolyl residue which is well-known to produce kink when it occurs in the middle of a long α-helical region. However, it is not clear why Ser and Thr with high intrinsic preference for α_R region are not favoured in α-helical regions. In general, this data seems to suggest that the formation of α-helical regions in proteins is more an attribute of the local sequence rather than strong tendency of individual residues to occur in the helix.

3.2 Correlation between propensities of residues in loops for β-region and those in β-sheet secondary structure

Figure 4a shows scatter plot depicting correlation between these Chou-Fasman propensities for β-sheet regions and propensities of residues in loops with β-conformation in the Ramachandran map.

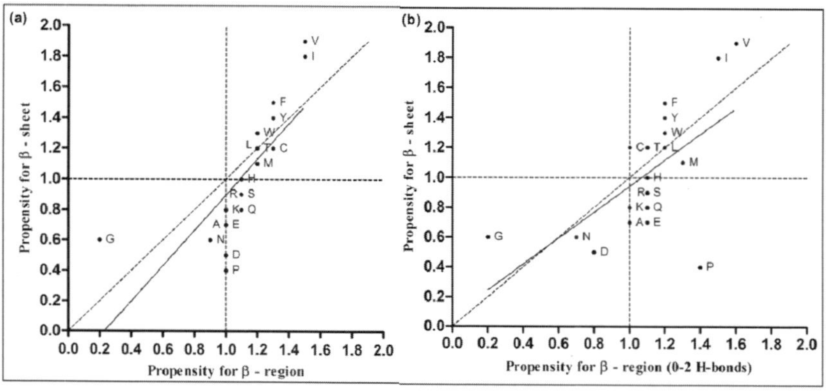

Figure 4: (a) Propensities of 20 residues in loops to adopt conformations in β-region of the Ramachandran map plotted against Chou-Fasman propensities of residues to occur in β-sheet regions in proteins. The three dashed lines represent propensity value of 1 along X and Y axes and 45° line. The solid line represents the best fit straight line for the points in the plot. (b) Same as in (a) but by choosing a sub-set of residues in loops with β-conformation with 0 to 2 intra-molecule hydrogen bonds present involving amide and carbonyl groups in the main chain flanking the residue of interest.

It is obvious from Figure 4a that almost all the residues with high propensity values for β-conformation in loops also show high Chou-Fasman propensity values for β-sheet regions. Therefore all the residues which are known to be favourable for their occurrence in β-sheet in proteins seem to have intrinsic tendency to adopt β-conformation even when they occur in loops. In fact the two residues (Val and Ile) which have highest propensities for β-sheet also have highest propensities to adopt β-conformation when they occur in loops. Most of the residues which show a low propensity values for their occurrence in β-sheet show neutral tendency to occur in the β-region of the Ramachandran map. Indeed the correlation coefficient for the data shown in Figure 4a is 0.95.

Thus, in general, the strong correlation between intrinsic tendency of amino acid residues to adopt β-conformation and their Chou-Fasman propensities to occur in β-sheet seem to suggest that β-sheet formation in proteins is favoured by high preponderance of residues which have intrinsic tendency to adopt β-conformation. This contrasts with the inference derived in the context of formation of α-helical regions.

3.3 Analysis of propensities of residues in loops with lower intra-molecular influence on their (ϕ, ψ) values

The main reason why we chose to consider loop regions in order to get an idea on intrinsic preferences of amino acid residues is that the formation of helical and β-structures could be more a sequence effect rather than intrinsic preferences. The spontaneous formation of series of N-H...O=C hydrogen bonds both in helical and β-sheet structures might be governed by the local sequence. Involvement of main chain polar groups in hydrogen bonding could have a direct bearing on the (ϕ,ψ) values adopted at individual residue positions. For example, it is well known that the twist in β-strands is likely to optimise the N-H...O=C hydrogen bond geometry and van der Waal's interactions.

In loop regions too it is possible to have some influence of intra-molecular interactions (example. β-turn) that could affect the intrinsic tendency of a given residue to adopt main chain conformations. In order to minimise such possible influences we re-calculated the intrinsic propensity values by considering only a sub-set of residues in loops whose main chain amide and carbonyl groups (4 in all flanking either side of a C^{α} position) are involved in no more than 2 hydrogen bonds.

The correlation plots of these propensity values are depicted in Figures 3b and 4b for α_R and β-conformations respectively. Comparison of Figures 3a and 3b and Figures 4a and 4b suggests that the general distribution of points are largely similar with spread of points being greater in Figures 3b and 4b. The correlation coefficient values corresponding to the data presented in Figures 3b and 4b are 0.26 and 0.64 respectively confirming that influence of sequence in helix formation is greater.

4. Conclusions

We have employed Ramachandran map[1] as a tool to investigate intrinsic preferences[12] of amino acid residues in loops of protein structures to adopt main chain conformations in the local regions of the Ramachandran map. The tendencies displayed by residue types to adopt β-conformation are well correlated with Chou-Fasman propensities for β-sheet. The same does not hold good for the correlation between propensities of residues for α_R conformation and for α-helical regions. Consideration of residues in loops with low influence due to intra-molecular interactions confirms these findings. It appears that the influence of local sequence is more in the formation of α-helical conformation than in β-sheet.

Acknowledgement

BL is supported in a project funded by DBT-AIST, New Delhi. This research is supported by DBT.

References

1. G.N. Ramachandran, C. Ramakrishnan and V. Sasisekharan, *J. Mol. Biol.* **7**, 95 (1963).
2. C. Ramakrishnan and G.N. Ramachandran, *Biophys. J.* **5**, 909 (1965).
3. C.B. Anfinsen, *Science* **181**, 223 (1973).
4. H. M. Berman, J. Westbrook, Z. Feng, G. Gilliland, T. N. Bhat, H. Weissig, I. N. Shindyalov and P. E. Bourne. *Nucleic Acids Res* **28**, 235 (2000).
5. S. Parthasarathy and M. R. N. Murthy. *Protein Sci* **6**, 2561 (1997).
6. W. Kabsch and C. Sander, *Biopolymers* **22**, 2577 (1983).
7. K. Mizuguchi, C. M. Deane, T. L. Blundell, M. S. Johnson and J. P. Overington. *Bioinformatics* **14**, 617 (1998).
8. N. Eswar, H.A. Nagarajaram, C. Ramakrishnan and N. Srinivasan, *Proteins* **49,** 326 (2002).
9. J. S. Richardson, *Adv. Prot. Chem.* **34,** 167 (1981).
10. V. Ravichandran and E. Subramanian. *Int. J. Pept. Prot. Res.* **18,** 121 (1981).
11. N. Srinivasan, V.S. Anuradha, C. Ramakrishnan, R. Sowdhamini and P. Balaram. *Int. J. Pept. Prot. Res.* **44,** 112 (1994).
12. M.B. Swindells, M.W.MacArthur and J.M. Thornton. *Nature Str. Biol.* **2,** 596 (1995).

STEREOCHEMISTRY OF POLYPEPTIDE CONFORMATION IN COARSE GRAINED ANALYSIS

ANIL KORKUT[1,4] AND WAYNE A. HENDRICKSON[1,2,3]

[1]*Department of Biochemistry and Molecular Biophysics*
[2]*Department of Physiology and Cellular Biophysics, and*
[3]*Howard Hughes Medical Institute*
Columbia University, New York, NY 10032 USA
[4]*Computational Biology Center*
Memorial Sloan-Kettering Cancer Center, New York, 10065, USA
E-mail: wayne@xtl.cumc.columbia.edu

The conformations available to polypeptides are determined by the interatomic forces acting on the peptide units, whereby backbone torsion angles are restricted as described by the Ramachandran plot. Although typical proteins are composed predominantly from α-helices and β-sheets, they nevertheless adopt diverse tertiary structure, each folded as dictated by its unique amino-acid sequence. Despite such uniqueness, however, the functioning of many proteins involves changes between quite different conformations. The study of large-scale conformational changes, particularly in large systems, is facilitated by a coarse-grained representation such as provided by virtually bonded Cα atoms. We have developed a virtual atom molecular mechanics (VAMM) force field to describe conformational dynamics in proteins and a VAMM-based algorithm for computing conformational transition pathways. Here we describe the stereochemical analysis of proteins in this coarse-grained representation, comparing the relevant plots in coarse-grained conformational space to the corresponding Ramachandran plots, having contoured each at levels determined statistically from residues in a large database. The distributions shown for an all-α protein, two all-β proteins and one α+β protein serve to relate the coarse-grained distributions to the familiar Ramachandran plot.

1. Introduction

Accurate understanding of the relationship between protein structure and biological function has been the central problem in biophysics since the dawn of structural biology[1]. Macromolecular crystallography has provided thousands of unique protein structures and that have helped to explain, at an atomic level of detail, diverse biological functions such as oxygen transport[2], protein folding[3], ATP synthesis[4], ion conduction across membranes[5], protein translation[6-8], RNA transcription[9], viral entry to host cells[10], and many more. Such triumphs of structural biology also demonstrate diverse tertiary structures, each one folded into a well-defined structure determined by its unique amino-acid sequence[11]. Although the overall foldings of proteins prove to be highly varied, locally it is the polypeptide conformations of helices and sheets that predominate as organized into regular secondary structures.

The actuality of polypeptides folding into regularly repeated conformations was anticipated brilliantly by Pauling and coworkers who proposed detailed structures of the α-helix[12] and β-pleated sheets[13] based on dimensions from crystal structures of amino acids and short peptides and principles of hydrogen bonding. The spatial distribution of hydrogen bonds between backbone amide and carboxyl groups and the planar nature of the peptide bonds determined the recurrent local structural elements of helices and strands in sheets. In return, the secondary structure elements have well defined geometries with energetically preferred dihedral angles along the polypeptide main chain. The analysis of this polypeptide geometry (i.e. the φ and ψ dihedral angles of polypeptide main chain, Figure 1A) by G. N. Ramachandran and colleagues resulted in the determination of stereochemical criteria for polypeptide chains and one of the most widely used structural analysis tools in biology, the Ramachandran plot[14]. Subsequently, the stereochemistry of polypeptide conformation was placed on a firm theoretical foundation by considering the energetic consequences of the interdependent restraints from bond distances, bond angles, dihedral angles, non-bonded interactions and the spatial organization of hydrogen bonds[15].

The static description of protein structures is insufficient to explain protein function. Descriptions of conformational transitions and allosteric regulation are essential to understand the relation between protein structure and function. For this purpose, macromolecular crystallography provided snapshots of proteins in distinct conformational (and activity) states and computational biophysics provided methods such as molecular dynamics simulations. However, atomic-level analysis of large conformational transitions, particularly in large macromolecular complexes such as ribosome particles[6-8], is still computationally infeasible. Thus, a reduction in the structural complexity is required. Based on this motivation, coarse grained approaches that provide normal mode evaluation of local protein fluctuations have become popular[16]; however, in order to follow the time course of large-scale molecular events, more is required. A successful coarse-grained approach for the simulation of protein dynamics has three crucial requirements: (i) an accurate representation of polypeptides with reduced complexity, (ii) a force field that captures the molecular restraints relevant to the coarse-grained representation, and (iii) a molecular mechanics algorithm that uses the reduced representation and the potential function to accurately and efficiently compute molecular dynamics.

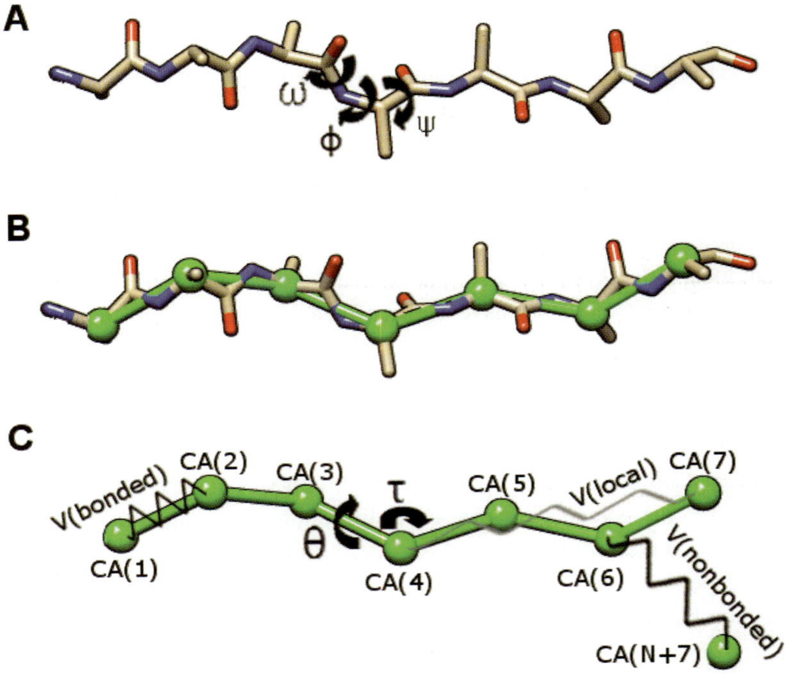

Figure 1. Polypeptide main chain and coarse grained modeling. A. The full-atom polypeptide main chain with dihedral angles (φ,ψ,ω). B. Conversion of the full-atom main-chain model to the Cα-atom coarse-grained representation. C. The coarse-grained representation in VAMM force field with virtual angles θ and τ.

2. Virtual atom molecular mechanics

For our analyses we have used the Cα virtual atom representation[17] to reduce complexity (Figure 1B). Then, to facilitate simulations, we developed virtual atom molecular mechanics (VAMM) force field[18] for coarse-grained analyses and a VAMM-based algorithm to compute conformational transition pathways for protein molecules[19]. VAMM is currently based on Cα-only representation of protein molecules similar to that in elastic network models[16] and can be extended to include side-chain centroids as in other coarse-grained models[20]. The VAMM force field restrains a given protein structure through use of virtual (i.e. Cα- Cα) bond length, angle bending, dihedral, non-bonded, and local interaction terms (Figure 1C, Equation 1). Each restraint is parameterized statistically against crystallographic data with the Boltzmann conversion method.

$$V_{VAMM} = V_{bond} + V_{angle} + V_{dihedral} + V_{nonbonded} + V_{local} \quad (1)$$

In VAMM, we determined secondary structure specific virtual dihedral (θ) and bond angle (τ) restraints based on the individual distributions of every possible configuration of dihedral angle quartets (e.g. HHHH for four consecutive residues with helical structure) and bond angle triplets (e.g. EEE for three consecutive residues with β-sheet structure). The analysis of the virtual dihedral and bond angle distributions also revealed that Cα backbones of proteins are specified with (θ, τ) virtual angles as the main chain is defined by (φ,ψ) dihedrals. In analogy to the Ramachandran plot, we described probability contours of (θ, τ) and (θ-, θ+) at levels of 95% (favored) and 99% (allowed) based on the statistical information in the Top500 structure database (See methods).

Here, we further analyze the virtual (θ, τ) angle distributions in different proteins to demonstrate how such contour plots differ from each other for distinct folds and how they relate to the Ramachandran plot. The analysis reveals well defined regions on the (θ, τ) and (θ-, θ+) plots, which are populated by distinct secondary structure elements similar to the case on Ramachandran plot. The overall analysis demonstrates the coarse-grained stereochemical properties of a peptide chain are directly related to the stereochemical properties of the main chain.

3. Results

In order to determine the stereochemical properties of polypeptides in coarse-grained representation, we constructed probability contour plots of virtual angle parameters and analyzed a set of protein structures with diverse secondary structure elements (Figure 2). Analyzed structures are $TorS_S$ (all α-helices), the sensor domain of a histidine kinase receptor taken from the complex of $TorS_S$ with the ligand-binding co-receptor TorT [21], histidine kinase sensor domain $HK3_S$ [22] (all β-strands), glycoprotein CD4 [23] (D1D2 immunoglobulin-like domains, all β-strands), HIV envelope glycoprotein gp120 [23] (mixture of α-helices/β-strands). For each protein molecule, the Cα (θ, τ) and (θ-, θ+) distributions are plotted on a contour plot as in Ref. 19 and compared to the Ramachandran plot.

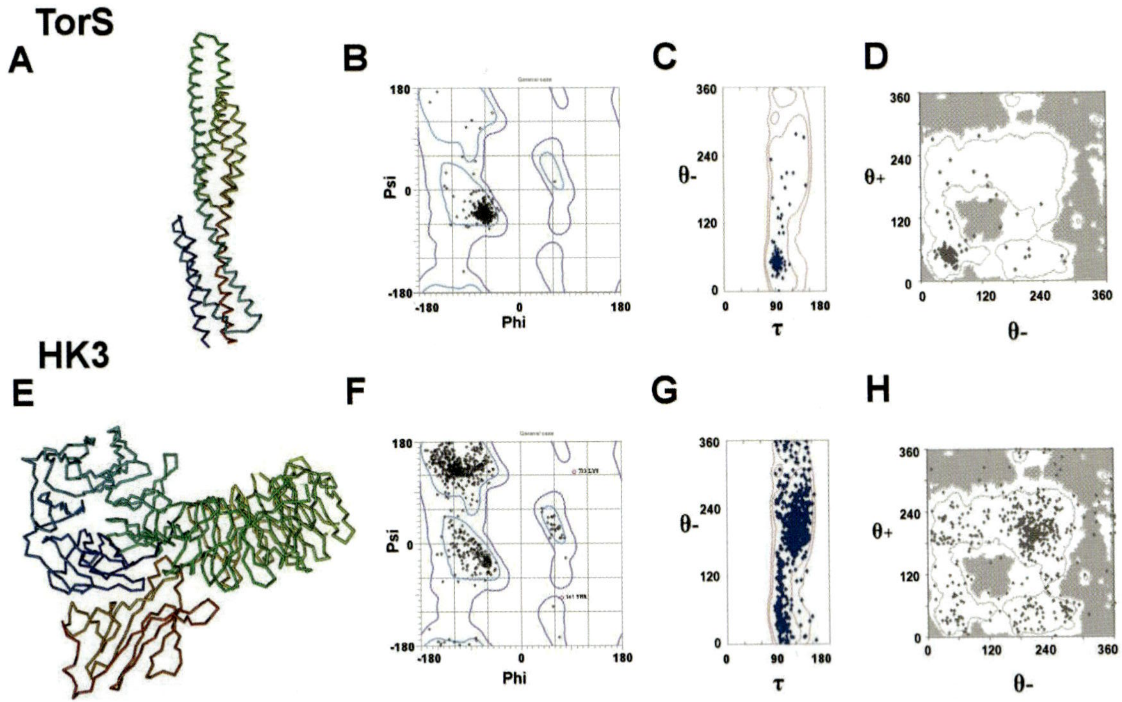

Figure 2. Stereochemical analysis of protein structures. A. The structure of $TosS_S$. B. Ramachandran plot for $TorS_S$. C. (θ, τ) plot for Tor_S. D. (θ-, θ+) plot for $TorS_S$. E. The structure of HK3. F. Ramachandran plot for $HK3_S$. G. (θ, τ) plot for $HK3_S$. H. (θ-, θ+) plot for $HK3_S$. I. The structure of CD4-D1D2. J. Ramachandran plot for CD4. K. (θ, τ) plot for CD4. L. (θ-, θ+) plot for CD4. M. The structure of HIV gp120. N. Ramachandran plot for gp120. O. (θ, τ) plot for gp120. P. (θ-, θ+) plot for gp120.

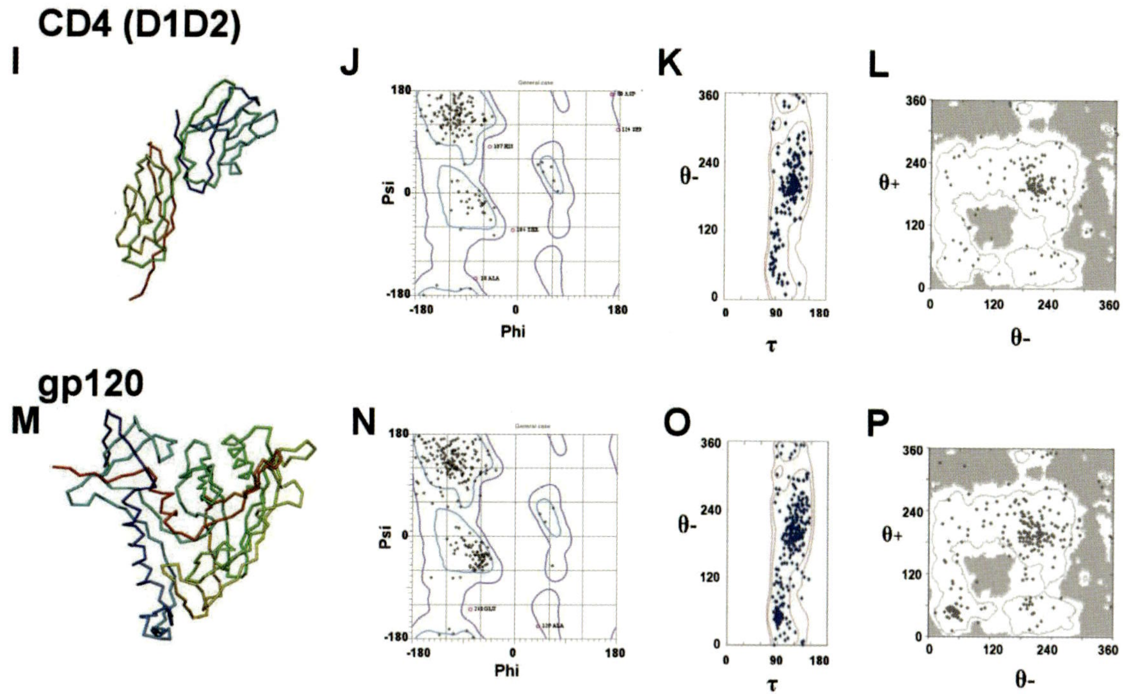

Figure 2. *Continued.*

3.1. *Virtual bond angles are more constrained than virtual dihedrals*

The analysis of the (θ, τ) contour plots reveals that the virtual bond angles are more constrained than the virtual dihedral angles. 99% of the measured virtual dihedral angles are confined between 75° and 148°, regardless of the secondary structure in their vicinity. On the other hand, the virtual dihedral angles sampled in protein structures span all possible angles between 0° and 360°. Thus, in the (θ, τ) distribution only a fraction of the plot is populated in the τ dimension, whereas there is a continuous distribution in the θ dimension. Nevertheless, each virtual dihedral is still constrained by the local structural preferences of its immediate vicinity. This constraint is depicted in the (θ-, θ+) contour plots such that the regions (300° < θ- < 360°, 300° < θ+ <360°) and (75° < θ- < 150°, 75° < θ+ < 150°) are predominantly disallowed according to the distribution of virtual dihedrals sampled in Top500 database. Note that polypeptide segments involving glycine residues more frequently sample the space 75° < θ- < 150° compared to the segments without glycine (data not shown).

3.2. *Proteins have characteristic (θ, τ) and (θ-, θ+) distributions*

The analysis of the θ and τ angles shows that each protein has a unique (θ, τ) and (θ-, θ+) distribution (Figure 2). In the all α-helical protein, $TorS_S$ (Figures 2A-D), the (θ, τ) plot is mainly populated around the peak (θ = 50°, τ = 90°) and few points are located outside this peak. Similarly, the peak around (θ- = 50°, θ+ = 50°) singles out on the (θ-, θ+) plot. Not surprisingly, the region corresponding to the α-helix on Ramachandran plot is most populated for $TorS_S$. The all β-strand protein CD4 (Figures 2I-L) shows a remarkably different profile with a major peak centered around (θ = 195°, τ = 117°) and a significant number of points extending outside this region to lower dihedral angle values. The structural analysis shows that these points correspond to those residues residing on the loops. The (θ-, θ+) distribution has a sharp peak around (θ- = 195°, θ+=195°). The significantly larger $HK3_S$ structure (all β-strand, 739 residues, Figures 2E-H) shares the same peaks with CD4. Most interestingly, a large number of points are observed outside this region extending to the helical zone on both (θ, τ) and (θ-, θ+). The same characteristics are also observed in the Ramachandran plot (Figure 2F). These points correspond to residues on the loops connecting the strands of the β-propeller domains, which retain short helical or turn structures (e.g. a 3_{10} helix at Pro776-His781 and an α-helix at Asp165-Asn168 according to a DSSP analysis). Not surprisingly, the (θ, τ) and (θ-, θ+) plots of the α/β mixed gp120 displays both of the peaks observed in all helical and β-sheet proteins.

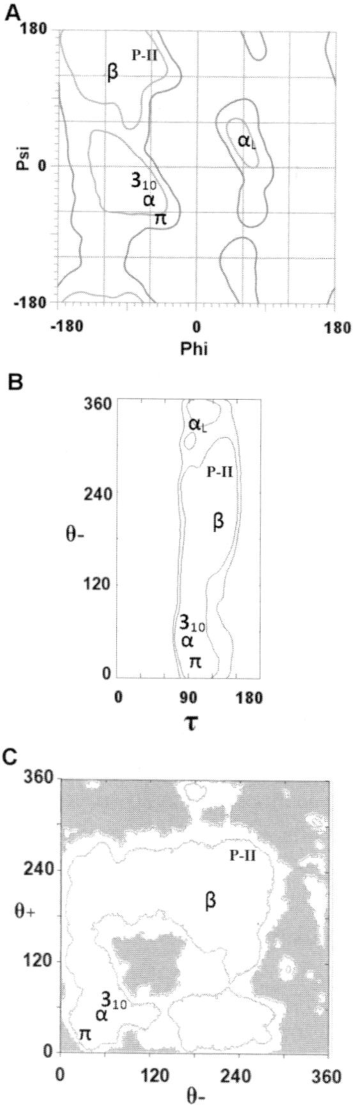

Figure 3. Distribution of secondary structure elements. Each secondary structure element (β-strand, α-helix, α_L, 3_{10} helix, and π-helix) populates a specific region in actual and virtual dihedral and bond angle spaces. The position of each element is shown on each plot: A. The Ramachandran plot. B. The (θ, τ) plot. C. The (θ-, θ+) plot. Note that the (θ, τ) distributions of peptides in α_L configurations were computed using short polypeptide segments (4 residues) extracted from the PDB as given by Novotny & Kleywegt[25]. There are no long α_L segments, and short fragments do not allow for computation of (θ-, θ+).

3.3. *Each secondary structure element populates a distinct region on θ/τ plots*

The unique θ/τ distributions for each protein and the VAMM force field parameterization[18] have suggested that we can define stereochemical criteria for coarse grained analysis of polypeptide chains. Such criteria are analogous to those seen in Ramachandran plots (Figure 3A), where positions of regularly repeated secondary structure are now refined from those identified initially[14] through careful analyses of the PDB[24-26] (Table 1). A detailed analysis of θ/τ distributions demonstrates how each secondary structure element occupies a distinct area on contour plots (θ, τ) and (θ-, θ+) (Table 1, Figures 3B-C).

Table 1. Approximate preferred virtual angles for secondary structure elements.

Secondary Structure	(φ, ψ)	(θ, τ)	(θ-, θ+)
α-helix	(-63°, -43°)[24]	(50°, 90°)	(50°, 50°)
β-strand	(-116°, 129°)[24]	(194°, 120°)	(194,194°)
Polypeptide-II	(-65°, 145°)[24]	(249°, 118°)	(249°, 251°)
3_{10}-helix	(-62°, -22°)[24]	(67°, 90°)	(67°, 67°)
α_L-helix	(59°, 42°)[25]	(320°, 90°)	(X, 320°)
π-helix	(-76°, -41°)[26]	(22°,103°)	(22°, 22°)

4. Discussion

Here we have analyzed the stereochemistry of polypeptides for coarse grained analysis. The distribution of virtual dihedral and bond angles in Cα-traces shows that there are a set of stereochemical criteria that distinguishes each secondary structure type in coarse grained representation. In analogy to the Ramachandran plot, (θ, τ) and (θ-, θ+) probability contour plots define favored and allowed regions for polypeptides. These contour plots can serve to determine the quality of structures in their energy-minimum conformations and transition states computed with coarse-grained simulations. Such plots will be useful in analysis of conformation transition pathways and guide us to develop more accurate coarse grained modeling strategies.

5. Methods

5.1. Protein structures

All protein structures were obtained from the Protein Data Bank (www.pdb.org). The TorS$_S$ chain A is stripped from the TorS$_S$-TorT complex structure at 2.80 Å resolution (PDB id: 3O1I). The HK3$_S$ structure is chain A of the dimeric structure of the extracellular domain of putative one component system BT4673 at 2.3 Å resolution (PDB id: 3OTT). The gp120 and CD4(D1D2) structures are extracted from the structure of the gp120 core complexed with the membrane-distal immunoglobin-like (Ig) domains D1 and D2 of CD4 and an Fab fragment of antibody 17b at 2.2 Å resolution (PDB id: 1RZJ).

5.2. Construction of the (θ, τ) and (θ-, θ+) plots

The favored and allowed regions of the virtual dihedral (θ) and bond angles (τ) were computed by a procedure similar to that used in MolProbity software to compute the contours of the Ramachandran plot[27]. The contour levels were determined statistically against the Top500 database, which was also used to construct Ramachandran plots in MolProbity. Different from the Ramachandran plots, the Gly and Pro residues were not analyzed separately. The Top500 database contains structures determined at better than 1.8 Å resolution and residues with B-factors higher than 40 are not included in the statistical analysis. A density dependent smoothening function[28] was used to generate contour plots such that the sparse regions of the distributions are represented smoothly and continuous, and the sharp transitions are preserved. All of the scripts and templates for generating the (θ, τ) and (θ-, θ+) plots are available at http://www.virtualatom.org.

Acknowledgement

This work was supported in part by National Institutes of Health Grant GM56550 (to W. A. H.).

References

1. F. H. White, Jr., and C. B. Anfinsen. *Ann. N. Y. Acad. Sci.* **81**, 515 (1959).
2. M. F. Perutz, H. Muirhead, J. M. Cox and L. C. Goaman. *Nature* **219**, 131 (1968).
3. K. Braig, Z. Otwinowski, R. Hegde, D. C. Boisvert, A. Joachimiak, A. L. Horwich and P. B. Sigler. *Nature* **371**, 578 (1994).

4. Abrahams JP, Leslie AG, Lutter R and Walker JE. *Nature* **370**, 621 (1994).
5. D. A. Doyle, J. Morais Cabral, R. A. Pfuetzner, A. Kuo, J. M. Gulbis, S. L. Cohen, B. T. Chait and R. MacKinnon. *Science* **280**, 69 (1998).
6. N. Ban, P. Nissen, J. Hansen, P. B. Moore and T. A. Steitz. *Science* **289**, 905 (2000).
7. F. Schluenzen, A. Tocilj, R. Zarivach, J. Harms, M. Gluehmann, D. Janell, A. Bashan, H. Bartels, I. Agmon, F. Franceschi F and A. Yonath. *Cell* **102**, 615 (2000).
8. B. T. Wimberly, D. E. Brodersen, W. M. Clemons, Jr., R. J. Morgan-Warren, A. P. Carter, C. Vonrhein, T. Hartsch and V. Ramakrishnan. *Nature* **407**, 327 (2000).
9. P. Cramer, D. A. Bushnell and R. D. Kornberg. *Science* **292**, 1863 (2001).
10. I. A. Wilson, J. J. Skehel and D. C. Wiley. *Nature* **289**, 366 (1981).
11. C. B. Anfinsen. *Science* **181**, 223 (1973).
12. L. Pauling, R. B. Corey and H. R. Branson. *Proc. Natl. Acad. Sci. USA* **37**, 205 (1951).
13. L. Pauling and R. B. Corey. *Proc. Natl. Acad. Sci. USA* **37**, 251 (1951).
14. G. N. Ramachandran, C. Ramakrishnan and V. Sasisekharan. *J. Mol. Biol.* **7**, 95 (1963).
15. G. Némethy and H. A. Scheraga. *Biopolymers* **3**, 155 (1965).
16. I. Bahar, A. R. Atilgan and B. Erman. *Fold. Des.* **2**, 173 (1997).
17. M. Levitt and A. Warshel. *Nature* **253**, 694 (1975).
18. A. Korkut and W. A. Hendrickson. *Proc. Natl. Acad. Sci. USA* **106**, 15667 (2009).
19. Korkut A, and Hendrickson WA. *Proc. Natl. Acad. Sci. USA* **106**, 15673 (2009).
20. A. Liwo, S. Oldziej, R. Kazmierkiewicz, M. Groth and C. Czaplewski. *Acta Biochim. Pol.* **44**, 527 (1997).
21. J. O. Moore and W. A. Hendrickson. *Structure* **20**, 729 (2012).
22. Q. Liu, Z. Zhang and W. A. Hendrickson. *Acta Crystallogr. D Biol. Crystallogr.* **67**, 45 (2011).
23. P. D. Kwong, R. Wyatt, J. Robinson, R. W. Sweet, J. Sodroski and W. A. Hendrickson. *Nature* **393**, 648 (1998).
24. S. A. Hollingsworth, D. S. Berkholz and P. A. Karplus. *Prot. Sci.* **18**, 1321 (2009).
25. M. Novotny and G. J. Kleywegt. *J. Mol. Biol.* **347**, 231 (2005).
26. M. N. Fodje and S. Al-Karadaghi. *Prot. Eng.* **15**, 353 (2002).

27. I. W. Davis, A. Leaver-Fay, V. B. Chen, J. N. Block, G. J. Kapral, X. Wang, L. W. Murray, W. B. Arendall, 3rd, J. Snoeyink, J. S. Richardson and D. C. Richardson. *Nucleic Acids Res.* **35**, W375 (2007).
28. S. C. Lovell, I. W. Davis, W. B. Arendall, 3rd, P. I. de Bakker, J. M. Word, M. G. Prisant, J. S. Richardson and D. C. Richardson. *Proteins* **50**, 437 (2003).

MULTIPROTEIN ASSEMBLIES: MODULATING CELL ACTIVITY THROUGH TARGETING PROTEIN-PROTEIN INTERFACES

ALICIA P. HIGUERUELO, HARRY JUBB AND TOM L. BLUNDELL
Department of Biochemistry, University of Cambridge
Cambridge CB2 1GA, UK
E-mail: alicia@cryst.bioc.cam.ac.uk

The central role of multiprotein assemblies in living organisms makes them attractive targets for therapeutic intervention. Here, we briefly describe the classification of protein complexes and the available databases for their study. The chapter explores the current findings for peptide and small molecule binding to protein interfaces and presents insights for modulating the protein-protein interactions with synthetic agents.

1. Multiprotein assemblies and cell regulation

Multiprotein assemblies are central to the regulation of biochemical pathways involved in cell signaling, growth and survival. They include not only molecular machines such as the ribosome and proteome, but also non-obligate and transient assemblies mediating cell recognition, signal transduction, DNA replication, translation and repair. Indeed, high-throughput interactomics suggest that there may be ~130,000 interactions between proteins in humans[1]. Many of these will be essential and likely constitute useful targets for modulation of cell activity.

There is now increasing interest in targeting protein-protein interfaces (PPI) of multiprotein complexes in the design of chemical tools and therapeutic agents[2]. Evidence is accumulating that targeting protein-protein interactions may offer great opportunities in improving specificity and selectivity, but at the same time poses new challenges, particularly because the protein-protein interfaces tend to be less ligandable than the active sites of enzymes or receptors.

Understanding protein-protein interactions is reductionist at the level of defining stability, cooperativity, assembly and disassembly of individual protein complexes. However, understanding interactions in molecular assemblies is also holistic in focus; it is central to the systems biology of cells and living organisms. The study of protein-protein interactions in subcellular molecular assemblies presents a challenge in defining forms and functions that Ramachandran would have found as fascinating as understanding conformations of individual proteins. He would undoubtedly have also shared in the pleasure of realizing that it may be useful in advancing new methods of therapeutic intervention.

2. Classification of multiprotein assemblies: continuous and discontinuous epitopes

Nooren and Thornton[3] have presented a useful characterization of protein-protein interactions and assemblies. They distinguished between transient/permanent interactions that are largely defined by affinities, and non-obligate/obligate interactions that are defined by whether individual protomers are or are not observed independently *in vivo*. A further basis for classification is whether the interaction region of a subunit is comprised of a discontinuous epitope[4-7], where strands of polypeptide from different regions of the sequence contribute, or a continuous epitope, often comprising a single secondary structure element (α-helix, ß-strand or turn). Many continuous epitopes arise from flexible polypeptides or disordered regions of polypeptide chains that undergo concerted folding and binding as discuss by Wright and colleagues[8,9]. Schwyzer[10] earlier visited this idea for polypeptide hormones in the 1970s and proposed it as a mechanism of receptor recognition for glucagon and other polypeptide hormones[11-14]. The importance of these continuous segments has been recognized in recent years, not only for the crucial role they play in cell function but also for being attractive drug targets[4,15].

3. Databases of protein-protein interactions

Protein-protein interactions have become of increasing interest to both systems biologists seeking to define all interactions in a cell as well as those interested in individual assemblies and their modulation. There are as a consequence many databases for identifying interacting proteins as well as those that describe individual assemblies and the interactions between component protomers.

Examples from the first class include the Database of Interacting Proteins (DIP[16]), which contains manually verified data regarding experimentally determined protein-protein interactions from 469 organisms. IntAct[17], an open source molecular interaction database from literature curation and user submission is also a software suite from the European Bioinformatics Institute (EBI). Another interacting protein resource, which focuses primarily on PPIs from mammalian genomes, is the Molecular Interaction Database (MINT[18]). Combining both experimental and predicted data, the STRING[19] database contains direct physical and indirect functional PPIs. The International

Molecular Exchange Consortium[20,21] is a non-redundant set of protein-protein interaction data from a broad taxonomic range of organisms.

A useful review of databases holding structurally characterized PPIs and tools for studying them has been prepared by Nussinov and co-workers[22]. Databases of this type include Domain Annotated Protein-protein Interaction Database (DAPID[23]), 3D Interaction Domains (3DID [24]), which describes intra- and inter-molecular interactions between Pfam domains from high-resolution crystal structures, and includes functional annotations, based on Gene Ontology and predicted similarity of interactions across families. Atomic interactions of the protein-protein interfaces are stored and classified in resources such as 3D Complex, which classifies protein-protein complexes by sequence, structure and contact topology, and allows 2D visualization of 3D complex topology[25].

PICCOLO[26] is a comprehensive database of structurally characterized PPIs derived from quaternary assemblies generated by the EBI's PISA resource[27]. Interactions are described at the level of interacting pairs of atoms, residues and polypeptide chains. PICCOLO distinguishes 12 different interaction types using distance and angle terms, which are used to analyze physico-chemical properties of protein-protein interfaces including residue propensity, hydropathy, polarity, residue contact preference and sequence entropy. PICCOLO was developed in our laboratory in conjunction with sister databases CREDO[28] (protein-small-molecule interactions) and BIPA[29] (protein-nucleic acid interactions).

4. Peptide binding to protein interfaces

Peptides bound to proteins have been studied mainly in the context of short linear motifs (SLM, LM or SLiMs). These motifs are defined as short regulatory modules, around ten contiguous residues, which are recognized by globular domains in a transient manner with typically low affinities[30]. They are often part of a larger usually disordered structure, but can bind to their globular targets adopting an organized structure upon binding, described as concerted folding and binding.

Stein and Aloy[31] analyzed SLiMs found in the PDB using computational alanine scanning. The authors differentiated between the residues in the motifs as defined by the Eukaryotic Linear Motif database (ELM[30]) and the residues (they called them context) that interact with the globular domain but are not defined in the motif pattern. Computing the contribution to binding for each residue, Stein and Aloy found that the amino acids in the motifs are optimized for maximal affinity while the residues in "the context" form suboptimal interactions, however they are most likely to be crucial for specificity. Their argument was that motifs are large enough to secure binding but too small to justify the exquisite specificity in vivo.

5. Databases of NCE-binding to PPI

There are now increasing numbers of examples of structurally defined protein-protein interfaces that have been targeted with New Chemical Entities (NCE). A resource bringing these data together is TIMBAL[32]. The original version was, a hand-curated relational database containing small molecules that inhibit protein-protein interactions. The database gave insights into which types of molecules are favored by protein interfaces. The data contained in TIMBAL highlighted the importance of structural information for these challenging targets. Virtually all PPIs that have been successfully disrupted by small molecules have crystallographic or NMR structural data for the protein-protein complex such as IL2/IL2Ra, Bcl-XL/Bad, MDM2/p53, MDM4/p53, S100B/p53, TNF trimer, XIAP/Smac, B-catenin/Tcf4, ZipA/FtsZ, cMyc/Max, E1/E2, iNOS dimer, CD80/CTLA-4, CMR1/Rev or UL42/HSV-Pol. The growth of data in the past years makes hand-curated databases a very time-consuming task. The current maintenance of TIMBAL is through automated searches on the ChEMBL database[33]. The new TIMBAL database contains more than 3,900 small molecules for 28 PPI targets, including almost 700 PDB entries of protein-protein and protein-small molecule complexes.

The 2P2I resource[7,34] is another hand-curated database of the structures of protein-protein complexes with known inhibitors, which now has 12 protein-protein complexes with 39 non-redundant protein-small molecules. Only targets with structural information for both the protein-protein complex and the protein-inhibitor complex are included in the database.

6. Concavity and pockets at protein-protein interfaces

A critical element defining interactions at protein-protein interfaces is the existence of complementary concave and convex regions on the two partners at an interface, and perhaps more importantly the existence of small concave pockets that can selectively accept an amino acid from the partner. Fuller *et al.*[35] examined these features using the program QSiteFinder[36], comparing binding pockets of protein-protein interfaces with those of protein-ligand interfaces of currently marketed drugs. They found that the ligands bind to proteins using fewer pockets of greater volume, compared to protein-protein and PPI inhibitor binding which use more pockets of smaller volume.

Fuller *et al.*[35] argued that the energetic approach of QSiteFinder for determining druggable pockets is more successful at identifying small cavities than geometric methods. However, we have had success in analyzing the concavity of different classes of protein-protein interfaces using Ghecom[37]. This grid-based method exploits operations from "mathematical morphology" to define concavity using a multiple sphere based approach. Ghecom calculates a property called Rinaccess, the minimum spherical probe radius that is unable to fit into a cavity, based on the solvent accessibility of a given atom under consideration. Figure 1 illustrates the Rinaccess concept using the apoptosis-related protein BCL-XL in the context of its interaction with the full-length BAD peptide[38] and the Rad51 interaction with BCR4 repeat of BRCA2[39]. We have also performed a large-scale (27,876 interfaces) comparative analysis of the concavity/pocket forming potential at pairwise PPI interfaces. We analyzed proportions of residues within given Rinaccess thresholds at different kinds of interface, finding the most striking differences between residues in the solvent inaccessible interface core vs. solvent accessible interface periphery, and between protein-peptide and protein-protein interactions (Figure 2; Jubb and Blundell, Unpublished).

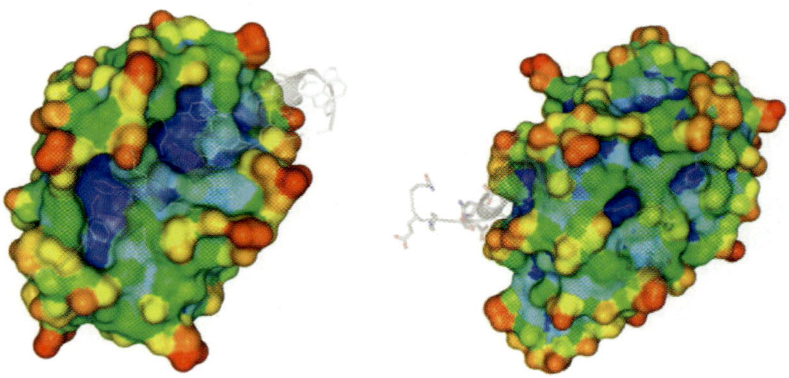

Figure 1: Interactions between BCL-XL and BAD peptide (left) and RAD51 and a BRCA2 BRC repeat (right), illustrating interface concavity. The BAD peptide and BRC repeat are translucent and shown in cartoon/sticks representation. BCL-XL and RAD51 is shown in surface representation and is colored by atomic Rinaccess values. Deeper concavities (with lower Rinaccess values, i.e. inaccessible to a small probe) are bluer in color. Knob regions protruding from the surface, which have a large Rinaccess value (as they are accessible to the larger spheres used by the program) are redder in color. In BCL-XL, two distinct clefts with low Rinaccess values can be seen in the binding site, occupied by large and aromatic residues from the BAD peptide. In RAD51, two small pockets with low Rinaccess can be seen binding the phenylalanine and alanine of the BRC repeat FxxA binding motif. The proportion of interface residues within given ranges of Rinaccess can be used to gain a comparative, whole-interface understanding of interface concavity (Figure 2).

Figure 2 indicates that a greater proportion of interactions in concave environments (lower Rinaccess) occur in the core of the solvent inaccessible interface for all interfaces. It is also of interest that in the bins of protein-peptide interactions, the high proportions of residues in more concave environments from the longer protein partner of the interaction implies that these interfaces are overall more concave than the protein-protein interfaces.

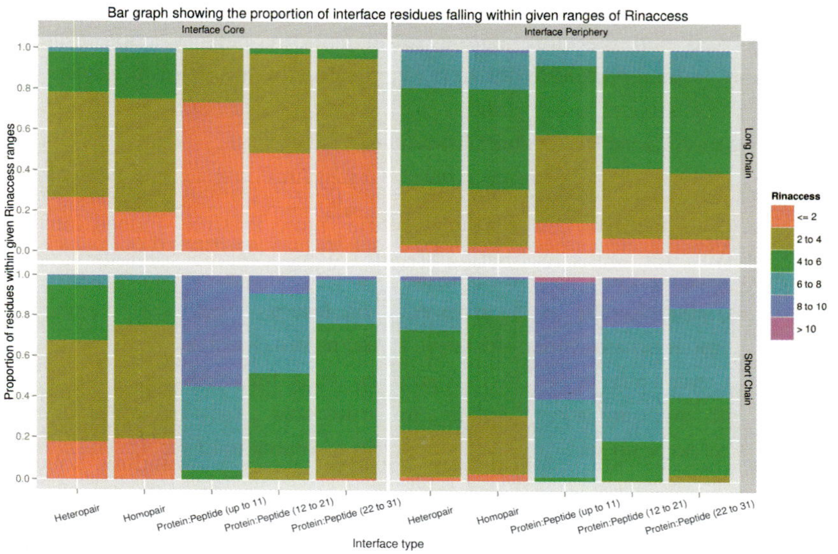

Figure 2: Interface concavity at pairwise PPI interfaces as measured by Rinaccess, a property calculated for each chain in its bound conformation in the context of a protein-protein interface. Smaller Rinaccess indicates a smaller probe radius that is inaccessible to a cavity, hence a deeper cavity. Rinaccess values are binned into ranges of two ångströms, and are shown as the proportion of residues meeting a given Rinaccess threshold in each residue environment. The plot is faceted vertically by the solvent exposure of the interface residue, and horizontally by whether the longer or shorter polypeptide chain is being considered with respect to primary structure. Interfaces are classified by whether the same protomers are interacting or not (homo or heteropair) or by whether a protein (> 31 residues) is interacting with a short polypeptide binned by different lengths of peptide.

7. Hotspots

Protein-protein interacting surfaces have long been recognized to exhibit "hot spots"[40] that are responsible for most of the interaction energy of the complex. This is a common characteristic of almost all interfaces of protein complexes[41]. Experimental[42] and computationally predicted hotspots[43,44] are described and systematically organized in databases. The early work of Bogan and Thorn[45] showed that the hot spot residues are usually clustered at the center of the interface and are surrounded by energetically neutral residues, giving rise to the "O-ring" model, the role of the neutral residues being to shield the hot spots from the solvent. The findings described in the previous section, regarding the location of the pockets are also consistent with this "O-ring" model. The most frequent hot spot residues Trp, Tyr and Arg are capable of both hydrophobic and electrostatic interactions. Nussinov and co-workers[46] have shown that hot spot residues are more evolutionarily conserved. They are more rigid as they are densely packed and therefore pay a lower entropy penalty upon binding, whereas non-optimal packed regions are responsible for site flexibility.

8. Binding of NCE at the protein interfaces

The existence of small and localized hotspots at protein-protein interfaces makes the design of competitive new chemical entities more feasible. Indeed location of binding sites for multiple small molecules using organic solvent mapping studies using X-ray crystallography[47,48] and NMR[48,49] as well as chemical fragment binding using NMR methods[50] or X-rays[51] can be used to define hotspots. Wells and co-workers suggest that protein-protein interaction hotspots mean that fragments can find ligand-efficient 'footholds', but higher affinity is often achieved by binding 'cryptic' binding sites within the binding interface[2,52]. It is clear that, where protein-protein hotspots and ligand-binding hotspots coincide, competitive small molecule inhibitors can be developed, as illustrated by the data described in section 5.

9. Comparisons of peptides and synthetic molecules

We can now begin to compare protein complexes of small peptides, natural and synthetic molecules many of which have been made by medicinal chemists. We show that small peptides (synthetic but simulating natural molecules) and other natural small molecules that evolved in living organisms engage on average more polar contacts with their targets than synthetic molecules. Analysis of the number of oxygens and nitrogens as a proportion of the total number of atoms interacting shows that the more polar interaction profile presented by natural molecules is due to a higher content of oxygens and nitrogens compared to carbon atoms in these molecules. We have also shown that the greater number of rotatable bonds in peptides and other natural molecules allows conformations to be adopted that optimize interactions with their protein target. Table 1 contains selected examples for proteins where there are structural data for both complexes, i.e. the protein target to the natural molecule or bound to a small peptide, as well as bound to a synthetic molecule. Figure 3 shows an example for each of these two classes.

Table 1: Examples of polar/sum Contacts ratio for proteins that bind to both natural (small molecules and peptides) and synthetic molecules.

Target	Natural molecules			Synthetic molecules			Refs
	PDB	Ratio	Aff	PDB	Ratio	Aff	
Visfatin	2G96 (A&B)	0.29	53uM	2G97 (A&B)	0.07	0.15uM	53
Abl Kinase	2G2I (A)	0.39	-	2HZI (B)	0.06	70nM	54 55
MK2	1NY3 (A)	0.61	-	2PZY (A)	0.28	34nM	56
HIV Protease	1A94 (D&E)	0.37	14nM	1D4Y (A&B)	0.29	8pM	57 58
Phospo lipase A2	2O1N (A)	0.20	-	2B17 (A)	0.13	620nM	59
Alpha-Thrombin	1NY2 (2)	0.33	1.75mM	1BCU (H)	0.20	0.53mM	60 61

Note: The PDB code includes the interacting chains, for example 2G96(A&B) denotes chains C and D interacting with the small molecule. When available, affinity measure (Aff) and units is specified in table. The three first target proteins bind to natural molecules and synthetic ones. The last three target proteins bind to small peptides and to synthetic molecules.

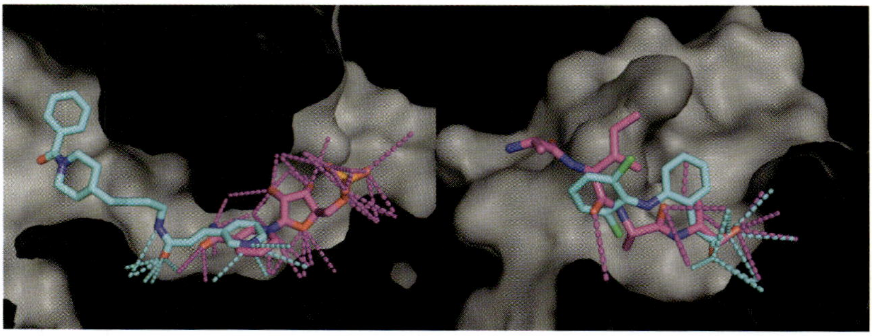

Figure 3: Examples of natural molecules (magenta) and synthetic molecules (cyan) binding to the same protein target. Only polar contacts are shown for clarity. Colours of the contacts (in dotted lines) are the same as the molecules making them. LEFT: Visfatin with Nicotinamide Mononucleotide (2G96) and FK-866 (2G97). RIGHT: Phospholipase A2 with a tetrapeptide (2O1N) and Diclofenac (2B17). In both cases synthetic molecules use only a fraction of the polar contacts available for the natural counterparts.

Comparison of the residue compositions of the binding sites of different classes of small molecules and also different protein-protein interfaces highlights the more hydrophobic and aromatic character of the binding sites selected by synthetic small molecules (drug-like), as found previously by Soga *et al.*[62].

Figure 4 shows the proportions of charged, polar and hydrophobic residues for all subsets. The proportion of charged residues in protein-protein complexes is significantly greater ($P < 0.05$) than that of the small molecule subsets. In contrast, the proportion of hydrophobic residues for small molecule inhibitor complexes with proteins is significantly greater than that for complexes with small peptides and proteins. Small peptide complexes are similar to transient dimers, having a greater proportion of polar residues than other subsets. In particular, transient dimers have more polar and fewer hydrophobic residues than obligate dimers.

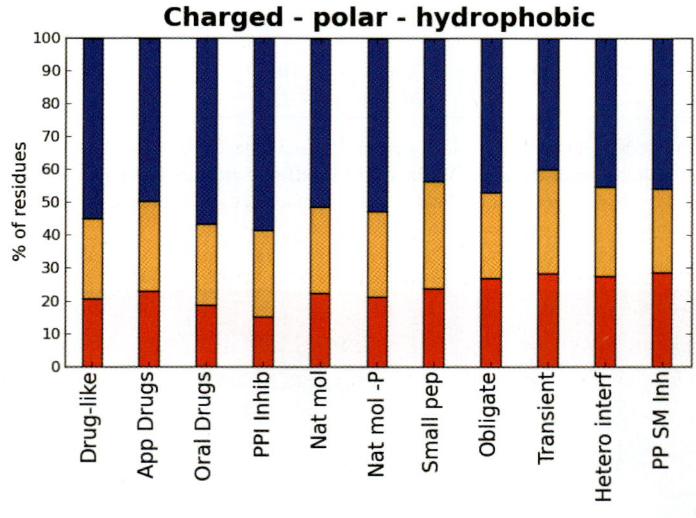

Figure 4: Average proportions of charged (red), polar (orange) and hydrophobic (blue) residues at the interfaces for each molecular subset: Drug-like (synthetic), Approved drugs, Oral drugs, small molecule protein-protein (PP) interaction inhibitors, natural molecules, natural molecules without phosphorous, small peptides, PP obligate dimers, PP transient dimers, PP hetero- quaternary interfaces and PP complexes successfully inhibited by small molecules. For the PP complexes, only the long chain is considered.

10. Lessons for modulating protein-protein interfaces in multiprotein assemblies

Amongst the protein-protein interfaces, the transient complexes appear to be, from the structural point of view, the most amenable to targeting by small molecule chemical tools and candidate drug molecules. These complexes have on average deeper pockets at the interface than obligate and quaternary interfaces. In addition, they often involve a short peptide chain binding to a larger protein, sometimes exhibiting concerted folding and binding.

Contrary to the views often expressed that some degree of site adaptability has to occur at the interface of globular protein complexes in order to bind to a small molecule[2], Fuller *et al.*[35] and Bourgeas and co-workers[7] find that for most complexes there are few conformational changes between the protomers in the complex and in isolation, and when bound to small molecules. Indeed Swapna *et al.*[63] find that around 65% of interface residues are rigid and likely play roles of contributing to the stability of the unbound protomer and in reducing the entropic cost of binding.

Higueruelo *et al.*[32] have made comparisons of small molecules targeting PPI with drugs and synthetic drug-like molecules used in classical screening and shown that inhibitors of protein-protein interfaces are on the average bigger, more lipophilic and have fewer hydrogen bonding features. Interestingly, although synthetic drug-like molecule binding sites are on average more hydrophobic and have higher aromatic content than those binding small natural molecules, these sites have a higher proportion of unmatched polar atoms, suggesting that in principle the polar interaction profile for drug-like molecules could be improved.

This is probably the most encouraging observation for those wishing to design more effective new chemical entities that bind and modulate protein-protein interfaces. More attention to optimizing hydrogen bonds, perhaps by introducing a little more flexibility, could lead to greater selectivity and less lipophilic character, a feature that often leads to off-site binding and failure of candidate drugs in clinical trials[64].

References

1. K. Venkatesan *et al.*, *Nat Methods,* **6**, 83 (2009).
2. J. Wells and C. McClendon, *Nature,* **450**, 1001 (2007).
3. I. M. Nooren and J. M. Thornton, *EMBO J,* **22**, 3486 (2003).
4. T. L. Blundell et al., Philos Trans R Soc Lond B Biol Sci, **361**, 413 (2006).

5. S. Jones and J. M. Thornton, Proceedings of the National Academy of Sciences, **93**, 13 (1996).
6. A. Pal, P. Chakrabarti, R. Bahadur, F. Rodier and J. Janin, *J Biosci,* **32**, 101 (2007).
7. R. l. Bourgeas, M.-J. Basse, X. Morelli and P. Roche, *PLoS ONE,* **5**, e9598 (2010).
8. P. E. Wright and H. J. Dyson, *J Mol Biol,* **293**, 321 (1999).
9. P. E. Wright and H. J. Dyson, *Current Opinion in Structural Biology,* **19**, 31 (2009).
10. R. Schwyzer *et al.*, *Pure and Applied Chemistry,* **51**, 831 (1979).
11. K. Sasaki, S. Dockerill, D. A. Adamiak, I. J. Tickle and T. Blundell, *Nature,* **257**, 751 (1975).
12. W. Braun, G. Wider, K. Lee and K. Wuthrich, *J Mol Biol,* **169**, 921 (1983).
13. T. Blundell, *Trends in Biochemical Sciences,* **4**, 80 (1979).
14. T. Blundell and S. Wood, *Annual review of biochemistry,* **51**, 123 (1982).
15. V. Neduva and R. B. Russell, *Curr Opin Biotechnol,* **17**, 465 (2006).
16. I. Xenarios *et al.*, *Nucleic Acids Research,* **28**, 289 (2000).
17. H. Hermjakob *et al.*, *Nucleic Acids Research,* **32**, D452 (2004).
18. A. Chatr-Aryamontri *et al.*, *Nucleic Acids Research,* **35**, D572 (2007).
19. D. Szklarczyk *et al.*, *Nucleic Acids Res,* **39**, D561 (2011).
20. S. Orchard *et al.*, *Proteomics,* **7 Suppl 1**, 28 (2007).
21. A. L. Turinsky, S. Razick, B. Turner, I. M. Donaldson and S. J. Wodak, *Nat Biotech,* **29**, 391 (2011).
22. N. Tuncbag, G. Kar, O. Keskin, A. Gursoy and R. Nussinov, *Briefings in bioinformatics,* **10**, 217 (2009).
23. Y. C. Chen, H. C. Chen and J. M. Yang, *Genome Inform,* **17**, 206 (2006).
24. A. Stein, A. Ceol and P. Aloy, *Nucleic Acids Res,* **39**, D718 (2011).
25. E. D. Levy, J. B. Pereira-Leal, C. Chothia and S. A. Teichmann, *PLoS Comput Biol,* **2**, e155 (2006).
26. G. R. Bickerton, A. P. Higueruelo and T. L. Blundell, *BMC Bioinformatics,* **12**, 313 (2011).
27. E. Krissinel and K. Henrick, *J Mol Biol,* **372**, 774 (2007).
28. A. Schreyer and T. Blundell, *Chem Biol Drug Des,* **73**, 157 (2009).
29. S. Lee and T. L. Blundell, *Bioinformatics,* **25**, 1559 (2009).
30. H. Dinkel *et al.*, *Nucleic Acids Res,* **40(Database issue)**, D242 (2012).
31. A. Stein and P. Aloy, *PLoS ONE,* **3**, e2524 (2008).
32. A. P. Higueruelo *et al.*, *Chem Biol Drug Des,* **74**, 457 (2009).
33. A. Gaulton *et al.*, *Nucleic acids research,* **40**, D1100 (2011).
34. X. Morelli, R. Bourgeas and P. Roche, *Curr Opin Chem Biol,* **15**, 475 (2011).
35. J. C. Fuller, N. J. Burgoyne and R. M. Jackson, *Drug Discov Today,* **14**, 155 (2009).
36. A. T. Laurie and R. M. Jackson, *Bioinformatics,* **21**, 1908 (2005).
37. T. Kawabata, *Proteins,* **78**, 1195 (2010).
38. K. Lee, W. Han, K. Kim and B. Oh, *PDB: 2BZW,* (2007).
39. L. Pellegrini *et al.*, *Nature,* **420**, 287 (2002).
40. T. Clackson and J. A. Wells, *Science,* **267**, 383 (1995).

41. D. Reichmann, O. Rahat, M. Cohen, H. Neuvirth and G. Schreiber, *Curr Opin Struct Biol,* **17**, 67 (2007).
42. K. S. Thorn and A. A. Bogan, *Bioinformatics,* **17**, 284 (2001).
43. E. Guney, N. Tuncbag, O. Keskin and A. Gursoy, *Nucleic acids res,* **36**, D662 (2008).
44. J. Segura and N. Fernandez-Fuentes, *Nucleic acids res,* **39**, D755 (2011).
45. A. A. Bogan and K. S. Thorn, *J Mol Biol,* **280**, 1 (1998).
46. O. Keskin, B. Ma and R. Nussinov, *J Mol Biol,* **345**, 1281 (2005).
47. P. A. Fitzpatrick, A. C. Steinmetz, D. Ringe and A. M. Klibanov, *Proceedings of the National Academy of Sciences,* **90**, 8653 (1993).
48. C. Mattos and D. Ringe, *Nat Biotech,* **14**, 595 (1996).
49. E. Liepinsh and G. Otting, *Nat Biotech,* **15**, 264 (1997).
50. P. J. Hajduk, J. R. Huth and S. W. Fesik, *J Med Chem,* **48**, 2518 (2005).
51. A. Ciulli, G. Williams, A. G. Smith, T. L. Blundell and C. Abell, *Journal of Medicinal Chemistry,* **49**, 4992 (2006).
52. C. D. Thanos, W. L. DeLano and J. A. Wells, *Proc Natl Acad Sci U S A,* **103**, 15422 (2006).
53. M. K. Kim *et al.*, *J Mol Biol,* **362**, 66 (2006).
54. N. M. Levinson *et al.*, *PLoS Biol,* **4**, e144 (2006).
55. S. W. Cowan-Jacob et al., Acta Crystallogr D Biol Crystallogr, **63**, 80 (2007).
56. K. W. Underwood *et al.*, *Structure,* **11**, 627 (2003).
57. J. Wu *et al.*, *Biochemistry,* **37**, 4518 (1998).
58. S. Thaisrivongs *et al.*, *J Med Chem,* **39**, 4349 (1996).
59. N. Singh et al., Acta Crystallogr D Biol Crystallogr, **62**, 410 (2006).
60. B. Pillai *et al.*, *J Mol Biol,* **365**, 343 (2007).
61. E. Conti, C. Rivetti, A. Wonacott and P. Brick, *FEBS Lett,* **425**, 229 (1998).
62. S. Soga, H. Shirai, M. Kobori and N. Hirayama, *J Chem Inf Model,* **47**, 400 (2007).
63. L. S. Swapna, R. M. Bhaskara, J. Sharma and N. Srinivasan, *Scientific Reports,* **2**, (2012).
64. M. M. Hann, *MedChemComm,* **2**, 349 (2011).

PREDICTION OF PROTEIN-PROTEIN COMPLEX STRUCTURES

EIJI KANAMORI*

Japan Biological Informatics Consortium, 2-3-26 Aomi, Koto-ku Tokyo 135-0064, Japan

YOICHI MURAKAMI

National Institute of Biomedical Innovation, 7-6-8 Asagi, Saito Ibaraki, Osaka567-0085, Japan

JOY SARMIENTO, SHIDE LIANG, DARON M. STANDLEY

Immunology Frontier Research Center, Osaka University, 3-1Yamadaoka Suita, Osaka 565-0871, Japan

MATSUYUKI SHIROTA, KENGO KINOSHITA

Graduate School of Information Sciences, Tohoku University, 6-3-9 Aoba Aramaki, Sendai, Miyagi980-8579, Japan

YUKO TSUCHIYA, JUNICHI HIGO, HARUKI NAKAMURA

Institute for Protein Research, Osaka University, 3-2 Yamadaoka Suita, Osaka 565-0871, Japan

A docking method, *surFit*, was developed that automatically and semi-automatically docks a pair of protein molecular surfaces. It performs the following six procedures: *Binding site prediction, Rigid surface docking, Coarse scoring, Refinement, Precise scoring, and Re-refinement by molecular dynamics simulation*. The first four procedures have been automated and implemented in the webserver *surFit*. The current protocol successfully built many acceptable predicted complex structures with high qualities for the recent CAPRI targets, and accurately estimated the solvent water positions at the interface for CAPRI Target 47. To reveal the complex structure of an intrinsically disordered protein (IDP) with its partner receptor protein, enhanced sampling computations were performed to simulate the free energy landscapes of the IDP with and without the receptor. Consequently, both induced fitting and population shift mechanisms were observed for the NRSF-Sin3 system.

1. Introduction

Protein-protein interactions govern almost all biological phenomena, and analyses at an atomic level have been performed to reveal the precise

*Present address: Hitachi Solutions Ltd., 1-1-43 Suehiro-cho, Tsurumi, Yokohama 230-0045, Japan.

mechanisms for individual, specific recognition at protein interfaces[1, 2] In particular, Kinjo and Nakamura[3] performed an exhaustive structural classification study, and find that almost all structural motifs at the interfaces are confined within single protein families, with exceptions only between α-helices. Namely, for a pair of proteins, the specific interface structures are well designed, and are often conserved in the two interacting families as homologous interactions.

Although the numbers of complex structures in the PDB are increasing, there are still many more protein-protein interactions[4]. Thus, precise prediction of protein complex structures is required for further experiments to understand the recognition mechanisms. For this purpose, we developed a computational procedure to dock a pair of protein molecular surfaces[5, 6]. It consists of several procedures arranged as a linear pipeline: binding site prediction, surface docking, scoring the complex structures, and refinement for promising candidates. Finally, a molecular dynamics simulation with explicit water molecules is performed to relax the complex structures and allow flexibility for the interfaces and the solvent molecules. Here, our procedures are described with several results from the recent CAPRI (Critical Assessment of Predicted Interactions: http://www.ebi.ac.uk/msd-srv/capri/) blind predictions for protein complex structures [7, 8].

In contrast to this rigid body docking with some flexibility at the interface, a completely different approach is required to analyze and predict coupled folding and binding, which is often observed in intrinsically disordered proteins (IDPs)[9]. Although the interfaces are localized for IDPs, the free energy landscapes of complex structures should be detected to correctly understand the interactions coupled with the folding of an IDP. Our recent approach[10] is also described.

2. Methods

2.1 SurFit: Docking procedure for protein domains

The docking server, *surFit*, was developed to automatically and semi-automatically dock a pair of protein molecular surfaces. It consists of several procedures arranged as a linear pipeline, in a multistage docking manner[11].

(i) Prediction of binding interfaces

The binding propensities are computed for each residue in each monomer, by using either the empirical method Pinup[12] or the real-value evolutionary trace (ET) method[5, 13]. Manually assigned interface information for particular residues from the literature can also be added. The propensities are then mapped onto the molecular surfaces of the proteins.

(ii) Docking two rigid molecular surfaces

Approximately one million candidate complex structures of protein A as a receptor and protein B as a ligand are rapidly produced, such that the two interface shapes are complementary. In the pre-processing phase of this procedure, all possible pairs of vertices with their associated normals on the molecular surfaces of the protein pair A and B are selected as the basis pair to build the reference coordinates (RC) systems, respectively, by satisfying the geometric constraints of the distances and directions of the normals. Here, all of the normals on the molecular surface of protein B (ligand) are inversed before computation. For every basis pair, the reference frame (RF) is defined along with the rotation and translation matrix, which change all the atomic coordinates from the original coordinate system to the RC system, as proposed by Fischer *et al.*[14]. For efficiency, we first prepare all of the RFs in protein A, which are selected to be nearly equally distributed on its surface, and then the atomic coordinates of protein Aare stored in a Hash table.

In the matching phase shown in Fig. 1, when a set of two RF pairs in the RC system, one pair from protein A and the other from protein B, have similar sets of normals with similar spatial arrangements, all of the atomic coordinates and the surface vertices of the ligand protein B are mapped onto the receptor protein A, using the rotation and translation matrices. If an atomic collision between protein A and B is observed, as shown in Fig. 1, the corresponding RF pair is rejected. The coarse score, S_{coarse}, computed from the matched vertex pairs of proteins A and B, located within 5 Å, is calculated according to eqs. 1-3, as shown below. This procedure is repeated for all of the possible RF pairs of proteins A and B, thus producing a large number of possible poses. Using a single core computer, a million complex structures can be produced in about one hour. Finally, similar poses are clustered, which reduce the candidates to about 1,000 high-scoring complex structures within about 10 min.

(iii) Coarse scoring

The following coarse score, S_{coarse}, is applied for assessing the shape complementarities at vertices,

$$S_{coarse} = \sum_{ij} I_{ij}^{2} W_{ij} D_{ij} \qquad \text{eq. 1}$$

$$I_{ij} = \frac{1 - \boldsymbol{n}_i \boldsymbol{n}_j}{2} \quad \text{eq. 2}$$

$$D_{ij} = \min\left(1, \frac{0.25}{d_{ij}^{2}}\right) \quad \text{eq. 3}$$

Here, $\boldsymbol{n}_i$ is the normal vector at the i'th vertex, and d_{ij} is the distance between the i'th and j'th vertices. W_{ij} is the weight incorporating the empirical interface probability[12], the ET score[5], and/or any other biological and experimental information. The candidates of the complex structures are automatically selected and sorted based on S_{coarse}.

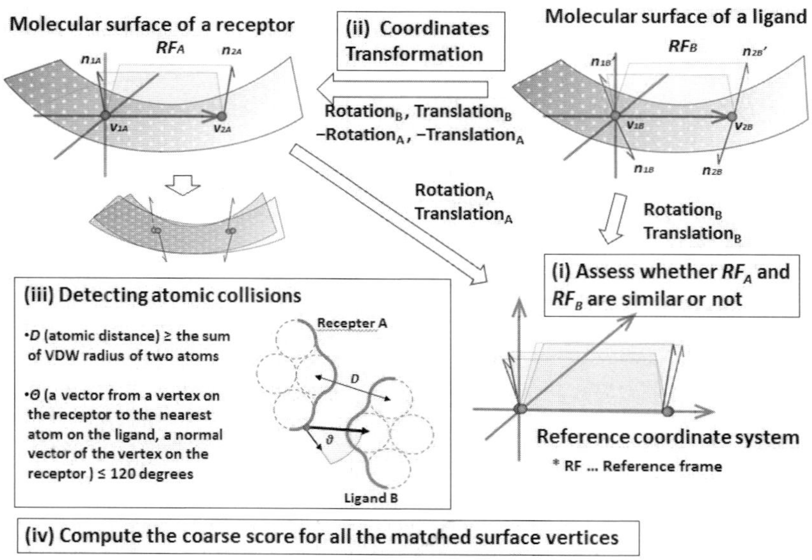

Figure 1: Rapid docking procedures for two protein molecular surfaces. (i) The similarity of a pair of two normals on the corresponding vertices (RFs: reference frames) of a receptor protein A and a ligand protein B are assessed in the common references coordinates (RC) system. The direction of all the normals on the vertices of protein B are first inversed. Here, RF frames are composed of the two normals, the distance of which was within 3 Å to 12 Å, and the direction angle difference is less than 50 degrees. (ii) For the RF pairs having the similar features (the distance difference between the two normals is less than 2 Å, and the direction difference of the normals is less than 60 degrees), the coordinates of all the vertices and atoms of protein B are transformed to the coordinate system of protein A, so that the RF pairs are superimposed. (iii) After superimposing the two structures, the atomic collisions are examined. If such collisions are observed, such RF pairs are rejected. (iv) Coarse score, S_{coarse}, was computed for the selected vertex pairs.

(iv) Refinement

About 500 promising candidates selected in the above automatic procedures are then refined by energy minimization, by weakly restraining the entire backbone structure. This procedure improves the surface geometry to avoid any overlapping of atoms between the ligand and receptor proteins.

(v) Precise scoring

The geometry-optimized docked surfaces are scored by a function that considers shape, hydrophobicity, and electrostatic complementary at the interface, as well as binding site propensity[6, 8]. We use a linear combination of the degrees of complementarities for the three properties as the complementarity score of an interface,

$$SCR_{comp} = W_H \times H_{cmp} + W_E \times E_{cmp} + W_S \times S_{cmp} \qquad \text{eq. 4}$$

where H_{cmp}, E_{cmp}, and S_{cmp} represent the degrees of complementarities for the hydrophobicity features calculated from the hydration free energies with the parameters by Ooi *et al.*[15] using the accessible surface area, the electrostatic potentials obtained by solving the Poisson-Boltzmann equations[16], and the surface curvatures[17], respectively. W_H, W_E, and W_S are their weights, which were optimized by using the complex structures for the 74 representative hetero-dimers[6]. We prepared up to 500 models for each hetero-dimer by our docking method as a training dataset, and optimized the three weights so that the discrimination between native-like models with an rmsd (root mean square displacement) less than 10 Å from the native complexes and other models in the training set could be performed with maximum accuracy. W_H, W_E, and W_S are determined as 0.343, 0.544, and 0.112, respectively.

(vi) Re-refinement for flexible docking

For the high-scoring complex structures, molecular dynamics simulation by the *myPresto* program[18] with explicit water molecules is performed to relax the complex structures and allow flexibility for the interfaces and the solvent molecules. In addition, this procedure can identify stable complex structures with satisfactory solvent interfaces.

The above steps (i) to (iv) have been automated and implemented in the web server *surFit* (http://sysimm.ifrec.osaka-u.ac.jp/surFit/). The procedures for Steps (v) and (vi) are performed manually.

2.2 *Detection of the free energy landscape for the analysis of coupled folding and binding*

In order to draw correct free energy landscapes for an IDP and its partner protein, we perform enhanced conformation sampling by using the multicanonical molecular dynamics (McMD) method[19-22], as described in detail elsewhere[10, 23]. McMD can explore a broad conformational space in a wide temperature range, and a thermodynamic conformational ensemble at any temperature can be reconstructed from the sampled conformation by using a reweighting technique. Here, our target IPD is neural restrictive silencer factor (NRSF), and its receptor is the paired amphipathic helix domain (PAH1) of Sin3. The minimal structural size of the former is the N-terminal 15 residues, and that of the latter is the N-terminal 83 residues, observed by NMR (PDBID 2CZY)[24]. We performed McMD simulations to obtain the ensemble from the single-chain NRSF system and that from the NRSF with the Sin3 system embedded in the explicit TIP3P water solvent, using the computer program *PRESTO* [10].

3. Results and Discussion

3.1 *Prediction of complex structures at CAPRI*

We participated in a series of the CAPRI blind prediction to predict the complex protein structures[8, 25]. Although our results were essentially obtained by rigid body docking, some flexibility was incorporated at the interface during the molecular dynamics simulations, and this flexibility contributed to some of the predictions.

In round 18, Target 40, the complex structures were predicted with structural information from the unbound form of trypsin and its inhibitor API-A in its bound form[26]. Interestingly, this inhibitor (chain C) simultaneously binds two trypsin molecules (chain A and B) at different interfaces, and one of them is a homologous interaction designated as the CA complex, and observed in the PDB. However, the recognition mode of the CB complex was new. Our prediction was very successful for this Target 40 with 10 submitted structures: 5 for CA complexes and 5 for CB complexes (Fig. 2). In particular, M09 for CA was the 7th best model, and M03 for CB was the highest quality structure among all of the 368 structures submitted by the CAPRI participants. The criteria of quality assessment are described in the caption of Fig. 2[25].

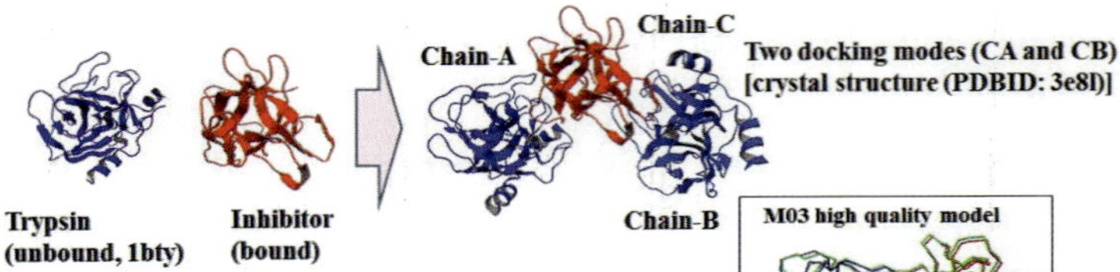

Model	surFit Rank	f_nat	L_rmsd	I_rmsd	classification
for CB					
M01	1	0.95	0.84	0.44	high
M02	1	0.76	1.79	0.64	high
M03	1	0.88	0.67	0.47	high
M04	1	0.45	5.34	2.06	acceptable
M05	1	0.76	1.23	0.80	high
for CA					
M06	41	0.63	5.95	1.31	medium
M07	41	0.88	2.82	1.26	medium
M08	41	0.83	4.55	1.00	medium
M09	41	0.89	1.00	0.45	high
M10	41	0.86	2.31	0.71	high

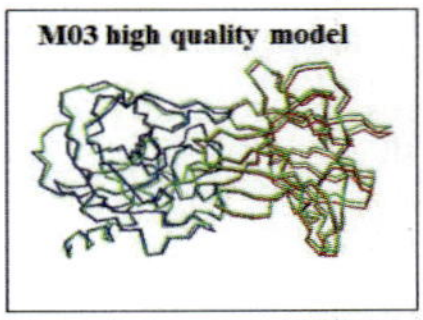

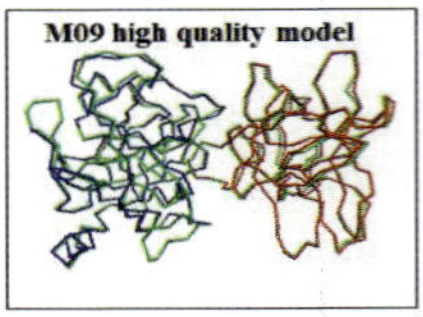

Figure 2: Blind prediction of CAPRI Target 40, composed of unbound trypsin and two inhibitor API-A molecules in their bound forms[26]. Two interaction forms CB and CA should be predicted, and our models, five of each, were very closed to the crystal structure (PDBID 3e8l). In the Table, *surFit* rank was the S_{coarse} order. The followings are parameters introduced in CAPRI[25]. f_nat is the fraction of native interface atom pairs, L_rmsd is the rmsd between backbone heavy atoms of the ligand protein, after superimposing the receptor proteins. Here, a larger protein is called as a receptor, and a smaller one a ligand. I_rmsd is the rmsd of the backbone atoms of the interface residues only, which at least have one atoms within 10 Å of an atom on the other molecule.When f_nat $\geq$ 0.5, L_rmsd $\leq$ 1Å and I_rmsd $\leq$ 1Å, the model has high precision.f_nat $\geq$ 0.3, 1Å < L_rmsd $\leq$ 5Å or 1Å < L_rmsd $\leq$ 2Å, medium precision.f_nat $\geq$ 0.1, 5Å < L_rmsd $\leq$ 10Å or 2Å < L_rmsd $\leq$ 4Å, acceptable precision. When f_nat < 0.1, then the model is incorrect. Our model M03 for the CB form (red and blue lines) was the most precise complex model among 368 predicted structures. M09 for the CA form (red and blue lines) was the 7'th best model. In both superimposed pictures, green lines show the crystal structure.

In round 19, Target 41, an homologous interaction was predicted for the DNase domain of colicin E9 and immunity protein2 (Im2)[27], for which the complex structure of the DNase domain of colicin E9 with immunity protein 9 (Im9) (PDBID 1emv) serves as a good template. Thus, all 10 deposited structures were correct, with 2 medium quality models. Target 42, also in round 19, was the prediction of the homodimer of a designed TPR motif[28], where homology modeling was performed for the protomer using our modeling engine, Spanner[29], based on the template, PDBID 1na0. Since the complex was an artificially designed one, homologous interactions were excluded in this case.

Among our 10 deposited structures, the M04 complex structure (Fig. 3) had the highest quality among all of the 278 structures deposited by the CAPRI participants.

Model (AB)	surFit Rank	L-rmsd (Å)	Interface complementarity		Contact surface area (Å²)
			H_{cmp}	E_{cmp}	
M02	7	34.81	best	7th best	519.4
M04	11	1.29	low	2nd best	764.2
M08	21	41.16	best	modest	590.3
M09	23	15.85	modest	best	455.9
M10	25	36.11	low	6th best	714.8

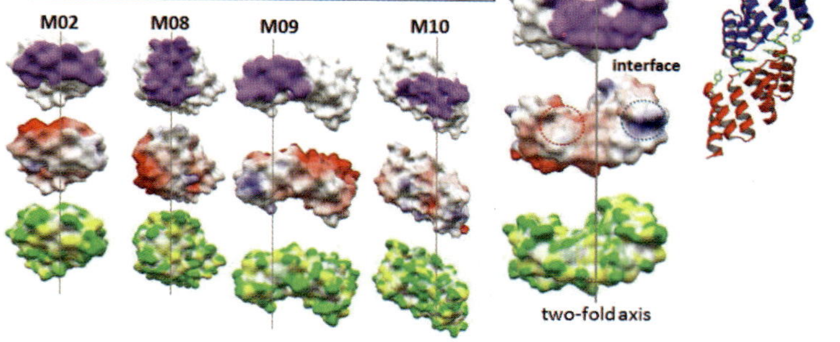

Figure 3: CAPRI Target 42 to predict an artificial homo-dimer, which was different form known natural oligomers. In this case, interface complementarities were independently assessed, and our model M04 considering only the electrostatic complementarity had the highest precision among 278 deposited structures. Here, the contact surface area was also considered. Purple color shows the interface of the homo-dimer. Blue and red colors show the positive and negative electrostatic potentials on the molecular surfaces. Yellow and green colors show the hydrophobic and hydrophilic residues.

In round 23, Target 47 was also a prediction of the homologous interaction between the DNase domain of colicin E2 and Im2[30], and the positions of the water molecules around the interface were assessed. With the above modeling procedure, we could predict many high quality models not only for the complex structures but also for the water positions (Table 1).

In round 24, Target 51 was a difficult problem involving a xylanase from Clostridium thermocellum. We were required to model GH5-CBM6 (glycoside hydrolase catalytic domain) with carbohydrate-binding modules/CBM13/Fn3 from the following 3 parts: A chain GH5-CBM6 (unpublished unbound), B chain CBM13 (homology model using PDBID 1knl as the template), and C chain Fn3 (unbound, PDBID 3mpc). One of our predicted structures for CBM13 with Fn3 had acceptable quality, which was highly significant because only 2 other structures were acceptable out of all the predicted structures from other groups. For the interaction between GH5-CBM6 and CBM13, only a single structure by one group was successful, and none of the participants for this round succeeded in modeling the complex of GH5-CBM6 with CBM13-Fn3.

Table 1: Assessment results for CAPRI Target 47, where not only the complex structure but also the positions of water molecules around the interface should be predicted. Here, fw_nat is the fraction of water-mediated contacts, which are indirect contacts where both ligand and receptor residues have at least one heavy atoms within 3.5Å of the same water molecule as that of the native one.

Model	Complex model assessment				Water position assessment	
	f_nat	L_rmsd (Å)	I_rmsd (Å)	classification	f^w_nat	classification
M01	0.821	3.658	1.131	medium	0.571	excellent
M02	0.839	3.794	1.036	medium	0.543	excellent
M03	0.821	2.730	1.011	medium	0.600	excellent
M04	0.839	4.144	1.195	medium	0.600	excellent
M05	0.839	2.882	1.005	medium	0.543	excellent
M06	0.893	3.676	1.114	medium	0.457	good
M07	0.839	3.387	1.074	medium	0.486	good
M08	0.786	4.210	1.255	medium	0.514	excellent
M09	0.768	3.777	1.056	medium	0.543	excellent
M10	0.804	2.365	0.963	high	0.486	good

In round 26, the information for Targets 53 and 54are not accessible as of this publication time. Both targets involved docking homologous models with their partner proteins in the unbound forms. In Target 53, the 3D structure of the former designed protein was modeled by Spanner[29]. After MD simulation (as described in Step 6 of Methods), two complex structures were acceptable out of our chosen 10 stable complex structures. This suggests that stability under MD simulation is a potentially powerful tool for assessing the quality of a complex. In Target 54, the top 500 models generated by *surFit* were evaluated and the top 10 models with favorable scores were submitted. This time, the MD calculations did not provide a clear evaluation. Among them, models M02 and M04 turned out to have acceptable accuracy. This prediction was difficult, and only 6 out of 377 models were submitted with acceptable accuracy.

In conclusion, our procedure for building an ensemble of complex structures exhibited encouraging performance. Our scoring method can often successfully extract the correct complex structures among the ensemble, however it is difficult to distinguish the best model using only the score. Homologous interactions are always well predicted with the aid of our homology modeling engine, Spanner. MD simulations with explicit water molecules at the very last stage occasionally identify more promising model complex structures. In the future, starting from the putative and acceptable complex structures, longer MD simulations with explicit water molecules may yield more reliable and precise complex structures, which are essentially beyond the capability of rigid body docking.

3.2 Ab initio simulation of coupled folding and binding

Figure 4 shows the distribution of NRSF structures at 300 K for the NRSF-Sin3 system (blue dots) and the single-chain NRSF system (red dots), mapped on three-dimensional space by the principal component (3DPC) analysis of the single-chain NRSF structure ensemble 10. The figure corresponds to the free energy landscapes of NRSF with and without Sin3. It is clear that the single-chain NRSF is intrinsically disordered in solution at 300 K, and lacks a dominant structure. It is interesting that structural variety is also observed in the ensemble of NRSF with Sin3.

The picture in Fig. 5A shows the conformational distributions for the NRSF-Sin3 system at 300 K[10]. The dots are projections of structures on the space with the first major three PC axes. The black sphere corresponds to the native complex (NMR model 1[24]). The black dashed lines highlight the free-energy barriers (labeled B1 and B2) separating clusters C1, C2 and C3.The tertiary structures chosen from the distributions are displayed, where "N" indicates the N-terminus of NRSF.

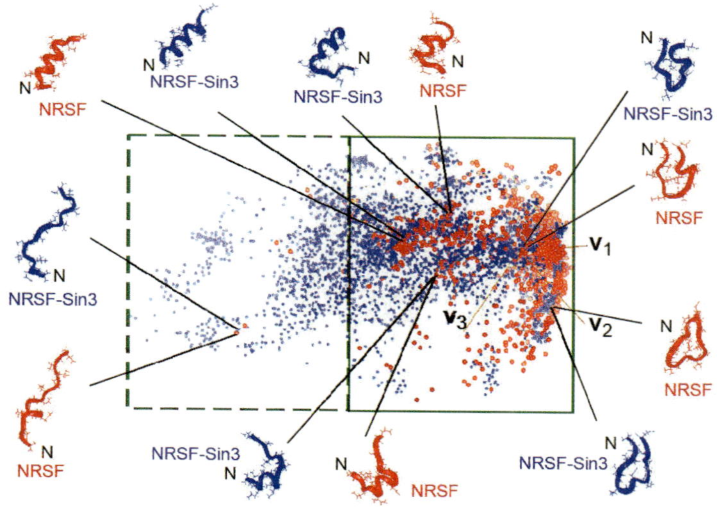

Figure 4: Distribution of NRSF structures from the NRSF-Sin3 system (blue dots with the representative structures identified with NRSF-Sin3) and the single-chain NRSF system (red dots with the representative structures identified with NRSF). The three major PC axes, v1, v2, and v3, are from the single-chain NRSF distribution. A majority of the structures are from the single-chain NRSF distributed in the solid-line rectangle. On the contrary, structures in the dashed-line rectangle originated from the NRSF-Sin3 system, and more various conformations such as the extended conformations were observed.

This free-energy landscape in Fig. 5A for the NRSF-Sin3 system consisted of non-native complexes as well as the native-like complex[10]. The structures from the single-chain NRSF system couple with those from the NRSF-Sin3 system (Fig. 4). Therefore, the various single-chain NRSF structures are adaptable to the Sin3 groove. Since the helix is not the most stable structure in the single-chain NRSF system, it is statistically rare for the native complex to be formed in the first approach to the Sin3 groove. Thus, we conclude that multiple encounter complexes (not a single encounter complex) are probably formed prior to the native complex. The picture in Fig. 5B shows a schematic overview of coupled folding and binding, based on the current study. White arrows represent conformational motions in each of the single-chain NRSF and NRSF-Sin3 systems. In the single-chain NRSF system, population shifts occur in both directions. The black arrows represent NRSF binding to Sin3. The arrow thickness represents transition probabilities assigned to the binding. The arrows in the NRSF-Sim3 complex indicate the conformational changes from the non-native to native complexes. In conclusion, population shifts in the disordered state and induced fit at the binding state were both observed [10].

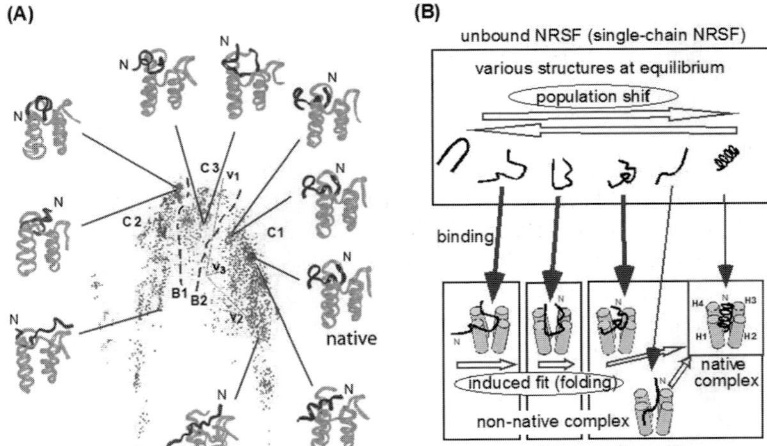

Figure 5: (A) Conformational distributions for the NRSF-Sin3 system at 300 K. Dots are projections of structures of NRSF-Sin3 at 300 K on the 3DPC subspace constructed by the first three major PC axes v1, v2, and v3, which includes the native complex (NMR model 1) as designated by "native". Dashed lines highlight the free-energy barriers with the labels of B1 and B2 separating clusters C1,C2, and C3. Tertiary structures picked from the distributions are displayed, where "N" indicates the N-terminus of NRSF. Black indicates NRSF, and grey, Sin3. (B) Schematic overview of coupled folding and binding. White arrows represent conformational motions in each of the single-chain NRSF and NRSF- Sin3 systems. Black arrows represent NRSF binding to Sin3. The arrow thickness represents the transition probability assigned to the binding.

In conclusion, our *surFit* procedure is able to provide many putative candidates of protein-protein interaction having protein flexibility, to some extent, by applying MD simulations with explicit water molecules as shown by the CAPRI blind prediction exercises. When the ligand molecule shows a large conformational change upon binding like the coupled folding and binding in IDP, the complex structural ensemble can be given by the free energy landscapes of the ligand with the receptor protein. In the future, both approaches can be integrated to provide more reliable and rapid prediction of protein complex structures to reveal the mechanisms of biological signal transduction.

Acknowledgments

J.H. was supported by a Grant-in-Aid for Scientific Research on Innovative Areas (21113006) from MEXT. H. N. was supported by Grants-in-Aid for Scientific Research (B) (23370071) from the Japan Society for the Promotion of Science. J.H. and H.N. were supported by grants from the New Energy and Industrial Technology Development Organization (NEDO), Japan.

References

1. J. Janin, R. P. Bahadur and P. Chakrabarti, *Quart. Rev. Biophys.* **41**, 133 (2008).
2. O. Keskin, A. Gursoy, B. Ma and R. Nussinov, *Chem. Rev.* **108**, 1225 (2008).
3. A. R. Kinjo and H. Nakamura, *J. Mol. Biol.* **399**, 526 (2010).
4. T. Yamada and P. Bork, *Nat. Rev. Mol. Cell Biol.* **10**, 791 (2009).
5. E. Kanamori, Y. Murakami, Y. Tsuchiya, D. M. Standley, H. Nakamura and K. Kinoshita, *Proteins* **69**, 832 (2007).
6. Y. Tsuchiya, E. Kanamori, H. Nakamura and K. Kinoshita, *Adv. Appl. Bioinfo. Chem.* **2**, 79 (2009).
7. J. Janin, K. Henrick, J. Moult, L. T. Eyck, M. J. E. Sternberg, S. Vajda, U. Vakser and S. Wodak, *Proteins* **52**, 2 (2003).
8. S. J. Fleishman *et al. J. Mol. Biol* **414**, 289 (2011).
9. H. J. Dyson and P. E. Wright, *Nat. Rev. Mol. Cell Biol* **6**, 197 (2005).
10. J. Higo Y. Nishimura and H. Nakamura, *J. Ame. Chem. Soc* **133**, 10448 (2011).
11. S. Vajda and D. Kozakov, *Curr. Opin. Struct. Biol* **19**, 164 (2009).
12. S. Liang, C. Zhang and Y. Zhou, *Nucl. Acids Res* **34**, 3698 (2006).
13. I. Mihalek, I. Res and O. Lichtage, *J. Mol. Biol* **336**, 1265 (2004).
14. D. Fischer, S. L. Lin, H. L. Wolfson and R. Nussinov, *J. Mol. Biol* **248**, 459 (1995).

15. T. Ooi, M. Oobatake, G. Nemethy and H. Scheraga, *Proc. Natl. Acad. Sci. USA* **84**, 3086 (1987).
16. K. Kinoshita and H. Nakamura, *Bioinformatics* **20**, 1329 (2004).
17. Y. Tsuchiya, K. Kinoshita and H. Nakamura, *Proteins* **55**, 885 (2004).
18. Y. Fukunishi, Y. Mikami and H. Nakamura, *J. Phys. Chem. B* **107**, 13201 (2003).
19. B. A. Berg and T. Neuhaus, *Phys. Rev. Let* **68**, 9 (1992).
20. U. H. E. Hansmann, Y. Okamoto and F. Eisenmenger, *Chem. Phys. Lett* **259**, 321 (1996).
21. N. Nakajima, H. Nakamura and A. Kidera, *J. Phys. Chem* **101**, 817 (1997).
22. J. Ikebe, K. Umezawa, N. Kamiya, T. Sugihara, Y. Yonezawa, Y. Takano, H. Nakamura and J. Higo, *J. Comput. Chem* **32**, 1286 (2011).
23. J. Higo, J. Ikebe, N. Kamiya and H. Nakamura, *Biophys. Rev* **4**, 27 (2012).
24. M. Nomura, H. Uda-Tochio, K. Murai, N. Mori and Y. Nishimura, *J. Mol. Bio.* **354**, 903 (2005).
25. M. F. Lensink and S. J. Wodak, *Proteins* **78**, 3073 (2010).
26. R. Bao, C-Z. Zhou, C. Jiang, S-X. Lin, C-W. Chi, and Y. Chen, *J. Biol. Chem.* **284**, 26676 (2009).
27. N. A. G. Meenan, A. Sharma, S. J. Fleishman, C. J. MacDonald, B. Morel, R. Boetzel, G. R. Moore, D. Baker and C. Kleanthous, *Proc. Natl. Acad. Sci. USA* **107**, 10080 (2010).
28. A. M. Krachler, A. Sharma and C. Kleanthous, *Proteins* **78**, 2131 (2010).
29. M. Lis, T. Kim, J. J. Sarmiento, D. Kuroda, H. V. Dinh, A. R. Kinjo, K. Amada, S. Devadas, H. Nakamura and D. M. Standley, *Immunome Res* **7**, 1 (2011).
30. A. Wojdyla, S. J. Fleishman, D. Baker and C. Kleanthous, *J. Mol. Biol.* **417**, 79 (2012).

NEW PARADIGMS IN ANTIBODY SPECIFICITY: STRUCTURAL BIOLOGY OF ANTIGEN RECOGNITION BY GERMLINE ANTIBODIES

DINAKAR M. SALUNKE[1,2], HARMEET KAUR[1] AND JASMITA GILL[1]
[1]*Regional Centre for Biotechnology, Gurgaon, India*
[2]*National Institute of Immunology, New Delhi, India*
E-mail: dmsalunke@rcb.res.in

Among the fundamental issues critical for body's defense is the maintenance of self-nonself discrimination. How does the immune system handle a limitless antigenic space and efficiently discriminate within a biological energy 'budget'? Is there a unique antibody for every new antigen that is encountered? Are there strategies built into the immune system to cope with the pathogenic intelligence? Crystallographic studies on antibody diversity have given newer insights into the mechanism of repertoire amplification at the germ line stage offering interesting physiological perspectives on primary immune response. Structural data pertaining to the interactions between germline antibodies and their corresponding antigens have been analysed. These studies redefine antigen recognition within the established norms of structural chemistry bringing out intriguingly new aspects of antigen recognition in humoral antibody response.

1. Introduction

The immune system in all vertebrates is composed of two arms, innate and adaptive, which together act in perfect synchrony to protect an individual from infections and foreign invaders. Hallmark of the adaptive immune system is the remarkable specificity in its recognition repertoire that not only counters the invader foreign substances but also ensures critical distinction of self with respect to the nuances of the non-self. The dynamic nature exhibited by the humoral immune system, a branch of adaptive immunity, and its molecular entity, the antibody, acts as an important arsenal for the surveillance of the body. Antibodies are produced by the plasma cells of the B lymphocyte lineage and then are circulated throughout the body and into secretions, where they bind strongly and specifically to invading organisms and dispose them off through a variety of effector functions.[1]

Antibodies are Y-shaped glycosylated molecules of 150 kDa, composed of two heavy and two light chains of about 50 kDa and 25 kDa, respectively.[2,3] Each arm of the antibody - called the Fab (Fragment antigen binding) because the antigen binding sites are located on them - is ellipsoidal in shape whereas the heart shaped stem has been named the Fc (Fragment crystallizable) because it was found to be highly amenable to crystallization at low ionic strength. The junction of the arm and the stem is known as the hinge region. There are five classes of heavy chains- α, δ, μ, γ and ε. Correspondingly, the antibodies can be classified into different isotypes IgA, IgD, IgM, IgG and IgE on the basis of the class of heavy chain utilized. In case of light chains, there are only two classes – the kappa (κ) and lambda (λ). The Fab consists of the Fv region containing the Complementarity Determining Regions (CDR). Each arm of the Y-shaped antibody molecule, comprising of heavy and light chains, has three CDRs namely, CDR1, CDR2 and CDR3.[4] CDR3 has been seen to play a major role in interaction with antigen. However, the size of an antigen is the primary factor governing the use of various CDR combinations used for interaction.[5]

It is well known that the antigen recognition potential does not reside on the whole of an antibody sequence, but is restricted to the CDRs. While the antibody region defined by the CDRs and involved in antigen recognition is called paratope, that of the antigen interacting with the paratope is called epitope. Crystallographic studies over the past couple of decades have gone a long way in providing insights into epitope-paratope interaction. Hundreds of antigen-antibody complex structures determined so far, delineate the full extent of antibody-antigen recognition. The general principles that have emerged indicate that the complementarity of charge and shape between the paratope and the epitope may be the key determinant of antigen recognition.

2. Dynamics and diversity of host-pathogen interactions

The physico-chemical basis for antigen recognition could be perfectly valid if immune system was designed to be recognizing the well-defined inanimate chemical entities having invariant shapes. However, the antigens or the real foreign invaders are not dead objects but living entities or the components of the living organisms, which are equipped to deal with the host immune system. Since the host defense is against the living objects and not against the inanimate molecular matter, the intelligence, diversity and dynamics of the host-pathogen interaction is therefore important. While the invading organisms will naturally attempt ways by which they are not to be detected or neutralized, the immune system would have also evolved to effectively counter the foreign invasion. It would be interesting to analyse how the host immune system handles the complexities and dynamics of the pathogen encounter.

One of the issues that matters is the handling of the limitless antigenic space. Virtually any chemical entity, ranging from the simplest of the molecules to the most complex of macromolecular structures, can serve as antigenic determinants. If one adds to this the fact that even relatively simple molecules can provide a multiplicity of recognition sites, the latent paradox of a finite receptor repertoire coping with a potentially infinite epitope repertoire becomes self-evident. While the process of affinity maturation can handle this to a great extent, the initial recognition encounter by the naive B cell receptors or the germline antibodies is very critical. Also, one of the ways pathogens tend to evade immune response is to mutate after being recognized by the antibody. If the physico-chemical basis was to be followed, every new shape will have to have different antibody, which is not practical. How does the immune system handle a situation like this? Third issue is that the pathogens tend to use disguise as one of the ways to evade the immune response. If the foreign antigen mimics the self, can the immune system still distinguish them to be different? In other words, is the immune system capable of recognizing subtle differences, if any, between self and non-self at molecular level and still mount an immune response without generating self reactive antibodies or without showing any manifestation of autoimmunity. In order to generate an effective long term response each antigen has to be recognized in the first encounter. Considering that the specificity in terms of the physico-chemical principles is critical for this, the number of antibodies required to neutralize all of them will also have to be infinite, which is not possible.

3. Infinite antigenic space

The issue of the initial encounter while dealing with the infinite antigenic space was extensively addressed at the genetic level. Diversity in the germline antibodies is generated by the imprecise re-arrangement of V, D, J gene segments.[6] Random shuffling and processing of the gene segments by the dynamic genetic system further expands the antibody repertoire to ~ 10^8. Indeed, the genetic recombination event that amplified the BCR repertoire, as suggested by Tonegawa, could not alone encompass the entire antigenic universe, particularly if we were addressing the initial encounter by the "un-exposed" immune system. So, how do the primary antibodies, which have not seen any antigen *a priori*, specifically recognize diverse antigens and initiate subsequent processes for affinity maturation? A possible explanation is that each antibody is capable of binding more than one antigen. Such a scenario was envisaged in the 1940s.[7] Based on computational and thermodynamic studies, it was earlier suggested that diversity is further provided by the conformational variability of the CDRs although it remained to be structurally demonstrated.[8,9] Thus, to generate an ensemble of diverse antibodies to cater to these antigens, degeneracy in antibody specificity is a necessary requirement.

The studies during past few years have elegantly illustrated that the degeneracy of specificity could be handled at structural level providing effective ways to detect the invading antigens at the onset of the immune action. While somatic hypermutations confer the germline antibody an increased affinity and specificity for an optimized response subsequent to the initial encounter, the studies on the structural basis of antibody diversity[10-15] have given newer insights into the mechanism of repertoire amplification at the germ line stage offering interesting physiological perspectives on primary immune response. We have analyzed the critical observations from such structural studies pertaining to the interactions between germline antibodies and their corresponding antigens. All these studies have demonstrated new ways of appreciating the antigen recognition within the established norms of structural chemistry.

4. Paratope flexibility

The antibody combining site needs to be as multifaceted as possible to accommodate the varied stereochemical and electronic features of the antigen. In comparison to numerous crystallographic datasets on mature antibody-antigen complexes, fewer studies have provided insights into the initial encounter of foreign antigens by naive primary antibodies or the germ line antibodies. A study on germline antibody BBE6.12H3 has elucidated the ability of the CDR to adopt different structural states to accommodate the antigen via induced fit mechanism.[10] The structural data comprising of four independent synthetic peptide antigens bound to a germline antibody BBE6.12H3 have provided a rich source of information reflecting correlation of paratope conformational flexibility and polyreactivity (Figure 1).

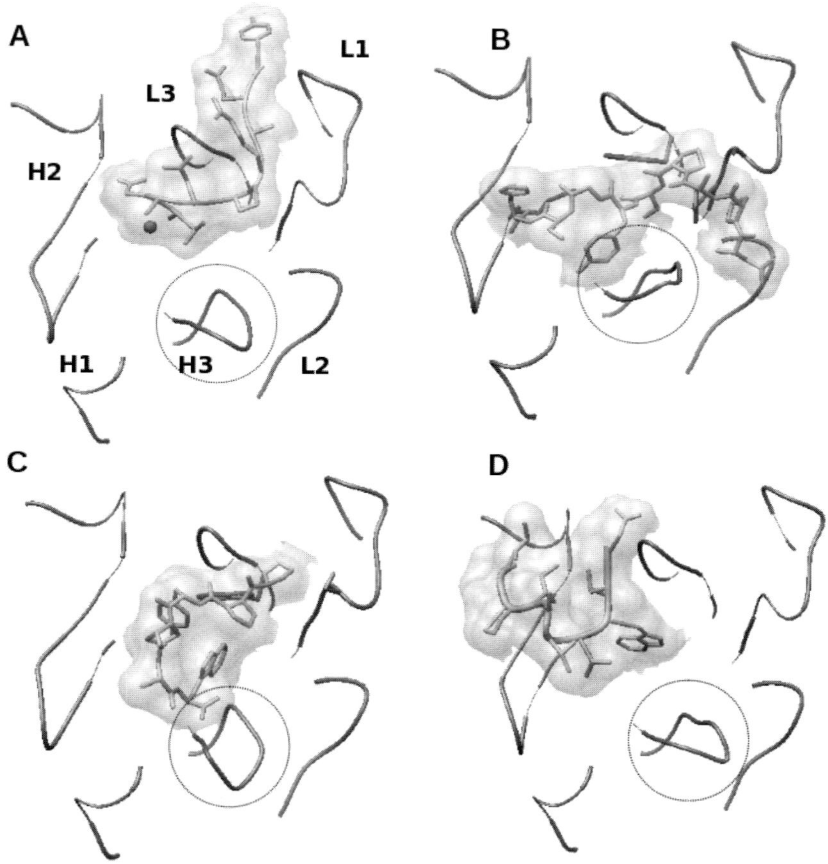

Figure 1: Germline antibody BBE6.12H3 binds to independent peptides via flexibility in CDR H3: A. antibody-Dlw peptide (PDB ID: 2Y36) B. antibody-Gdp peptide (PDB ID: 2Y06) C. antibody-Ppy peptide (PDB ID: 2Y07) and, D. antibody-Yql peptide (PDB ID: 2XZQ). The peptides are shown as sticks and translucent surface.

It was revealed that the hapten binding pocket at the centre of the antigen-combining site can tolerate significant variety in the sequence of the peptide antigens that share no structural similarity in their epitopes though few peptide residues have been seen to bind at common paratope sites. Three such interacting residues form a minimal conserved motif essential for recognition of these multiple antigens.[10] Thus, the germline antibody BBE6.12H3 has a predesigned binding site containing both conserved and non-conserved interacting residues for recognizing multiple antigens, specifically. Despite the extraordinary differences between cognate antigen hapten and the selected peptides, the paratope could accommodate diverse peptide residues by forming new hydrogen bonds. Thus, the previously conceived idea for the requirement of conformational flexibility as a rationale for the discrepancy that exists between the finite number of germline antibodies and infinite number of antigens has been very well elucidated in this work.

Further, conformational rearrangement of CDRH3 has been shown to facilitate hapten binding to germline antibodies 28B4, 48G7, and AZ28[13,16-18]. However, none of these studies have associated conformational flexibility of CDRs with degenerate specificity. Also, in most of the crystallographic studies reported on antigen recognition by germline antibodies involves extensive interactions of the antigen with mainly CDR L3 and CDR H3.[19] However, another recent study[15] has shown an important and exclusive role for CDR L1 in achieving CDR flexibility in germline antibody S64-4 to be able to bind to different antigens (Figure 2). These studies show that a germline antibody utilizes the inherent flexibility of CDRs including Cα and residue side-chain flexibility to accommodate different antigens.

5. Differential juxtaposition

While the adaptation of the paratope through conformational changes in the CDRs generating different binding potentials is an obvious structural solution in order to cope with the diversity of antigenic challenges, an antibody has been seen to adopt an altogether different mechanism, which does not harness the conformational flexibility of the paratope. Crystallographic studies involving germline antibody 36-65 in complex with different synthetic peptide antigens have illustrated a unique mechanism called "differential ligand positioning".[11] It was observed that three diverse peptide epitopes were bound to the germline antibody 36-65 in a surprisingly common paratope conformation (Figure 3). However, each peptide revealed spatially different footprints within the antigen-combining site. The most interesting aspect was the observed variety in the modes of epitope juxtapositioning while interacting with the single conformational state of the paratope.

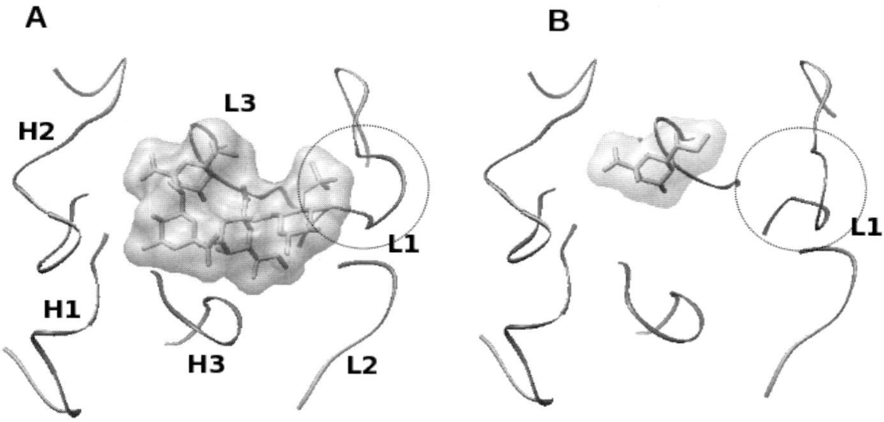

Figure 2: Germline antibody S64-4 binds to independent peptides via flexibility in CDR L1: A. antibody-Kdo derivative (PDB ID: 3PHO) B. antibody-Kdo derivative (PDB ID: 3PHQ). The carbohydrate ligands are shown as sticks and translucent surface.

This has also been observed in another germline antibody interacting with two distinct ligands. Further, germline antibody S25-2 have been shown to utilize only one portion of the antigen-combining site to recognize the terminal Kdo (3-deoxy-D-manno-oct-2-ulosonic acid) a synthetic oligosaccharide, on each of the 12 different epitopes in a similar manner with high specificity, whereas the rest of the combining site adapts to or is excluded from the remainder of the different epitopes.[12] Thus, despite the significant conformational flexibility available to both these antibodies, little change in CDR conformation mainly attributed to varied residue side-chain orientations in the paratope is observed upon binding to a wide range of similar but not identical carbohydrate antigens (Figure 4).

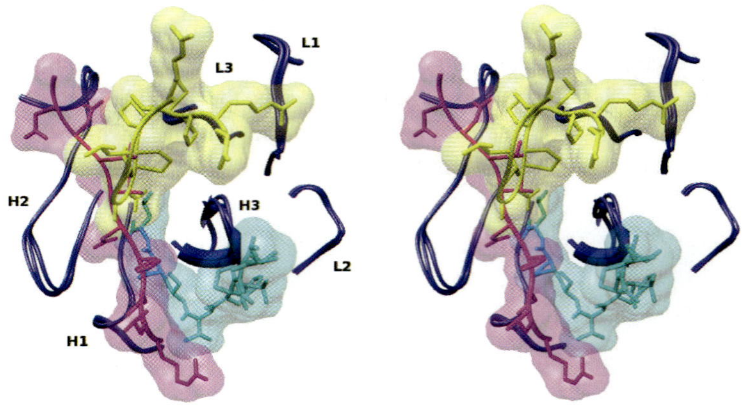

Figure 3: Germline antibody 36-65 binds to independent antigens in common conformation: Stereoscopic view of the superposition of CDRs. The peptides are shown as sticks and translucent surface. Kla (PDB ID: 2A6I) , RII (PDB ID: 2A6D) and Slg (PDB ID: 2A6K) peptides are colored red, yellow and cyan respectively and shown as sticks and translucent surface.

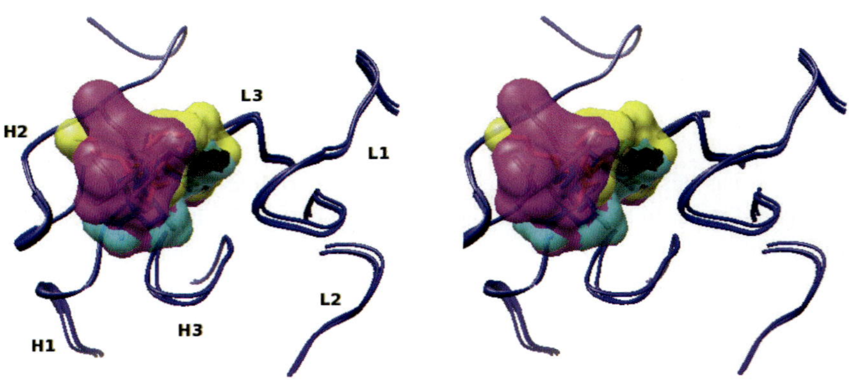

Figure 4: Differential binding of germline antibody S25-2 to four different ligands: Stereoscopic view of the superposition of CDRs. The ligands are shown as sticks and translucent surface and colored red, cyan, yellow and magenta (PDB ID: 3SYO, 3T4Y, 3T65 and 3T77).

These studies substantially contribute to the understanding of degeneracy in antibody specificity by showing that recognition of diverse antigens by a single paratope structure can be possible by compromising on the stringency in the nature of interactions otherwise associated with specific recognition. This study brings to light that a germline antibody paratope need not always act as a malleable template to accommodate diverse antigens.

6. Conclusions

The exclusively demonstrated phenomenon of antibody degeneracy in germline antibodies holds profound significance that beautifully addresses the paradox central to the dynamics of host-pathogen interaction. The crystallographic studies on how a germline antibody can use the different antigen 'read out' modes to productively interact and neutralize a myriad of distinct antigens which it did not have prior cognizance of. The data analyzed above sheds light on the very functioning of the primary antibody repertoire. In order to address the milieu of antigens that intelligently try to evade the surveillance system of the body, antibodies naturally prefer to explore all possible mechanisms that facilitate them with requisite binding affinities towards an encountered antigen. The antibody thus tries to explore all possible ways to expand its germline repertoire suggesting also a paradigm shift in the established notion of antigen recognition. It can be stated that mere physical interaction is not sufficient to cater to the diversity exhibited in the antibody-antigen recognition. Degenerate specificity or polyspecificity in this context is a rescue in real earnest.

It was envisaged that pluripotency of the germline antibody substantially serves the purpose of enhancement in recognition potential in the primary response. With the plethora of antigens in the environment this may not be necessarily limited to germline antibodies alone but can transcend to their matured counterparts as well. Indeed, we now have several instances of polyreactivity structurally demonstrated.[7,20] Molecular mimicry seen in affinity matured antibodies is also a manifestation of this.[21-23] Thus, while expanding the antibody repertoire to account for diverse antigens which have not been exposed to the host immune system, the pluripotency arising due to similar structural plasticity post antibody maturation helps countering the immune evasion.

Acknowledgments

We thank Department of Biotechnology, Government of India for financial support. DMS is a recipient of J. C. Bose National Fellowship of Department of Science and Technology, Government of India.

References

1. N. Maizels, *Annu. Rev. Genet.* **39**, 23 (2005).
2. J.B. Fleischmann, R.R. Porter and E. Press, *Biochem. J.* **88**, 220 (1963).
3. A. Nisonoff, F.C. Wissler and L.N. Lipman, *Science* **132**, 1770 (1960).
4. E.A. Kabat, T. T. Wu and H. Bilofsky, *J. Biol. Chem.* **19**, 6609 (1977).
5. J.A. Caravella, D. Wang, S.M. Glaser and A. Lugovskoy, *Curr. Comput. Aided Drug Des.* **6**, 128 (2010).
6. S. Tonegawa, *Nature* **302**, 575 (1983).
7. L. C. James, P. Roversi and D.S. Tawfik, *Science* **299**, 1362 (2003).
8. L.T. Chong, Y. Duan, L. Wang, I. Massova and P.A. Kollman, *Proc. Natl. Acad. Sci. USA* **96**, 14330 (1999).
9. V. Manivel, N. C. Sahoo, D. M. Salunke and K. V. S. Rao, *Immunity* **13**, 611 (2000).
10. T. Khan and D.M. Salunke, *J. Immunol.* **188**, 1819 (2012).
11. D.K. Sethi, A. Agarwal, V. Manivel , K.V.S. Rao and D.M. Salunke, *Immunity* **24**, 429 (2006).
12. H.P. Nguyen, N.O. Seto, C.R. MacKenzie, L. Brade, P. Kosma, H. Brade and S.V. Evans, *Nat. Struct. Biol.* **12**, 1019 (2003).
13. G. J. Wedemayer, L. H. Wang, P. A. Patten, P. G. Schultz and R. C. Stevens, *J. Mol. Biol.* **268**, 390 (1997).
14. J. Yin, A. E. Beuscher IV, S. E. Andryski, R. C. Stevens and P. G. Schultz, *J. Mol. Biol.* **330**, 651 (2003).
15. D.W. Evans, S. Müller-Loennies, C. L. Brooks , L. Brade, P. Kosma, H. Brade and S. V. Evans, *Glycobiology* **8**, 1049 (2011).
16. G. J. Wedemayer, P. A. Patten, L. H. Wang, P. G. Schultz and R. C. Steven, *Science* **276**, 1665 (1997).
17. J. Yin, E. C. Mundorff, P. L. Yang, K. U. Wendt, D. Hanway, R. C. Stevens and P. G. Schultz, *Biochemistry* **40**, 10764 (2001).
18. L. C. Hsieh-Wilson, P. G. Schultz and R. C. Stevens, *Proc. Natl. Acad. Sci. USA* **93**, 5363 (1996).
19. M. Babor and T. Kortemme, *Proteins* **75**, 846 (2009).
20. Y. Li, H. Li, F. Yang, S.J. Smith-Gill and R.A. Mariuzza, *Nat. Struct. Biol.* **10**, 482 (2003).

21. D. Jain, K. J. Kaur, B. Sundaravadivel and D. M. Salunke, *J. Biol. Chem.* **275**, 16098 (2000).
22. M. Goel, L. Krishnan, S. Kaur, K. J. Kaur and D. M. Salunke, *J. Immunol.* **173**, 7358 (2004).
23. L. Krishnan, S. Lomash, B.P.J. Raj, J. Kaur and D.M. Salunke, *J. Immunol.* **178**, 7923 (2007).

PROTEOMICS, PROTEIN FOLDING AND SELF-ASSOCIATION: NEW INSIGHTS FROM RECENT ADVANCES IN PROTEIN NMR METHODOLOGIES

RAMAKRISHNA V. HOSUR[1,2], DINESH KUMAR[1], SWAGATA CHAKRABORTY[1], MANOJ KUMAR ROUT[1], JITHENDER G. REDDY[1] AND PUSHPA MISHRA[1]

[1]*Department of Chemical Sciences, Tata Institute of Fundamental Research, 1-Homi Bhabha Road, Colaba, Mumbai-400005, India*

[2]*UM-DAE Centre for Excellence in Basic Sciences, University of Mumbai, Kalina Campus, Santa Cruz Mumbai-400098, India*

Nuclear Magnetic Resonance (NMR) spectroscopy has been undergoing unparalleled development, much of which has been driven by applications in structural biology, especially, proteins. We have designed new NMR methods and protocols for rapid chemical shift assignment of protein backbone resonances. These have enabled rapid determination of protein structures, elucidation of folding transitions, and getting valuable insights into protein-protein interaction and self association. These have significant implications for understanding structure-dynamics-function relationships, mechanisms of protein mis-folding, and thus may help in designing of drugs against specific diseases.

1. Introduction

In the current post genomic era, where genome sequences are being determined very rapidly, there is a high demand on establishing structure-dynamics-function relationships in all the proteins constituting the proteome in every species, because of the hope that these may lead to design of drugs and cures for diseases, which have hitherto remained elusive. This constitutes the area of research called as ‘proteomics’ and that has thrown up many challenges: Firstly, a proteome in any species may have anything like 100000 proteins and the structures of all of these will have to be determined, and that too has to be done in a high-throughput manner in a practical time frame for the efforts to yield some tangible benefits. The speed becomes particularly important in cases where the protein is less stable and gets degraded in a matter of days. Secondly, the destiny of a freshly synthesized protein inside a cell can be very uncertain. It may fold to the proper native state, or it may get misfolded leading to various kinds of ordered or disordered aggregates and the mechanisms of these processes are yet to be understood. Finally, unraveling structure-function relationship in proteins often requires study of various protein-protein complexes, and here the difficulties do not scale linearly. The structure-function paradigm has also undergone revision in recent years, and it is now believed that the function of a protein arises as a consequence of interplay between different kinds of species such as unfolded chain, molten globule, pre-molten globule and the folded chain1-6. Each of these species exhibits high degree of internal dynamics at different time scales ranging from milli-seconds to pico-seconds and these have decisive roles in protein function.

One of the most outstanding problems in structural biology is to understand how proteins fold. Post translation, a protein exists as an unfolded/denatured polypeptide chain lacking any well-defined organized three-dimensional structure, though it may possess residual structural propensities. How these polypeptide chains adopt their functional, highly structured states in the intracellular environment has remained one of the most remarkable challenge for scientists over the past few decades. Currently, the most widely accepted view of protein folding is the so-called 'funnel view', also called as the 'energy landscape view'[7-11] (Fig. 1).

This depicts protein folding as a parallel, diffusion-like motion of conformational ensembles on the folding energy landscape inclined towards the native state. The depth in the funnel represents the free energy of the polypeptide chain in fixed conformations and the width indicates the chain entropy. The broad end of the funnel represents the heterogeneous ensemble of unfolded polypeptide chains which rapidly interconvert, while the narrow end represents the homogeneous native state.[12,13] Different members of the ensemble may follow different trajectories to fold/unfold and their energy profiles are different. According to the landscape view, protein folding starts from the unfolded state and may encompass a range of topologies like the pre-molten globule, the molten globule and various other ordered or disordered forms as it gradually folds down the funnel.[14,15]

Nuclear Magnetic Resonance (NMR) spectroscopy has proved to be the technique with the greatest potential to address many of these issues, and the methodology continues to evolve rapidly to handle several of those challenges. Major efforts in this context have focused on (i) enhancing the speed of data acquisition, (ii) designing of new NMR pulse sequences to obtain more instructive data with minimum of experimentation, (iii) development of new strategies and methods to speed up data analysis, remove ambiguities, and be able to handle small as well as large systems, and (iv) automation of data analysis. It is beyond the scope of this article to review the entire vast literature on this subject. Therefore, we limit our discussion here to the work carried out in our own laboratory, but will try to place it in the right perspective.

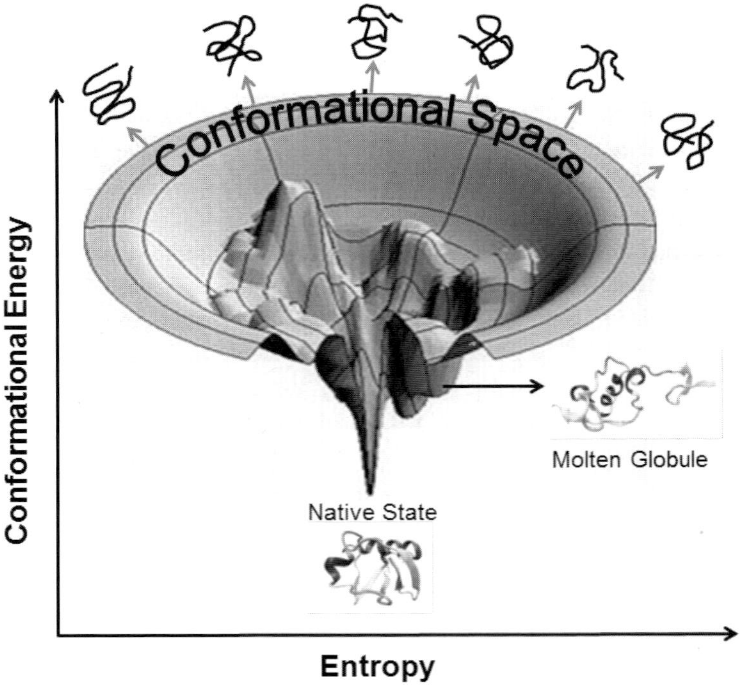

Figure 1: Funnel landscape view or energy landscape view of protein folding, where the broad end reflects the unfolded state which diffuses energetically downhill toward its native state (represented by the narrow end of the funnel) through an unlimited number, essentially a continuum, of intermediates and paths. (Folding Funnel taken from Dill and Chan 1997)

2. High throughput protein NMR

The development of triple resonance multidimensional NMR experiments for proteins in the 1990's revolutionized NMR applications in protein research, and ever since methodological developments are continuing unabated. The Protein Data Bank (PDB), the sole repository of macromolecular structures, has nearly 9500 depositions from NMR data till date. However, as of now, protein structure determination by NMR is a very tedious and arduous process requiring an enormous investment of time and hard-work. The most laborious and time-consuming step in protein structure determination by NMR is the assignment of NOESY cross peaks to particular side chain proton resonances. The process not only requires additional data collection to establish the sequence specific assignment of side chain proton resonances, but also extensive human intervention. This is a serious limitation in the context of proteomics research which demands high-throughput determination of protein structures. Substantial efforts have recently been made both in terms of development of novel NMR methods (for rapid data collection)[16-27] and automation for data analysis to alleviate these difficulties. The NMR methods developed are based on direct sequential correlation of three consecutive residues exploiting good dispersion of 15N chemical shifts, on one hand, and on optimization of various time delays in the pulse sequences, on the other, to minimize experimental acquisition time and to generate specific patterns of peaks in the spectra. In this background we focus here on the methods (HNN, HN(C)N and their variants) developed in our laboratory during the last decade and describe below their salient features in some detail.

The HNN, HN(C)N experiments are three dimensional triple resonance experiments and their pulse sequences employ magnetization transfer pathways along the backbone atoms, as is done in many other pulse sequences, but the implementation of the transfers is such that it results in substantial new features and information in the spectra. The transfer pathways and the schematic spectral patterns are shown in Figure 2.

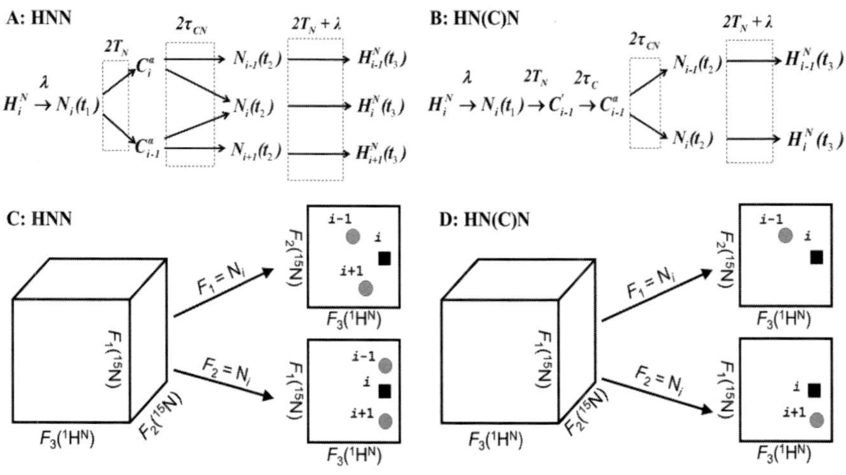

Figure 2: Schematic diagram showing the magnetization transfer pathway in the HNN (A) and HN(C)N (B) experiments. (C) and (D): Schematic three-dimensional HNN and HN(C)N spectra and the correlations observed in the $F_1(^{15}N)$–$F_3(^1H^N)$ and $F_2(^{15}N)$–$F_3(^1H^N)$ planes at the $F_2(^{15}N)$ and $F_1(^{15}N)$ chemical shifts of residue ***i***, respectively. Squares and circles represent diagonal ($F_1 = F_2$) and sequential peaks, respectively. Black and grey colors represent positive and negative peak signs, respectively. Note that HNN spectrum shows ***i*** to ***i+1*** and ***i–1*** correlations in both $F_1(^{15}N)$–$F_3(^1H^N)$ and $F_2(^{15}N)$–$F_3(^1H^N)$ planes, whereas HN(C)N spectrum shows ***i*** to ***i+1*** correlation in $F_1(^{15}N)$–$F_3(^1H^N)$ plane and ***i*** to ***i–1*** correlation in the $F_2(^{15}N)$–$F_3(^1H^N)$ plane.

The most significant feature of these spectra is the occurrence of special patterns of positive and negative peaks in the different planes of the three dimensional spectra, depending upon whether the triplet of consecutive residues correlated by the peaks includes a Gly residue or not (Fig. 3A, B). This makes sequential assignment of backbone atoms along the polypeptide chain very easy and straightforward (illustrated in Fig. 3C).

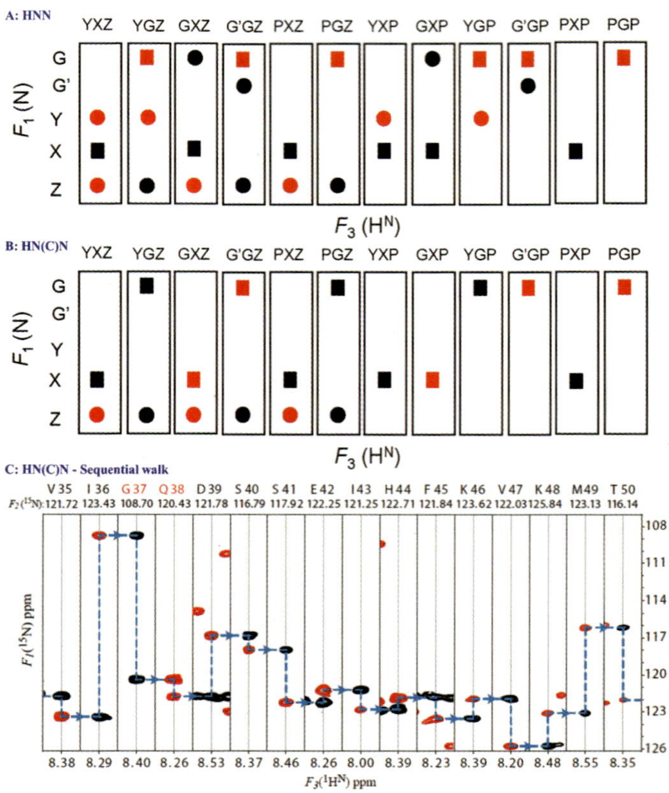

Figure 3: (A) and (B) are the schematic patterns in the F_1–F_3 planes of HNN and HN(C)N spectra, respectively, for various special triplet sequences at the F_2 chemical shift of the central residue in the triplets mentioned on the top of each panel. In all cases, the peaks are aligned at the F_3 ($^1H^N$) chemical shift of the central residue. The letters X and Y represent any amino acid other than glycine and proline, the letter Z represents any amino acid other than proline, G/G' represent glycine, and P represents proline. (C) Sequential walk along the F_1(^{15}N) dimension of HN(C)N spectra recorded on urea denatured SUMO-1 protein. Glycine and its following residue (highlighted in red) are acting as check points during the sequential assignment.

The glycine residues act as anchor points. Several improvisations of these sequences have been developed to make Ser/Thr and Ala look like Glycine, and together they increase the number of anchor points quite substantially, so much so that side chain assignment may even become unnecessary to be able to complete backbone assignment of a given protein. Two dimensional versions of these experiments enable direct identification of Gly/Ser/Thr/Ala peaks and also of peaks of residues following them along the sequence. All the pulse sequences derived thus are summarized in Table 1. Together these have enabled rapid assignment of a large number of proteins, folded as well as unfolded or partially folded under of variety of conditions.[42-48]

Because of the very useful features of the above spectra, it has become possible and has been demonstrated that for small and medium sized proteins (~100 – 120 residues) complete backbone assignment, which includes data acquisition and analysis, can be accomplished in less than a day. Further, analysis of the above spectra is so straight forward that it can be readily automated and this resulted in the algorithm AUTOBA.[49] Success rate of AUTOBA assignment has been demonstrated to be >95% using published assignments in BMRB data base. Extending these applications further, and in conjunction with other algorithms such as TALOS+, CS-ROSETTA[50,51], which allow structure calculation on the basis of backbone chemical shifts only, we have shown as a proof of principle that the structure of a small well-folded protein can be determined in less than two days (Kumar et al, *manuscript under review*). This is indeed a significant achievement in the context of proteomics research.

One of the concerns for the application of the above pulse sequences to large systems is the relatively low sensitivity of these experiments compared to the more conventional and routinely used sequences such as HNCA[29]; transverse relaxation during the long evolution periods causes significant loss of the signal. However, some improvements based on TROSY[52], application of selective excitation strategies (BEST sequences[53]), and non uniform sampling strategies[54] have been suggested to circumvent these limitations.

3. Equilibrium folding transitions

Protein folding is a process by which a polypeptide chain folds into its characteristic and functional three-dimensional structure. A newly synthesized protein chain inside a cell is in some kind of denatured state and this can vary from one cell type to another depending upon the conditions in the intracellular medium. How the chain adopts a functional, highly structured state in the intracellular environment has remained one of the most remarkable challenge for scientists over the past few decades. In many instances, Chaperones have been invoked to be playing roles in directing the protein to its folded state. However, there have been innumerable cases where this does not seem to be necessary; in in-vitro conditions the proteins have been found to be folding to their functionally active states.

Table 1: All the pulse sequences derived from basic HNN & HN(C)N experiments.

Experiments	Information
3D-HNN[28,29] 3D-HNN-A[30] 3D-HNN-ST[31] 3D-BEST-HNN[32]	$(\boldsymbol{i-1}) \leftarrow (\boldsymbol{i}) \rightarrow (\boldsymbol{i+1})$ $^1H^N$ and ^{15}N sequential correlations and generate check points of Glycines/Alanine/Serine and Threonines and following residues.
2D-(HC)NH[33]	Check points highlighted $^1H^N$-^{15}N HSQC type spectra for Alanines, Serines and Threonines.
2D-(HN)NH[32]	$^1H^N$-^{15}N HSQC type spectra showing only Alanines, Serines, Threonines and following residues.
3D-HNN & 2D-NCa (double detection)[34]	$(\boldsymbol{i-1}) \leftarrow (\boldsymbol{i}) \rightarrow (\boldsymbol{i+1})$ $^1H^N$ and ^{15}N correlations in 3D-HNN and $(\boldsymbol{i}) \rightarrow (\boldsymbol{i-1})$ ^{15}N and ^{13}Ca Correlations in 2D-NCa.
3D-HNCAN[35]	$(\boldsymbol{i-1}) \leftarrow (\boldsymbol{i}) \rightarrow (\boldsymbol{i+1})$ $^1H^N$ and ^{13}Ca edited ^{15}N sequential correlations.
3D-HN(C)N[28,29,36] 3D-HN(C)N-A[30] 3D-HN(C)N-J(N-CA)[37]	$(\boldsymbol{i}) \rightarrow (\boldsymbol{i+1})$ $^1H^N$ and ^{15}N sequential correlations and generate check points of Glycines/Alanine/Serine and Threonines and following residues. $^1\boldsymbol{J(N-C\alpha)}$ Coupling Constant information.
3D-hnCOcaNH & 3D-hncoCANH [38]	Provide $(\boldsymbol{i-1}) \rightarrow (\boldsymbol{i})$ sequential correlation along ^{13}CO and $^{13}C\alpha$ dimensions along with $^1H^N$ and ^{15}N chemical shifts.
3D-hNCOcanH[39]	Provide $(\boldsymbol{i-1}) \rightarrow (\boldsymbol{i})$ sequential correlation along amides ^1H dimension along with ^{15}N and ^{13}CO chemical shifts.
2D-hNcocanH & 2D-hncocaNH-G/A/S&T [40]	$^1H^N$-^{15}N HSQC type spectra showing $(\boldsymbol{i-1}) \rightarrow (\boldsymbol{i})$ sequential correlations and check points highlighting only Alanines, Serines, Threonines and following residues.
2D-hnCOcanH & 2D-hncoCAnH [41]	$(\boldsymbol{i}) \rightarrow (\boldsymbol{i+1})$ $^1H^N$ - ^{13}CO and $^1H^N$ - $^{13}C\alpha$ sequential Correlations and generate check points of Glycines, Alanines, Serines, Threonines and following residues.

In the landscape view of protein folding described above, what pathway a particular chain in the ensemble will follow depends upon the energy barriers along the way, and such kinetically driven structural transitions may or may not coincide with thermodynamically driven transitions, that is, those transitions where the final state is an energetically more stable state. In those fortuitous situations where the energy barriers to the thermodynamically stable state are low, the protein will approach that state and such equilibrium transitions can be probed by NMR methods. Intuitively, it is not unreasonable to expect that a large fraction of the molecules in the ensemble would undergo such transitions, since that way the thermodynamic principles will be most easily satisfied.

Even when the protein chain goes through equilibrium structural transitions during the folding process, the various intermediates along the pathway may be short-lived for detailed structural characterization. There is no easy way to trap the molecule in such intermediate states so that a detailed structural characterization can be carried out. An approximation that can be adopted is to use to some denaturants and by using them in a controlled manner the molecules can be stabilized via some states midway between the fully unfolded and native states. While these states would correspond to some point on the folding funnel, one hopes that these may represent the intermediates in some pathways, which, any way are innumerable.

Using the above strategy, we have been able to characterize by NMR methods[44] the most probable equilibrium structural transitions that occur in SUMO-1 protein when starting from a urea denatured state step-wise folding is caused by dilution of urea concentration. This is schematically shown in Figure 4. At 8M urea the polypeptide chain has largely β preferences. With urea dilution, the protein progressively loses majority of the broad β-domain structural preferences, acquires some helical propensities at 5 M urea and then loses some of them again on further dilution of urea. At 3M urea the chain has lost all preferences. These observations contrast the conventional notion that proteins fold with monotonously increasing native type preferences. For folding below ~ 3 M urea, the region around the α1 helix appears to be a potential folding initiation site. The folding seems to start with a collapse into native like topologies, at least in parts, and is followed by secondary and tertiary structure formation, perhaps by cooperative rearrangements. This is evidenced by the motional characteristics of the protein which show maximum conformational exchange contributions in those areas which are proximal in the final folded state. Similar detailed characterization of equilibrium folding transitions occurring along the acetic acid dilution–driven folding coordinate of HIV-1 protease (PR) has also been recently reported.[55] Here, a comparative study of the wild type protein and a mutant protein, PR_{D25N}, was carried out to investigate the role of mutations in the folding characteristics of the protein. It was observed that though the conformational and dynamics preferences of PR and PR_{D25N} are fairly similar in 9M acetic acid, they seem to undergo different folding transitions when acetic acid concentration is slowly reduced. The differences are seen in the active site, in the flap, and in the hinge of the flap regions. One can extrapolate that such differences, though different in detail, would occur for other mutations as well, and also for different initial denatured states. These observations would have important implications for understanding effects of mutations on efficacies of drug binding to the protein and thus to design of new drugs.

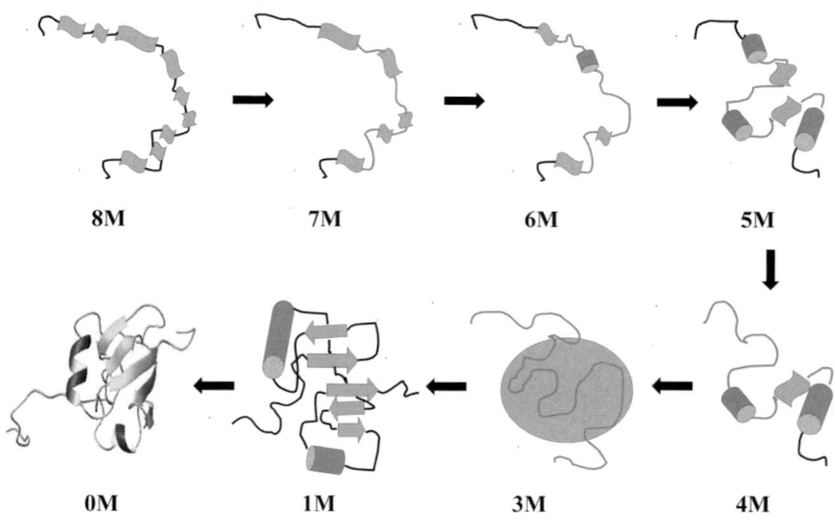

Figure 4: Schematic representation of most probable equilibrium structural transitions of SUMO-1 protein caused by step-wise dilution of urea concentration from 8M Urea- denatured state.

4. Protein-protein association

Protein-protein association to form large molecular species is an important step in many cellular functions.[56,57] Examples include, ribosome, chaperone, proteasome, chromatin, multi-subunit proteins, dynamin assembly, dynein motor assembly, to name a few. The biological functions arise as a consequence of controlled and coordinated assembly formation, and thus association-dissociation processes would serve as regulatory mechanisms. There are other instances where uncontrolled associations lead to dysfunctions which turn out to be the causes of certain diseases. Recently, it has been recognized that proteins, in general, can form fibrils – which are huge self-assemblies – under certain experimental conditions.[58-60] Large soluble oligomers are precursors to such fibrils and presumably, it is the precursors which are responsible for the disease conditions.[61,62]

Structural characterization of such large assemblies, which holds the clues for understanding their biological functions, has remained a difficult task. From the NMR point of view, the primary difficulty arises from the fact that the line widths are extremely large, so much so that the signals are hardly visible above the noise. The largest assembly from which signals have been observed in NMR till date is the 900 kDa GroEL-GroES chaperone complex.[63]

In this background we recently described an approach to characterize self association pathways and thus obtain structural insights into very large megadalton-size molecular species. The basic idea here is to first dissociate the assembly into the smaller monomers by use of external agents, and then reassemble them in a controlled manner by removal of the external agent, and as this association progresses systematically, different segments of the protein undergo exchange processes which in turn lead to differential line broadening along the polypeptide chain; structural characterization of the intermediate species provides insights into the structure of the assembly. The advantage here would be that while monitoring the signals of the monomer, we would be deriving information on the multimeric species. This strategy is schematically shown in Figure 5A. Clearly, such a strategy would not depend on the size of the assembly, and moreover, would provide high resolution detail into the structures of the molecules in the assembly.

Following the above strategy, we reported the self association pathway of GTPase effector domain (GED) of dynamin, an important protein in the endocytic machinery[64-68]; GED is responsible for self association of dynamin into huge spiral ropes which wrap around the budding vesicles during the endocytosis process. In this investigation, the HNN and HN(C)N suite of experiments were extremely crucial for obtaining sequence specific assignments at every intermediate stage of association. Figure 5 (B and C) shows schematically the structural changes that occur when association was induced starting from two different initial denatured states, one created by dimethyl

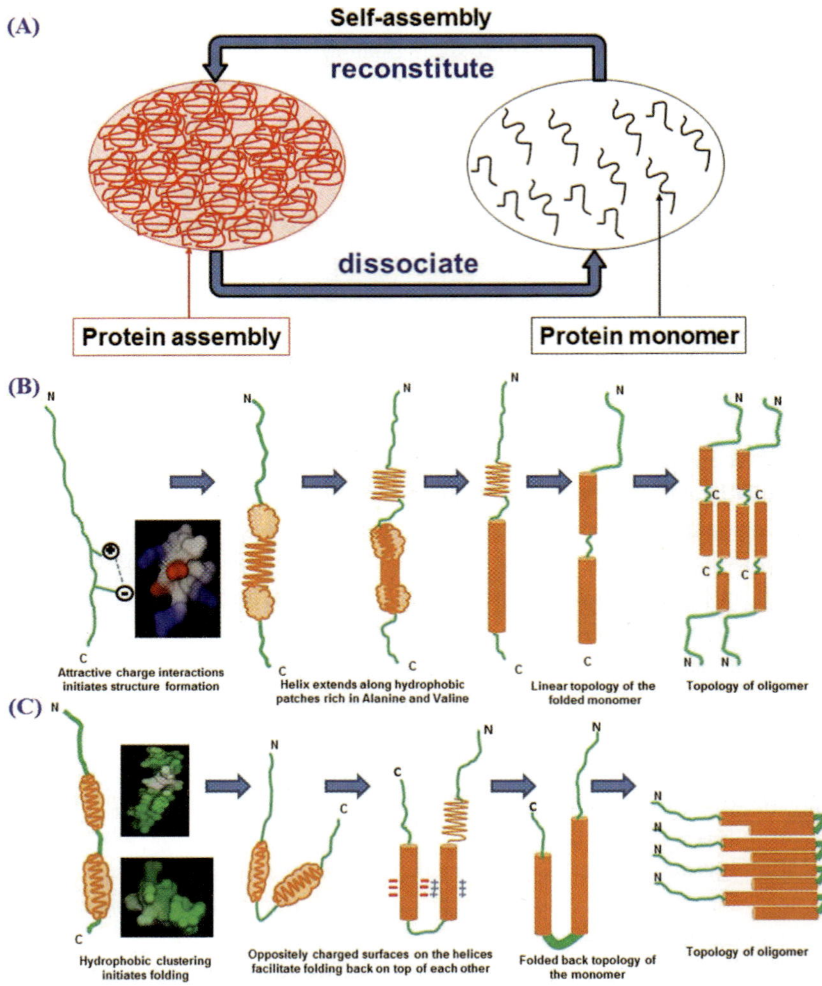

Figure 5: (A) Strategy for studying megadalton size aggregates. (B and C) Schematic representation of the structural changes that occur when association is induced starting from two different initial denatured states, one created by dimethyl sulfoxide (DMSO) (B) and the other created by guanidine hydrochloride (Gdn-HCl) (C)[64] (figure reproduced with permission).

sulfoxide (DMSO) (Fig. 5B) and the other by guanidine hydrochloride (Gdn-HCl) (Fig. 5C). It is observed that the initial transitions are different in the two cases, but finally the protein seems to reach the huge associated state in which the dominant structural feature is the α-helix. It is, however, not possible at this stage to know the details of packing of the helices. Even so, basing on available data from other biochemical results on inter-domain interactions in dynamin and on the results on GED association, some speculative suggestions have been made for dynamin self-assembly.[64]

Similarly, in the case of P2 protein from Plasmodium falciparum, which is seen to form higher order structures, insights into mechanism of folding starting from urea denatured state have been obtained, which in turn have yielded insights into the self-association mechanism of the protein.[47] It is observed that 'hydrophobic collapse' is the initial event in the folding of this protein.

5. Concluding remarks

This article is a survey of the exciting developments in NMR methodologies in our laboratory, which have enabled application of NMR to problems which had remained inaccessible for the last several decades. Structures of small proteins can be determined in less than two days. Folding transitions can be elucidated at residue level detail which show that structural changes occur continuously and one cannot truly talk about two-state, three-state models etc. Large molecular assemblies and aggregates can be studied and the process of association can be investigated to unravel mechanisms and the driving forces. All of these would have significant implications for understanding protein function and take NMR applications in structural biology to greater heights than have been achieved before.

References

1. Schulz, G. E., *Nucleotide Binding Proteins*, 79 (1979).
2. Pontius, B. W., *Trends in biochemical sciences* **18**, 181 (1993).
3. Plaxco, K. W.; Gross, M., *Nature* **386**, 657 (1997).
4. Spolar, R. S.; Record, M. T., Jr., *Science* **263**, 777 (1994).
5. Dobson, C. M., *Nature* **426**, 884 (2003).
6. Dobson, C. M., *Nature reviews. Drug discovery* **2**, 154 (2003).
7. Bryngelson, J. D.; Onuchic, J. N.; Socci, N. D.; Wolynes, P. G., *Proteins* **21**, 167 (1995).
8. Onuchic, J. N.; Luthey-Schulten, Z.; Wolynes, P. G., *Annual review of physical chemistry* **48**, 545 (1997).
9. Shoemaker, B. A.; Wang, J.; Wolynes, P. G., *Journal of molecular biology* **287**, 675 (1999).
10. Wolynes, P. G., *Philosophical transactions. Series A, Mathematical, physical, and engineering sciences* **363**, 453 (2005).
11. Dill, K. A.; Chan, H. S., *Nature structural biology* **4**, 10 (1997).
12. Dobson, C. M.; Karplus, M., *Current opinion in structural biology* **9**, 92 (1999).
13. Dyson, H. J.; Wright, P. E., *Methods in enzymology* **394**, 299 (2005).

14. Dunker, A. K.; Brown, C. J.; Obradovic, Z., *Advances in protein chemistry* **62**, 25 (2002).
15. Uversky, V. N., *European journal of biochemistry / FEBS* **269**, 2 (2002).
16. Atreya, H. S., *Journal of the Indian Institute of Science* **90**, 87 (2009).
17. Atreya, H. S.; Szyperski, T., *Methods in enzymology* **394**, 78 (2005).
18. Kupce, E.; Freeman, R., *J Am Chem Soc* **125**, 13958 (2003).
19. Kupce, E.; Freeman, R., *J Magn Reson* **163**, 56 (2003).
20. Kupce, E.; Freeman, R., *Journal of biomolecular NMR* **25**, 349 (2003).
21. Lee, K. M.; Androphy, E. J.; Baleja, J. D., *Journal of biomolecular NMR* **5**, 93 (1995).
22. Lescop, E.; Schanda, P.; Brutscher, B., *J Magn Reson* **187**, 163 (2007).
23. Pervushin, K.; Riek, R.; Wider, G.; Wuthrich, K., *Proceedings of the National Academy of Sciences of the United States of America* **94**, 12366 (1997).
24. Riek, R.; Pervushin, K.; Wuthrich, K., *Trends in biochemical sciences* **25**, 462 (2000).
25. Tugarinov, V.; Muhandiram, R.; Ayed, A.; Kay, L. E., *J Am Chem Soc* **124**, 10025 (2002).
26. Jaravine, V. A.; Zhuravleva, A. V.; Permi, P.; Ibraghimov, I.; Orekhov, V. Y., *J Am Chem Soc* **130**, 3927 (2008).
27. Bonvin, A. M.; Houben, K.; Guenneugues, M.; Kaptein, R.; Boelens, R., *Journal of biomolecular NMR* **21**, 221 (2001).
28. Bhavesh, N. S.; Panchal, S. C.; Hosur, R. V., *Biochemistry* **40**, 14727 (2001).
29. Panchal, S. C.; Bhavesh, N. S.; Hosur, R. V., *Journal of biomolecular NMR* **20**, 135 (2001).
30. Chatterjee, A.; Kumar, A.; Hosur, R. V., *J Magn Reson* **181**, 21 (2006).
31. Chugh, J.; Kumar, D.; Hosur, R. V., *Journal of biomolecular NMR* **40**, 145 (2008).
32. Kumar, D.; Paul, S.; Hosur, R. V., *J Magn Reson* **204**, 111 (2010).
33. Chugh, J.; Hosur, R. V., *J Magn Reson* **194**, 289 (2008).
34. Chakraborty, S.; Paul, S.; Hosur, R. V., *Journal of biomolecular NMR* **52**, 5 (2012).
35. Rout, M. K.; Mishra, P.; Atreya, H. S.; Hosur, R. V., *J Magn Reson* **216**, 161 (2012).
36. Chatterjee, A.; Bhavesh, N. S.; Panchal, S. C.; Hosur, R. V., *Biochemical and biophysical research communications* **293**, 427 (2002).
37. Bhavesh, N. S.; Chatterjee, A.; Panchal, S. C.; Hosur, R. V., *Biochemical and biophysical research communications* **311**, 678 (2003).
38. Kumar, D.; Reddy, J. G.; Hosur, R. V., *J Magn Reson* **206**, 134 (2010).
39. Kumar, D.; Hosur, R. V., *Magnetic resonance in chemistry : MRC*, (2011).
40. Kumar, D.; Hosur, R. V., *Research Communication:Current Science* **99**, 1581 (2010).
41. Kumar, D.; Borkar, A.; Hosur, R. V., *Magnetic resonance in chemistry : MRC* **50**, 357 (2012).
42. Juneja, J.; Bhavesh, N. S.; Udgaonkar, J. B.; Hosur, R. V., *Biochemistry* **41**, 9885 (2002).
43. Chatterjee, A.; Mridula, P.; Mishra, R. K.; Mittal, R.; Hosur, R. V., *The Journal of biological chemistry* **280**, 11369 (2005).
44. Kumar, A.; Srivastava, S.; Kumar Mishra, R.; Mittal, R.; Hosur, R. V., *Journal of molecular biology* **361**, 180 (2006).

45. Chugh, J.; Chatterjee, A.; Kumar, A.; Mishra, R. K.; Mittal, R.; Hosur, R. V., *The FEBS journal* **273**, 388 (2006).
46. Mohan, P. M.; Hosur, R. V., *Biochimica et biophysica acta* **1784**, 1795 (2008).
47. Mishra, P.; Das, S.; Panicker, L.; Hosur, M. V.; Sharma, S.; Hosur, R. V., *PloS one* **7**, e36279 (2012).
48. Rout, M. K.; Hosur, R. V., *Archives of biochemistry and biophysics* **482**, 33 (2009).
49. Borkar, A.; Kumar, D.; Hosur, R. V., *Journal of biomolecular NMR* **50**, 285 (2011).
50. Shen, Y.; Lange, O.; Delaglio, F.; Rossi, P.; Aramini, J. M.; Liu, G.; Eletsky, A.; Wu, Y.; Singarapu, K. K.; Lemak, A.; Ignatchenko, A.; Arrowsmith, C. H.; Szyperski, T.; Montelione, G. T.; Baker, D.; Bax, A., *Proceedings of the National Academy of Sciences of the United States of America* **105**, 4685 (2008).
51. Shen, Y.; Vernon, R.; Baker, D.; Bax, A., *Journal of biomolecular NMR* **43**, 63 (2009).
52. Pervushin, K.; Riek, R.; Wider, G.; Wuthrich, K., *J Am Chem Soc* **120**, 6394 (1998).
53. Schanda, P.; Van Melckebeke, H.; Brutscher, B., *J Am Chem Soc* **128**, 9042 (2006).
54. Rovnyak, D.; Frueh, D. P.; Sastry, M.; Sun, Z. Y. J.; Stern, A. S.; Hoch, J. C.; Wagner, G., *J Magn Reson* **170**, 15 (2004).
55. Rout, M. K.; Reddy, J. G.; Phillips, M.; Hosur, R. V., *Journal of Biomolecular Structure and Dynamics* **(in press)**, (2012).
56. Hinshaw, J. E., *Annual review of cell and developmental biology* **16**, 483 (2000).
57. Kentsis, A.; Borden, K. L., *Current protein & peptide science* **5**, 125 (2004).
58. Kumar, S.; Mohanty, S. K.; Udgaonkar, J. B., *Journal of molecular biology* **367**, 1186 (2007).
59. Goto, Y.; Yagi, H.; Yamaguchi, K.; Chatani, E.; Ban, T., *Current pharmaceutical design* **14**, 3205 (2008).
60. Hosur, R. V.; Chugh, J., *Journal of Indian Chemical Socity* **87**, 43 (2010).
61. Ono, K.; Hirohata, M.; Yamada, M., *Current pharmaceutical design* **14**, 3247 (2008).
62. Yamin, G.; Ono, K.; Inayathullah, M.; Teplow, D. B., *Current pharmaceutical design* **14**, 3231 (2008).
63. Fiaux, J.; Bertelsen, E. B.; Horwich, A. L.; Wuthrich, K., *Nature* **418**, 207 (2002).
64. Chakraborty, S.; Hosur, R. V., *Protein and peptide letters*, (2012).
65. Chakraborty, S.; Pratihar, S.; Hosur, R. V., *PloS one* **7**, e30109 (2012).
66. Chakraborty, S.; Hosur, R. V., *Journal of molecular biology* **405**, 1202 (2011).
67. Chakraborty, S.; Hosur, R. V., *Biomolecular NMR assignments* **5**, 59 (2011).
68. Chugh, J.; Sharma, S.; Hosur, R. V., *Protein science : a publication of the Protein Society* **17**, 1319 (2008).

COMPUTATIONAL APPROACHES FOR UNDERSTANDING THE RECOGNITION MECHANISM OF PROTEIN COMPLEXES

M. MICHAEL GROMIHA

Department of Biotechnology, Indian Institute of Technology Madras, Chennai 600036, India

Understanding the recognition mechanism of proteins with other molecules is a challenging task in molecular and computational biology. Experiments provide a wealth of data on binding specificity in terms of dissociation constant and binding free energy change upon protein complex formation. Computationally, the three-dimensional structures of protein complexes have been used to delineate the binding sites and to understand their recognition mechanisms. We have developed an unified energy based approach for identifying the binding site residues in protein–protein, protein-RNA and protein-DNA complexes. In protein-protein complexes, the residues and residue-pairs with charged and aromatic side chains are found to be important for binding. These residues influence to form cation–π, electrostatic and aromatic interactions. In protein-RNA complexes, the positively charged, polar and aromatic residues are important for binding. These residues influence to form electrostatic, hydrogen bonding and stacking interactions. The positive charged and polar residues prefer to bind with DNA in protein-DNA complexes. Our observations showed a good agreement with the experimental binding specificity of protein-protein and protein-nucleic acid complexes.

1. Introduction

The interactions of proteins with other molecules, such as proteins, DNA, RNA, carbohydrate and ligand are important for most of the cellular processes such as signal transduction, gene expression and control, antibody-antigen complex, gene regulation, DNA replication and repair, packaging and so on. Understanding the recognition mechanism of protein complexes is a challenging task in computational and molecular biology, which would aid for function prediction and drug design.

The availability of protein complex structures in Protein Data Bank[1] enables researchers to identify, analyze and develop methods to predict the binding sites using various physicochemical and energetic features of amino acid residues. Protein-protein interactions at molecular level have been analyzed in terms of amino acid residues at the interface, conservation of residues, electrostatic interactions, hydrogen bonds, solvent accessibility, conformational changes and docking simulations[2-10]. The development of methods for understanding the factors, which govern protein-protein interactions have been reviewed from different perspectives such as experimental techniques, development of databases, identification of binding sites and prediction of protein-protein and domain interactions[11-20]. In addition, protein-nucleic interactions have been viewed from the interacting pattern between amino acid residues and nucleotides. The tendency of amino acid residues to form non-covalent interactions with nucleotide bases in protein-DNA/RNA complexes have been brought out[21-28]. The interactions are classified into two different types, namely, (i) intermolecular interactions (or direct readout) due to specific interactions between amino acid side chains and DNA bases as well as non-specific interactions involving main chain atoms of proteins and backbone of DNA[29], and (ii) intramolecular interactions (or indirect readout), which are mainly due to the conformational changes of DNA and/or water mediated contacts[30-32]. Gromiha et al.[33] revealed the contributions of both inter and intramolecular interactions using different types of protein-DNA complexes.

The binding sites in protein complexes have been identified mainly with a criteria based on the contacts between the partners in protein complexes[6,34-41]. The atomic contacts between any heavy atoms in a pair of proteins or protein-RNA/DNA within a specific distance between 3.5 and 6.0Å have been used to assign the contacts. These criteria include the repulsive interactions in which two residues or residue-nucleotide pairs are close to each other. In addition, these pairs with different distances have been treated as the same manner. On the other hand, solvent accessibility based methods have also been proposed for identifying the binding sites[3,42].

We have proposed an approach based on interaction energy for defining the binding sites in protein-protein, protein-RNA and protein-DNA complexes[43-47]. The binding sites are identified with the same cutoff energy in all these complexes. We observed that the binding sites are dominated with aromatic and charged residues in protein-protein complexes, indicating the importance of electrostatic, aromatic and cation-π interactions. In protein-RNA complexes, the binding sites are dominated with positively charged, aromatic and polar residues, showing the importance of electrostatic, hydrogen bonding and aromatic interactions. The preference of charged and polar residues are noticed in protein-DNA complexes. The computational results have been compared qualitatively with the experimental binding free energy change upon complex formation and found a good agreement between them. Further, specific examples will be discussed in detail.

2. Materials and methods

2.1. *Dataset of protein complex structures*

Three different datasets have been developed for protein-protein, protein-RNA and protein-DNA complexes with 153, 81 and 212 structures, respectively. These complexes were obtained from the Protein Data Bank (1) with two major conditions: (i) non-redundant (sequence identity is less than 25% and 35% in protein-protein and protein-nucleic acid complexes, respectively) and (ii) the structures are solved at high resolution. In protein-protein complexes we considered only heterodimers. The missing regions are less than 10% and there should be a minimum of 50 amino acid residues and 5 nucleotides in proteins and nucleic acids, respectively. In addition, all structures were manually checked to construct clean datasets[43-49].

2.2. *Computation of interaction energy*

We have calculated the interaction energy between atoms in protein complexes using AMBER potential[50]. It is given by:

$$E_{\text{inter}} = \Sigma\ [(A_{ij}/r_{ij}^{12} - B_{ij}/r_{ij}^{6}) + q_i q_j/\varepsilon r_{ij}] \qquad (1)$$

where $A_{ij} = \varepsilon_{ij}^{*}(R_{ij}^{*})^{12}$ and $B_{ij} = 2\ \varepsilon_{ij}^{*}(R_{ij}^{*})^{6}$; $R_{ij}^{*} = (R_i^{*} + R_j^{*})$ and $\varepsilon_{ij}^{*} = (\varepsilon_i^{*}\ \varepsilon_j^{*})^{1/2}$; R^{*} and ε^{*} are, respectively, the van der Waals radius and well depth and these parameters are obtained from Cornell et al.[50]; q_i and q_j are, respectively, the charges for the atoms i and j, and r_{ij} is the distance between them. We have used the distant dependent dielectric constant ($\varepsilon = r_{ij}$) to take account of the dielectric damping effect of the Coulomb interactions, as used in other studies on protein complexes[51].

2.3. *Binding propensity*

The binding propensity for the 20 amino acid residues in protein-protein, protein-RNA and protein-DNA complexes has been developed as follows: we have computed the frequency of occurrence of amino acid residues in binding sites (f_b) and in the protein as a whole (f_t). The binding propensity (P_{bind}) is calculated using the equation[52]:

$$P_{\text{bind}}(i) = f_b(i)/f_t(i) \qquad (2)$$

where, i represents each of the 20 amino acid residues.

2.4. *Experimental binding free energy change*

We have constructed thermodynamic databases for proteins and mutants, ProTherm[53,54], protein-nucleic acid interactions, ProNIT[54,55] and protein-protein interactions, PINT[56] using experimental thermodynamic data reported in the literature along with sequence and structural information, experimental conditions and literature information. The experimental binding free energy change upon complex formation for protein-nucleic acid and protein-protein complexes has been taken from ProNIT and PINT databases, respectively.

3. Results and discussion

3.1. *Binding site residues in protein complexes*

In a protein complex, we have computed the interaction energy (Eqn. 1) of each atom in a protein with all atoms in the partner molecule (protein, RNA or DNA). The contributions of all atoms in a residue have been summed up to obtain the interaction energy of each residue. We used a cutoff of -1 kcal/mol and the residues, which have the interaction energy of less than -1 kcal/mol are identified as binding site residues. This common criteria showed the presence of 10.8%, 7.6% and 8.7% of residues as binding site residues in protein-protein, protein-RNA and protein-DNA complexes, respectively[43,45,47].

3.2. *Binding propensity of residues in protein complexes*

We have computed the binding propensity of protein-protein, protein-RNA and protein-DNA complexes using Eqn. 2. We observed that the residues, Phe, His, Arg, Trp and Tyr have remarkably high contribution in protein-protein complexes, showing the preference to form electrostatic, hydrophobic, cation-π and aromatic interactions. On the other hand, the residues Arg, His, Lys, Asn, Tyr, Gln and Ser highly contribute for the interaction between protein and RNA. Interestingly, these residues belong to positively charged, polar and aromatic groups, which form electrostatic, hydrogen bonds and aromatic interactions with RNA. In protein-DNA complexes, positive charged residues are more dominant than protein-protein and protein-RNA complexes. Further, polar residues prefer to interact with DNA in the form of hydrogen bonds[46].

3.3. *Atomic contribution in protein complexes*

In proteins, we classified the atoms into seven groups, three for the main chain (C, N, and O) and four for the side chain (C, N, O, and S), and computed their energetic contributions (Figure 1a). The contributions of side chain atoms are twice to that of main chain atoms in proteins, which indicates the importance of specific interactions such as electrostatic, aromatic and cation-π interactions formed by the side chain/base atoms of partner proteins/DNA/RNA[46]. In DNA/RNA, the atoms are grouped into P, O, and C in main chain (backbone) and N, O, and C in side chain (base). The contribution of phosphate is remarkably higher in DNA than RNA. Interestingly, the contribution of main chain atoms is stronger than side chain atoms in both protein-DNA and protein-RNA complexes (Figure 1b).

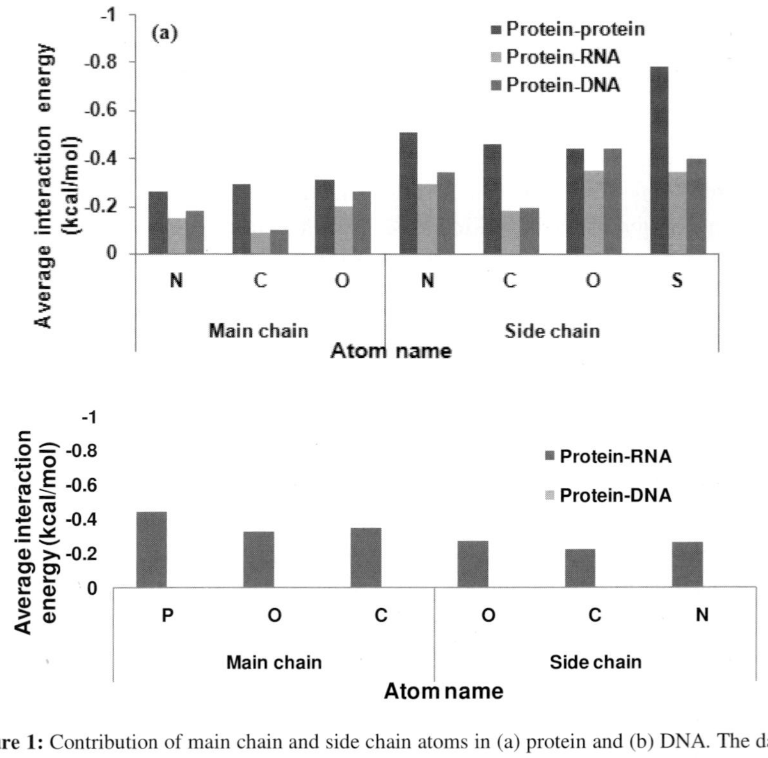

Figure 1: Contribution of main chain and side chain atoms in (a) protein and (b) DNA. The data for protein-protein, protein-RNA and protein-DNA complexes are shown.

3.4. *Comparison with experiments*

We have compared the results obtained in this work with experimental binding free energies of protein-protein and protein-nucleic acid complexes deposited in ProNIT and PINT databases[55,56]. The number of mutants with specific interactions in protein complexes are presented in Table 1.

Table 1
Comparison of experimental binding specificity in protein complexes

Condition	Protein-protein	Protein-RNA	Protein-DNA
Binding free energy change	>2kcal/mol	>1kcal/mol	>1kcal/mol
Total interactions	217	424	405
Unique interactions	67	72	244
Specific interactions	charged: 38 +ve charge and aromatic: 32 hydrophobic: 7	charged: 19 aromatic: 19 polar: 16 others: 18	+ve charged: 83 aromatic: 34 polar: 34 others: 93
Comments	cation-π aromatic electrostatic	electrostatic aromatic HB	electrostatic, HB, aromatic DNA conformation

HB: hydrogen bond; others: mutations with similar chemical behavior.

In protein-protein interactions, we observed that majority of the mutants, which severely change the binding free energy are due to the replacements with aromatic and positive charged residues. On the other hand, only seven mutants are hydrophobic residues. This result demonstrates the importance of electrostatic, cation-π and aromatic interactions for the recognition of protein-protein complexes[43].

In protein-RNA complexes, majority of the mutants are due to the replacement of aromatic, charged and polar residues, indicating the importance of electrostatic, hydrogen bonds and aromatic interactions for the recognition of protein-RNA complexes[45].

The influence of electrostatic, hydrogen bonding and aromatic interactions are prevalent in protein-DNA interactions[47]. On the other hand, appreciable number of amino acid residues with similar chemical behavior also showed the binding free energy change of 1 kcal/mol, which is mainly due to the conformational change of DNA. The results obtained with the computational analysis revealed the importance of these interactions, showing good agreement with experiments.

3.5. *Specific examples*

We have explored the importance of specific interactions for the recognition of protein-protein, protein-RNA and protein-DNA complexes and specific examples are discussed below.

(a) Protein-protein

Eletr and Kuhlman[57] measured the binding free energies of 49 mutants in E6AP-UbcH7 complex (PDB code: 1C4Z) and showed that the replacement of Phe63Ala altered the binding free energy of 3 kcal/mol. Our analysis showed that Phe63 in UbcH7 makes a strong aromatic interaction with Tyr694 in E6AP and the interaction free energy is -1.2 kcal/mol, indicating the importance of aromatic-aromatic interactions (Figure 2a).

Zhang et al.[58] carried out binding experiments on interleukin4 receptor binding protein complex (1IAR) and most of the hotspots involved cation-π interaction forming residues (aromatic and positively charged). An example for this interaction between Tyr127 in interleukin4 and Arg85 in receptor protein is shown in Figure 2b and the substitution of Tyr127Ala changed the binding free energy to 2.2 kcal/mol.

Kiel et al.[59] studied the thermodynamic behavior of binding in Ras-Rap complex and the hotspot residues involved mainly the charged residues, Lys, Arg and Asp. Figure 2c shows the importance of electrostatic interactions formed between the residues Lys32 and Asp238 in Ras and Rap, respectively.

(b) Protein-RNA

Shiels et al.[60] measured the binding free energies of mutants with the replacement of Phe56 with other aromatic and hydrophobic residues in U1A protein-RNA complex (1URN). The substitutions with non-aromatic hydrophobic residue Leu heavily lost the binding affinity whereas with other aromatic residues (Trp/Tyr) did not show any significant effect in binding affinity, which demonstrates the importance of aromatic interactions. Our analysis showed that Phe56 in U1A protein makes a strong aromatic interaction with A11 in RNA and the major contribution is between the side chain atoms of Phe and base atoms of A (Figure 3a).

Robert and Brakier-Gingras[61] carried out binding experiments on ribosomal protein S7-RNA complex (1VS5) and reported that the affinity is decreased with the mutations of charged residues. Our analysis showed the contribution of electrostatic interactions between Arg3 in ribosomal protein S7 and C932 in RNA (Figure 3b). We have also noticed that the electrostatic interactions made by the charged residues Lys35 and Lys113 with RNA are reported to play important roles in the recognition of ribosomal protein S7-RNA complex[61].

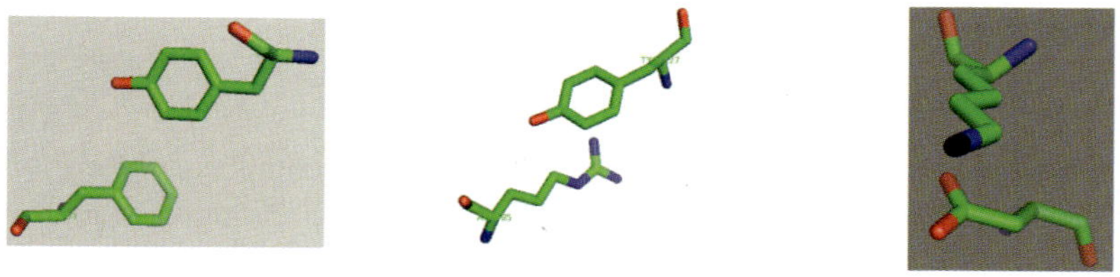

Figure 2: Specific interactions in protein-protein complexes[43]: (a) E6AP-UbcH7 complex, (b) Interleukin4 receptor binding protein and (c) Ras-Rap.

Figure 3: Interactions between amino acid residues and nucleotides in protein-RNA complexes[45]: (a) U1A protein-RNA and (b) ribosomal protein S7-RNA.

(c) Protein-DNA

Larkin et al.[62] measured the binding free energies of wild type and 26 mutants in F factor relaxase-DNA complex (2A0I) and showed that R237A changed the binding free energy of 4.39 kcal/mol. Our analysis revealed a strong electrostatic interaction between the side chain atoms of Arg237 and main chain atoms in T8 (Figure 4a).

Figure 4: Interactions between amino acid residues and nucleotides in protein-DNA complexes[47]: (a) F factor relaxase-DNA complex and (b) Telomere binding proteins Cdc13-DNA complex.

Anderson et al.[63] carried out binding experiments on telomere binding protein Cdc13-DNA complex (1S40) and reported that the replacement of Tyr85 caused severe loss of binding. We observed the presence of aromatic interactions between Tyr85 and T4 as seen in Figure 4b.

3.6. *Probable mechanism for recognition*

The contributions of different interactions in protein complexes along with experimental binding free energy change prompted to propose a probable mechanism for recognition[43,45,47]. The available experimental and computational data suggested that the recognition may be initiated mainly with specific residues as binding with single residue segments are dominant in protein complexes (data not shown). It has been reported that the initiation may take place through charged and aromatic residues in the form of electrostatic and aromatic interactions. The role of conformational changes of DNA is also vital in the case of protein-DNA complexes. Once the initiation takes place the polar residues tend to form hydrogen bonds between the partners. It is observed that the hydrogen bonds are more prevalent in protein-nucleic acid complexes than protein-protein complexes. Finally, the complexes are stabilized with hydrophobic and other interactions. It may be noted that binding is also altered with concentrations of ions, salt conditions, environment etc. which are not considered in the present study.

Conclusions

We have proposed a common method to identify the binding sites residues in protein-protein, protein-RNA and protein-DNA complexes using an energy based approach. The identified residues have been interpreted in terms of their importance to form non-covalent interactions such as electrostatic, hydrogen bond, cation-π, aromatic-aromatic and hydrophobic interactions. The results obtained with computational analysis have been compared with experimental binding free energy change upon mutations in protein-protein, protein-RNA and protein-DNA complexes. We observed a good agreement between computational analysis and experimental data. Based on the results obtained in the study, we proposed plausible mechanisms for understanding the recognition of protein complexes.

Acknowledgments

The author thanks Prof. B. Jayaram, Dr. K. Fukui and Dr. S. Selvaraj for helpful discussions.

References

1. H.M. Berman, G. J. Kleywegt, H. Nakamura and J. L. Markley, *Structure* **20**, 391 (2012).
2. Y. Ofran and B. Rost, *J Mol Biol.* **325**, 377 (2003).
3. M. Guharoy and P. Chakrabarti, *Proc. Natl. Acad. Sci. USA,* **102**, 15447 (2005).
4. F. B. Sheinerman and B. Honig, *J Mol Biol.* **318**, 161 (2002).
5. M. F. Lensink and R. Méndez, *Curr Pharm Biotechnol.* **9**, 77 (2008).
6. M. Sikić, S. Tomić and K. Vlahovicek, *PLoS Comput Biol.* **5**, e1000278 (2009).
7. A. Shulman-Peleg, R. Nussinov and H. J. Wolfson, *J Mol Biol.* **339**, 607 (2004).
8. M. M. Gromiha, N. Saranya, S. Selvaraj, B. Jayaram and K. Fukui, *Proteome Sci.* **9**, Suppl 1:S13 (2011).
9. M. Tyagi, R. R. Thangudu, D. Zhang, S. H. Bryant, T. Madej and A. R. Panchenko, *PLoS One.* **7**, e28896 (2012).
10. S. Ahmad and K. Mizuguchi, *PLoS One.* **6**, e29104 (2011).
11. A. Valencia and F. Pazos, *Curr Opin Struct Biol.* **12**, 368 (2002).
12. L. Salwinski and D. Eisenberg, *Curr Opin Struct Biol.* **13**, 377 (2003).
13. S. J. Wodak and R. Mendez, *Curr Opin Struct Biol.* **14**, 242 (2004).
14. P. Aloy, M. Pichaud and B. B. Russell, *Curr Opin Struct Biol.* **15**, 15 (2005).
15. B. A. Shoemaker and A. R. Panchenko, *PLoS Comput. Biol.* **3**, e42 (2007).
16. B. A. Shoemaker and A. R. Panchenko, *PLoS Comput. Biol.* **3**, e43 (2007).
17. O. Keskin, A. Gursoy, B. Ma and R. Nussinov, *Chem. Rev.* **108**, 1225 (2008).
18. N. Tuncbag, G. Kar, O. Keskin, A. Gursoy and R. Nussinov, *Brief. Bioinform.* **10**, 217 (2009).
19. I. Ezkurdia, L. Bartoli, P. Fariselli, R. Casadio, A. Valencia and M.L. Tress, *Brief. Bioinform.* **10**, 233 (2009).
20. R. D. Gorham Jr, C. A. Kieslich and D. Morikis, *Ann. Biomed. Eng.* **39**, 1252 (2011).
21. S. Jones, D. T. Daley, N. M. Luscombe, H. M. Berman and J. M. Thornton, *Nucleic Acids Res.* **29**, 943 (2001).
22. M. Treger and E. Westhof, *J. Mol. Recognit.* **14**, 199 (2001).
23. J. Allers and Y. Shamoo, *J. Mol. Biol.* **311**, 75 (2001).
24. A. Sarai and H. Kono, *Annu. Rev. Biophys. Biomol. Struct.* **34**, 379 (2005).
25. B. Jayaram, K. McConnell, S. B. Dixit, A. Das and D. L. Beveridge, *J. Comput. Chem.* **23**, 1 (2002).
26. M. M. Gromiha, C. Santhosh, and S. Ahmad, *Int. J. Biol. Macromol.* **34**, 203 (2004).
27. R. Rohs, S. M. West, A. Sosinsky, P. Liu, R. S. Mann and B. Honig, *Nature* **461**, 1248 (2009).

28. P. Zhou, P. Tian, Y. Ren, J. Zou and Z. Shang, *J. Chem. Inf. Model.* **50**, 1476 (2010).
29. H. Kono and A. Sarai, *Proteins: Struct. Funct. Genet.* **35**, 114 (1999)
30. W. K. Olson, A. A. Gorin, X. J. Lu, L. M. Hock, and V. B. Zhurkin, *Proc. Natl. Acad. Sci. USA*, **95**, 11163–11168. (1998).
31. M. M. Gromiha, *J. Biotechnol.* **117**, 137 (2005).
32. B. Jayaram and T. Jain, *Annu. Rev. Biophys. Biomol. Struct.* **33**, 343 (2004).
33. M. M. Gromiha, J. G. Siebers, S. Selvaraj, H. Kono and A. Sarai, *J. Mol. Biol.* **337**, 285 (2004).
34. S. Ahmad, M. M. Gromiha and A. Sarai, *Bioinformatics.* **20**, 477 (2004).
35. N. Bhardwaj and H. Lu, *FEBS Lett.* **581**, 1058 (2007).
36. J. Wu, H. Liu, X. Duan, Y. Ding, H. Wu, Y. Bai, and X. Sun, *Bioinformatics.* **25**, 30 (2009)
37. A. Koike and T. Takagi, *Protein Eng Des Sel* **17**, 165 (2004).
38. Y. Ofran and B. Rost, *Bioinformatics.* **23**, e13 (2007).
39. M. Terribilini, J. D. Sander, J. H. Lee, P. Zaback, R. L. Jernigan, V. Honavar and D. Dobbs, *Nucleic Acids Res.* **35**, W578 (2007).
40. L. Wang and S. J. Brown, *Nucleic Acids Res* **34**, W243 (2006).
41. M. Kumar, M. M. Gromiha and G. P. S. Raghava, *Proteins* **71**, 189 (2008).
42. P. Chakrabarti and J. Janin, *Proteins.* **47**, 334 (2002).
43. M. M. Gromiha, K. Yokota and K. Fukui, *Mol. Biosyst.* **5**, 1779 (2009).
44. M. M. Gromiha, K. Yokota and K. Fukui, *Int. J. Biol. Macromol.* **46**, 187 (2010).
45. M. M. Gromiha, K. Yokota and K. Fukui, *Curr. Protein Pept. Sci.* **11**, 629 (2010).
46. M. M. Gromiha, S. Selvaraj, B. Jayaram and K. Fukui, *Lecture Notes Comp. Sci.* **6215**, 626 (2010).
47. M. M. Gromiha and K. Fukui, *J Chem Inf Model.* **51**:721 (2011).
48. R. P. Bahadur, M. Zacharias and J. Janin, *Nucleic Acids Res.* **36**, 2705 (2008).
49. B. Xu, Y. Yang, H. Liang and Y. Zhou, *Proteins.* **76**, 718 (2009).
50. W. D. Cornell, P. Cieplak, C. I. Bayly, I. R. Gould, K. M. Merz, D.M. Ferguson, D. C. Spellmeyer, T. Fox, J. W. Caldwell and P. A. Kollman, *J Amer Chem Soc* **117**, 5179 (1995).
51. F. Pichierri, M. Aida, M. M. Gromiha and A. Sarai A, *J. Amer. Chem. Soc.* **121**, 6152 (1999).
52. M. M. Gromiha, Protein Bioinformatics: From Sequence to Function, Academic Press/Elseviers, (2010).
53. M. M. Gromiha, J. An, H. Kono, M. Oobatake, H. Uedaira and A. Sarai, *Nucleic Acids Res.* **27**, 286 (1999).
54. M. D. Kumar, K. A. Bava, M. M. Gromiha, P. Prabakaran, K. Kitajima, H. Uedaira and A. Sarai, *Nucleic Acids Res.* **34**, D204 (2006).

55. P. Prabakaran, J. An, M. M. Gromiha, S. Selvaraj, H. Uedaira, H. Kono and A. Sarai, *Bioinformatics*. **17**, 1027 (2001).
56. M. D. Kumar and M. M. Gromiha, *Nucleic Acids Res.* **34**, D195 (2006)
57. Z. M. Eletr and B. Kuhlman, *J Mol Biol*. **369**, 419-28 (2007)
58. J. L. Zhang, I. Simeonowa, Y. Wang and W. Sebald, *J. Mol. Biol.* **315**, 399 (2002).
59. C. Kiel, L. Serrano and C. Herrmann, *J Mol Biol*. **340**: 1039 (2004).
60. J. C. Shiels, J. B. Tuite, S. J. Nolan and A. M. Baranger, *Nucleic Acids Res.* **30**, 550 (2002).
61. F. Robert and L. Brakier-Gingras, *Nucleic Acids Res*. **29**:677 (2001).
62. C. Larkin, S. Datta, M. J. Harley, B. J. Anderson, A. Ebie, V. Hargreaves and J. F. Schildbach, *Structure*. **13**, 1533 (2005).
63. E. M. Anderson, W. A. and D. S. Wuttke, *Biochemistry*. **42**, 3751 (2003).

UNIVERSALITIES IN PROTEIN TERTIARY STRUCTURES: SOME NEW CONCEPTS

B.JAYARAM[1,2*], ADITYA MITTAL[1*], AVINASH MISHRA[1], CHANCHAL ACHARYA[1] AND GARIMA KHANDELWAL[2]

[1]*Kusuma School of Biological Sciences, Indian Institute of Technology Delhi Hauz Khas, New Delhi 110016, India.*

[2]*Department of Chemistry & Supercomputing Facility for Bioinformatics & Computational Biology, Indian Institute of Technology Delhi Hauz Khas, New Delhi 110016, India.*

[*]*E-mail: bjayaram@chemistry.iitd.ac.in; amittal@bioschool.iitd.ac.in*

We discuss the recent extraction of signatures of stoichiometry driven universal spatial organization of backbones of folded proteins regardless of their size, shape/structure and function. We present further evidence for secularity of amino acids in protein structures from the perspectives of surface area and energy. While conceptual fragmentation to gain insights into the diversity of protein structures appears to be a popular approach, we believe that the secrets to solving the protein folding problem lie in appreciating concepts that are universally applicable.

1. Introduction

Historically, the ideas of Pauling[1-3] and Ramachandran[4] proposed in 1950s and 1960s established universality among secondary structures in proteins. Pauling's work led to our understanding that proteins, irrespective of their structure and function, are made up of regular secondary structural elements called alpha helices and/or beta sheets and irregular regions connecting these called loops with Ramachandran's work providing a raison d'etre in terms of a stereochemical interpretation for these. Each secondary structural element (alpha helix or beta sheet), is characterized by a well-defined allowed region in the dihedral angle (ϕ, ψ) space of the backbones of proteins. Not surprisingly, even the so called "irregular regions", i.e. the loops, assume either helical or sheet like dihedral values[5]. A protein consisting of n peptide linkages shows up as n points in the 2D- (ϕ, ψ) Ramachandran plot, exhibiting a clustering of points as per its secondary structural composition. Structural studies via crystallography and NMR (RCSB[6]) have verified the hypotheses of Pauling and Ramachandran time and again.

How to extend these ideas of commonalities among proteins to tertiary structures remains a pending question however. A specification of the 2n Ramachandran angles (n ϕs and n ψs) leads to a coarse-grained description of the tertiary structure of a protein. The overall conformation of a protein thus corresponds to a single point in this 2n dimensional Hyper-Ramachandran plot. Of course, one could add more dimensions to account for the degrees of freedom associated with the side chains. If free energy is added to this 2n-dimensional surface, native structure of the protein, according to Anfinsen[7], corresponds to the bottom most point or the global minimum in free energy on this (2n+1) dimensional surface. Thus ensued several proposals on energy landscapes which overall conform to the concept of minimum free energy for the native structure[8-10]. Several other physico-chemical parameters based on size, shape, area, energy (intrinsic as well as transfer) have been investigated[11-14] but few universal ideas applicable to all proteins have emerged. This has led to extensive classifications of both secondary and tertiary structures of proteins such as various flavors of helices, turns, super-secondary structural motifs, folds etc.[15-19]. The corollary of all such classifications which chronicle the architectural splendor of proteins is to give up on universality.

We revisit proteins from a new perspective here and show evidence for very compelling clues to the existence of some universal principles, not on the folding pathways but, on the organization of protein tertiary structures. We hope that the perspective presented here paves the way for re-embracing unifying principles of protein structures rather than develop numerous subtle fragmentations.

2. Stoichiometry

It has been demonstrated recently[20-24] that amino acid space of proteins is not infinite rather proteins have well defined stoichiometries with bounds set on amino acid compositions (Table 1) as seen from the sequence data available in Swissprot/Uniprot[25]. The compositions are non-random and the deviations from the averages (called the margin of life[20]) account for the diversity of proteins.

It is logical to expect that these stoichiometries are the essence of protein size, shape, structure and function. In fact, a sequence analysis reveals the requirement of all 20 amino acids in all known protein sequences, as shown in Figure 1. This, coupled with the complete absence of naturally occurring anagramic protein sequences, strongly supports the idea that protein structure and function is governed by the stoichiometric ratios of amino acids.

Table 1. Margin of Life.

AMINO ACID	Protein sequences confirmed by annotation and experiments (Mean ± Std. dev.; n = 131855)
A	7.2 ± 3.0
V	6.3 ± 2.1
I	5.1 ± 2.2
L	9.6 ± 2.9
Y	3.0 ± 1.5
F	3.9 ± 1.8
W	1.2 ± 0.9
P	5.4 ± 2.6
M	2.2 ± 1.1
C	1.9 ± 2.3
T	5.5 ± 1.8
S	7.9 ± 2.8
Q	4.3 ± 2.0
N	4.2 ± 1.9
D	5.2 ± 1.9
E	6.8 ± 2.8
H	2.4 ± 1.3
R	5.3 ± 2.9
K	6.0 ± 2.9
G	6.6 ± 2.8

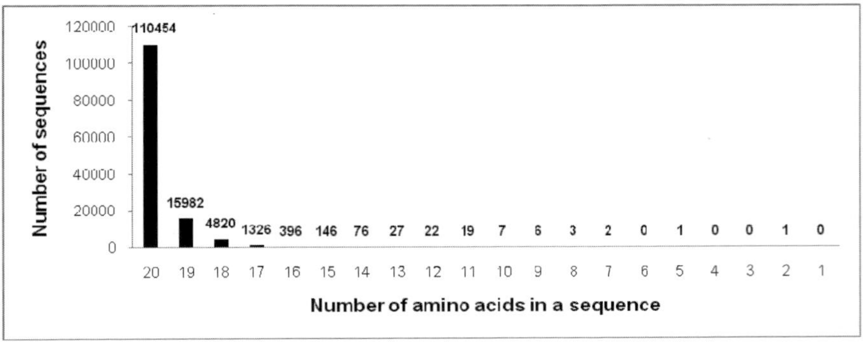

Figure 1: In the 133388 sequences (> 75 amino acids) examined, all the 20 amino acids occur in 82.87%, only 19 in 11.99%, 18 in 3.62% and 17 in 0.99% of the sequences.

3. Sizes and Shapes

3.1 Radius of gyration

Physical chemistry of polymers is rich with scaling laws, the most celebrated one being that of Flory[26-29]. Root mean square end to end distances and equivalently the radius of gyration, R_G, varies as N^{ν} where N is the number of monomeric units and ν is the scaling exponent. It is now accepted that ν = 0.60 for homopolymers in good solvents, ν = 0.33 for homopolymers in poor solvents, while for proteins, values of ν ranging from 0.30 to 0.40 have been observed (Figure 2).). It may be noted that ν = 1.00 for a fully extended chain, ν = 0.50 for random chains, the lower the ν value, the higher the compaction or exclusion by solvent[20, 21]. The low values of ν seen for proteins clearly point to a 'precipitation' type phenomenon upon folding.

Many of these early discoveries on radii of gyration transited to advancing hydrophobicity and fractal dimensions for proteins and did not stay focused on molecular origins of such high levels of compaction.

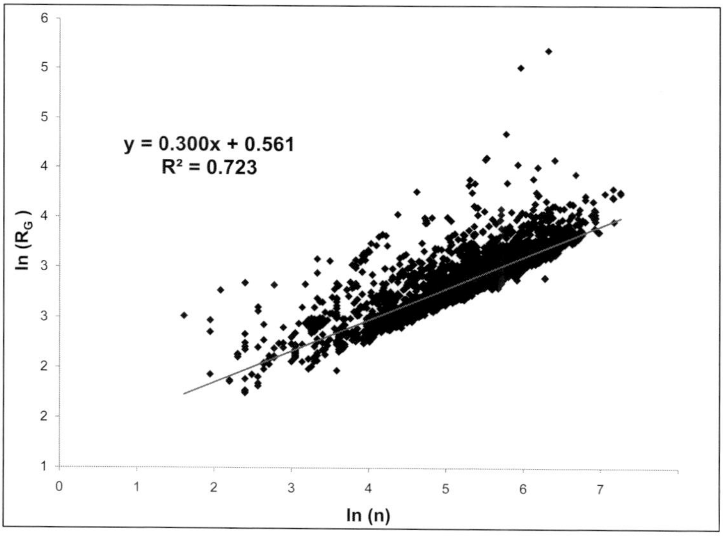

Figure 2: Radius of gyration plotted against number of residues as a log-log plot for ~ 6750 proteins. Proteins are seen to be extremely compact compared to random chains and synthetic polymers in good solvents. In the parlance of Flory, water is not a "good solvent" for proteins.

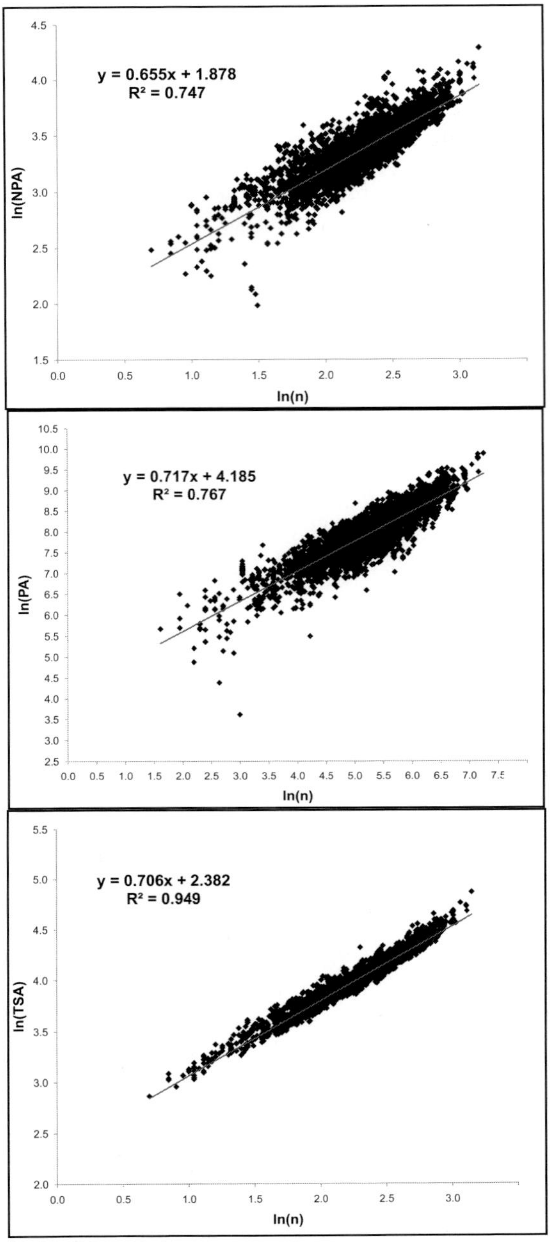

Figure 3: Solvent accessible surface areas Nonpolar (top panel), polar (middle panel), total (bottom panel) versus number of residues (n) in ~6750 proteins shown as log-log plots.

3.2 Surface area

It is believed for long that folded proteins follow a simple axiom: hydrophobic residues 'in' and hydrophilic 'out'. Structural analyses of 1000 globular proteins[30-31] suggested that the ratio of loss in surface area of nonpolar residues to that of polar residues is close to unity (~ 1.1). The axiom is thus obeyed more as an exception than as a rule. The nonpolar and polar areas of proteins have been examined hundreds of times.

Figure 3 demonstrates a simple fact that it does not profit to divide the areas into nonpolar or polar to understand the molecular phenomena. Total area (with a very high correlation coefficient of 0.95) serves as a simple and elegant variable to estimate the solvent exposed surface area of proteins. Furthermore, it is strongly suggestive of an inherent shape/area conservation principle or an invariant area per residue metric.

4. Spatial Distribution of Backbone Cα Atoms

A recent study[20], reported that regardless of the size, shape and composition of proteins, the spatial distribution of their backbone Cα atoms obey a simple sigmoidal equation: $Y = Y_{Max}(1-e^{-kX})^n$, Y is the cumulative number of Cα neighbors, X is the radial coordinate and, n and k are two parameters characterizing the sigmoid. It may be expected just as the (ϕ, ψ) space of proteins for secondary structures, that n and k should display allowed regions for tertiary structures of proteins.

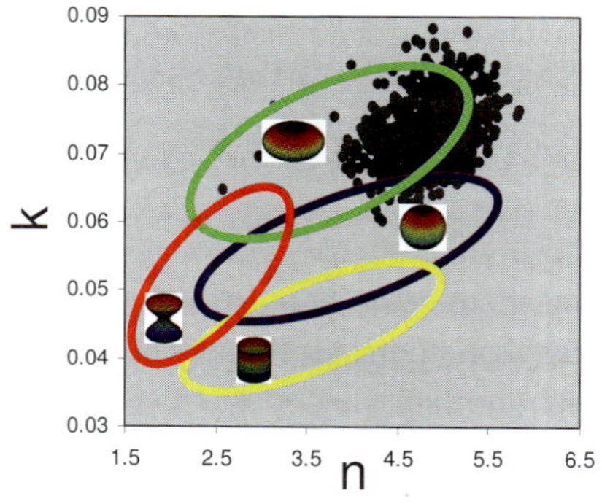

Figure 4: n, k values of 13,550 proteins superposed on allowed n, k values for some regular solid geometrical objects. Regardless of the structural classification of the proteins, and regardless of the amino acid neighbourhoods investigated, Cα neighbourhoods in folded proteins form a definite cluster in the n-k space. Proteins structures fit ellipsoidal geometries better[32-33].

A computer simulation was carried out on solid objects with idealized geometries (sphere, cylinder, ellipsoids, hourglass etc.)[32]. The n,k space of each was characterized (Figure 4). This was followed by an analysis of 13550 crystal structures. Considering each of the amino acids in the crystal structures individually, the geometrical analysis gave rise to neighbourhood data for a given protein backbone (i.e. Cα coordinates) in the form of a 20x20 matrix at each neighbourhood distance and the n, k values were extracted from this data. The above simulation results and the crystal structure analyses provide a direct and definitive proof confirming the earlier observations[20-22] that soluble protein indeed have a universal spatial organization regardless of size, fold, structure and function.

5. Energetics: Secularity

Empirical energy functions with an implicit consideration of solvent have been in vogue for quite some time in the protein structure prediction field. The functional form and parameters therein are commonly anchored to a well-established force field. One such all atom based energy function originating in Amber force field reported by us earlier[34] accounts for all non-bonded interactions in proteins including van der Waals, electrostatics and solvation. It was seen that this function was able to separate the native from thousands of decoys in 67 of the 69 protein systems studied[34], thus demonstrating that it captures the essential features of a free energy function. The total energies of 6750 proteins are shown in Figure 5. The high correlation obtained (0.99) and the stability of the energy per residue clearly imply the compensatory nature of the diverse intra- and inter-molecular energy components and the resultant secularity or energetic equivalence of amino acids across a large number of tertiary structures of proteins of varying sizes and sequence compositions.

6. Perspectives and Conclusions

A preliminary application of the above ideas was implemented (BhageerathH www.scfbio-iitd.res.in/bhageerath/bhageerath_h.jsp) during the first ever successful completion of an Indian entry for tertiary structure prediction of proteins in CASP10 (1st May to 17th July, 2012). CASP - Critical Assessment of Techniques for Protein Structure Prediction (http://predictioncenter.org/casp10/) refers to a worldwide biennial experiment to gauge the current abilities in the area of protein tertiary structure prediction. Initiated in the year 1994 by the University of Maryland under the leadership of John Moult, CASP has provided

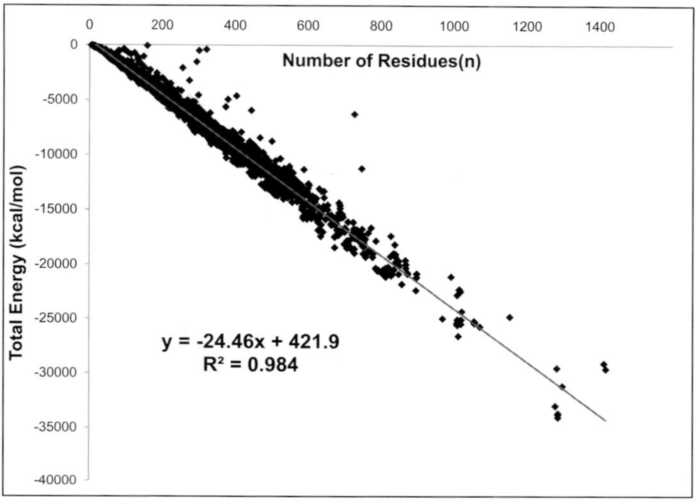

Figure 5: Total energy of 6750 proteins shown as a function of number of residues.

a platform to the prediction community all over the world to test their methodologies, ideas and "true predictions" on soon to be known/released proteins structures. Over the years, CASP has managed to evolve major benchmarks and computational breakthroughs in the field of protein structure prediction. Further, in spite of CASP being almost the equivalent of 'Olympics' of protein folding, no group in India has managed to participate in this competition for predicting tertiary structures of proteins. Till the time this article has been compiled, 38 experimental structures have been released since the beginning of CASP10. Our group (BhageerathH) has been able to predict a structure within 3Å rmsd from the native in 20 cases. These are remarkably promising results especially considering the participation of 261 prediction groups which include 125 registered servers comprising some of the best structure prediction groups for 113 blind prediction targets.

We are encouraged by the fact that universal insights into the rich diversity of protein structures, rather than development of numerous micro-subclassifications to account for the face value of the observed diversity, have led to the biggest breakthroughs in protein folding. We take our inspiration, in large part, from the universally applicable work of G. N. Ramachandran published 50 years ago, and, continue our march towards addressing one of the most important grand challenges in science today, i.e. protein folding.

Acknowledgments

BJ acknowledges the programme support from the Department of Biotechnology (DBT), Govt. of India, to the supercomputing facility for bioinformatics & Computational biology. CA is a senior research fellow of the Council of Scientific and Industrial Research (CSIR), Govt. of India. This manuscript is dedicated to the 50th anniversary of the first publication on "Ramachandran maps", a ubiquitous text-book concept now, by the great Indian scientist Prof. G. N. Ramachandran with his colleagues.

References

1. L. Pauling, R. B. Corey and H. R. Branson, *Proc. Natl. Acad. Sci. U.S.A.*, **37**, 205 (1951).
2. L. Pauling and R. B. Corey, *Proc. Natl. Acad. Sci. U.S.A*, **37**, 251 (1951).
3. L. Pauling and R. B. Corey, *Proc. Natl. Acad. Sci. U.S.A*, **37**, 235 (1951).
4. G. N. Ramachandran, C. Ramakrishnan and V. Sasisekharan, *J. Mol. Biol.*, **7**, 95 (1963).
5. L. Thukral, S. R. Shenoy, K. Bhusan and B. Jayaram, *Journal of Biosciences*, **32(1)**, 71 (2007).
6. H. Berman, K. Henrick, H. Nakamura and J. L Markley, *Nucl. Acid Res.*, **35**, D301 (2007).
7. C. B. Anfinsen, *Science*, **181**, 223 (1973).
8. K. A. Dill, *Biochemistry*, **29(31)**, 7133 (1990).
9. H. Frauenfelder, S. Sligar and P. G. Wolynes, *Science,* **254**, 1598 (1991).
10. S. Auer, M. A. Miller, S. V. Krivov, C. M. Dobson, M. Karplus and M. Vendruscolo, *Phys. Rev. Lett.*, **26**, 178104 (2007).
11. D. W. Bolen and G. D. Rose, *Annual Review of Biochemistry*, **77**, 339 (2008).
12. D. Thirumalai, E. P. O'Brien, G. Morrison and C. Hyeon, *Annu. Rev. Biophys.,* **39**, 159 (2010).
13. H. S. Chan and K. A. Dill, *Annu. Rev. Biophys. Biomol. Struct.,* **27**, 425 (1997).
14. J. Janin, *Nature*, **277**, 491 (1979).
15. A. G. Murzin, S. E. Brenner, T. Hubbard and C. Chothia, *J. Mol. Biol.*, **247**, 536 (1995).
16. A. Andreeva, D. Howorth, J. M. Chandonia, S. E. Brenner, T. J. Hubbard, C. Chothia and A. G. Murzin, *Nucl. Acid Res.*, **36**, D419 (2008).
17. Structural Classification of Proteins [SCOP : http://scop.mrc-lmb.cam.ac.uk/scop/]
18. J. S. Richardson, *Nature,* **268**, 495 (1977).
19. P. D. Sun, C. E. Foster and J. C. Boyington, in *Curr. Protoc. Prot. Sci.*, (John Wiley & Sons, 2004), p. 17.1.1.

20. A. Mittal, B. Jayaram, S. R. Shenoy and T. S. Bawa, *J. Biomol. Struc. Dyn.*, **28(2)**, 133 (2010).
21. A. Mittal and B. Jayaram, *J. Biomol. Struc. Dyn.*, **28(4)**, 669 (2011).
22. A. Mittal and B. Jayaram, *J. Biomol. Struc. Dyn.*, **28(4)**, 443 (2011).
23. R. H. Sarma, *J. Biomol. Struct. Dyn.*, **28**, 587 (2011).
24. P. S. Agutter, *J. Biomol. Struct. Dyn.*, **28**, 643 (2011).
25. Swissprot/Uniprot.[ftp://ftp.uniprot.org/pub/databases/uniprot/current_release/knowledgebase/complete/uniprot_sprot.fasta.gz].
26. P. J. Flory, *Statistical Mechanics of Chain Molecules* (Wiley, New York, 1969).
27. H. S. Chan and K. A. Dill, *Annu. Rev. Biophys. Biophys. Chem.*, **20**, 447 (1991).
28. L. Hong and J. Lie, *J. Polymer Science: Part B: Polymer Physics*, **47**, 207 (2009).
29. C. F. Tsai and K. J. Lee, *Int. J. Mol. Sci.*, **12**, 8449 (2011).
30. B. Jayaram, K. Bhushan, et al., *Nucl. Acids Res.*, **34**, 6195 (2006)
31. P. Narang, K. Bhushan, S. Bose and B. Jayaram, *Phys. Chem. Chem. Phys.*,**7**, 2364 (2005)
32. A. Mittal [http://precedings.nature.com/documents/6038/version/1/files/npre20116038-1.pdf].
33. B. D. Silverman, *Proc. Natl. Acad. Sci. U.S.A.*, **98**, 4996 (2001).
34. P. Narang, K. Bhushan, S. Bose and B. Jayaram, *J. Biomol. Str. Dyn.*, **23**, 385(2006).

COLD-SHOCK DOMAINS – VERSATILE MOLECULAR MODULES FOR SINGLE-STRANDED RNA BINDING AND REMODELING

FLORIAN MAYR AND UDO HEINEMANN
Crystallography, Max-Delbrück Center for Molecular Medicine
Robert-Rössle-Straße 10, 13125 Berlin, Germany, and
Institute for Chemistry and Biochemistry, Freie Universität Berlin
Takustraße 6, 14195 Berlin, Germany

The cold-shock domain (CSD) is a structurally conserved nucleic acid-binding domain that is present in all kingdoms of life from bacteria to vertebrates. It associates with a wide range of single-stranded nucleic acids via a conserved, hydrophobic platform located on one side of its β-barrel fold. The limited sequence specificity in nucleic-acid binding, in line with its RNA-remodeling function, makes the CSD a powerful molecular module for regulating RNA metabolism at various levels. Moreover, combinations of CSDs with other protein domains result in an enormous functional versatility of CSD-containing proteins that comprises cold acclimation in bacteria as well as regulation of development, differentiation and stress tolerance in higher eukaryotes. This review summarizes the role of CSDs in prokaryotic and eukaryotic proteins with a special emphasis on their nucleic acid-binding and domain-swap capabilities.

1. Introduction

CSDs are highly abundant nucleic acid-binding domains that are widely distributed in bacteria, animals and plants. The SMART protein database[1] currently lists 6052 CSD-containing proteins. Most of these proteins have been detected in bacteria (90%) and eukaryotes (9%), while only 51 archaeal proteins contain CSDs. Interestingly, all archaeal and the majority of the bacterial proteins are composed of a single CSD. These so called major cold-shock proteins (Csps) are known to bind single-stranded RNA and DNA with K_D values in the nanomolar to micromolar range[2-7] and work as RNA chaperones that help to destabilize RNA secondary structures[8,9]. Some members of this family are strongly induced at low temperatures and help to keep mRNA in a translation-competent folding state[10].

In the course of molecular evolution, processes like domain duplication and domain shuffling have led to 62 different domain architectures in CSD-containing proteins[11] with versatile biological functions. Especially in eukaryotes, auxiliary domains and signal peptides in addition to the CSD have broadened the biological function of these proteins. Consequently, CSD-containing proteins regulate several cellular processes like cold-shock adaption and development via direct regulation of transcription, miRNA biogenesis and translation.

2. Bacterial cold-shock domain proteins

Csps were first discovered in bacteria, as members of this family are induced during the bacterial response to cold-shock conditions[10,12]. In *E. coli*, after exposure to low temperatures, the expression of 4 out of 9 *csp*-family genes (*cspA, cspB, cspG* and *cspI*) is transiently induced, with the major cold-shock protein *Ec*-CspA accounting for up to 10% of total protein in the cell[13-17]. These proteins are known to bind single-stranded DNA and RNA oligonucleotides and to dissolve local secondary structures that may impede transcription or translation. Moreover, *Ec*-CspA, *Ec*-CspC and *Ec*-CspE are characterized as transcription anti-terminators by destabilizing RNA hairpins in mRNAs that otherwise act as Rho-independent transcriptional terminators[18].

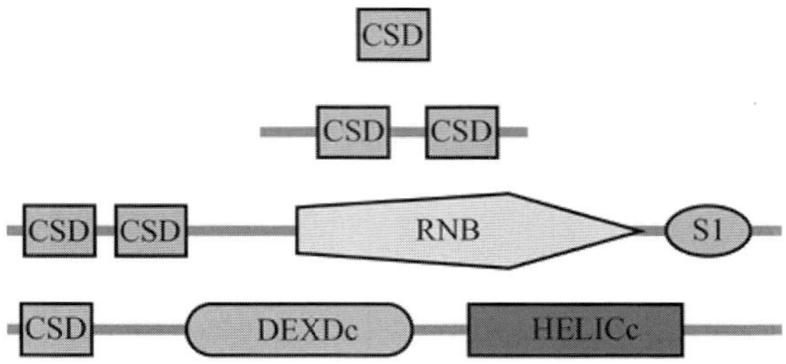

Figure 1. Domain architecture of bacterial CSD-containing proteins. S1: Ribosomal protein S1-like RNA-binding domain, RNB: catalytic domain of ribonuclease II, HELICc: helicase superfamily C-terminal domain, DEXDc: DEAD-like helicase superfamily.

Next to major Csps, bacteria also possess proteins in which a CSD is associated with other domains. Among them are Csp-like proteins that contain two CSDs, exoribonuclease R and DEAD-box RNA helicases (Fig. 1). Like *Ec*-CspA, the RNA helicase CsdA and exoribonuclease R are also induced under cold-shock conditions in *E. coli*[19]. *Ec*-CspA is essential only at low temperatures and has been linked to ribosome biogenesis, promotion of translation initiation and mRNA stability and degradation. It was suggested that the major task of this protein lies in unwinding RNA secondary structures within mRNAs that otherwise impair their degradation[19]. The exoribonuclease R, on the other hand, possesses two different activities: a helicase and a ribonuclease activity. The helicase activity is independent of its ribonuclease activity and is crucial for cold acclimation[20]. Moreover, the protein is composed of two N-terminal CSDs, and at least the second CSD is essential for RNA remodeling and helicase activity.

3. Eukaryotic cold-shock domain proteins

In contrast to bacterial Csps, eukaryotic CSD-containing proteins are often larger proteins in which a CSD is associated with auxiliary domains, like multiple CSDs (UNR), A/P domains and basic-acidic repeats (Y-box proteins), retroviral-like CCHC Zn-finger domains (Lin28) and CCHC Zn finger-containing glycine-rich repeats (*A. thaliana (At)* Csps, Fig. 2). Frequently, natively unfolded peptide regions are present in proteins alongside CSDs. With this architecture, CSD-containing proteins can fulfill pleiotropic functions mainly related to RNA metabolism and thereby regulate complex biological processes such as stress tolerance, development and differentiation[21].

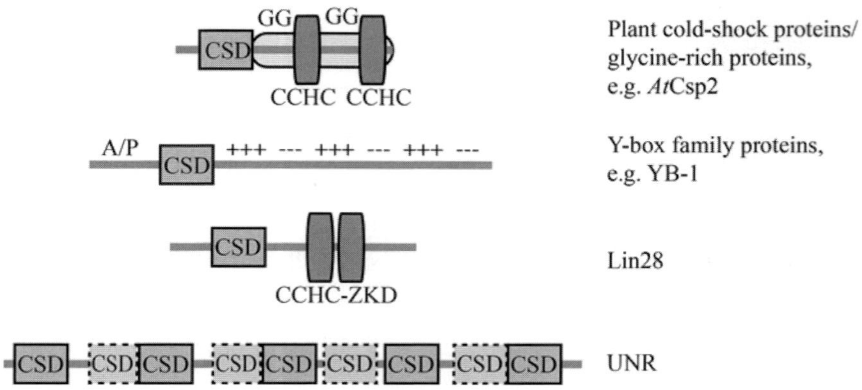

Figure 2. Domain architecture of eukaryotic CSD-containing proteins. Plant cold-shock proteins typically comprise a CCHC Zn-finger containing glycine-rich sequence C-terminal to the CSD. In animals, CSD-containing proteins are larger proteins containing additional CSDs (e.g. UNR) CCHC Zn-knuckle domains (ZKD, e.g. Lin28) or regulatory sequences like alanine/proline (A/P)-rich regions or basic/acidic repeats (+++ ---, e.g. YB-1). CSDs shown in dashed lines are either truncated or only predicted by homology.

3.1. Plant cold-shock domain proteins

Like bacterial Csps, plant CSD proteins are important for conferring freezing tolerance[22,23]. These proteins are typically composed of a highly conserved N-terminal CSD followed by a variable number of glycine-rich repeats (GRPs) that are interspersed by CCHC Zn fingers. The number of GRP/Zn-finger repeats can vary between one and seven. A recent study showed that both the CSD and Zn-finger containing GRPs contribute to the RNA-binding and RNA-chaperone activity of *At*-Csp1 and *At*-Csp2 during cold adaption[24]. Furthermore, the different modular arrangement of CSDs and GRP/CCHC Zn fingers might determine the nucleic acid-binding specificity and RNA-remodeling properties of plant Csps enabling the regulation of multiple processes like cold adaption, embryo development, flowering time and fruit development.

3.2. *Lin28*

The domain architecture of Lin28 (cell lineage abnormal 28) resembles those of plant CSD proteins. Originally discovered in *C. elegans* where mutations within the gene disturbed the developmental progression of the worm[25-27], it was soon found that Lin28 inhibits the biogenesis of *let-7* miRNAs. By this means Lin28 can maintain pluripotency and promote cell transformation and tumorigenesis, as *let-7* controls multiple oncogenes[28,29]. Lin28 binds to the terminal loop of pri- and pre-*let-7* and interferes with cleavage by the ribonucleases III Drosha and Dicer. Moreover, Lin28 was shown to promote the 3'-terminal uridylation of pre-*let-7* via TUT-4 leading to a subsequent degradation by cellular nucleases[30,31]. Recent X-ray structures showed that both the CSD and the two retroviral-type CCHC Zn fingers (comprising a Zn-knuckle domain, ZKD) bind to single-stranded parts of the terminal loop of pre-*let-7*. However, only the ZKD mediated sequence-specific binding to a conserved GGAG motif while the CSD showed limited sequence specificity[32]. Recently, Nam *et al.* observed the binding of a pre-*let-7* stem loop-derived RNA to mLin28A CSD[33]. In this structure, an RNA single strand of up 9 nucleotides completely wraps around the CSD. However, the closing of the stem loop as well as additional elements from a basic linker following the CSD contribute to binding. Furthermore, the CSD was shown to have an RNA-remodeling activity that re-structures the terminal loop of pre-*let-7*, thereby enabling a subsequent binding of the ZKD to the conserved GGAG motif[32]. By this means Lin28 can specifically bind to all 11 structurally different pre-*let-7* family members with diverse secondary structures and inhibit their biogenesis.

3.3. *UNR*

The UNR (upstream of N-ras) protein is an 85 kDa transcription factor, which is comprised of up to 9 consecutive CSDs. However, only 5 of these CSD are structurally and biochemically verified. The *UNR* gene was named after its location immediately upstream of the *N-ras* loci in several vertebrates[34,35] and was shown to be essential for certain stages in differentiation and development[36]. UNR is ubiquitously expressed in cell lines and tissues and shows a high binding affinity towards single-stranded nucleic acids with K_D values in the low nanomolar range[37]. Further SELEX experiments suggested that UNR preferentially binds to 11-14 nt long purine-rich sequences with a conserved core motif of AAGUA/G or AACG.

Unlike the Lin28 CSD, UNR CSDs show only a low sequence match to bacterial CSPs (25% to 30% identity compared to *Ec*-CspA, see Fig. 4). Interestingly, from the 7 conserved aromatic residues that are present in classical Csps and involved in nucleic-acid binding only three are recognizable in UNR CSDs (see Fig. 4). Moreover, the UNR CSDs themselves differ considerably with respect to their nucleic acid-binding capabilities. While the first CSD bound to a ssRNA oligonucleotide with a dissociation constant of 55 nM, the fifth CSD showed for the same ssRNA only a K_D value of 1500 nM[37]. Recent NMR solution structures of all domains suggested that the conformation of residues F43 and H45 in UNR CSD1 might be responsible for the observed difference in binding affinity[38]. These two residues are highly conserved in CSDs (see Fig. 4) and were shown to contribute considerably to nucleic-acid binding via base stacking. In UNR CSD1 the conformation of H45 is further stabilized by a unique cysteine (C46) that is not present in other CSDs. According to the authors, this cysteine probably leads to an increased intrinsic stability and triggers an optimal conformation of H45 and F43 for nucleic-acid binding.

4. Structure of monomeric and domain-swapped cold-shock domains

From an evolutionary and structural perspective, the CSD is an ancient ssDNA/RNA-binding module with a close relationship to the ribosomal protein S1-like RNA-binding domain (S1 domain). Both domains belong to the oligo-saccharide-oligonucleotide-binding (OB) fold and consist of a β-barrel, composed of five antiparallel β-strands (Fig. 3). The CSD typically comprises 70 amino-acid residues and is characterized by a prominent amphipathic surface. One side the protein exposes a negatively charged surface that does not favor RNA binding. On contrary, the opposite surface forms a hydrophobic platform surrounded by individual basic side chains. This side accommodates the moderately conserved ribonucleoprotein motifs RNP1 (GF/YGFI) an RNP2 (DV/LFVH) and interacts with ssDNA/RNA via extensive stacking interactions between bases and exposed aromatic protein sidechains (see Fig. 4).

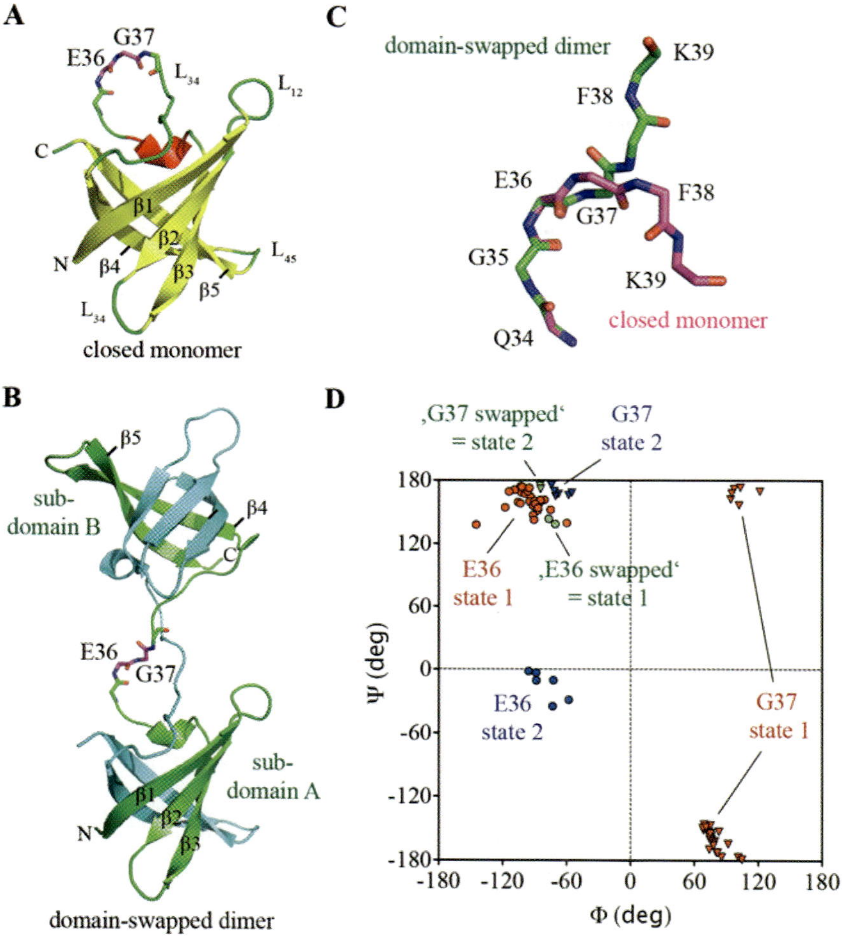

Figure 3. Structure of monomeric and domain-swapped *Bc*-Csp. (A) Normally, *Bc*-Csp forms a closed β-barrel with two subdomains that are connected by the flexible loop L_{34}. (B) In the domain-swapped form, a specific E36Ψ/G37Φ torsion-angle combination within this loop leads to an effective 180° rotation of the main chain and an opening of the molecule. Subdomain 2 of one molecule can consequently re-associate with subdomain 1 of the other molecule forming a closed β-barrel. (C) Superimposition of the L_{34} loop-forming residues Q34 to K39 in the closed and domain-swapped form. (D) Ramachandran diagram for E36 and G37 in closed and domain-swapped forms. Loop L_{34} allows a two-state conformational variability with respect to the torsion angles of E36Ψ and G37Φ. Based on 21 closed models for *Bc*-Csp and *Bs*-CspB, E36Ψ/G37Φ either adopt torsion angles of 163 ± 11° and 91 ± 13° (state 1) or -16 ± 15° and -66 ± 7° (state 2). In the domain-swapped dimer of *Bc*-Csp however, E36Ψ adopts a torsion angle according to state 1, whereas G37Φ shows a torsion angle similar to state 2.

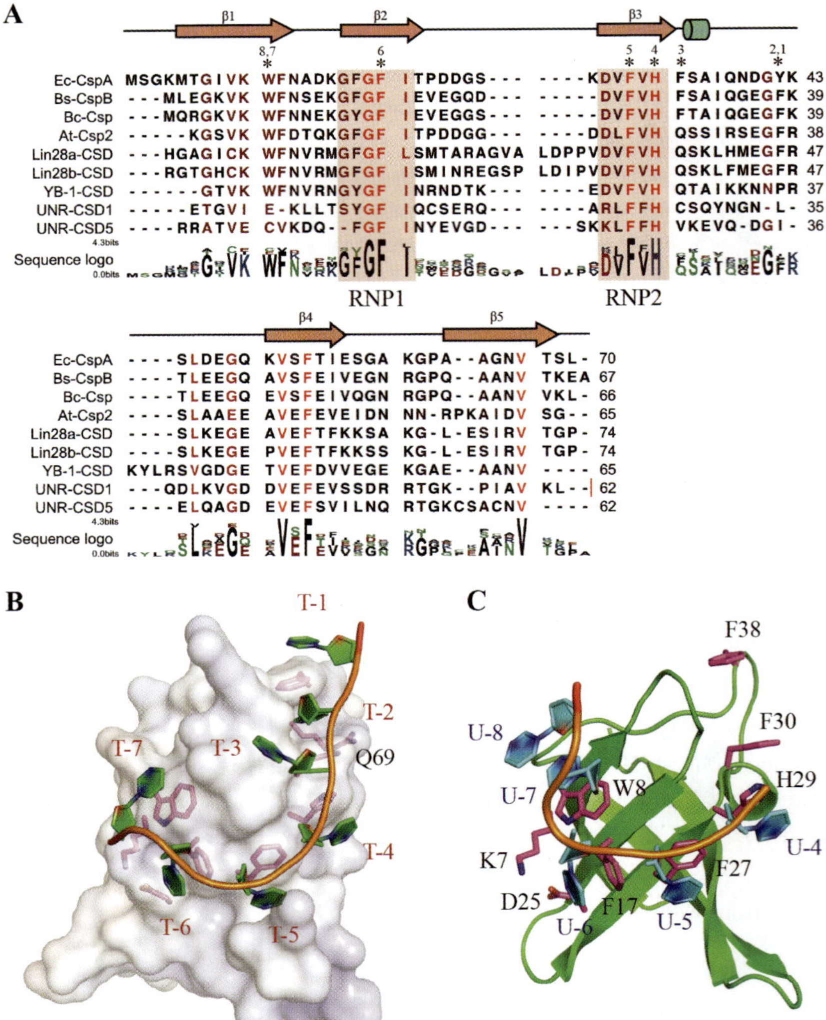

Figure 4. (A) Structure-based sequence alignment of CSD-containing proteins. Residues that are conserved at a level of least 75% are highlighted with a red foreground color. The conserved RNP1 and RNP2 motifs are shaded in red. (B, C) Structural comparison of DNA- and RNA-bound CSDs. Binding to single-stranded nucleic acids occurs via a pre-formed nucleic acid-binding platform harboring the conserved, surface-exposed aromatic residues. These residues interact with the bases of RNA and DNA mainly through base stacking while the sugar-phosphate backbone remains solvent exposed. The binding mode and polarity of RNA and DNA oligonucleotides is similar and conserved in bacterial (right: *Bs*-CspB:rU$_6$) and eukaryotic proteins (left: *Xtr*Lin28b CSD:dT$_7$).

Another interesting aspect of CSDs is their potential to form domain-swapped dimers. In the crystal structure of *B. caldolyticus* Csp (*Bc*-Csp) with bound hexathymidine (dT_6), β-strands 1-3 (subdomain 1) associate with β-strands 4 and 5 (subdomain 2) of another protein and *vice versa* (Fig. 3)[5]. Interestingly, the globular units of this dimer strongly resemble the functional unit of the monomeric apo protein[39]. Even the binding of dT_6 is not affected and similar to that observed for Lin28B CSD:dT_6 [32]. The swapped chains are correlated by a non-crystallographic two-fold symmetry and arise from a strictly localized conformational change within the peptide linkage between residues E36 and G37 (Fig. 3). In the closed form of *Bc*-Csp, the torsion angles E36Ψ and G37Φ adopt a two-state conformational variability with either 163 ± 11° and 91 ± 13° (state 1) or -16 ± 15° and -66 ± 7° (state 2)[5,39]. In the domain-swapped dimer, however, E36Ψ adopts a state 1-like torsion angle (141 ± 4°) whereas G37Φ shows a torsion angle similar to state 2 (-85 ± 1°). As a result, the main chain rotates about 180° at E36, causing an opening of *Bc*-Csp in the loop region between β-strands 3 and 4 (Fig. 3). The association of subdomain 2 with subdomain 1 of another molecule finally leads to the re-formation of closed β-barrels.

5. Binding of nucleic acids to cold-shock domains

Crystal structures of hexa- or heptathymidine bound to bacterial (*Bc*-Csp, *Bs*-CspB) and eukaryotic CSDs (Lin28B CSD) revealed a curved conformation of the bound oligonucleotide[4,5,32]. Binding is dominated by stacking interactions between aromatic protein sidechains and thymine bases. The 5'-end of the oligonucleotide is located at the loop region between β-strands 3 and 4, the single strand then proceeding over a groove formed by residues from β-strands 5 and 3 of the CSD, crossing β-strand 2 until making a final kink over β-strand 1. The binding site is already preformed and can accommodate up to 8 nucleotides (Fig. 4). Sequence-specific binding is only observed at positions 2 (T/U/G) and 6 (U/T), while the lack of specific hydrogen bonds with bases at other positions should allow binding of a broad spectrum of potential nucleic acids. Various biochemical data support these observations and emphasize the overall limited sequence specificity of CSDs in nucleic-acid binding[5-7,32].

For RNA oligonucleotides a similar binding mode as for DNA was observed[40]. As the sugar-phosphate backbone is surface exposed and hardly involved in protein:RNA interactions, no substantial difference in binding mode and strength between ssDNA and ssRNA oligonucleotides is observable (Fig. 4). Only the additional hydrophobic contribution of the 5-methyl group from thymidine compared to uridine led to higher binding affinities of dT_7 compared to rU_7. Recently, Nam *et al.* observed the binding of a more complex RNA oligonucleotide derived from the pre-*let-7* stem loop to mLin28A CSD[33]. In this structure, a single strand with up to 9 nucleotides completely wraps around the CSD. However, the closed RNA stem-loop as well as additional elements from a basic linker following the CSD contribute to binding.

References

1. Letunic, I. et al., *Nucleic Acids Res.* **30**, 242 (2002).
2. Lopez, M.M. and Makhatadze, G.I., *Biochim. Biophys. Acta* **1479**, 196 (2000).
3. Lopez, M.M., Yutani, K. and Makhatadze, G.I., *J. Biol. Chem.* **276**, 15511 (2001).
4. Max, K.E.A., Zeeb, M., Bienert, R., Balbach, J. and Heinemann, U., *J. Mol. Biol.* **360**, 702 (2006).
5. Max, K.E.A., Zeeb, M., Bienert, R., Balbach, J. and Heinemann, U., *FEBS J* **274**, 1265 (2007).
6. Morgan, H.P. et al., *Nucleic Acids Res.* **35**, e75 (2007).
7. Zeeb, M. et al., *Nucleic Acids Res.* **34**, 4561 (2006).
8. Phadtare, S. and Severinov, K., *Nucleic Acids Res.* **33**, 5583 (2005).
9. Phadtare, S., Tyagi, S., Inouye, M. and Severinov, K., *J. Biol. Chem.* **277**, 46706 (2002).
10. Horn, G., Hofweber, R., Kremer, W. and Kalbitzer, H.R., *Cell. Mol. Life Sci.* **64**, 1457 (2007).
11. Bateman, A. et al., *Nucleic Acids Res.* **30**, 276 (2002).
12. Ermolenko, D.N. and Makhatadze, G.I., *Cell. Mol. Life Sci.* **59**, 1902 (2002).
13. Brandi, A., Spurio, R., Gualerzi, C.O. and Pon, C.L., *EMBO J.* **18**, 1653 (1999).
14. Goldstein, J., Pollitt, N.S. and Inouye, M., *Proc. Natl. Acad. Sci. USA* **87**, 283 (1990).
15. Lee, S.J. et al., *Mol. Microbiol.* **11**, 833 (1994).
16. Nakashima, K., Kanamaru, K., Mizuno, T. and Horikoshi, K., *J. Bacteriol.* **178**, 2994 (1996).
17. Wang, N., Yamanaka, K. and Inouye, M., *J. Bacteriol.* **181**, 1603 (1999).
18. Bae, W., Xia, B., Inouye, M. and Severinov, K., *Proc. Natl. Acad. Sci. USA* **97**, 7784 (2000).
19. Phadtare, S., *RNA Biol.* **8**, 394 (2011).
20. Phadtare, S. and Severinov, K., *RNA Biol.* **7**, 788 (2010).
21. Mihailovich, M., Militti, C., Gabaldón, T. and Gebauer, F., *BioEssays* **32**, 109 (2010).
22. Chaikam, V. and Karlson, D.T., *BMB Rep.* **43**, 1 (2010).

23. Sasaki, K. and Imai, R., *Front. Plant Sci.* **2**, 116 (2011).
24. Park, S.J., Kwak, K.J., Jung, H.J., Lee, H.J. and Kang, H., *Phytochemistry* **71**, 543 (2010).
25. Ambros, V. and Horvitz, H., *Science* **226**, 409 (1984).
26. Moss, E.G., Lee, R.C. and Ambros, V., *Cell* **88**, 637 (1997).
27. Seggerson, K., Tang, L. and Moss, E.G., *Dev. Biol.* **243**, 215 (2002).
28. Rybak, A. et al., *Nat. Cell Biol.* **10**, 987 (2008).
29. Viswanathan, S.R., Daley, G.Q. and Gregory, R.I., *Science* **320**, 97 (2008).
30. Heo, I. et al., *Mol. Cell* **32**, 276 (2008).
31. Heo, I. et al., *Cell* **138**, 696 (2009).
32. Mayr, F., Schütz, A., Döge, N. and Heinemann, U., *Nucleic Acids Res.* **40**, 7492 (2012).
33. Nam, Y., Chen, C., Gregory, R.I., Chou, J.J. and Sliz, P., *Cell* **147**, 1080 (2011).
34. Jeffers, M., Paciucci, R. and Pellicer, A., *Nucleic Acids Res.* **18**, 4891 (1990).
35. Doniger, J. and DiPaolo, J.A., *Nucleic Acids Res.* **16**, 969 (1988).
36. Nicolaiew, N., Triqueneaux, G. and Dautry, F., *Oncogene* **6**, 721 (1991).
37. Triqueneaux, G., Velten, M., Franzon, P., Dautry, F. and Jacquemin-Sablon, H., *Nucleic Acids Res.* **27**, 1926 (1999).
38. Goroncy, A.K. et al., *J. Struct. Funct. Genom.* **11**, 181 (2010).
39. Delbrück, H., Mueller, U., Perl, D., Schmid, F.X. and Heinemann, U., *J. Mol. Biol.* **313**, 359 (2001).
40. Sachs, R., Max, K.E.A., Heinemann, U. and Balbach, J., *RNA* **18**, 65 (2012).

INSIGHTS INTO DNA HELICAL TRANSITIONS FOUND IN PROTEIN-DNA COMPLEXES

ANDREW V. COLASANTI, YUN LI[*], GAUTAM SINGH, XIANG-JUN LU[†] AND WILMA K. OLSON

Department of Chemistry & Chemical Biology, BioMaPS Institute for Quantitative Biology, Rutgers, the State University of New Jersey
Wright-Rieman Laboratories, 610 Taylor Road
Piscataway, New Jersey 08854 USA

Many proteins essential for the viability of the cell induce large-scale deformations in the DNA double helix. Some of these deformations include transitions of regular B DNA to alternate helical forms, such as A DNA, where base pairs are unwound, displaced, and inclined with respect to the helical axis. Here we examine structural features of the A-like DNA base-pair steps found in a diverse set of high-resolution protein-DNA complexes. The protein-bound DNA is more deformable than the A DNA crystallized in the absence of protein, and the protein microenvironment surrounding A DNA differs in subtle ways from that associated with other types of DNA. The kinds of amino-acid atoms in immediate contact with the sugars, phosphates, and bases hint of ways in which small molecules may cluster around A DNA in solution and how changes in solvent may contribute to alterations of helical structure.

1. Introduction

The proteins that bind double-helical DNA present various microenvironments that sense and/or induce signals in the genetic material. The high-resolution structures of protein-DNA complexes reveal the nature of both the microenvironments and the conformational responses in DNA and protein. Complex networks of interactions within the structures somehow tie the protein and DNA together and induce the observed spatial forms. Understanding how the local chemical environment presented by a bound protein influences the three-dimensional structure of DNA is fundamental to the prediction of protein function and to the design of DNA-binding ligands for specific purposes.

Investigations of the contacts between protein and DNA in high-resolution structural complexes have provided numerous insights into the mechanisms of

[*]Current address: Department of Chemistry & Biochemistry, Delaware Valley College, Doylestown, Pennsylvania 18901 USA.
[†]Current address: Department of Biological Sciences, Columbia University, New York, New York 10027 USA.

macromolecular interaction and recognition. The amino acids that associate with DNA typically cluster on the protein exterior, with patches of positively charged residues at the binding interface[3-7]. The contacts to DNA are dominated by, but not limited to, interactions with the negatively charged phosphate groups. The associations of proteins with the DNA bases[8-13] include many of the specific amino-acid contacts anticipated from the patterns of hydrogen-bond donor and acceptor atoms on the major-groove edges of the Watson-Crick base pairs and the protein side groups[16]. The contacts to the sugar depend upon the puckering and the exposed surface of the five-membered ring, *i.e.*, the groove in which the protein approaches the sugar[17].

The ionic character of the sugar-phosphate backbone makes DNA especially sensitive to its local physicochemical environment. Changes in solvation lead to changes of helical state. For example, variation in the ion composition or water content of DNA fibers introduces a large-scale conformational transformation of the classic 10-fold B-DNA double helix[18] to the compacted 11-fold A-DNA structure[19]. The change of state rearranges the paired bases and the intervening chemical moieties. Furthermore, the ease of conformational transition depends upon sequence[20-27], with certain base-pair steps standing out as elements that are more easily converted to A-type structures in solution[28,29]. The deformations of B DNA induced by protein include similar transitions to the A form. The localized distortions alter the exposure of specific atoms and introduce bends in the double helix[30].

The forces that dominate protein-induced conformational changes in DNA appear to reside at the binding interface. The sequence-specific coordination of monovalent and divalent cations in the grooves of ultra-high-resolution crystal structures[31] along with the asymmetric build-up of counterions around DNA in solution[32-34], hint of the critical role played by electrostatic interactions in sequence-dependent structural changes of DNA. The nonuniformity of counterions at individual base-pair steps can bend the double helix along the same lines as site-specific phosphate neutralization, whereby DNA collapses into a chemically modified site[35]. The bending that results from all-atom energy minimization of DNA helices that contain patches of phosphate neutralization supports these predictions and hints of the similar role of cationic amino acids in guiding protein-induced deformations of DNA[36,37].

Here, in an effort to decipher the forces that induce B→A transitions of DNA, we focus on the organization of amino-acid and nucleotide atoms at the sites of A-type structure in high-resolution protein-DNA complexes. Our work takes advantage of methods that we have developed to distinguish A DNA from

B DNA in molecular complexes and to characterize the chemical microenvironments of protein-bound DNA. The types of protein atoms in close contact with the A-like base-pair steps differ from those associated with other kinds of DNA, and the sites of greatest protein build-up around DNA undergo a subtle redistribution upon the change of conformational state. The data are suggestive of ways in which the double helix might respond to changes in the solvent environment and provide new benchmarks for the simulation of such processes.

2. Computational Approach

The information presented below is based on a set of 239 protein-DNA crystal complexes of 2.5 Å or better resolution taken from the Nucleic Acid Database[38]. The dataset includes 101 structures of double-helical DNA bound to enzymes, 121 duplexes in the presence of regulatory proteins, 16 complexes with structural proteins, and one DNA bound to a multi-functional protein. The structures are filtered to exclude over-represented complexes in order to obtain a balanced sample of spatial and functional forms. Complete details will be reported elsewhere.

The A-DNA elements within the protein-DNA complexes are distinguished by the value of z_P, the mean out-of-plane displacement of the phosphorus atoms at each base-pair step[30]. The normal symmetric spacing of phosphorus on B DNA, roughly mid-way between the planes of neighboring base pairs, becomes skewed in A DNA, with the atoms shifted toward the 3′-ends of the base-pair steps. The limits used here correspond to the range of z_P characteristic of pure A-DNA structures ($z_P \geq 1.5$ Å). Other conformational parameters used to identify A-like DNA, such as the sugar puckering[17], are less discriminatory. Large positive values of z_P also occur at highly deformed base-pair steps, termed TA-DNA[39], in some high-resolution structures. These steps, which differ from A DNA in terms of the extreme unwinding and inclination of the base pairs, are excluded in the present study with the classification tools in the 3DNA suite of programs[14,15]. Also omitted from the dataset are steps with extreme values of the six rigid-body parameters used to describe the spatial arrangements of neighboring base pairs[40], *i.e.*, values more than three standard deviations beyond the mean. The remaining base-pair steps, which are predominantly B-like, are referred to as other DNA. The rigid-body parameters and the torsion angles along the chain backbones in both types of DNA are determined with the 3DNA software[14,15].

The identities of the protein atoms in the immediate vicinity of the DNA in a given structural complex are found by partitioning the coordinate space into a set of 5-Å cubes, finding the distances between all pairs of protein and DNA atoms within each cube and in adjacent cubes, and choosing the atom pairs within a desired distance. The selected coordinates are then re-expressed in standard reference frames on the DNA, *i.e.*, in established frames on the four bases[41] and/or in newly defined frames on the phosphate and sugar, depending upon the chemical context of the atom. This procedure makes it possible to examine the patterns of atomic interaction in different structures from a common perspective. The frame on the phosphate is defined so that the origin lies on the P atom, the *y*-axis bisects the OP1-P-OP2 valence angle, and the *x*-axis runs along the normal to the OP1-P-OP2 plane, with positive values taken in the O3´→O5´ direction. The frame on the sugar is similarly described with the origin at the midpoint of the C2´-C3´ bond, the *x*-axis pointed along the C2´→C3´ direction, the *y*-axis in the plane of C2´, C3´, and O4´, and the positive direction of *y* pointed toward O4´. The *z*-axes of the two frames follow from the right-handed rule.

The selected protein atoms are characterized in terms of the types of amino acids in which they occur, the secondary structures of the peptide backbone to which they are attached, the locations of the atoms on the chemical structure, and the family to which the protein belongs. The contacted atoms on DNA are grouped into similar chemical categories, and the solvent-accessible surface areas of the atoms are determined with the MSMS software[42]. The amino-acid atoms are further divided into four chemical categories — cation, anion, water, and hydrocarbon — that reduce the protein to a simple 'solvent' mixture. The 'cations' include the positively charged nitrogen atoms on arginine, lysine, and histidine and the 'anions' the negatively charged oxygen atoms on aspartic acid and glutamic acid. The 'water' corresponds to any other atom with the capacity to act as a hydrogen-bond donor or acceptor, and the 'hydrocarbon' to atoms, mainly carbon, that do not fit the preceding categories.

3. Results

The base-pair steps identified as A-like occur in 42 structures. Of these, 29 are complexes with enzymes, 12 with regulatory proteins, and one with a structural protein. The proteins adopt a variety of folds and perform diverse cellular tasks (Table 1). Many of the proteins introduce bends in the DNA. In some examples the proteins encircle the DNA and in others they merely touch a small patch on the double-helical surface (Figure 1). Although many of the proteins introduce isolated distortions in 1-2 base-pair steps, the A-like DNA associated with other proteins extends over longer nucleotide stretches.

Table 1. Protein-DNA complexes with A-like base-pair steps.

Protein	PDB_ID/citation	Contacts ≤ 3.6 Å	A-like steps
DNA endonucleases			
homing endonuclease I-CreI	1G9Y[43]	14	1
homing endonuclease I-PpoI	1IPP[44]	79	8
restriction endonuclease BsobI	1DC1[45]	148	2
restriction endonuclease EcoO1091	1WTE[46]	244	8
restriction endonuclease EcoRV	1BGB[47] 1RV5[48] 1EOO[49]	152	3
restriction endonuclease Mspi	1SA3[50]	88	2
restriction endonuclease PvuII	1EYU[51] 3PVI[52]	265	13
DNA excisionases			
excisionase (bacteriophage λ)	1RH6[53]	–	1
DNA glycosylases			
formamidopyrimidine glycosylase	1K82[54]	–	4
DNA lyases			
deoxyribonuclease I	1DNK[55] 2DNJ[56]	55	3
DNA polymerases			
DNA polymerase I (*T. aquaticus*)	1QSS[57] 2KTQ[58] 3KTQ[58] 4KTQ[58]	148	6
DNA polymerase I (*Bacillus*)	1L3S[59] 1L3U[59] 1L3V[59] 1LV5[59] 2BDP[60] 4BDP[60]	333	13
DNA polymerase iota	1T3N[61]	66	
T7 phage DNA polymerase	1SKR[62] 1TK0[63]	60	2
Glucocorticoid receptor-like domains			
ecdysone receptor-ultraspiracle heterodimer	1R0O[64]	25	1
retinoid X receptor	1DSZ[65]	24	1
High-mobility-group proteins			
HMG1 +cisplatin	1CKT[66]	–	1
Non-specific nucleases			
periplasmic nuclease Vvn	1OUP[67]	8	1
TBP-like domains			
TATA-box binding protein	1QN5[68] 1QNB[68] 1QNE[68]	4	3
TFIIA + TBP	1YTF[69]	7	3
TFIIB + TBP	1AIS[70]	–	1
Transcription factors			
catabolite gene activator protein	1J59[71]	4	1
HoxA9/Pbx1 homeodomains	1PUF[72]	11	3
NarL response regulator	1JE8[73]	32	4
regulatory factor X	1DP7[74]	16	3
Transposases			
DNA transposase Tc3	1TC3[75]	39	6
Tumor suppressor proteins			
tumor suppressor p53	1TUP[76]	1	1

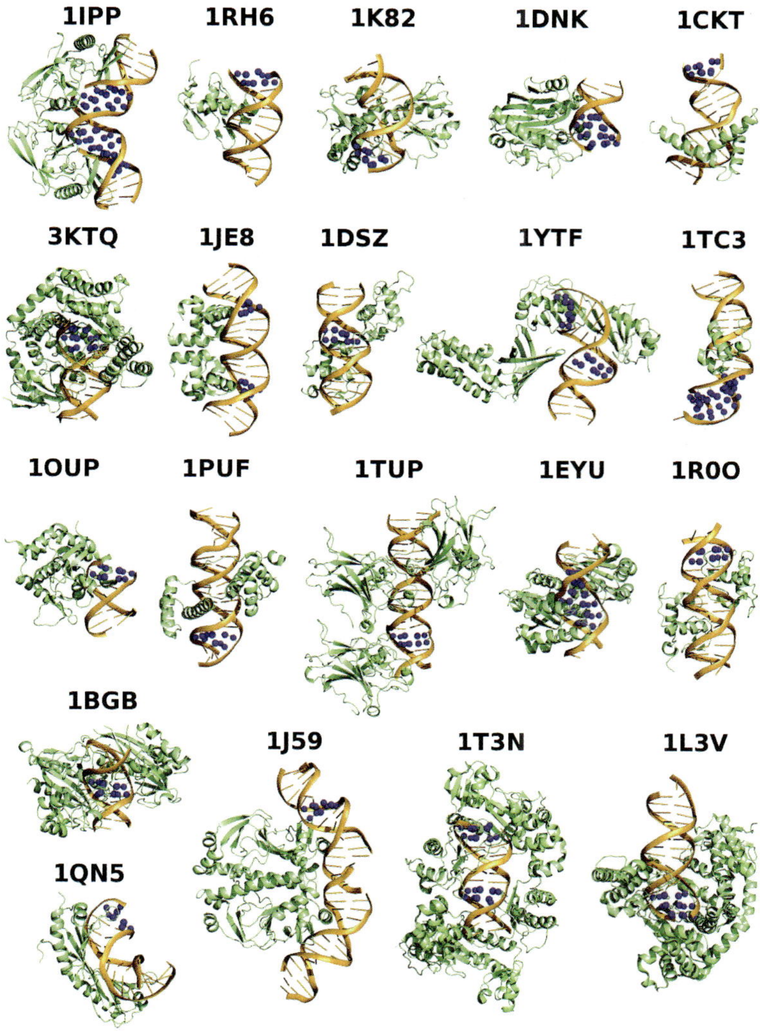

Figure 1. Gallery of representative protein-DNA structures highlighting, in blue, the sites of A-like base-pair steps. The DNA is depicted in gold with the phosphorus atoms on complementary strands joined by ribbons and the bases drawn as stick figures. The proteins are shown as pale green ribbons and the complexes are labeled with the Protein Data Bank[2] identifiers described in Table 1. Views looking down the shortest principal axis of each DNA with the longest principal axis pointed upward.

3.1. *Protein-bound A-like DNA*

The selected A-like steps bear close resemblance to those found in high-resolution structures of protein-free A DNA[41]. The angle of bending about the long axes of successive base pairs, the so-called Roll angle[40], is slightly lower on average in the protein-bound compared to the pure structures[41] (6.1 *vs.* 8.0°). The accompanying lateral displacement of base pairs, the so-called Slide[40], and the average Twist are remarkably close in the two types of A DNA (–1.51 *vs.* –1.53 Å and 30.9 *vs.* 31.1° in the protein-bound and free DNA, respectively). Like the sugars in pure A-DNA structures[77] and in classic fiber-diffraction models[1], most of the sugars at the A-like steps in the protein-DNA complexes adopt C3′-endo, or N, puckered forms[78]. The distributions of angles about the backbone and side-group chemical bonds in the protein-bound steps also resemble those found in pure A-DNA structures[77] (data not shown).

As suggested by the molecular images, almost all of the A-like steps come into direct contact with protein. Of the 102 identified steps, only seven contain atoms that do not come within 3.6 Å of a heavy (non-hydrogen) atom on protein. The proteins make ~16 contacts per step on average and over threefold more in a few structures. The reported set of contacts, however, does not include the interactions of DNA with the small molecules and ions thought to work in combination with protein to induce the changes in DNA structure found in many complexes[79]. The omission is justified by the fact that the sites of hydration on DNA often coincide with the binding sites at protein- and drug-DNA interfaces[80,81]. The predominance of A-like steps made up of only G·C and C·G base pairs and the lack of steps with only A·T and T·A pairs follows the known sequence-dependent free energies of B→A conformational transitions[21,28,29]. The limited number of base-pair steps of different chemical composition, however, precludes detailed analysis of the sequence-dependent occurrences of A DNA in the presence of proteins.

The superimposed structures in Figure 2 illustrate the variability in the A-like steps. Each dimer is represented in a common reference frame[14,15] and the deformations of A DNA are suggested by the range of positions of corresponding atoms in the different structures. The illustrated steps differ more from one another than do those found in pure A DNA. The protein-bound steps contain a much higher proportion of correlated, crankshaft-like rotations[82] about chemical bonds in the sugar-phosphate backbone than do those in pure A DNA. These non-canonical states occur in almost all cases in one strand of the protein-bound A DNA and at isolated steps or the ends of longer A-DNA stretches.

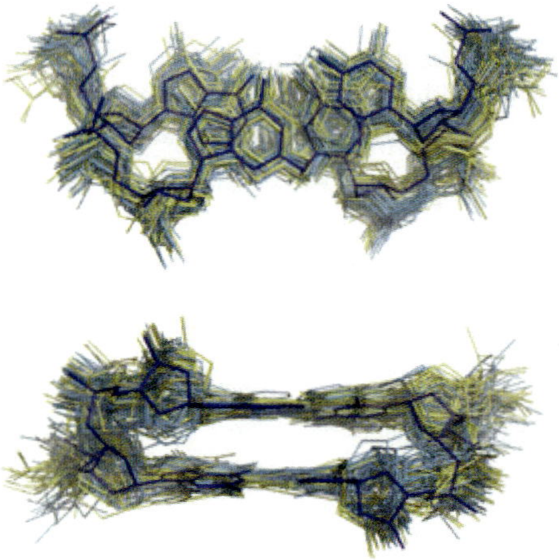

Figure 2. Superposed stick images illustrating the variation in A-like base-pair steps (light blue) found in high-resolution protein-DNA complexes compared to those in canonical A DNA[1] (deep blue). Composite structure obtained by superposition of the mean coordinate frames between the base pairs in each dimer. Steps highlighted in gold adopt the non-canonical backbone arrangements discussed in the text. Views perpendicular to the mean base-pair plane (top) and into the minor groove (bottom).

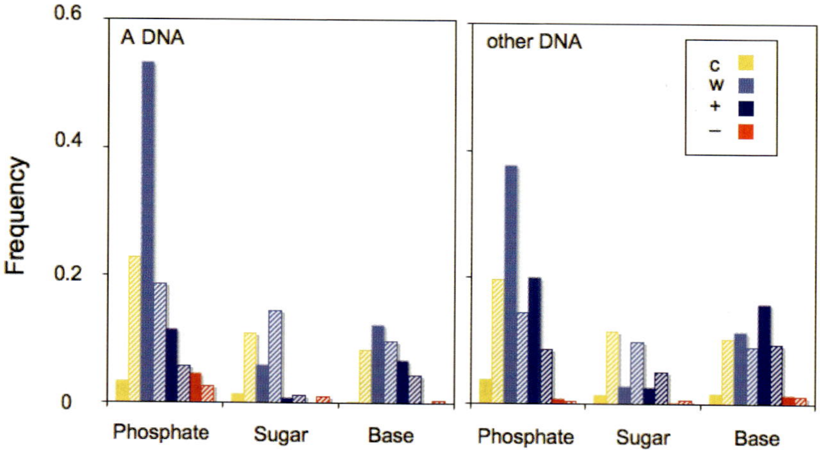

Figure 3. Distribution of the amino-acid atoms in close contact with the chemical moieties of A-like base-pair steps (left) compared to those around other types of DNA (right). Amino-acid atoms grouped by chemical type: c 'hydrocarbon'; w 'water'; + 'cation'; – 'anion'. Contacts at close range (≤ 3 Å) denoted by deeper shading and those more distant (≤3.6 Å) by lighter shading in the color-coded pair of entries for each 'solvent' type.

3.2. *Protein Interactions with A DNA*

The proteins bound to A-like DNA bring different kinds of amino acids to the double helix compared to the proteins that bind other types of DNA. Unlike the vast majority of protein-DNA complexes, where proteins associate preferentially with DNA through charged amino acids, most of the close contacts of protein with A DNA involve polar amino acids (Table 2). These preferences persist at the level of bases, sugars, and phosphates although there are sizable populations of charged amino acids in close contact with A DNA and large numbers of polar amino acids that bind other types of DNA.

Table 2. Characteristics of protein atoms in contact with A DNA *vs*. other DNA steps.

Step	A DNA		Other DNA	
	≤ 3.0 Å	≤ 3.6 Å	≤ 3.0 Å	≤ 3.6 Å
Contacts within specified limit				
	375	1900	4203	22665
Amino acid types				
charged	0.29	0.27	0.53	0.47
polar	0.60	0.61	0.41	0.39
hydrophobic	0.12	0.12	0.06	0.14
Chemical microenvironment				
'hydrocarbon'	0.05	0.42	0.07	0.41
'water'	0.71	0.43	0.52	0.33
'cation'	0.19	0.11	0.39	0.23
'anion'	0.05	0.04	0.03	0.02
Peptide backbone				
coil	0.24	0.19	0.26	0.27
helix	0.15	0.21	0.42	0.36
sheet	0.22	0.20	0.14	0.18
turn	0.39	0.40	0.18	0.19
Atom position				
backbone	0.35	0.29	0.18	0.17
ε	0.07	0.13	0.13	0.15
ζ	0.09	0.07	0.12	0.11

The chemical microenvironment around A DNA depends upon the distance between amino-acid and nucleotide atoms (Table 2). Most of the contacts to A-like dimers at close range (≤ 3 Å) entail water-like amino-acid atoms and at greater distances (≤ 3.6 Å) a mix of 'water' and 'hydrocarbon' atoms sprinkled with 'cations'. The amount of 'water' around A DNA consistently exceeds that around other DNA, and the proportion of 'cations' falls below the levels in other

DNA. The fraction of 'anions' is similarly low in both types of DNA and nearly independent of the distance between amino-acid and nucleotide atoms.

The proteins seemingly approach A DNA with different types of secondary structures from those used to contact other base-pair steps. The proportion of contacts to A-like DNA from atoms in helices is substantially lower than that found in the vicinity of other DNA steps. The contacts to A DNA at very close range entail a large proportion of residues from sheets, and at greater distances a large number of contacts from turns. The large fraction of sheets in the protein-DNA complexes that bind A-like DNA is clear from the depicted structures (Figure 1).

Interestingly, about a third of the contacts to A DNA come from peptide backbone atoms, nearly tenfold the fraction found very close to other kinds of DNA steps and double the fraction found at greater distances. As expected from the lower 'cationic' content of the microenvironment around A DNA, there are fewer contacts from atoms at the tips of long amino-acid side groups, *e.g.*, the N^{ε} and N^{ζ} atoms of arginine, histidine, and lysine. A large fraction of the close contacts to other DNA steps entail these atoms.

3.3. *Interactions of A DNA with its Microenvironment*

A-like DNA resembles other types of DNA in the sense that it makes more contacts to protein via phosphate atoms than either sugar or base atoms, but it differs in terms of the relative populations (Figure 3). The likelihood of association to protein through the A-DNA phosphates exceeds that found at the phosphates in other types of DNA, and the chance of interaction through the bases is lower. The differences are more pronounced at close range but persist at greater separation.

The preponderance of phosphate interactions with protein involves the side group oxygens, OP1 and OP2 (Figure 4). The former atom is 25-33% more exposed than the latter in terms of its solvent-accessible surface area. Moreover, the propensity to contact OP1 over OP2 in A DNA exceeds the increase in surface area. The number of contacts to OP1 is more than double that to OP2. The enhanced association to OP1 over OP2 in other types of DNA is roughly proportional to the relative surface areas.

The O4′ atom makes more contacts to protein than any other sugar atom (Figure 4) and comprises nearly 10% of all the contacts to A-DNA steps at greater separation distances. As noted by others[17], the overwhelming majority of contacts to O4′ involve the approach of protein via the minor groove. The majority of contacts to the remaining sugar atoms also occur via the minor groove, and the proportion of minor-groove contacts to A DNA is greater than that for other kinds of DNA. The contacts of protein to the bases, by contrast, occur preferentially in the major groove. The propensity to bind major- over minor-groove base atoms is greater for A DNA than that for other types of DNA steps.

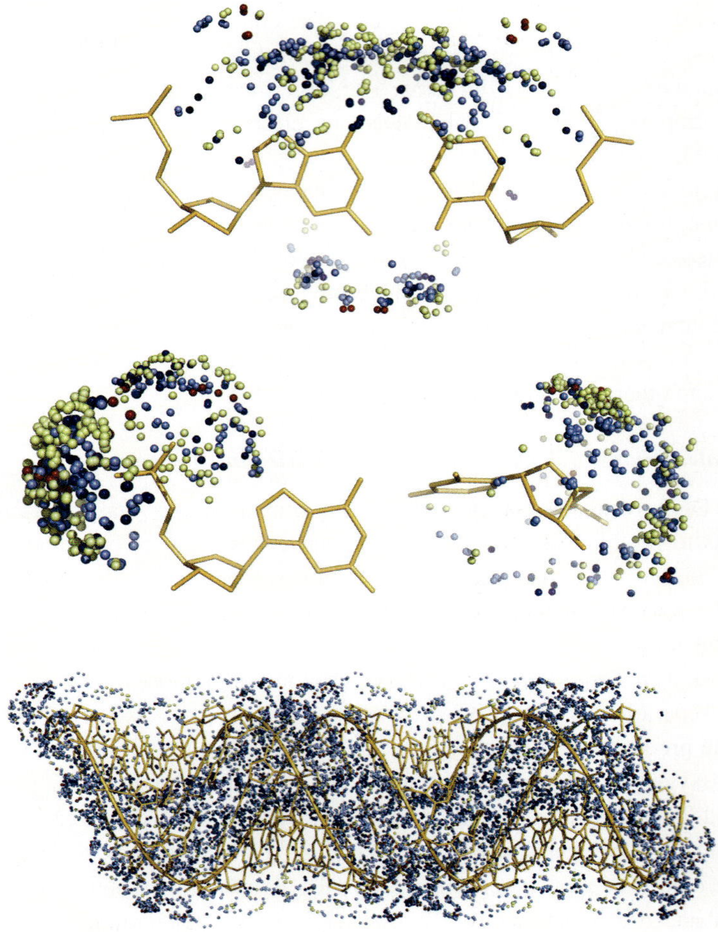

Figure 4. Composite microenvironment around the G·C base pairs (top), phosphates (middle left), and sugars (middle right) of A-like steps in high-resolution protein-DNA complexes and the inferred distribution of solvent around a $dG_{15}{\cdot}dC_{15}$ stretch of canonical A-DNA[1] (bottom). Protein atoms in contact with DNA are colored-coded by chemical type: 'hydrocarbon' (gold); 'water' (pale blue); 'cations' (dark blue); 'anions' (red). Illustrated points lie within 3.6 Å of the nucleotide fragments and 3.0 Å of the polymer. Contacts to OP1 on the phosphate and O4´ on the sugar are noted by larger spheres. Polymer model generated with 3DNA[14, 15].

4. Conclusions and Outlook

Interest in uncovering a molecular code, analogous to the genetic code, to account for the recognition of DNA by proteins has stimulated many studies of the binding interfaces in high-resolution protein-DNA complexes. Extensive examination of three-dimensional structure and composition has shown that there is no simple correspondence between amino acids and bases and no clear set of protein folds that account for the DNA sequences targeted by particular proteins. The various ways in which the amino acids interact with DNA point to the more fundamental level of protein-DNA association, considered here, involving specific types of atoms. From the perspective of either protein or DNA, its binding partner is a spatially organized collection of atoms that present features complementary to its chemical composition and architecture. Different amino-acid side groups and different protein folds can bring the same kinds of atoms to DNA and *vice versa*.

Here we show how the cumulative build-up of amino-acid atoms around the sugars, phosphates, and bases in different protein-DNA complexes produces a binding cloud around the double helix and how different kinds of atoms fill that cloud. Intriguingly, the distribution of atom types depends upon DNA conformation. Most of the atoms that accumulate around A-like DNA base-pair steps resemble water in that they are capable of forming hydrogen bonds. This local build-up is perplexing in that A-DNA formation in fibers[18,19] and in solution[20,21] is accompanied by a loss of water. The atoms at the binding sites of other types of DNA include a sizable proportion of cationic atoms in place of the water-like atoms.

When superimposed on a canonical A-DNA duplex (Figure 4), the observed binding sites create a ribbon-like cluster of 'waters' reminiscent of the spine of hydration seen in B DNA[83]. Here the clusters lie in the major rather than the minor groove and represent the hydrogen-bond donor and acceptor atoms from proteins in contact with the OP1 and OP2 atoms of phosphate rather than crystallized water. The redirection of these side-group oxygens, long known to accompany the B→A transition of DNA[84,85], underlies the unique pattern. The change in orientation places the preferred sites of phosphate binding across and into the major groove. The fragments of A-DNA determined at high resolution, in both protein-free and protein-bound structures, have not been long enough to detect this structural feature.

The sites of protein-binding found here differ from the positions of water seen in the first few well-resolved DNA crystal structures, which suggested that A DNA may be stabilized by an economy of hydration not possible in other helical forms.[86] The repuckering of the sugar ring in the crystal structures brings successive phosphate groups close enough that a single water molecule can form a hydrogen-bonded bridge between the side-group oxygens. There are few corresponding bridges in the A-like steps formed in the presence of protein. The lack of bridged phosphates may reflect the greater variability of the protein-bound steps compared to the A-DNA formed in crystalline arrays. As noted above, a number of the protein-bound A-like steps adopt non-canonical backbone arrangements. The rigid-body parameters relating adjacent base pairs in the protein-DNA complexes also span a broader range of values, with standard deviations roughly 1.5 times those reported for pure A-DNA crystals[41]. The build-up of water and ions around the phosphate edges of the major groove may play a role analogous to the water bridges found in crystals in stabilizing naturally fluctuating A-DNA helical structures in solution.

Finally, the treatment of DNA-bound proteins as ordered collections of solvent-like particles provides a new source of information about the likely distribution of water, ions, and other small molecules around DNA in solution. Aside from the ordered waters and ions found in collections of high-resolution DNA structures[87], current understanding of DNA solvation is based on the behavior of model compounds incorporated in atomic-level simulations[88-91]. The calculated properties are governed by force fields that are applied to DNA and the surrounding molecular species and thus are subject to the limitations of these treatments. Although recent improvements to nucleic acid force fields have incorporated structural information about DNA and RNA[92-94], they do not yet consider the unique information found at the binding interfaces of protein-DNA complexes. The inferences about solvation that can be drawn from these structures provide new stimuli for improvement of nucleic acid force fields and fresh ideas for exploration of the properties of DNA in solution.

Acknowledgments

Support of this work through USPHS Research Grants GM20861 and GM34809 to WKO and GM096889 to XJL is gratefully acknowledged.

References

1. R. Chandrasekaran and S. Arnott, in *Landolt-Börnstein Numerical Data and Functional Relationships in Science and Technology, Group VII/1b, Nucleic Acids*, Ed. W. Saenger, (Springer-Verlag, Berlin, 1989), p. 31.
2. H. M. Berman, J. Westbrook, Z. Feng, Gilliland, B. G., T.N., H. Weissig, I. N. Shindyalov and P. E. Bourne, *Nucleic Acids. Res.* **28**, 235 (2000).
3. D. H. Ohlendorf and J. B. Matthew, *Advan. Biophys.* **20**, 137 (1985).

4. Y. N. Chirgadze and E. A. Larionova, *Mol. Biol. (Engl. Tr.)* **37**, 232 (2003).
5. S. Jones, H. P. Shanahan, H. M. Berman and J. M. Thornton, *Nucleic Acids Res.* **31**, 7189 (2003).
6. E. W. Stawiski, L. M. Gregoret and Y. Mandel-Gutfreund, *J. Mol. Biol.* **326**, 1065 (2003).
7. H. Tjong and H.-X. Zhou, *Nucleic Acids Res.* **35**, 1465 (2007).
8. C. O. Pabo and R. T. Sauer, *Annu. Rev. Biochem.* **61**, 1053 (1992).
9. M. Suzuki, *Structure* **2**, 317 (1994).
10. Y. Mandel-Gutfreund, O. Schueler and H. Margalit, *J. Mol. Biol.* **253**, 370 (1995).
11. H. Kono and A. Sarai, *Proteins: Struct. Funct. Genet.* **35**, 114 (1999).
12. N. M. Luscombe, R. A. Laskowski and J. M. Thornton, *Nucleic Acids Res.* **29**, 2860 (2001).
13. T. W. Siggers, A. Silkov and B. Honig, *J. Mol. Biol.* **345**, 1027 (2005).
14. X.-J. Lu and W. K. Olson, *Nucleic Acids Res.* **31**, 5108 (2003).
15. X.-J. Lu and W. K. Olson, *Nature Protoc.* **3**, 1213 (2008).
16. N. C. Seeman, J. M. Rosenberg and A. Rich, *Proc. Natl. Acad. Sci., USA* **73**, 804 (1976).
17. M. Y. Tolstorukov, R. L. Jernigan and V. B. Zhurkin, *J. Mol. Biol.* **337**, 65 (2004).
18. J. D. Watson and F. H. C. Crick, *Nature* **171**, 964 (1953).
19. R. E. Franklin and R. G. Gosling, *Nature* **171**, 740 (1953).
20. F. M. Pohl and T. M. Jovin, *J. Mol. Biol.* **67**, 375 (1972).
21. V. I. Ivanov, L. E. Minchenkova, E. E. Minyat, M. D. Frank-Kamenetskii and A. K. Schyolkina, *J. Mol. Biol.* **87**, 817 (1974).
22. S. Arnott and E. Selsing, *J. Mol. Biol.* **88**, 509 (1974).
23. S. Arnott and E. Selsing, *J. Mol. Biol.* **88**, 551 (1974).
24. J. F. Burd, R. M. Wartell, J. B. Dodgson and R. D. Wells, *J. Biol. Chem.* **250**, 5109 (1975).
25. T. A. Early, D. R. Kearns, J. F. Burd, J. E. Larson and R. D. Wells, *Biochemistry* **16**, 541 (1977).
26. Y. Wang, G. A. Thomas and W. L. Peticolas, *J. Biomol. Struct. Dynam.* **5**, 249 (1987).
27. Y. Wang, G. A. Thomas and W. L. Peticolas, *J. Biomol. Struct. Dynam.* **6**, 1177 (1989).
28. V. I. Ivanov and L. E. Minchenkova, *Mol. Biol.* **28**, 780 (1995).
29. M. Y. Tolstorukov, V. I. Ivanov, G. G. Malenkov, R. L. Jernigan and V. B. Zhurkin, *Biophys. J.* **81**, 3409 (2001).
30. X.-J. Lu, Z. Shakked and W. K. Olson, *J. Mol. Biol.* **300**, 819 (2000).
31. L. D. Williams and L. J. Maher III, *Annu. Rev. Biophys. Biomol. Struc.* **29**, 497 (2000).
32. V. P. Denisov and B. Halle, *Proc. Natl. Acad. Sci., USA* **97**, 629 (2000).

33. N. V. Hud and M. Polak, *Curr. Opin. Struct. Biol.* **11**, 293 (2001).
34. N. C. Stellwagen, S. Magnusdottir, C. Gelfi and P. G. Righetti, *J. Mol. Biol.* **305**, 1025 (2001).
35. J. K. Strauss and L. J. Maher III, *Science* **266**, 1829 (1994).
36. R. Gurlie and K. Zakrzewska, *Theor. Chem. Accts.* **106**, 83 (2001).
37. K. M. Kosikov, A. A. Gorin, X.-J. Lu, W. K. Olson and G. S. Manning, *J. Am. Chem. Soc.* **124**, 4838 (2002).
38. H. M. Berman, W. K. Olson, D. L. Beveridge, J. Westbrook, A. Gelbin, T. Demeny, S.-H. Hsieh, A. R. Srinivasan and B. Schneider, *Biophys. J.* **63**, 751 (1992).
39. G. Guzikevich-Guerstein and Z. Shakked, *Nature Struct. Biol.* **3**, 32 (1996).
40. R. E. Dickerson, M. Bansal, C. R. Calladine, S. Diekmann, W. N. Hunter, O. Kennard, E. von Kitzing, R. Lavery, H. C. M. Nelson, W. K. Olson, *et al.*, *J. Mol. Biol.* **208**, 787 (1989).
41. W. K. Olson, M. Bansal, S. K. Burley, R. E. Dickerson, M. Gerstein, S. C. Harvey, U. Heinemann, X.-J. Lu, S. Neidle, Z. Shakked, *et al.*, *J. Mol. Biol.* **313**, 229 (2001).
42. M. F. Sanner, A. J. Olson and J.-C. Spehner, *Biopolymers* **38**, 305 (1996).
43. B. Chevalier, R. J. Monnat Jr. and B. L. Stoddard, *Nature Struct. Biol.* **8**, 312 (2001).
44. K. E. Flick, M. S. Jurica, R. J. Monnat Jr. and B. L. Stoddard, *Nature* **394**, 96 (1998).
45. M. J. van der Woerd, J. J. Pelletier, S. Xu and A. M. Friedman, *Structure* **9**, 133 (2001).
46. H. Hashimoto, T. Shimizu, T. Imasaki, M. Kato, N. Shichijo, K. Kita and M. Sato, *J. Biol. Chem.* **280**, 5605 (2005).
47. N. C. Horton and J. J. Perona, *J. Biol. Chem.* **273**, 21721 (1998).
48. N. C. Horton and J. J. Perona, *J. Mol. Biol.* **277**, 779 (1998).
49. N. C. Horton and J. J. Perona, *Proc. Natl. Acad. Sci., USA* **97**, 5729 (2000).
50. Q. S. Xu, R. B. Kucera, R. J. Roberts and H. C. Guo, *Structure* **12**, 1741 (2004).
51. J. R. Horton and X. Cheng, *J. Mol. Biol.* **300**, 1049 (2000).
52. J. R. Horton, H. G. Nastri, P. D. Riggs and X. Cheng, *J. Mol. Biol.* **284**, 1491 (1998).
53. M. D. Sam, D. Cascio, R. C. Johnson and R. T. Clubb, *J. Mol. Biol.* **338**, 229 (2004).
54. R. Gilboa, D. O. Zharkov, G. Golan, A. S. Fernandes, S. E. Gerchman, E. Matz, J. H. Kycia, A. P. Grollman and G. Shoham, *J. Biol. Chem.* **277**, 19811 (2002).
55. S. A. Weston, A. Lahm and D. Suck, *J. Mol. Biol.* **226**, 1237 (1992).
56. A. Lahm and D. Suck, *J. Mol. Biol.* **221**, 645 (1991).
57. Y. Li, V. Mitaxov and G. Waksman, *Proc. Natl. Acad. Sci., USA* **96**, 9491 (1999).

58. Y. Li, S. Korolev and G. Waksman, *EMBO J.* **17**, 7514 (1998).
59. S. J. Johnson, J. S. Taylor and L. S. Beese, *Proc. Natl. Acad. Sci., USA* **100**, 3895 (2003).
60. J. R. Kiefer, C. Mao, J. C. Braman and L. S. Beese, *Nature* **391**, 304 (1998).
61. D. T. Nair, R. E. Johnson, S. Prakash, L. Prakash and A. K. Aggarwal, *Nature* **430**, 377 (2004).
62. Y. Li, S. Dutta, S. Doublie, H. M. Bdour, J. S. Taylor and T. Ellenberger, *Nature Struct. Mol. Biol.* **11**, 784 (2004).
63. L. G. Brieba, B. F. Eichman, R. J. Kokoska, S. Doublie, T. A. Kunkel and T. Ellenberger, *EMBO J.* **23**, 3452 (2004).
64. S. Devarakonda, J. M. Harp, Y. Kim, A. Ozyhar and F. Rastinejad, *EMBO J.* **22**, 5827 (2003).
65. F. Rastinejad, T. Wagner, Q. Zhao and S. Khorasanizadeh, *EMBO J.* **19**, 1045 (2000).
66. U.-M. Ohndorf, M. A. Rould, Q. He, C. O. Pabo and S. J. Lippard, *Nature* **399**, 708 (1999).
67. C.-L. Li, L.-I. Hor, Z.-F. Chang, L.-C. Tsai, W.-Z. Yang and H. S. Yuan, *EMBO J.* **22**, 4014 (2003).
68. G. A. Patikoglou, J. L. Kim, L. Sun, S.-H. Yang, T. Kodadek and S. K. Burley, *Genes Dev.* **13**, 3217 (1999).
69. S. Tan, Y. Hunziker, D. F. Sargent and T. J. Richmond, *Nature* **381**, 127 (1996).
70. P. F. Kosa, G. Ghosh, B. S. DeDecker and P. B. Sigler, *Proc. Natl. Acad. Sci., USA* **94**, 6042 (1997).
71. G. Parkinson, C. Wilson, A. Gunasekera, Y. W. Ebright, R. E. Ebright and H. M. Berman, *J. Mol. Biol.* **260**, 395 (1996).
72. N. A. LaRonde-LeBlanc and C. Wolberger, *Genes Dev.* **17**, 2060 (2003).
73. A. E. Maris, M. R. Sawaya, M. Kaczor-Grzeskowiak, M. R. Jarvis, S. M. D. Bearson, M. L. Kopka, I. Schroeder, R. P. Gunsalus and R. E. Dickerson, *Nature Struct. Biol.* **9**, 771 (2002).
74. K. S. Gajiwala, H. Chen, F. Cornille, B. P. Roques, W. Reith, B. Mach and S. K. Burley, *Nature* **403**, 916 (2000).
75. G. van Pouderoyan, R. F. Ketting, A. Perrakis, R. H. A. Plasterk and T. K. Sixma, *EMBO J.* **16**, 6044 (1997).
76. Y. Cho, S. Gorina, P. D. Jeffrey and N. P. Pavletich, *Science* **265**, 346 (1994).
77. B. Schneider, S. Neidle and H. M. Berman, *Biopolymers* **42**, 113 (1997).
78. C. Altona and M. Sundaralingam, *J. Am. Chem. Soc.* **94**, 8205 (1972).
79. B. Jayaram and T. Jain, *Annu. Rev. Biophys. Biomol. Struct.* **33**, 343 (2004).
80. J. Woda, B. Schneider, K. Patel, K. Mistry and H. M. Berman, *Biophys. J.* **75**, 2170 (1998).
81. W. Ge, B. Schneider and W. K. Olson, *Biophys. J.* **88**, 1166 (2005).

82. W. K. Olson, in *Biomolecular Stereodynamics*, Ed. R. H. Sarma, (Adenine Press, New York, 1981), p. 327.
83. H. R. Drew and R. E. Dickerson, *J. Mol. Biol.* **151**, 535 (1981).
84. J. Pilet and J. Brahms, *Biopolymers* **12**, 387 (1973).
85. W. Pohle, V. B. Zhurkin and H. Fritzsche, *Biopolymers* **23**, 2603 (1984).
86. W. Saenger, W. N. Hunter and O. Kennard, *Nature* **324**, 385 (1986).
87. P. Auffinger and Y. Hashem, *Bioinformatics* **23**, 1035 (2007).
88. M. Feig and B. M. Pettitt, **77**, 1769 (1999).
89. K. J. McConnell and D. L. Beveridge, *J. Mol. Biol.* **304**, 803 (2000).
90. T. E. Cheatham III and M. A. Young, *Biopolymers* **56**, 232 (2001).
91. N. Pastor, *Biophys. J.* **88**, 3262 (2005).
92. A. Pérez, I. Marchán, D. Svozi, J. Sponer, T. E. Cheatham III, C. A. Laughton and M. Orozco*, *Biophys. J.* **92**, 3817 (2007).
93. I. Yildirim, S. D. Kennedy, H. A. Stern, J. M. Hart, R. Kierzek and D. H. Turner*, **8**, 172 (2012).
94. K. Hart, N. Foloppe, C. M. Baker, E. J. Denning, L. Nilsson and A. D. MacKerell Jr., *J. Chem. Theor. Comp.* **8**, 348 (2012).

D-AMINO ACIDS: OCCURRENCE, STEREOCHEMISTRY AND EXCLUSION FROM TRANSLATIONAL MACHINERY

SATYA BRATA ROUTH*, SADEEM AHMAD* AND RAJAN SANKARANARAYANAN

CSIR-Centre for Cellular and Molecular Biology
Uppal Road, Hyderabad-500 007, India

The biological macromolecular world is chiral. How homochirality at the molecular level crept into the living system is an aspect which still remains unresolved. The presence of D-amino acids in organisms apart from bacteria was not appreciated until recently. Several studies in the last three decades have established the ubiquity of D-amino acids in the living world. Some of these D-amino acids have now been implicated in several important physiological functions, including neurotransmission and hormone regulation. However, proteins are comprised of only L-amino acids; a protein containing both L- and D-enantiomers loses its functionality owing to the fact that it fails to take its proper conformation. Hence, it becomes imperative for the cell to maintain "enantiomeric fidelity" during every step of translation. To this end, the cell employs several molecular machineries which, in their own capacity, help in excluding D-amino acids from polypeptides. An interesting case is presented by D-aminoacyl-tRNA deacylase which removes a D-amino acid that is mischarged onto the corresponding tRNA. Further discrimination against D-amino acids at the level of elongation factors and ribosomes has been noted. All these cellular systems play their part in perpetuation of homochirality in proteins.

1. Introduction

All biological macromolecules are composed of basic units which have one particular handedness or chirality. Nucleic acids (DNA and RNA) have only D-ribose as the sugar moiety; likewise, only L-amino acids (except glycine which is achiral) are found in proteins. It is well known that a protein will not be able to take a proper conformation if it were to be made from a mixture of D- and L-amino acids, and hence will lose its functionality. However, what has been a subject of much speculation is why and how only L-amino acids were selected to form peptides/proteins during the prebiotic era. It is, in general, considered that chiral-selection in the primordial environment was a mere chance event, but research in this area over the past few decades has brought to the fore several new and interesting possibilities which alone or in combination might have played crucial role in perpetuation of biological homochirality.

---

*Contributed equally.

The parity-violating energy difference (PVED) model[1,2] proposes that there exists extremely small difference in energy between enantiomers[3,4] which result in slight excess (enantiomeric excess), thereby leading to stabilization of one of the enantiomeric products during reactions of achiral or racemic substrates. This small energy difference is then amplified through preferential stabilization by electroweak interactions.[5-7] Certain studies have shown that at least some amino acids exhibit preferential adsorption of one enantiomer to faces of the same handedness of certain inorganic crystals such as calcite, while the other enantiomer adheres preferentially to the mirror-related faces of the crystals. Thus, this preference in adsorption might have led to sustenance of the L-configuration and elimination of the D-isomer from proteins.[8] Another proposition has implicated the opposite polarities of circularly polarized UV light at dawn and dusk owing to asymmetric rotation of the earth to be a factor behind exclusion of D-amino acids from proteins.[9] An interesting finding suggests that homochirality of L-amino acids was dictated by the prior evolution of D-ribose RNA, since the latter displays preference for L-amino acids during non-enzymatic aminoacylation reactions.[10-13] Evidence also exists for asymmetric amplification in autocatalysis (a phenomenon whereby a product acts as a catalyst to enhance its own production, while suppressing the production of its enantiomer) as a mechanism for the evolution of homochirality.[14]

If we assume that fortuity did not play a role, it was the interplay of several of these physico-chemical events, if not all, which consequently gave rise to homochirality in biological systems. How large and significant was the effect of each of these individual forces is a question that remains to be answered.

In this review article, we will focus on the occurrence of D-amino acids in biological systems, their roles and stereochemical considerations and molecular mechanisms of their exclusion from the translational machinery.

2. Distribution and Roles of D-amino acids in Living Organisms

It has long been known that some D-amino acids are components of certain peptides that are a part of bacterial cell wall,[15] or those which form antibiotics such as valinomycin, actinomycin, gramicidin S and tyrocidines.[16,17] These amino acids are incorporated through the actions of non-ribosomal peptide synthetases (NRPSs) and polyketide synthases (PKSs).[18] However, the occurrence of D-amino acids in higher organisms was not known mainly because of lack of adequate analytical procedures.[17] The discovery of D-amino acid oxidase in mammals in 1935 came as a bit of a surprise since the presence of

D-amino acids outside of the bacterial kingdom was not obvious at that time.[19] Again, it was not until the advent of advanced analytical techniques such as chiral-phase gas chromatography that research through the 1980's and the 1990's helped establish the ubiquitous occurrence of D-amino acids in the living world.[17]

The formerly underappreciated and largely overlooked D-amino acids are now known to play some critical roles in the cellular physiology. Table 1 summarizes the distribution of several D-amino acids in a variety of organisms and their physiological roles. Certain D-amino acids, especially D-Asp, have been reported from several proteins of many aged tissues,[20] and have been implicated in diseases such as Alzheimer's.[21,22] It is also interesting to note that in some cases, certain D-amino acids can be present at very high concentrations comparable to their L-enantiomers. For example, D-Asp and D-Ser in mammalian brains[23,24] and D-Ser in silkworm blood[25] are found in high D- to L-ratios.

3. Biosynthesis of D-amino acids

Two principal mechanisms are known for the synthesis of free D-amino acids *in vivo* — racemisation and transamination.[26] We now know of several amino acid racemases which convert L-forms to their antipodes and vice versa. Since all the racemases have an equilibrium constant near unity, they catalyse the interconversion of L- and D-enantiomers with almost equal efficiency. However, depending on regulatory and other factors as well as cellular requirement, the enzyme drives the reaction towards one direction. The racemases fall in two categories, based on their cofactor requirement — pyridoxal 5'-phosphate (PLP)-dependent racemases and PLP-independent racemases.[27,28] The members of PLP-dependent racemases include serine racemase, alanine racemase and aspartate racemase of both bacterial and eukaryotic origin. The second class of racemases comprises of aspartate racemase, glutamate racemase and proline racemase, the enzymes being exclusively present in bacteria and archaea (except for proline racemase which was also reported from *Trypanosoma*[29]). Both classes of racemases follow a common mode for effecting D-L interconversion — extraction of C^{α}-hydrogen as a proton from the substrate, leading to the formation of a carboanion transition state, and subsequent transfer of a proton to the substrate from the other side. Nevertheless, the mechanism employed by the two is totally different.[28]

Table 1. Distribution and roles of D-amino acids in various organisms/tissues.

Amino Acid	Organism/ Tissue	State	Physiological Significance	References
D-Ala, D-Glu	Bacteria	Peptide-bound; part of peptidoglycan	Peptidoglycan provides mechanical strength to cell wall and resistance to protease digestion	78
D-Leu, D-Met, D-Phe, D-Tyr	Bacteria	Free	Regulation of peptidoglycan synthesis	78
D-Leu, D-Val	Bacteria	Peptide-bound; part of gramicidin	Gramicidin is an antibiotic	79-81
D-Phe	Bacteria	Peptide-bound; part of gramicidin S	Gramicidin S is a cyclic antibiotic	82,83
D-Ala, D-Leu, D-Met	Frog (eg, *Phyllomedusa*)/ Skin	Peptide-bound; part of opioid peptides — dermorphins and deltorphins	Opioid peptides are neuroactive compounds, some of them probably acting as hormones as well	84,85
D-Asn, D-Leu, D-Phe	Snail (eg, *Achatina fulica*)/Ganglia, atrium	Peptide-bound; part of neuropeptides such as achatin and fulicin	These neuropeptides control muscle contraction	86,87
D-Ser	*Agelenopsis aperta* (funnel web spider)	Peptide-bound; part of omega-agatoxin IVB	Agatoxin blocks voltage-sensitive Ca^{2+}-channels	32,33
D-Ser	Earthworm (eg, *Lumbricus terrestris*)	Bound; part of lombricine (a phosphodiester of guanidine)	Being a high-energy phosphate compound, lombricine acts as an energy storage compound	88
D-Cys	Firefly (eg, *Photinus*)	Bound; part of D-luciferin	D-luciferin acts as the precursor to oxyluciferin, which is bioluminescent	88,89
D-2,3-diamino-propionic acid, D-Ser	Insect (eg, *Bombyx mori*)	Free	Role in metamorphosis	88
D-Ala	Aquatic crustaceans and bivalve molluscs	Free	Cell volume modulation through osmoregulation	90
D-Asp	Mammals/ Central nervous, neuroendocrine and endocrine systems like brain, pituitary gland, pineal gland, adrenal gland and testis (Leydig cells)	Free	Neuromodulator of NMDA receptor signalling, regulates neurogenesis, suppresses melatonin synthesis and secretion, modulates production of hormones like prolactin, oxytocin, vasopressin and testosterone	91-101
D-Ser	Mammals/ Brain	Free	Neurotransmitter; acts as a co-agonist of NMDA receptor	94,102-104

D-amino acid transaminases (also called D-amino acid aminotransferases and abbreviated as DATs) are another class of enzymes which mediate biosynthesis of D-amino acids. The enzyme catalyses transfer of the α-amino group from a donor D-amino acid to a recipient α-keto acid. Consequently, the donor is converted to the corresponding α-keto acid, whereas the recipient to its D-amino acid. DATs differ from L-amino acid transaminases (LATs) in their requirement for D-amino acid as the substrate. DATs have been discovered in various bacterial genera, *Bacillus* being the first and the principal among them.[26] *In vitro*, *Bacillus* DATs have been shown to produce D-Glu, D-Leu, D-Phe and D-Tyr.[30] Very recently, one such transaminase has been reported to catalyse D-Asp synthesis from D-Glu and oxaloacetate in *Arabidopsis thaliana*.[31] Whether such D-amino acid-specific transamination occurs in other organisms too is an aspect that needs to be explored.

Besides these two enzymes probably playing a major role in D-amino acid biosynthesis, a novel peptide isomerase, discovered in the venom of American funnel-web spider (*Agelenopsis aperta*), has been shown to convert a number of L-amino acids to their D-enantiomers when the former are present in peptides.[32,33] Moreover, certain D-amino acids such as D-Asp can form by spontaneous isomerisation of its antipode in the peptide-bound state.[31] Furthermore, D-aspartyl endopeptidase can liberate D-Asp through degradation of proteins which contain D-Asp.[34]

4. Stereochemical Aspects of D-amino acids

There are 22 genetically encoded, proteinogenic amino acids, and many others which have their own biological significance. Chemically, these amino acids differ in the nature of their side chain; some are basic, some acidic, some polar and some non-polar. The chemistry of the side chain of these amino acids confers on them their specific chemical properties which in turn determine their physiological functionality. However, it is not just the chemical nature of the side chain that defines the role of an amino acid; the spatial arrangement of the different groups (–H, –COOH, $–NH_2$ and –R) at C^{α} relative to one another, *i.e.* whether the amino acid is L-enantiomer or D-enantiomer, also has marked effect on how the amino acid will behave in multiple cellular contexts. For example, D-Ser acts as co-agonist or co-activator of N-Methyl-D-Aspartate (NMDA) receptors, whereas L-Ser lacks this ability[35]. Thus, metabolically, stereochemistry of amino acids is as important as their chemistry.

In a polypeptide, an amino acid residue is not allowed to take all possible conformations, since certain conformations will engender steric clashes, rendering them energetically unfavourable. The different conformations in a peptide backbone can arise because of rotation around N–C$^{\alpha}$ and C$^{\alpha}$–C bonds, termed φ and ψ torsion angles respectively. Rotation around the peptide bond, called ω, is restricted because of the partial double bond nature of the peptide bond. It was through painstaking yet meticulous calculations that Ramachandran and his group were able to determine the allowed values of φ and ψ.[36] When φ and ψ values are plotted in an *x-y* coordinate system, the plot/map is referred to as Ramachandran plot/map. Though Ramachandran generated the map using peptides comprising L-amino acids, the same criteria can be used to plot a map for peptides containing all D-enantiomers. In the latter case, the φ and ψ values will remain the same as those for the L-enantiomers; however, the signs will become opposite (*i.e.*, positive φ or ψ value will become negative and *vice versa*) and the map will get an "inverted" appearance with the quadrants flipping from 1st to 3rd and *vice versa*, and from 2nd to 4th and *vice versa*[37,38] (Figure 1).

D-amino acids are usually not found in proteins. Nevertheless, they do occur in many naturally synthesised peptides such as gramicidin S. A DL or LD sequence in an oligopeptide can lead to the formation of "hairpin or β-bend", which reverses the direction of the peptide, thereby bringing about a sharp turn. This helps in greater compaction of the oligopeptide. Further, such a sequence

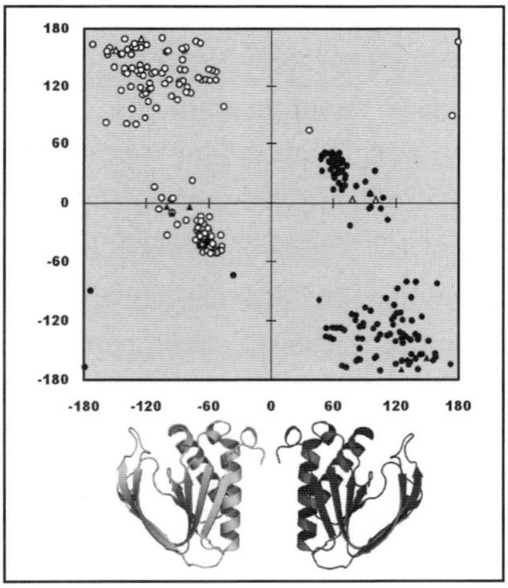

Figure 1. Ramachandran plot for a protein with inverted chirality. Shown on the left (in light grey) is threonyl-tRNA synthetase editing domain from *Pyrococcus abyssi* (PDB entry 3pd2). The same protein, with inverted chirality of amino acids, is shown on the right (in dark grey). Ramachandran plot shows the dihedral angles with open spots corresponding to the protein with L-amino acids and closed spots corresponding to the protein with D-amino acids.

also enables closure of the ring in cyclic peptides.[39] Therefore, the occurrence of D-enantiomers generates such physical or structural characteristics in these peptides that impart greater stability to them and become almost essential for their function.

5. The Problem of Enantioselectivity

The cellular pool and the extracellular environment contain myriad optically active compounds (amino acids, sugars, etc.) of which usually only one of the enantiomers acts as the substrate or ligand for their respective proteins (enzymes, receptors, etc.). These proteins thus face the problem of discrimination between D- and L-isomers of their substrate or ligand. How this enantioselectivity is achieved by the proteins is a matter of debate. The "three-point attachment" (TPA) model, invoked independently by Easson and Ogston, says that three distinct sites on a protein molecule must interact with the complementary points on the substrate or ligand molecule for efficient chiral discrimination.[40,41] Many examples of protein-ligand interaction were then found that validated the TPA model. Later on, another model — the "four-location" model[42,43] — was proposed according to which in addition to a minimum of three distinct complementary interactions between a protein and its ligand, another condition or constraint such as steric hindrance or restriction on the approach route of the ligand to its receptor's binding site must exist for adequate and efficient enantioselection (Figure 2). In the absence of the additional constraint, at least four points of interaction between a receptor and its ligand becomes absolutely essential. It is, therefore, seen that the "three-point attachment" model is a special case of the "four-location" model.[44] It is noteworthy to mention that the interactions these models talk about are not exclusively attractive forces; rather they are a combination of both attractive and repulsive forces.

6. Defence against D-amino acids: Exclusion of D-amino acids from the Translational Machinery

The process of information transfer in the cell is tightly regulated, and all the steps comprising this process — right from replication to translation — undergo strict quality control. The cellular machinery has to ensure that the genetic code is faithfully translated to the amino acid sequence of the polypeptide without incorporating D-amino acids. At the level of translation, aminoacyl-tRNA synthetases (aaRSs) are one of the principal players which maintain fidelity of information transfer. They are responsible for charging of tRNAs with cognate L-amino acids. However, aaRSs face the challenge of discrimination between

D- and L-enantiomers of amino acids. Although most aaRSs are capable of achieving this discrimination at a level that is sufficient for the viability of a cell, some such as tyrosyl-tRNA synthetase (TyrRS), aspartyl-tRNA synthetase (AspRS) and tryptophanyl-tRNA synthetase (TrpRS)[45-47] are much less efficient in doing so. In such cases, D-enantiomers get charged onto tRNAs, and the cell is afflicted with toxicity of accumulation of these dead-end products, and might even face the risk of D-amino acid incorporation in proteins, an event which becomes lethal for the cell. In order to prevent such mishaps, the cell is endowed with enzymes such as D-aminoacyl-tRNA deacylase (DTD) and D-amino acid oxidase (DAAO).

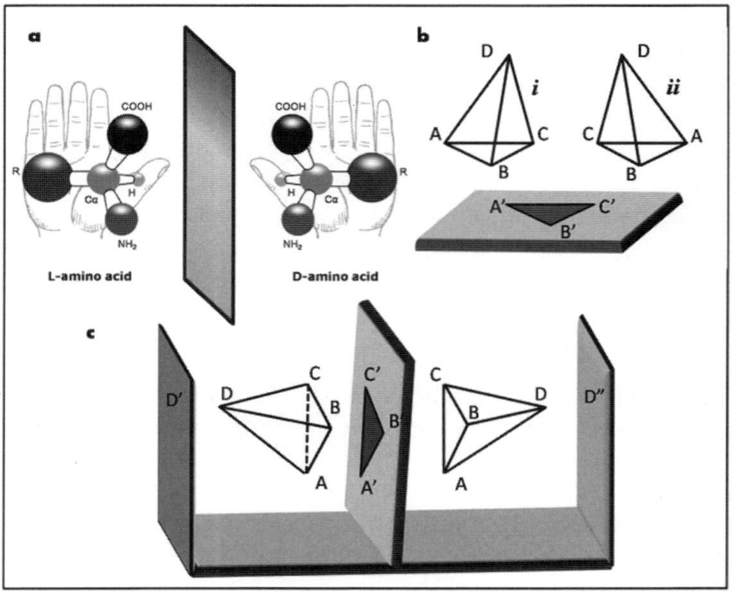

Figure 2. Models for stereospecificity of proteins: a) *Chirality in amino acids.* C^{α} is the chiral centre in amino acids; b) *Three-point attachment model.* If A, B, C from the substrate interact with A', B' and C' from the protein respectively, then only enantiomer (*i*) will bind and not enantiomer (*ii*). The two main assumptions made in this model are coplanarity of the binding sites and fixed approach of the ligand with respect to the binding surface (In this case, from above the plane of the binding site); c) *Four-location model.* This states that in addition to three interaction points, a fourth site is essential to fix the approach of the ligand. Here, D' and D" can interact with D from ligand. However, D' and D" would bind ligands of opposite chirality (Adapted from Ref. 41).

6.1. *D-Aminoacyl-tRNA Deacylase*

In 1967, Calendar and Berg for the first time reported the existence of an enzyme in certain cell extracts which was capable of hydrolyzing the ester linkage between D-Tyr and $tRNA^{Tyr}$, but was not active towards L-Tyr-$tRNA^{Tyr}$. They could partially purify this protein, and they named it "D-tyrosyl RNA deacylase" (DTD).[48] However, with the emerging knowledge of the broad specificity of this enzyme for various D-amino acids charged onto tRNAs such as D-Phe,[48] D-Asp and D-Trp,[47] it later came to be known as D-aminoacyl-tRNA deacylase. Calendar and Berg also made another interesting observation that D-Tyr, once charged onto $tRNA^{Tyr}$, could be incorporated into polypeptides by an *in vitro* protein-synthesis machinery. These key findings by Calendar and Berg had two major consequences — firstly, it underlined the importance of this enzyme in excluding D-amino acids from proteins, and secondly, it rekindled interest (after the discovery of DAAO in mammals in 1935) in addressing the question as to why such enzymes must exist in the cell if the cellular pool lacks the D-isomers. Since the discovery of DTD, active research in this field has firmly established the occurrence of D-amino acids throughout the living world with several of them now known to be playing crucial roles in the cellular physiology as mentioned earlier.

Three distinct types of DTDs have been reported so far. DTD1 is present in bacteria and its homologs are found in almost all eukaryotes; however, it could not be identified in any archaeal species, except *Methanosphaera stadtmanae* and *Methanobrevibacter ruminantium*.[49,50] A structurally different but functionally similar DTD2 was reported in archaea, and its homologs were shown to occur in plants. DTD2 could rescue *E. coli* cells lacking DTD1 from D-amino acid toxicity.[51] Genomes of several organisms were shown to lack either *dtd1* or *dtd2*. However, yet another DTD, DTD3, was discovered in the cyanobacterium, *Synechocystis*. Inactivation of *dtd3* gene in *Synechocystis* led to marked reduction in the growth of the organism in the presence of external D-Tyr. Moreover, DTD3 could functionally substitute DTD1 in a Δ*dtd E. coli* strain.[52] All these studies indicate that nature has endowed the living systems with a ubiquitous enzymatic activity which denies the hijacking of the translational machinery by D-amino acids.[52]

DTD hydrolyses the ester bond between a D-amino acid and the terminal adenosine of the corresponding tRNA, thereby releasing free D-amino acid and tRNA as the products. It, therefore, prevents toxicity arising out of lack of adequate amounts of L-aminoacyl-tRNA species in the cell.[50] This view is in contrast to what had been proposed earlier by Calendar and Berg — that

D-amino acids, once charged onto tRNAs, can be incorporated into peptides by the ribosome. More recent findings suggest that cellular toxicity of D-amino acids is manifested because of the accumulation of metabolically inactive D-aminoacyl-tRNA species rather than the incorporation of D-enantiomers in polypeptides.[50] Moreover, the stereospecificity of the translational machinery (EF-Tu and ribosome) and the binding of aminoacyl-tRNA species to the ribosome favours the L-enantiomer by more than 1000-fold for its incorporation in proteins.[50,53-56] The evidence indicates that using multiple checkpoints, D-amino acids are largely excluded from proteins even if they are charged onto tRNAs. The first crystal structure of DTD from *E. coli* became available in 2001 (PDB id: 1JKE).[57] It showed that the homodimeric enzyme had a structure different from any of the known protein structures available at that time. However, since there was no ligand, the authors could not propose any mechanism of enzyme action. On the basis of multiple sequence alignment of DTDs from various bacteria and eukaryotes, they could only speculate that the active site pocket in each monomer comprised two conserved sequence motifs — SQFTL and NXGP(V/F)T. A little later, the structure of DTD from *Haemophilus influenzae* was solved, and based on modelling studies with yeast $tRNA^{Phe}$, it was proposed that the enzyme follows a substrate-assisted deacylation mechanism.[58] Nevertheless, the mechanism by which DTD removes D-amino acids from tRNAs is still not fully understood owing to the lack of complex structures with D-aminoacyl-tRNA-mimicking ligands, and whatever has been proposed in this regard needs further scrutiny.[59] Furthermore, the mechanistic basis of enantioselection by the enzyme is also not known; whether DTD selects D- or rejects L-enantiomers is a question that needs to be addressed.

In 2005, an interesting discovery was made wherein it was shown that *E. coli* DTD bears remarkable structural similarity (r.m.s.d. of 1.16 Å over 112 C^{α} atoms) with the N-terminal editing domain of threonyl-tRNA synthetase from the archaeon, *Pyrococcus abyssi* (Pab-NTD), though the two proteins share only 14% sequence identity.[60,61] This finding suggested that Pab-NTD and modern DTDs probably evolved from a common ancient D-amino acid-editing enzyme, which in a primordial scenario might have played a crucial role in the perpetuation of homochirality.[61] A model was proposed which suggested that functional or covalent coupling of a DTD-like enzyme to a primordial aaRS could have played a crucial role in the perpetuation of homochirality during the formation of early proteins (Figure 3). Based on structural and functional (hydrolysis of aminoacyl-tRNA ester) similarity between the two, it is believed that they share a common enzymatic mechanism as well, which, in the case of Pab-NTD, has been proposed to be RNA-assisted and water-mediated.[62,63] However, a proper ligand-bound structure of DTD would be more helpful in elucidating its unique mechanism of removal of any D-amino acid from any tRNA.

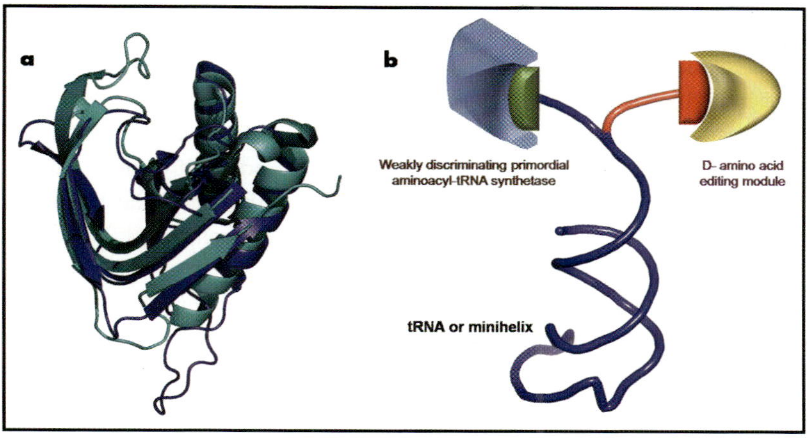

Figure 3. A model for perpetuation of homochirality in proteins: a) Superimposition of *E. coli* DTD (blue) with threonyl-tRNA synthetase editing domain (cyan) from *Pyrococcus abyssi*; b) A DTD-like fold coupled to primordial aminoacyl-tRNA synthetases may have helped in enantioselection of amino acids and thereby perpetuation of homochirality.

6.2. *D-Amino acid Oxidase*

D-amino acid oxidase (DAAO) was discovered by Hans Krebs, who, in 1935, could successfully show the presence of an enzymatic activity in kidney and liver of several mammals that oxidatively deaminates D-amino acids. Krebs was also able to show that this enzyme is distinct from the one that acts on the L-isomers.[19] The physiological significance of DAAO could not be ascertained immediately as living systems other than bacteria were believed to be devoid of D-amino acids. Research in the last two decades of the twentieth century established the presence of D-amino acids throughout the living world. It also became evident that the D-isomers have several important biological functions. All these findings led to the realization that DAAO, along with other molecular factors, helps maintain the intricate balance of D- and L-isomers in the cellular pool which is crucial for normal functioning of the cell.

DAAO is a peroxisomal homodimeric FAD-containing enzyme that catalyses oxidative deamination of neutral, polar and basic amino acids. It does not act on acidic amino acids and, therefore, the cell has D-aspartate oxidase which deaminates both D-Asp and D-Glu. DAAO is one of the most thoroughly studied enzymes, and the mechanism of its action is well understood. The enzyme extracts α-hydrogen from the substrate leading to the formation of α-imino acid and concomitant reduction of FAD. The imino acid spontaneously converts to α-keto acid, thereby releasing ammonia. The enzyme-bound reduced FAD is oxidised using molecular oxygen with the consequent formation of hydrogen peroxide.

Various important roles have been ascribed to DAAO. For instance, the enzyme helps yeast to utilize D-amino acids as carbon and energy source.[64-67] DAAO is also involved in regulating the levels of D-Ser in the brain, where D-Ser acts as a neuromodulator.[68-70] Hyperactivity of the enzyme in the brain has been implicated in schizophrenia.[71,72]

It is worth mentioning here that bacteria have a different enzyme, called D-amino acid dehydrogenase (DAAD), which helps them grow on D-amino acids, and also prevents accumulation-engendered toxicity of D-amino acids.[64] Biochemically, DAAD differs from DAAO in that it uses an electron acceptor (such as CoQ) other than oxygen as the second substrate.[73,74]

6.3. *Elongation Factor Tu and Ribosome*

It is quite evident that every step of translation acts as a checkpoint to prevent incorporation of wrong amino acids into proteins, including the D-enantiomers. In this orchestra of quality control, elongation factor Tu (EF-Tu) and ribosome also seem to play important roles in addition to the initial aaRS checkpoint.

As early as in 1980, it was shown through hydrolysis protection experiments that differences in binding affinity of various aminoacyl-tRNAs (aa-tRNAs) with EF-Tu.GTP depend on the nature of the aminoacyl moiety and not on the tRNA species.[75] Variations in the GC content of the acceptor stem, or changing the length of the variable arm of the tRNA did not influence the binding of aa-tRNA with EF-Tu.GTP. All non-acylated tRNAs showed similar, though weak, binding with EF-Tu.GTP. However, aa-tRNAs with different bound amino acids displayed a large variation in binding. For instance, Phe-$tRNA^{Phe}$ bound EF-Tu.GTP about 10 times as strongly as did Ser-$tRNA^{Ser}$. There was hardly any binding detectable with D-Tyr-$tRNA^{Tyr}$. More than 10-fold excess of D-Tyr-$tRNA^{Tyr}$ failed to competitively inhibit binding of L-Tyr-$tRNA^{Tyr}$ with EF-Tu.GTP.[75] Moreover, protection conferred by EF-Tu.GTP on the labile acyl (ester) bond between amino acid and tRNA moieties of aa-tRNA was absent in the case of D-Tyr-$tRNA^{Tyr}$.[53] These observations clearly indicated that stereospecificity of EF-Tu prevents, to a great extent, D-amino acid incorporation into polypeptides. This was in contrast to Calender and Berg's earlier postulate that D-amino acids coupled to corresponding tRNAs can be inducted into proteins by the ribosomal machinery.[48] However, a clear structural basis for such a selection by EF-Tu is not yet available.

The ribosome too has been shown to be capable of sifting out non-cognate aa-tRNA species during translation. In the presence of EF-Tu.GTP, *in vitro* dipeptide formation on the ribosome is 30 times faster with L-Tyr-$tRNA^{Tyr}$ than with D-Tyr-$tRNA^{Tyr}$.[53] This difference has been attributed to combined stereospecificity at various steps — binding of EF-Tu.GTP.aa-tRNA ternary complex with the ribosome, subsequent rate of GTP hydrolysis, and the fraction of aa-tRNA dissociating from the ribosome upon GTP hydrolysis.[53] Bhuta *et al.* demonstrated some degree of stereospecificity in peptidyl transferase activity of the ribosome.[76] They could show that the acceptor ($1/K_m^{app}$) activity of an aminoacyl-dinucleoside phosphate C-A(D-Phe) is between 1 and 2 orders of magnitude lower than that of C-A(L-Phe), while the V_{max} for C-A(D-Phe) is ~2 times lower than that for C-A(L-Phe). Extending the stereospecificity of peptidyl transferase activity further, a subsequent study showed that N-acetyl-D-phenylalanyl-$tRNA^{Phe}$ acted as a poor donor substrate in peptidyl transferase reaction when compared to N-acetyl-L-phenylalanyl-$tRNA^{Phe}$, the former being only 7% as efficient as a donor substrate than the latter.[54] Structural analysis of the peptidyl transferase centre (PTC) has revealed a possible role of a universal nucleotide U2585 in chiral discrimination against D-aminoacyl-tRNAs.[77] Thus, the ribosome further mitigates the problem of D-amino acid incorporation into polypeptides. Overall, it seems that the interplay of all the mechanisms discussed above ensures against the incorporation of D-amino acids in cellular proteins (Figure 4).

7. Perspective

During the course of biological evolution, only L-amino acids were selected for making polypeptides. What forces or factors were involved in making this selection is still a matter of speculation. Thus, as life evolved, the living systems acquired the ability of synthesising polypeptides exclusively from L-amino acids. However, this did not lead to the elimination of D-amino acids from the living systems. Instead they were assigned certain critical functions which were

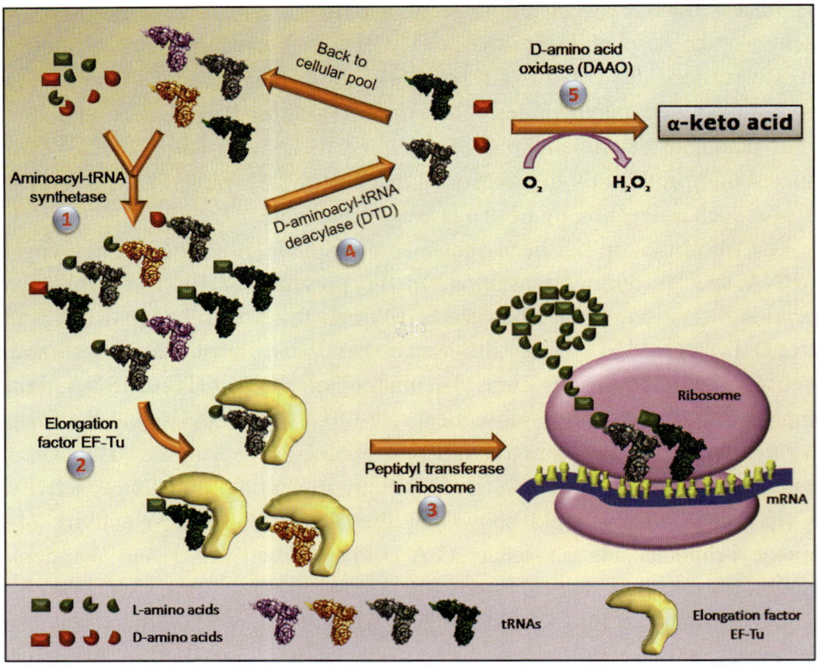

Figure 4. A schematic representation showing cellular mechanisms that exclude D-amino acids from protein synthesis.

systems. Instead they were assigned certain critical functions which were probably not possible for the L-enantiomers to carry out. Nevertheless, D-amino acids presented the cell with a challenging circumstance — that of preventing them from getting incorporated into polypeptides. Nature, therefore, evolved several mechanisms to exclude D-amino acids from proteins. As we now know, every step in translation acts as an "enantioselective" checkpoint to prevent the incorporation of D-amino acids into proteins. Homochirality in proteins was thus maintained and perpetuated during the course of biological evolution. However, structural and mechanistic insights into some of the key D-amino acid exclusionary processes remain to be revealed. Furthermore, mutations that lead to a reduced ability of these processes and their link to disease manifestations are unexplored territories.

Acknowledgements

S. B. R. and S. A. acknowledge a Junior Research Fellowship and a Senior Research Fellowship respectively from Council of Scientific and Industrial Research (CSIR), India. R. S. acknowledges funding from Swarnajayanti Fellowship of DST, India.

References

1. G. E. Tranter, *Biosystems* **20**, 37 (1987).
2. A. Szabo-Nagy and L. Keszthelyi, *Proc. Natl Acad. Sci. USA* **96**, 4252 (1999).
3. S. F. Mason and G. E. Tranter, *Mol. Phys.* **53**, 1091 (1984).
4. R. Berger and M. Quack, *Chem. Phys. Chem.* **1**, 57 (2000).
5. L. N. Johnson, *Eur. Rev.* **13**, 77 (2005).
6. S. F. Mason and G. E. Tranter, *Chem. Phys. Lett.* **94**, 34 (1983).
7. G. E. Tranter, *Chem. Phys. Lett.* **120**, 93 (1985).
8. P. Cintas, *Angewandte Chemie International Edition* **41**, 1139 (2002).
9. G. Goodman and M. E. Gershwin, *Exp. Biol. Med. (Maywood)* **231**, 1587 (2006).
10. J. M. Bailey, *FASEB J.* **12**, 503 (1998).
11. K. Tamura, *Biosystems* **92**, 91 (2008).
12. K. Tamura and P. Schimmel, *Science* **305**, 1253 (2004).
13. K. Tamura and P. R. Schimmel, *Proc. Natl Acad. Sci. USA* **103**, 13750 (2006).
14. D. G. Blackmond, *Proc. Natl Acad. Sci. USA* **101**, 5732 (2004).
15. M. J. Osborn, *Annu. Rev. Biochem.* **38**, 501 (1969).
16. M. Bodanszky and D. Perlman, *Science* **163**, 352 (1969).
17. N. Fujii and T. Saito, *Chem. Rec.* **4**, 267 (2004).
18. C. T. Walsh *et al.*, *Curr. Opin. Chem. Biol.* **5**, 525 (2001).
19. H. A. Krebs, *Biochem. J.* **29**, 1620 (1935).
20. N. Fujii *et al.*, *Biochem. Biophys. Res. Commun.* **265**, 746 (1999).
21. N. Fujii and Y. Kaji, *J. Chromatogr. B Analyt. Technol. Biomed. Life Sci.* **879**, 3141 (2011).
22. A. E. Roher *et al.*, *J. Biol. Chem.* **268**, 3072 (1993).
23. D. S. Dunlop *et al.*, *Biochem. Biophys. Res. Commun.* **141**, 27 (1986).
24. A. Hashimoto *et al.*, *FEBS Lett.* **296**, 33 (1992).
25. N. G. Srinivasan *et al.*, *J. Biol. Chem.* **237**, 3844 (1962).
26. H. K. Kuramitsu and J. E. Snoke, *Biochim. Biophys. Acta* **62**, 114 (1962).
27. T. Yoshimura and M. Goto, *FEBS J.* **275**, 3527 (2008).
28. P. Conti *et al.*, *Chem. Rev.* **111**, 6919 (2011).
29. B. Reina-San-Martin *et al.*, *Nat. Med.* **6**, 890 (2000).
30. P. P. Taylor *et al.*, *Trends Biotechnol.* **16**, 412 (1998).

31. M. Katane and H. Homma, *J. Chromatogr. B Analyt. Technol. Biomed. Life Sci.* **879**, 3108 (2011).
32. S. D. Heck *et al.*, *Science* **266**, 1065 (1994).
33. S. D. Heck *et al.*, *Proc. Natl Acad. Sci. USA* **93**, 4036 (1996).
34. T. Kinouchi *et al.*, *Biochem. Biophys. Res. Commun.* **314**, 730 (2004).
35. J. P. Mothet *et al.*, *Proc. Natl Acad. Sci. USA* **97**, 4926 (2000).
36. C. Ramakrishnan and G. N. Ramachandran, *Biophys. J.* **5**, 909 (1965).
37. R. Mahalakshmi and P. Balaram, in *D-Amino Acids: A New Frontier in Amino Acid and Protein Research* (Nova Science Publishers, Inc., 2007), p. 415.
38. C. Ramakrishnan, *Resonance* **6**, 48 (2001).
39. R. Chandrasekaran *et al.*, *Biochim. Biophys. Acta* **303**, 14 (1973).
40. L. H. Easson and E. Stedman, *Biochem. J.* **27**, 1257 (1933).
41. A. G. Ogston, *Nature* **162**, 963 (1948).
42. A. D. Mesecar and D. E. Koshland, Jr., *Nature* **403**, 614 (2000).
43. A. D. Mesecar and D. E. Koshland, Jr., *IUBMB Life* **49**, 457 (2000).
44. R. Bentley, *Arch. Biochem. Biophys.* **414**, 1 (2003).
45. R. Calendar and P. Berg, *Biochemistry* **5**, 1681 (1966).
46. R. Calendar and P. Berg, *Biochemistry* **5**, 1690 (1966).
47. J. Soutourina *et al.*, *J. Biol. Chem.* **275**, 32535 (2000).
48. R. Calendar and P. Berg, *J. Mol. Biol.* **26**, 39 (1967).
49. J. Soutourina *et al.*, *J. Biol. Chem.* **274**, 19109 (1999).
50. O. Soutourina *et al.*, *J. Biol. Chem.* **279**, 42560 (2004).
51. M. L. Ferri-Fioni *et al.*, *J. Biol. Chem.* **281**, 27575 (2006).
52. S. Wydau *et al.*, *J. Biol. Chem.* **284**, 14096 (2009).
53. T. Yamane *et al.*, *Biochemistry* **20**, 7059 (1981).
54. T. G. Heckler *et al.*, *Biochemistry* **27**, 7254 (1988).
55. J. R. Roesser *et al.*, *Biochemistry* **28**, 5185 (1989).
56. J. D. Bain *et al.*, *Biochemistry* **30**, 5411 (1991).
57. M. L. Ferri-Fioni *et al.*, *J. Biol. Chem.* **276**, 47285 (2001).
58. K. Lim *et al.*, *J. Biol. Chem.* **278**, 13496 (2003).
59. T. K. Bhatt *et al.*, *J. Biol. Chem.* **285**, 5917 (2010).
60. D. J. Rigden *RNA* **10**, 1845 (2004).
61. S. Dwivedi *et al.*, *Nat. Struct. Mol. Biol.* **12**, 556 (2005).
62. T. Hussain *et al.*, *EMBO J.* **25**, 4152 (2006).
63. T. Hussain *et al.*, *Proc. Natl Acad. Sci. USA* **107**, 22117 (2010).
64. L. Pollegioni *et al.*, *Appl. Microbiol. Biotechnol.* **78**, 1 (2008).
65. M. P. Simonetta *et al.*, *J. Gen. Microbiol.* **135**, 593 (1989).
66. G. Molla *et al.*, *Yeast* **20**, 1061 (2003).
67. Y. Sakai *et al.*, *Yeast* **14**, 1175 (1998).
68. K. Horiike *et al.*, *Brain Res.* **652**, 297 (1994).
69. K. Horiike *et al.*, *Brain Res. Bull.* **19**, 587 (1987).

70. L. Pollegioni *et al.*, *Cell Mol. Life Sci.* **64**, 1373 (2007).
71. S. V. Khoronenkova and V. I. Tishkov, *Biochemistry (Mosc)* **73**, 1511 (2008).
72. I. Chumakov *et al.*, *Proc. Natl Acad. Sci. USA* **99**, 13675 (2002).
73. P. J. Olsiewski *et al.*, *J. Biol. Chem.* **255**, 4487 (1980).
74. H. Jones and W. A. Venables, *FEBS Lett.* **151**, 189 (1983).
75. A. Pingoud and C. Urbanke, *Biochemistry* **19**, 2108 (1980).
76. A. Bhuta *et al.*, *Biochemistry* **20**, 8 (1981).
77. I. Agmon *et al.*, *FEBS Lett.* **567**, 20 (2004).
78. H. Lam *et al.*, *Science* **325**, 1552 (2009).
79. D. W. Urry, *Proc. Natl Acad. Sci. USA* **68**, 672 (1971).
80. B. A. Wallace, *Biophys. J.* **49**, 295 (1986).
81. B. M. Burkhart *et al.*, *Biopolymers* **51**, 129 (1999).
82. E. Conti *et al.*, *EMBO J.* **16**, 4174 (1997).
83. A. L. Llamas-Saiz *et al.*, *Acta Crystallogr. D Biol. Crystallogr.* **63**, 401 (2007).
84. A. Mor *et al.*, *FEBS Lett.* **255**, 269 (1989).
85. M. Broccardo *et al.*, *Br. J. Pharmacol.* **73**, 625 (1981).
86. Y. Kamatani *et al.*, *Biochem. Biophys. Res. Commun.* **160**, 1015 (1989).
87. N. Ohta *et al.*, *Biochem. Biophys. Res. Commun.* **178**, 486 (1991).
88. J. J. Corrigan, *Science* **164**, 142 (1969).
89. G. Meroni *et al.*, *ARKIVOC* **2009**, 265 (2009).
90. A. Hiroki and Y. Naoko, *Viva Origino* **30**, 221 (2002).
91. Y. Nagata *et al.*, *FEBS Lett.* **444**, 160 (1999).
92. A. Hashimoto *et al.*, *Eur. J. Neurosci.* **7**, 1657 (1995).
93. F. Errico *et al.*, *J. Neurosci.* **28**, 10404 (2008).
94. A. Hashimoto and T. Oka, *Prog. Neurobiol.* **52**, 325 (1997).
95. J. A. Lee *et al.*, *Biochem. Biophys. Res. Commun.* **231**, 505 (1997).
96. S. Ishio *et al.*, *Neurosci. Lett.* **249**, 143 (1998).
97. J. A. Lee *et al.*, *Brain Res.* **838**, 193 (1999).
98. Z. Long *et al.*, *Biochem. Biophys. Res. Commun.* **276**, 1143 (2000).
99. G. D'Aniello *et al.*, *Endocrinology* **141**, 3862 (2000).
100. H. Wang *et al.*, *J. Endocrinol.* **167**, 247 (2000).
101. M. Pampillo *et al.*, *Endocrine* **15**, 309 (2001).
102. A. Hashimoto *et al.*, *J. Neurochem.* **60**, 783 (1993).
103. M. J. Schell *et al.*, *Proc. Natl Acad. Sci. USA* **92**, 3948 (1995).
104. M. J. Schell *et al.*, *J. Neurosci.* **17**, 1604 (1997).

INDUCTION OF FOLDED STRUCTURES IN DESIGNED PEPTIDES USING CONFORMATIONALLY CONSTRAINED RESIDUES

SUBRAYASHASTRY ARAVINDA[1], NARAYANASWAMY SHAMALA[1] AND PADMANABHAN BALARAM [2]

[1] *Department of Physics*

[2]*Molecular Biophysics Unit, Indian Institute of Science*

Bangalore, 560 012, India

Folding into compact globular structures, with well-defined modules of secondary structure, appears to be a characteristic of long polypeptide chains, with a specific patterning of coded amino acid residues along the length of sequence. Cooperative hydrogen bond driven secondary structure formation and solvent forces, which contribute favorably to the entropy of folding, by promoting compaction of the polymeric chain, have long been discussed as major determinants of the folding process. First principles design approaches, which use non-coded amino acids, employ an alternative structure directing strategy, by using amino acid residues which exhibit a strong conformational bias for specific regions of the Ramachandran map. This overview of ongoing studies in the authors' laboratory, attempts to explore the use of conformationally restricted amino acid residues in the design of peptides with well-defined secondary structures. Short peptides composed of 20 genetically coded amino acids usually exist in solution as an ensemble of equilibrating conformations. Apolar peptide sequences, which are readily soluble in organic solvents like chloroform and methanol, facilitate formation of structures which are predominately driven by intramolecular hydrogen bond formation. The choice of sequences containing residues with a limited range of conformational choices strongly favors formation of local turn structures, stabilized by short range intramolecular hydrogen bonds. Two residue β-turns can nucleate either helical or hairpin folding, depending on the precise conformation of the turn segment. Restriction of the conformational space available to amino acid residues is easily achieved by introduction of an additional alkyl group at the Cα carbon atom or by side chain backbone cyclization, as in proline. Studies of synthetic sequences incorporating two prototype residues α-aminoisobutyric acid (Aib) and D-proline (DPro) illustrate the utility of the strategy in construction of helices and hairpins. Extensions to the design of conformationally switchable sequences and structurally defined hybrid peptides containing backbone homologated residues are also surveyed.

1. Introduction

The nature of the forces that contribute to the stability of the folded structures of globular proteins are reasonably well understood.[1,2] Non-bonded (van der Waals) interactions and hydrogen bonds, principally those involving backbone amide groups, are predominant contributors to favorable enthalpies of folding, while hydrophobic effects in aqueous media contribute entropically.[3,4] Spontaneous folding from an unstructured state requires that the effect of conformational entropy in the unfolded state must be more than adequately balanced by the enthalpy and entropy factors contributing to compaction of polypeptide chains.[5] Two opposing views of energetic contributions to the folded state have emerged. In one view, backbone hydrogen bonds play an important role in determining the formation of units of secondary structures, like helices and β-sheets.[6] Backbone-backbone interactions have been suggested to be energetically more favorable than backbone-water interactions. In an alternative view, solvent forces drive chain collapse, with the clustering of amino acid side chains resulting in favorable non-bonded interactions.[7,8] This view necessarily incorporates sequence information. Even as polypeptide chains move downhill towards a global energy minimum the possibility remains that kinetically trapped structures may be formed, which are separated by substantial barriers from escaping towards a more stable state. Practically, there is little evidence for the widespread occurrence of metastable structures, arguing that folding pathways facilitate an approach to the global minimum; a feature that matches folding rates with the stability of the folded structures.[9,10] Polypeptide conformations are most conveniently analyzed by considering two degrees of torsional freedom (ϕ,ψ) at each residue in the polypeptide chain. The Ramachandran map is a powerful means of reducing an analytical problem in three-dimensions to an exercise in two-dimensional space, in which the torsional variables ϕ and ψ are implicitly representative of three dimensional atomic arrangements.[11] The principle of steric overlap, in which two non-bonded atoms cannot approach one another at distances that are less than sum of their van der Waals radii, dramatically restricts the allowed regions of conformational space for residues which bear substituents at the C^{α} carbon atom. Protein and peptide design provides an opportunity to experimentally verify or validate our understanding of the principles by which folded structures are realized. In peptide design approaches, the role of solvent forces, primarily the entropically driven hydrophobic effect, can be eliminated by the choice of sequences which are appreciably soluble in relatively inert solvents like chloroform. In such media solvent competition for backbone hydrogen bond sites is also minimized, facilitating chain folding under conditions in which local non-bonded interactions and intramolecular hydrogen bonds provide the principal driving forces. The successful construction of peptides with designed structures is then a test of our understanding of the factors that determine polypeptide chain folding

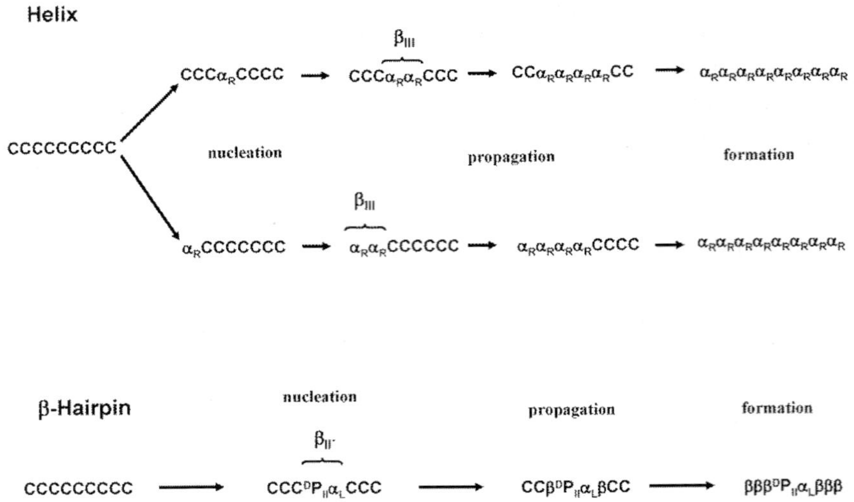

Figure 1: Schematic representation of structure nucleation by biasing local conformations in ϕ-ψ space. C: random; α_R/α_L: right/left-handed $3_{10}/\alpha$-helical; ${}^{D}P_{II}$: polyproline (right-handed, $\phi \sim +60°$, $\psi \sim -120°$); β: extended β-strand; $\beta_{III}/\beta_{II'}$: Type III/II′ β-turns.

in the absence of competing solvent forces. We describe in this article the use of conformationally constrained residues, which impose a local bias on the range of accessible conformations of an appropriately positioned residue, serving to nucleate the formation of peptide helices and hairpins.

In principle, helix nucleation may be achieved by placing residues which are largely constrained to the helical region of ϕ,ψ space. Local conformational restrictions can promote turn formation, with successive hydrogen bond formation resulting in the generation of incipient helical structures, which can further propagate to form a helical segment. Helix nucleating residues may be placed either in the centre of the sequence or at the N-terminus. β-hairpin generation may be achieved by placement of two residue tight turns, type I′ or II′ β-turn in the centre of the sequence (Figure 1). In both cases, internal hydrogen bond formation between acceptor CO groups and donor NH groups constitutes the major driving force for creation of the nucleus and subsequent propagation of the structures. Despite the restrictions imposed by the Ramachandran map, a fairly large region of conformational (ϕ,ψ) space remains available to individual amino acid residues. Figures 2 and 3 illustrate the formation of folding nuclei and subsequent propagation of the structure in the case of helical and hairpin segments. Nucleation may be formally viewed as a process in which contiguous residues cluster into a limited region of conformational space, which corresponds to a state where intramolecular backbone hydrogen bond formation becomes possible. For example, clustering of two successive residues in the right-handed helical region of ϕ,ψ space results in the formation of a type III β-turn structure. In a type III β-turn ϕ,ψ angles at both corner residues (i+1, i+2) lie in the helical region of ϕ-ψ space, -60°, -30°.

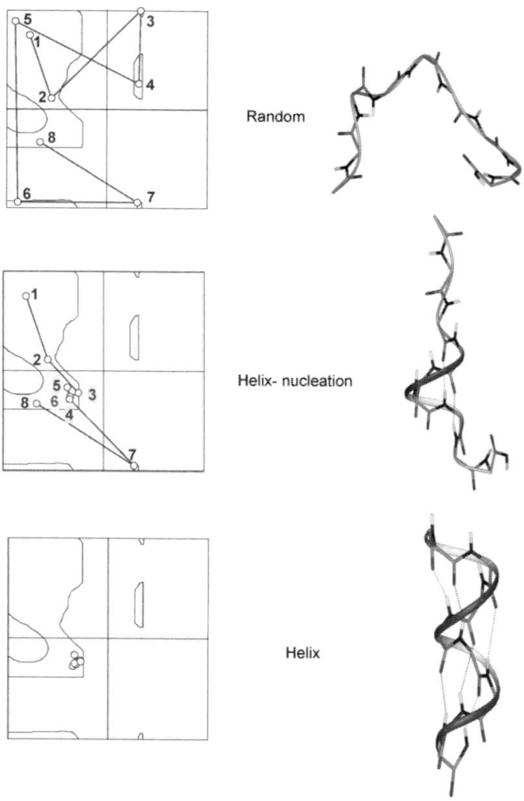

Figure 2: Pictorial representation of the formation of a helix from a random structure. The clustering of residues in the helical region of conformational space is shown. The outline of Ramachandran map is for L-Ala. A large region of ϕ,ψ space is accessible in random coils for Gly. Helix nucleation in the central segment is illustrated when residues 3 to 6 form a local cluster.

Repetition of the type III β-turn structures results in the generation of a 3_{10}-helix. Type II′ β-turns, are most frequently found in peptide β-hairpins. In this case, the turn segment stands isolated by the flanking strands, which recognize one another to form a registered, backbone hydrogen bonded, β-hairpin. Figure 4 illustrates the role of β-turn structures in nucleating both helices and hairpins. In the case of formation of helices and hairpins, cooperative hydrogen bond formation can drive rapid formation of a defined element of secondary structure. Structure nucleation can therefore be promoted by locally biasing the extent of conformational choices available to an individual residue in a polypeptide chain. Peptide design approaches, which attempt to direct the course of polypeptide chain folding using backbone hydrogen bonds as the major structure promoting force, are necessarily based on the use of non-protein residues in which the allowed regions of conformational space are severely limited.

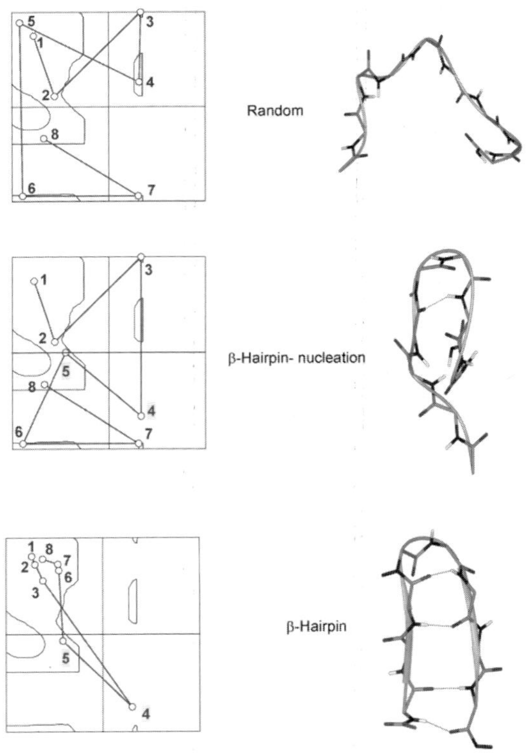

Figure 3: Pictorial representation of the formation of β-hairpins from a random structure along with Ramachandran plot showing the clustering of residues in the β-region of conformational space, with residues 4, 5 forming a Type II′ β-turn.

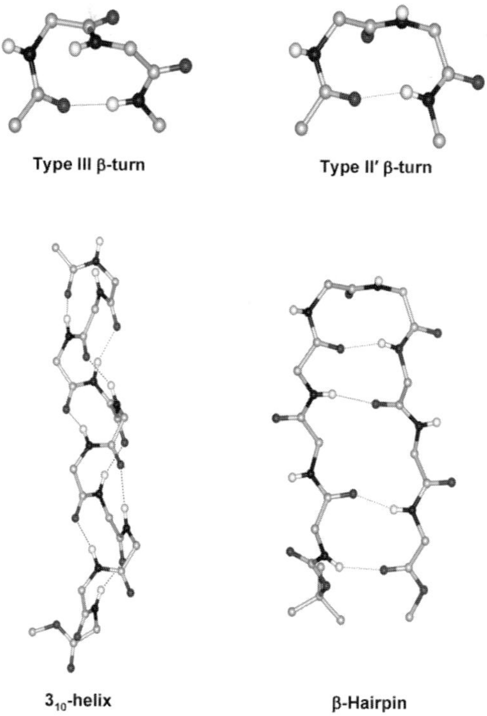

Figure 4: The role of β-turns in nucleating secondary structures. Type III β-turns nucleate helices (left) whereas a type II′ β-turn nucleates β-hairpins (right).

2.1 Aib and D-Proline residues

The Ramachandran allowed regions for α-aminoisobutyric (Aib) residue may be derived by overlap of the allowed regions for L and D-alanine residues, which are related by inversion about the origin of the φ-ψ map.[12] As seen from Figure 5, it is clear that the presence of geminal dialkyl groups at the C^{α} carbon atom results in a dramatic reduction in available conformational space, limiting Aib residues to the conformation which corresponds to right (α_R) and left (α_L) helices. The influences of tetra-substituted C^{α} atoms on polypeptide backbone conformations have been considering in terms of the classical Thorpe-Ingold effect.[13] Conformational energy calculations confirm that the left and right-handed helical regions are indeed significantly deep energy minima for Aib and related α,α-dialkyl residues.[14,15] The results of crystallographically determined conformations of nearly 1400 Aib residues in both synthetic and natural peptides clearly supports this expectation, as evident by the clustering of experimentally observed conformations. Insertion of an Aib residue into a peptide chain should therefore promote helix formation. A large body of experimental studies both in the solid state and in solution have established the overwhelming tendency of Aib containing sequences to form stable helices even in short peptides.[16-22] The example illustrated in Figure 5 demonstrates that centrally positioned Aib residue suffices to stabilize a helical structure over a ten residue peptide sequence. Aib containing peptides have a remarkable tendency to crystallize, especially in the case of relatively long sequences (10 to 16 residues) which form helices. The facility with which apolar helical peptides pack in crystals and the greater conformational homogeneity of the sequences in solution may be contributory factors in determining high crystallianity in these cases. Aib residues have been infrequently observed to adopt conformations in non-helical regions of ϕ-ψ space. A particularly interesting example, is the case of the peptide Boc-Cys-Val-Aib-Ala-Leu-Cys-NHMe, where the central Aib-^{L}Ala segmet adopts a type II′ β-turn conformation with Aib having backbone torsional angles $\phi = 49.9°$, $\psi = -131.2°$.[23] A recent survey reveals that nonhelical conformations of Aib residues have almost always been observed only in short linear sequences up to the level of tripeptides.[24] The overwhelming tendency of Aib residues to adopt either α_R or α_L conformations has led to a number of studies in which this strong conformational bias has been used to drive structure formation.

Of the 20 genetically coded amino acids, proline is unique in being the residue with the greatest degree of conformational restriction. The constraints of pyrrolidine ring formation limit available conformational space, with the torsional angle ϕ being restricted to values -60°±20° for L-Proline and +60°±20° for D-Proline.[19] Figure 5 highlights the available region of ϕ-ψ space for ^{D}Pro residues. Crystallographically observed values in model peptides, cluster predominately in the region $\phi \approx +60°$, $\psi \approx -120°$ ($^{D}P_{II}$). This conformation is favorable for nucleating type II′ β-turns, which in turn facilitate hairpin formation, as exemplified by the structure of the octapeptide shown in Figure 5.[25]

The stereochemical constraints imposed by strategically positioned Aib and ^{D}Pro residues have been extensively used in the synthetic design of secondary structure modules. The formation of ordered conformations in short oligopeptides promotes crystallization, permitting definitive characterization of three-dimensional structures by X-ray diffraction. In the case of Aib containing sequences both, 3_{10} and α-helical structures which differ marginally in ϕ,ψ

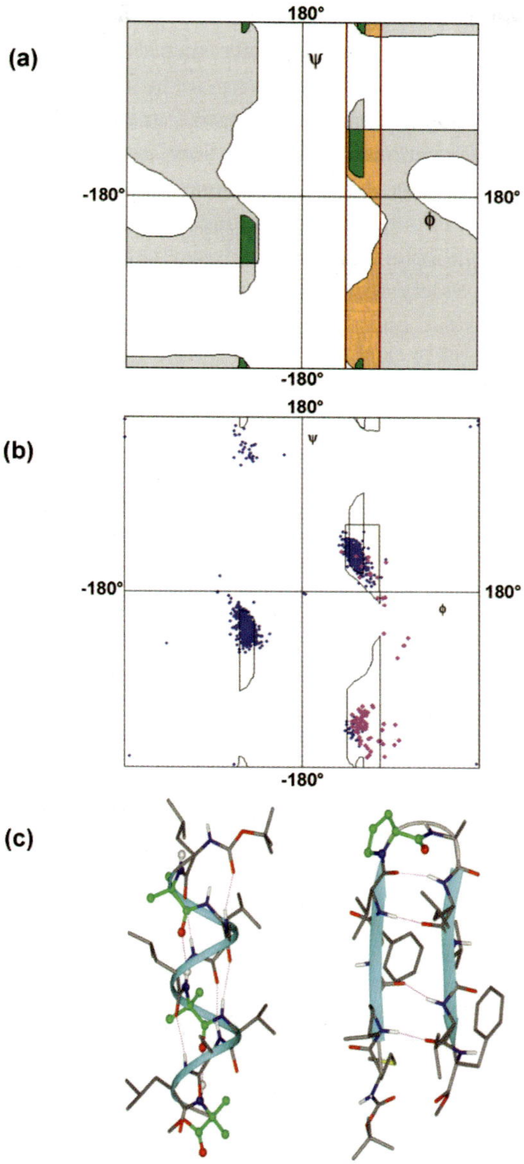

Figure 5: (a) Ramachandran map showing the allowed regions for the residues, L-Ala and D-Ala (gray) and ^{D}Pro (gold) and Aib (green). (b) The distribution of Aib residues (1430, blue) and ^{D}Pro (93, magenta) residues from peptide crystal structures. (c) Representative examples of a helix and a β-hairpin characterized in crystals. (Left) Boc-Leu-Aib-Val-Ala-Leu-Aib-Val-Ala-Leu-Aib-OMe.[26] (Right) Boc-Met-Leu-Phe-Val-^{D}Pro-Ala-Leu-Val-Val-Phe-OMe.[25]

values, have been characterized. While the former is stabilized by successive 4→1 hydrogen bonds, the latter exhibits successive 5→1 hydrogen bond formation.[26] There has been considerable debate in the literature on the relative stability on $3_{10}/\alpha$-helical structures and dynamic interconversion between these closely related conformations. Relatively low activation barriers may be anticipated for small changes in the backbone torsional angles in the helical region, suggesting that minor changes in sequence and environmental conditions can influence the equilibrium between the two helical forms. Indeed, mixed helical structures are widely observed in the crystal structures of Aib containing peptides.[27] A substantial body of experimental evidence suggests that Aib rich peptides, particularly Aib oligomers almost exclusive favor 3_{10}-helical conformations, while sequences with lower Aib content appear to tilt towards an α-helical conformation.[16] The strong helix nucleation ability of Aib is dramatically illustrated in Figure 6, which shows the experimentally determined crystal state conformations of four peptides containing only a single Aib residue positioned approximately in the middle of the sequence. Notably even a single Aib residue is able to promote a helical structure in peptides varying in length from 8 to 11 residues.[28-31] Interestingly, insertion of a β-amino acid residue, β^3-homophenylalanine, in the octapeptide[28] or the insertion of a hydroxy acid (lactic acid) in the 11- residue peptide does not perturb the overall helical conformation.[31] Thus helix construction can be readily achieved by incorporation of α,α-dialkyl residues, with helix propagation extending on either side of the inserted residue. Helical modules containing as many as 4 to 5 helical turns can therefore be prepared by synthetic design. The crystal structure of a 19-residue peptide containing three Aib and three D-residues (Boc-Leu-Aib-Val-Ala-Leu-Aib-Val-^{D}Ala-^{D}Leu-Leu-Val-Phe-Val-Aib-^{D}Val-Leu-Phe-Val-Val-OMe) reveals a helical conformation over the entire peptide length.[32]

Peptide hairpin formation requires a centrally positioned type I′ or II′ β-turn. In the 'prime turns' the flanking strands can be registered by cross-strand hydrogen bonds.[19] In the case of the type I β-turn there is considerable divergence of the flanking segments of the polypeptide chain. The nucleation of prime turns is readily achieved in ^{D}Pro-Xxx sequences where Xxx = L-residue for a II′ β-turn and D-residue for a I′ turn.[33-41] The ^{D}Pro-Gly sequence can, in principle, adopt both II′ and I′ conformations.[42-45] In a majority of crystallographically characterized β-hairpins, the ^{D}Pro-Xxx unit exists in a type II′ β-turn conformation.[25,42-46] A particularly stable hairpin can be generated with a heterochiral diproline unit, ^{D}Pro-^{L}Pro, as the hairpin nucleating segment. Both

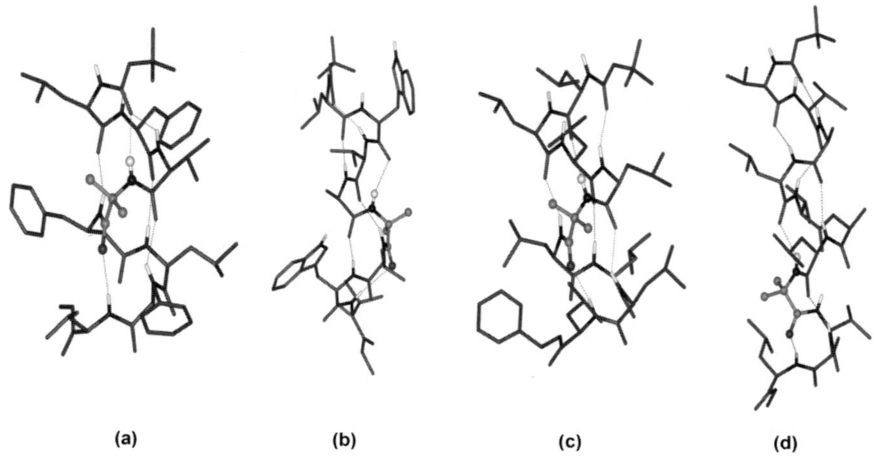

Figure 6: Example of crystal structures where a single Aib residue nucleates formation of helices in 8 to 11 residue peptides. The Aib residue shown as ball and stick. (a) Boc-Leu-Phe-Val-**Aib**-βPhe-Leu-Trp-Val-OMe (βPhe = β^3-homophenylalanine).[28] (b) Boc-Leu-Trp-Val-Ala-**Aib**-Leu-Trp-Val-OMe.[29] (c) Boc-Leu-Leu-Leu-Leu-**Aib**-Leu-Leu-Leu-Leu-OBzl.[30] (d) Boc-Val-Val-Ala-Leu-Val-Lac-Leu-**Aib**-Val-Ala-Leu-OMe.[31]

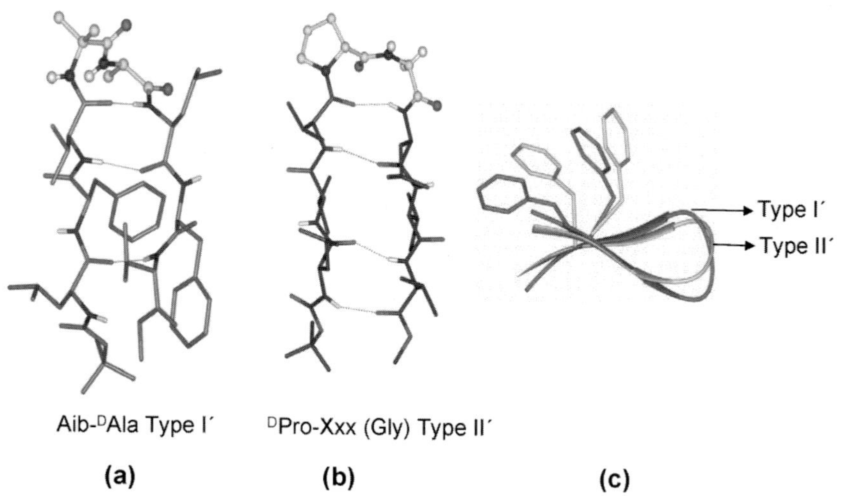

Figure 7: (a) Type I′ nucleated β-hairpin in Boc-Leu-Phe-Val-Aib-^{D}Ala-Leu-Phe-Val-OMe.[49] (b) Type II′ nucleated β-hairpin in Boc-Leu-Val-Val-^{D}Pro-Gly-Leu-Val-Val-OMe.[34] Turn segments are shown as ball and stick. (c) Superposition of type I′ and type II′ β-hairpins illustrating the difference in the strand twist.

NMR and crystal structure determination reveal that the octapeptide Boc-Leu-Phe-Val-^{D}Pro-^{L}Pro-Leu-Phe-Val-OMe favors a hairpin conformation in solution and in the solid state.[47] Indeed, the ^{D}Pro-^{L}Pro segment has been used in generating and stabilizing a synthetic cyclic peptide antigen for a potential malarial vaccine.[48] An alternative strategy for generating I′ turns is to use a centrally positioned Aib-^{D}Xxx sequence,[49-51] although in this case the possibility of a type I β-turn in which the D-residue adopts an α_R conformation is also to be considered. The Aib-^{D}Pro segment is an example of obligatory type I′ β-turn. Incorporation of this unit into the centre of a designed sequence results in the formation of well defined β-hairpins with a central type I′ β-turn.[51] Figure 7 shows two examples of designed β-hairpins nucleated by an Aib-^{D}Ala type I′ turn[49] and ^{D}Pro-Gly type II′ turn.[34] In both cases, the octapeptides form well registered hairpins which differ significantly in the extent of backbone twist.

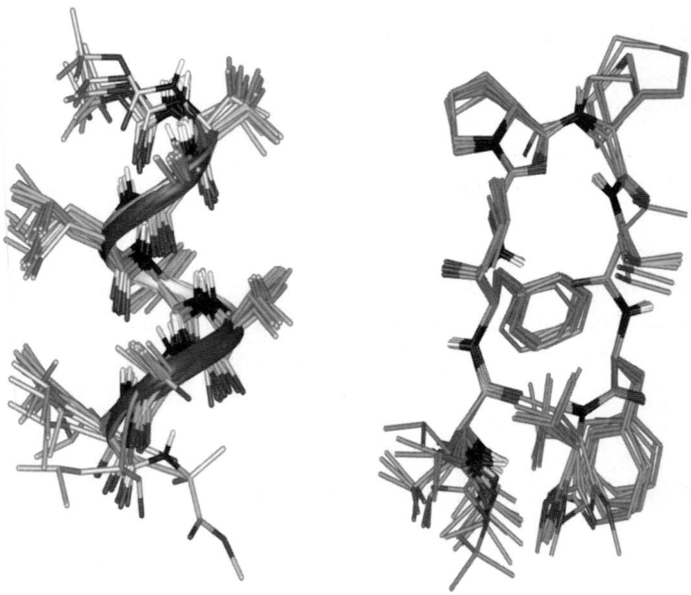

Figure 8: (a) Superposition of 15 structures containing the heptapeptide segment,-Val-Ala-Leu-Aib-Val-Ala-Leu-, all the structures adopt helical conformations for residues 1-6, demonstrating the intrinsic robustness of the helical fold nucleated by an Aib residue. RMSD for backbone atoms is 0.152Å. (b) Superposition of 4 β-hairpins structures in designed octapeptide sequence of Boc-Leu-Phe-Val-^{D}Pro-Xxx-Leu-Phe-Val-OMe (Xxx = Ala, Ac_6c, Ac_8c, Pro). RMSD is 0.145Å for resides 2-7 (backbone atoms).[25,46,47]

Type I′ β-turn nucleated hairpins reveal an extent of strand twist that is significantly greater than that of the type II′ hairpin. The angle of twist is conveniently defined as the virtual torsion angle formed by the four C^{α} atoms ($C^{\alpha}_{(i-1)}$-$C^{\alpha}_{(i)}$-$C^{\alpha}_{(i+1)}$-$C^{\alpha}_{(i+2)}$)[52] of the residues constituting the turn segments. A large body of work has established the structural characterization and stability of hairpins formed in sequences containing centrally positioned ^{D}Pro-Xxx segments. At present, the crystal structures of as many as 15 peptide hairpins are available (for a partial list see ref 46). The robustness of the approach of using local backbone constraints as a structure nucleating feature for both peptide helices and β-hairpins is exemplified in Figure 8, by the superposition of multiple structures determined by X-ray diffraction. The ability to design robust peptide hairpins has been widely exploited in analyzing cross strand side chain interaction in antiparallel sheet structures[53-57] and in dissecting the influences of thermodynamics and kinetic factors in the folding process.[58,59]

2.2 *Conformational switches: helix-hairpin transition*

In an early study the effect of a guest Aib-^{L}Pro sequence on an oligomeric L-Val host sequence was investigated. The peptide Boc-Val-Val-Aib-^{L}Pro-Val-Val-Val-OMe showed a helical conformation in the crystalline state with the central proline residue being accommodated, despite lose of an internal hydrogen bond.[60] However NMR studies in solution supported a possible transition between helical and hairpin structures, with the Aib-^{L}Pro segments, also facilitating a reversal of chain direction. The ability of Aib residues to populate distinct regions of conformational space (α_R, α_L and to a limited extents P_{II}) provides an opportunity to design peptides which can undergo conformational transitions between the two well-defined structural states, the helix and the hairpin. For example, if an Aib-Aib unit is positioned centrally in a peptide sequence two distinct possibilities exist for the local conformations of this segment. If both Aib residues adopt α_R conformations then a type I/III β-turn results, which should favor helix propagation over the sequence. If both Aib residues adopt α_L conformations then a type I′ β-turn results, a feature which facilitates β-hairpin formation when the flanking residues are composed of L-residues. Solvent dependent conformational switching is then possible. Investigations of octapeptides with such segments have revealed that helix-hairpin transitions can indeed be monitored in the peptide Boc-Leu-Phe-Val-Aib-Aib-Leu-Phe-Val-OMe.[50] Helices which contain a greater number of hydrogen bonds (six in the case of a 3_{10}-helix in an octapeptide) are favored in apolar non-hydrogen bonding solvents like $CHCl_3$. On the other hand, in a polar, solvating medium like methanol or dimethylsulfoxide, β-hairpin conformations are detectable as evidenced by characteristic inter-strand nuclear Overhauser effects. In these studies the tripeptide segment Leu-Phe-Val has been used as the test sequence to examine the effects of centrally positioned nucleating segments. It is relevant to point out that such a tripeptide segment exhibits a high degree of structural plasticity and can be induced to adopt either helical or β-strand conformations. Examples of both structures are available in designed peptides as exemplified in Figure 9.

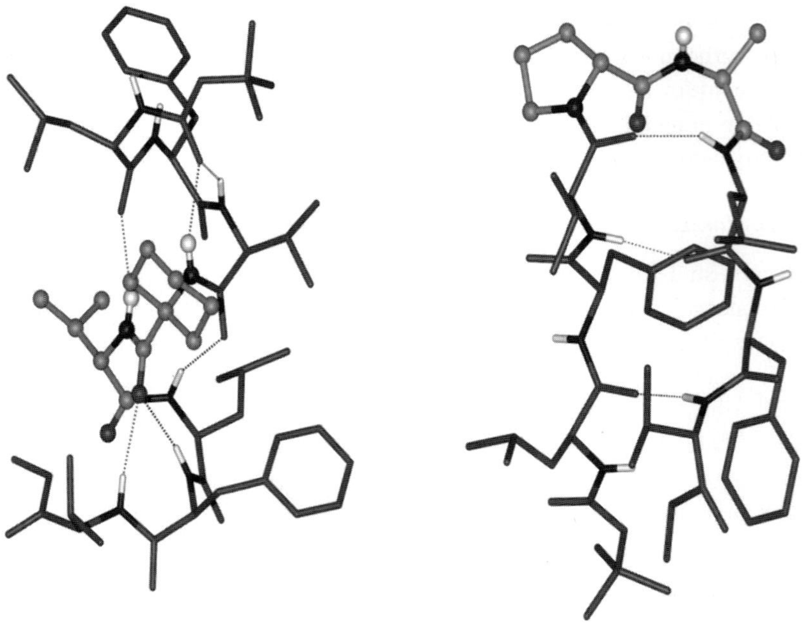

Figure 9: Structural variability of the tripeptide Leu-Phe-Val in octapeptides incorporating a central segment which is constrained to form a type I/III β-turn or type II′ β-turn. (Left) The helical conformation Boc-Leu-Phe-Val-Ac$_6$c-Val-Leu-Phe-Val-OMe in crystals. (Right) The β-hairpin conformation of Boc-Leu-Phe-Val-^{D}Pro-Ala-Leu-Phe-Val-OMe in crystals.[25] The central turn segment shown as ball and stick.

2.3 *Going beyond isolated helices and hairpins*

An interesting feature of peptide hairpins is that generation of multiple hairpin containing structures is readily possible in large sequences by insertion of ‘prime turn’ nucleating segments at appropriate positions. Successive reversals of chain direction results in the formation of an antiparallel β-sheet structures.

The construction of three, four and five stranded β-sheets formed with sequences containing multiple ^{D}Pro-Xxx segments has been demonstrated by two-dimensional NMR studies in solution.[61-63] Multi-strand sheet structures can be conveniently characterized by the observation of long range cross-strand NOEs between backbone and side chain protons and by the identification of strongly solvent shielded peptide NH groups. Thus far, designed multi-stranded sheets in synthetic peptides have not been characterized by X-ray diffraction. Mixed helix-hairpin structures have been characterized in crystals in a 17-residue peptide (Boc-Val-Ala-Leu-Aib-Val-Ala-Leu-Gly-Gly-Leu-Phe-Val-^{D}Pro-Gly-Leu-Phe-Val-OMe) generated by fusing isolated helical and hairpin segments using a two residue linker peptide unit.[64] A major issue in the design of polypeptides with super-secondary structure and organized tertiary structure is the problem of controlling the conformation of linking loop segments. In the present approach, which relies exclusively on biasing backbone conformations, it is not immediately possible to control linker conformation. A greater understanding of the conformations of irregular loops segments in globular proteins is emerging from an analysis of coil segments in large protein data sets.[65] In principle, the construction of defined connecting elements is possible: a goal which will be facilitated by expanding the repertoire of amino acid building blocks with restricted back bone conformations. While control of linker conformations is not immediately realizable, limited attempts to expand the turn segment in β-hairpins have been reported. For example, a registered β-hairpin containing a three residue loop has been established in the nonapeptide Boc-Leu-Phe-Val-^{D}Pro-^{L}Pro-^{D}Ala-Leu-Phe-Val-OMe (Figure 10a). In this case the central ^{D}Pro-^{L}Pro-^{D}Ala segment adopts an α-turn (C_{13} hydrogen bond) incorporating an internal γ-turn (C_7 hydrogen bond) as established by NMR studies.[47] Interestingly, the crystal structure of the isolated segment Piv-^{D}Pro-^{L}Pro-^{D}Ala-NHMe reveals an α-turn with bifurcated hydrogen bonds involving the Piv CO group with ^{D}Ala(3) NH group (C_{10} hydrogen bond) and methylamide NH group (C_{13} hydrogen bond, Figure 10b).[66] In the protein design area, alternative approaches in which solvent forces drive side-chain clustering have been more successful in creating synthetic mimics for protein structures.

Conformation directing residues which bias local folding in peptides have found significant application in the area of the design of analogs of biologically active peptides in which the range of accessible conformations is limited.[67-72] Multiple substitutions on backbone atoms and side chain backbone cyclization strategies have been extended to backbone homologated residues in imposing

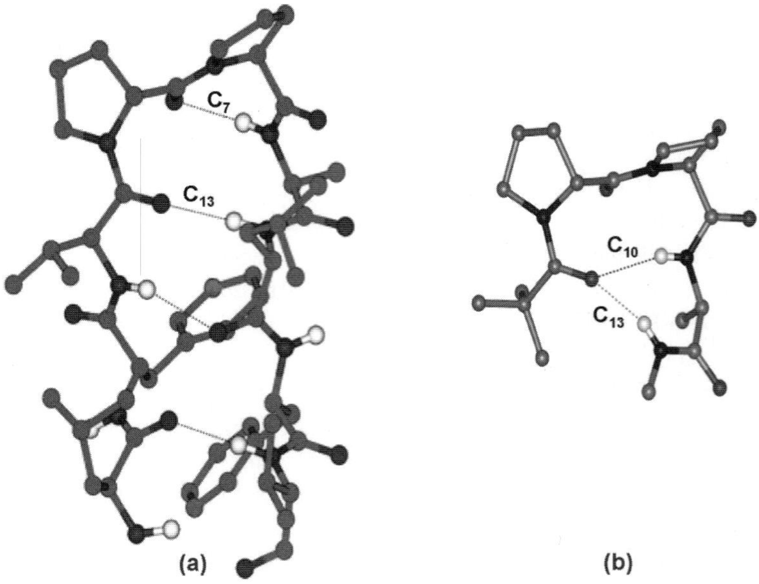

Figure 10: (a) NMR derived structure for the nonapeptide Boc-Leu-Phe-Val-^{D}Pro-^{L}Pro-^{D}Ala-Leu-Phe-Val-OMe.[47] (b) Conformation in crystals of the tripeptide segment ^{D}Pro-^{L}Pro-^{D}Ala illustrating α-turn (C_{13} hydrogen bond) formation.

local constraints on folding.[73-79] Conformationally rigid turn segments using ^{D}Pro-^{L}Pro sequences have been successfully employed in design of peptidomimetics.[48,80,81] D-Proline has also been introduced into ribonuclease.[82] With the methods of unnatural amino acid mutagenesis becoming more accessible there is the possibility that backbone constrained amino acids can be introduced, site specifically, into proteins. Analysis of appropriately designed mutants may then provide a measure of evaluating the role of local conformational features in nucleating protein folding.

References

1. R. L. Baldwin, *J Mol Biol* **371**, 283 (2007).
2. K. A. Dill, S. B. Ozkan, M. S. Shell and T. R. Weikl, *Annu Rev Biophys* **37**, 289 (2008).
3. G. D. Rose, and R. Wolfenden, *Annu Rev Biophys Biomol Struct* **22**, 381 (1993).
4. D. W. Bolen, and G. D. Rose, *Annu Rev Biochem* **77**, 339 (2008).
5. T. E. Creighton, *Biochem J* 270, 1(1990).
6. R. L. Baldwin, *J Biol Chem* **278**, 17581 (2003).

7. K. A. Dill, and H. S. Chan, *Nat Struct Biol* **4**, 10 (1997).
8. A. D. Miranker, and C. M. Dobson, *Curr Opin Struct Biol* **6**, 31 (1996).
9. R. L. Baldwin, *Nature* **369**, 183 (1994).
10. R. L. Baldwin, *Annu Rev Biophys* 37, 1 (2008).
11. G. N. Ramachandran and V. Sasisekharan, *Adv Protein Chem* **23**, 283 (1968).
12. B. V. V. Prasad and P. Balaram, *CRC Crit Rev Biochem* **16**, 307 (1984).
13. C. Toniolo, M. Crisma, F. Formaggio and C. Peggion, *Biopolymers (Peptide Science)* **60**, 396 (2001).
14. G. R. Marshall and H. E. Bosshard, *Circ Res* ***30/31***, Suppl. II: 143 (1972).
15. A. W. Burgess and S. J. Leach, *Biopolymers* **12**, 2599 (1973).
16. I. L. Karle and P. Balaram, *Biochemistry* **29,** 6747 (1990).
17. C. Toniolo and E. Benedetti, *Macromolecules* **24**, 4004 (1991).
18. R. Kaul and P. Balaram, *Bioorg. Med. Chem.,* **7**, 105 (1999).
19. J. Venkatraman, S. C. Shankaramma and P. Balaram, *Chem Rev* **101**, 3131 (2001).
20. S. Aravinda, N. Shamala, R. S. Roy and P. Balaram, *Proc Indian Acad Sci (Chem Sci)* **115**, 373 (2003).
21. M. Crisma, F. Formaggio, A. Moretto and C. Toniolo, *Biopolymers* **84**, 3 (2006).
22. I. L. Karle, J. L. Flippen-Anderson, K. Uma and P. Balaram, *Proteins: Struct Funct Genet* **7**, 62 (1990).
23. I. L. Karle, R. Kishore, S. Raghothama and P. Balaram, P. *J Am Chem Soc* **110**, 1958 (1988).
24. S. Aravinda, N. Shamala and P. Balaram, *Chem Biodivers* **5**, 1238 (2008).
25. S. Aravinda, V. V. Harini, N. Shamala, C. Das and P. Balaram, *Biochemistry* **43**, 1832 (2004).
26. S. Aravinda, S. Datta, N. Shamala and P. Balaram, *Angew Chem Int Ed* **43**, 6728 (2004).
27. I. L. Karle, J. L. Flippen-Anderson, R. Gurunath and P. Balaram, *Protein Sci* **3**, 1547 (1994).
28. K. Ananda, P. G. Vasudev, A. Sengupta, K. M. P. Raja, N. Shamala and P. Balaram, *J Am Chem Soc* **127**, 16668 (2005).
29. R. Mahalakshmi, A. Sengupta, S. Raghothama, N. Shamala and P. Balaram, *J Pept Res* **66**, 277 (2005).
30. K. Okuyama, Y. Saga, M. Nakayama and M. Narita, *Biopolymers* **31**, 975 (1991).
31. S. Aravinda, N. Shamala, C. Das and P. Balaram, *Biopolymers* **64**, 255 (2002).
32. I. L. Karle, H. N. Gopi and P. Balaram, *Proc Natl Acad Sci USA* **100**, 13946 (2003).
33. S. K. Awasthi, S. Raghothama and P. Balaram, *Biochem Biophys Res Commun* **216**, 375 (1995).

34. I. L. Karle, S. K. Awasthi and P. Balaram, *Proc Natl Acad Sci USA* **93**, 8189 (1996).
35. T. S. Haque, J. C. Little and S. H. Gellman, *J Am Chem Soc* **118**, 6975 (1996).
36. T. S. Haque and S. H. Gellman, *J Am Chem Soc* **119**, 2303 (1997).
37. S. R. Raghothama, S. K. Awasthi and P. Balaram, *J Chem Soc Perkin Trans 2*, 137 (1998).
38. S. H. Gellman, *Curr Opin Chem Biol* **2**, 717 (1998).
39. J. F. Espinosa and S. H. Gellman, *Angew Chem Int Ed* **39**, 2330 (2000).
40. H. E. Stranger and S. H. Gellman, *J Am Chem Soc* **120**, 4236 (1998).
41. C. Das, G. A. Naganagowda, I. L. Karle and P. Balaram, *Biopolymers* **58**, 335 (2001).
42. I. L. Karle, H. N. Gopi and P. Balaram, *Proc Natl Acad Sci USA* **98**, 3716 (2001).
43. I. L. Karle, H. N. Gopi and P. Balaram, *Proc Natl Acad Sci USA* **99**, 5160 (2002).
44. R. S. Roy, H. N. Gopi, S. Raghothama, R. D. Gilardi, I. L. Karle and P. Balaram, *Biopolymers (Peptide Science)* **80**, 787 (2005).
45. R. S. Roy, H. N. Gopi, S. Raghothama, I. L. Karle and P. Balaram, *Chem Eur J* **12**, 3295 (2006).
46. V. V. Harini, S. Aravinda, R. Rai, N. Shamala and P. Balaram, *Chem Eur J* **11**, 3609 (2005).
47. R. Rai, S. Raghothama and P. Balaram, *J Am Chem Soc* **128**, 2675 (2006).
48. J. A. Robinson, *Acc Chem Res* **41**, 1278 (2008).
49. S. Aravinda, N. Shamala, R. Rajkishore, H. N. Gopi and P. Balaram, *Angew Chem Int Ed* **41**, 3863 (2002).
50. R. Rai, S. Raghothama, R. Sridharan and P. Balaram, *Biopolymers* (*Peptide Science*), **88**, 350 (2007).
51. U. S. Raghavender, S. Aravinda, R. Rai, N. Shamala and P. Balaram, *Org Biomol Chem* **8**, 3133 (2010).
52. K. Gunasekaran, C. Ramakrishnan and P. Balaram, *Protein Eng* **10**, 1131 (1997).
53. C. D. Tatko, M. L. Waters, *J Am Chem Soc* **124**, 9372 (2002).
54. S. J. Russell, T. Blandl, N. J. Skelton, A. G. Cochran, *J Am Chem Soc* **125**, 388 (2003).
55. M. L. Waters, *Biopolymers* **76**, 435 (2004).
56. R. M. Hughes, M. L. Waters, *Curr Opin Struct Biol* **16**, 514 (2006).
57. L. Wu, D. McElheny, T. Takekiyo, T. A. Keiderling, *Biochemistry* **49**, 4705 (2010).
58. M. S. Searle, *J. Chem. Soc. Perkins Trans I* **2**, 1011 (2001).
59. M. S. Searle, B. Ciani, *Curr Opin Struct Biol* **14**, 458 (2004).
60. I. L. Karle, J. L. Flippen-Anderson, K. Uma, H. Balaram and P. Balaram, *Biopolymers* **29**, 1433 (1990).

61. C. Das, S. Raghothama and P. Balaram, *J Am Chem Soc* **120**, 5812 (1998).
62. C. Das, S. Raghothama and P. Balaram, *Chem Commun* 967 (1999).
63. J. Venkatraman, G. A. Naganagowda and P. Balaram, *J Am Chem Soc* **124**, 4987 (2002).
64. I. L. Karle, C. Das and P. Balaram, *Proc Natl Acad Sci USA* **97**, 3034 (2000).
65. L. L. Perskie, T. O. Street and G. D. Rose, *Protein Sci* **17**, 1151 (2008).
66. B. Chatterjee, I. Saha, S. Raghothama, S. Aravinda, R. Rai, N. Shamala and P. Balaram, *Chem Eur J* **14**, 6192 (2008).
67. S. Zikou, A. Koukkou, P. Mastora, M. Sakarellos-Daitsiotis, C. Sakarellos, C. Drainas and E. Panou-Pomonis, *J Pept Sci* **13**, 481 (2007).
68. J. R. Kumita, C. J. Weston, L. Choo-Smith, G. A. Woolley and O. S. Smart, *Biochemistry* **42**, 4492 (2003).
69. J. M. Conlon, R. Al-Kharrge, E. Ahmed, H. Raza, S. Galadari and E. Condamine, *Peptides* **28**, 2075 (2007).
70. M. Arduin, B. Spagnolo, G. Calò, R. Guerrini, G. Carrà, C. Fischetti, C. Trapella, E. Marzola, J. McDonald, D. G. Lambert, D. Regoli and S. Salvadori, *Bioorg Med Chem* **15**, 4434 (2007).
71. A. Beck, C. Klinguer-Hamour, M. Bussat, T. Champion. J. Haeuw, L. Goetsch, T. Wurch, M. Sugawara, A. Milon, A. V. Dorsselaer, T. Nguyen and N. Corvaïa, *J Pept Sci* **13**, 588 (2007).
72. S. Prasad, R. Mathur, N. Gupta, M. Jaggi, A. T. Singh, P. Rajendran, V. K. Sanna, K. Datta and R. Mukherjee, *J Pept Sci* **13**, 54 (2007).
73. R. P. Cheng, S. H. Gellman and W. F. DeGrado, *Chem Rev* **101**, 3219 (2001).
74. D. Seebach, A. K. Beck and D. J. Bierbaum, *Chem Biodivers* **1**, 1111 (2004).
75. W. S.Horne and S. H. Gellman, *Acc Chem Res* **41**, 1399 (2008).
76. S. Chatterjee, R. S. Roy and P. Balaram, *J R Soc Interface* **4**, 587 (2007).
77. C. M. Goodman, S. Choi, S. Shandler and W. F. DeGrado, *Nat Chem Biol* **3**, 252 (2007).
78. D. Seebach and J. Gardiner, *Acc Chem Res* **41**, 1366 (2008).
79. P. G. Vasudev, S. Chatterjee, N. Shamala and P. Balaram, *Acc Chem Res* **42**, 1628 (2009).
80. J. A. Robinson, *Synlett* **4**, 429 (2000).
81. C. E. Stotz and E. M. Topp, *J Pharm Sci* **93**, 2881 (2004).
82. B. L. Nilsson, M. B. Soellner and R. T. Raines, *Annu Rev Biomol Struct* **34**, 91 (2005).

CYCLIC BETA-AMINO ACIDS AS CONFORMATIONAL CONSTRAINTS

GURUNATH RAMANATHAN AND ANJU DULEY

Department of Chemistry, Indian Institute of Technology Kanpur
Kanpur 208016, India

β-amino acids are naturally occurring non-protein amino acids found in several fungal peptide antibiotics. β-amino acids can also tailor stable well-defined secondary structures in peptides. Cyclic β-amino acids can restrain peptide conformations further as the C^2 and C^3 bond is now part of a ring. Data analysis of the crystal structures of these β-amino acids, both for cyclic as well as acyclic amino acid residues, from the 2012 Cambridge structure database reveals that the (φ, ψ) values are spread in the allowed region of the glycine Ramachandran map.

1. Introduction

In protein structures, the polypeptide backbones can adopt folded structures[1-4] through hydrogen boding between the NH and C=O groups. It is therefore important to investigate if synthetic or non-coded amino acids can ensure similar type of conformations as α-amino acids in peptides. In the literature considerable effort has already been devoted to generate analogues of various biologically and structurally important molecules incorporating synthetic or non-coded amino acids[5,6]. β-amino acids are non-protein amino acids found in nature. β-amino acids have an extra carbon between the amine and the carboxyl group. As a consequence of having an extra carbon atom they are metabolically more stable than the α-amino acids themselves and thus result in pharmacologically more potent analogs when incorporated into biologically active peptides[7]. The pioneering work of Seebach[8] and Gellman[9] has shown that oligopeptides composed of β-amino acids can fold into stable secondary structures like those formed with α-amino acids. During the past few years, it has been shown that cyclic as well as acyclic peptides with beta amino acids can fold into well-defined secondary structures as well[10].

The simplest β-amino acid, β-alanine (Figure1), which is unsubstituted, conformationally flexible and gives extended sheet like structures that are disordered in solution[11,12]. Extra stereochemical constraints can be introduced by substituting the two carbons C^2 and C^3 in this β-amino acid.

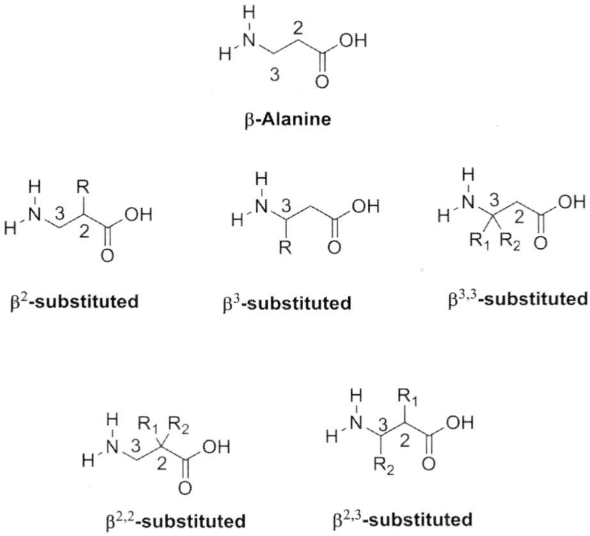

Figure 1: General structure of the various substituted β-amino acids.

Various β-amino acids can be obtained by substituting either of the carbons or both the carbons (Figure 1). Substituted β-amino acids are often generated by homologation of the corresponding α-amino acids[13]. There are various synthetic routes for the synthesis of these amino acids[14-17].

α-Amino acids with the gem dialkyl group, for example, Aib, have been introduced into peptides to nucleate helical conformations[18-21]. Keeping this observation in mind, the gem dialkyl β-amino acid homolog of the Aib was also introduced in peptides as a conformationally constrained β-amino acid[22,23]. Dialkyl substituted β-amino acids containing carbohydrate side chain was also explored where both the substitutions are present in the 2 position ($\beta^{2,2}$)[24] or in 3 position ($\beta^{3,3}$)[25]. The oligomers of dialkyl substituted $\beta^{2,2}$ amino acids where substituents at the 2,2 position was a part of the cyclopropane ring were also structurally investigated[26]. There are also conformationally constrained amino acids where C^2 and C^3 is part of a ring e.g. 2-amino cyclopropane carboxylic acid[27,] *Cis* and *trans*-2-amino cyclobutane carboxylic acid (2-AC^{4}C)[28-34,] *trans*-2-amino cyclopentane carboxylic (2-AC^{5}C)[35-37,] *trans*-2-amino-cyclohexanecarboxylicacid (2-AC^{6}C)[38,39] and *trans*-2,5-diamino-cyclohexanecarboxylicacid (2,5-DC^{6}C)[39] etc. (Figure 2). Some cyclic β-amino acids like pyrrolidine 3-carboxylic acid (PCA) and nipecotic acid (Nip) which are analogous to the α-amino acids proline and pipecolic acid respectively, also nucleate defined secondary structures[40-44] (Figure 2). There are reviews on structure and function of both acyclic and cyclic β-amino acids containing peptides found in literature[10, 45-48]. In this review we are going to limit our discussion to the cyclic β-amino acids and their conformational impact on peptides.

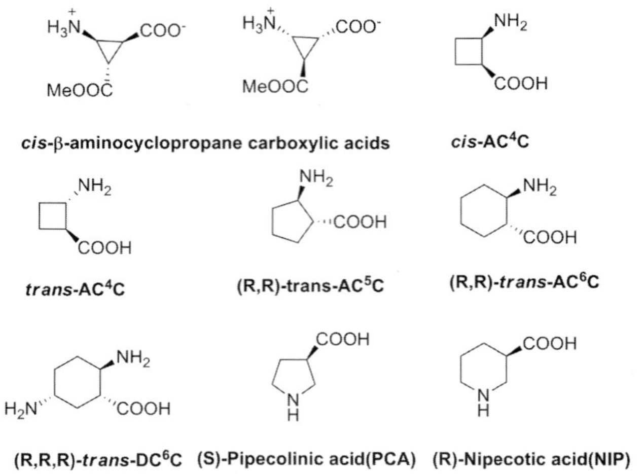

Figure 2: Structure of various conformationally constrained cyclic β-amino acids. The numbers in the superscript represent the ring size in the amino acid.

2. Conformational analysis of β-amino acids

For α-amino acids, the conformation of a residue can be described in terms of torsion angle φ and ψ. In beta amino acids, due to the presence of an extra carbon in the backbone, there is an additional torsion angle θ (Figure 3). The structures formed by the introduction of β-amino acids are favored when torsion angle θ between C^2 and C^3 adopts gauche/anti conformation.

We performed a data analysis of all beta amino acid structures in the 2012 Cambridge structure database[49] using conquest[50]. A total of 100 structures yielding 212 φ, ψ, θ values were extracted from these structures. A close scrutiny of these θ values as drawn in a circular plot is revealed that the θ values clustered around +40° to + 130° and -40° to -130°. As many as 155 examples out of the total 212 observations in the β amino acids crystal structures exhibited θ values in this range. Cyclization at C^2 and C^3 further imposes an additional constraint in five and six membered ring systems as theta is now constrained to ±60° as a result of sp^3 hybridization at these carbon atoms. Thus β amino acids have (φ, ψ)

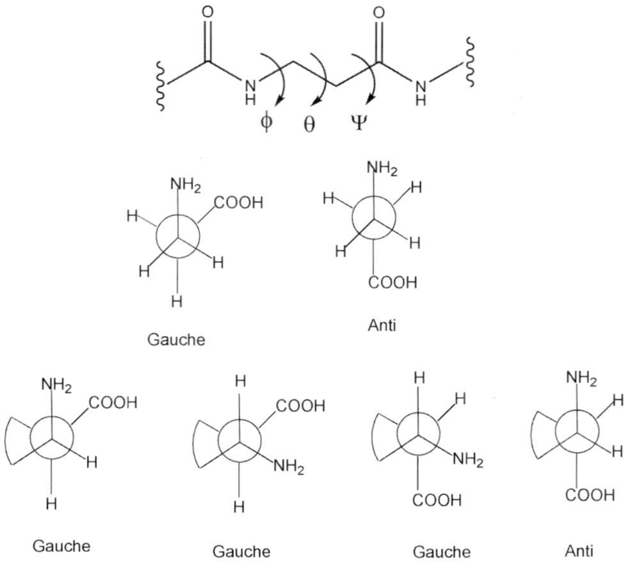

Figure 3: Representation of the torsion angle φ, θ and ψ in the peptide backbone and the Newman projection of acyclic and cyclic β-amino acid. Notice that in cyclic β-amino acids there can be four different stereoisomers arising out of the di axial, di equatorial, and axial-equatorial or equatorial-axial substitution of the amino and carboxyl groups.

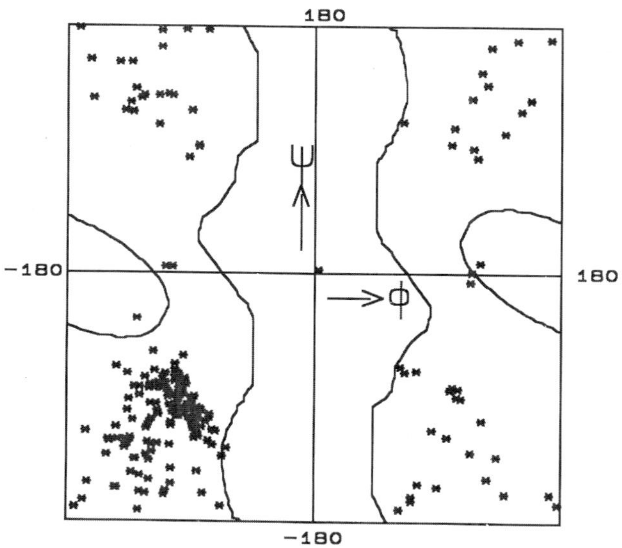

Figure 4: The Ramachandran-Gly map showing 212 β-amino acid (φ, ψ) observations from 100 independent crystal structures in the Cambridge structural database (CSD databank).

values that are spread on the Ramachandran space for the glycine residue[51,52] as shown in Figure 4. These results point to a tendency of these β-amino acids to adopt conformational space normally easily accessed only by glycine.

What is surprising from this result is that even though the two peptide units are separated by a third torsion θ, the glycine map could still be useful for its conformational analysis. There is also a complete absence of points in the disallowed central vertical band of the map *even* for beta amino acids! Thus a deeper investigation of their conformational preferences is warranted as further substituents on the backbone could easily skew this plot further. A full analysis of this work is also currently underway and will be published later.

3. Cyclic β-amino acids

3.1 β-amino acids with a three membered ring

The α/β peptides with 1:1 ratio (Figure 5) where the β-amino acid was either an enantiomer of *cis*-β-amino cyclopropane carboxylic acid adopted a helical conformation in solution[53]. Different chain lengths of peptides with alanine (A) (β-amino cyclopropane carboxylic acid) and entA (enantiomer of A)[27] were structurally investigated by 2D NMR studies. Peptides with one enantiomer gave a helical conformation and the opposite enantiomer gave a mixture of conformations that remained in equilibrium. So these type of α/β peptides generated new types of foldamers.

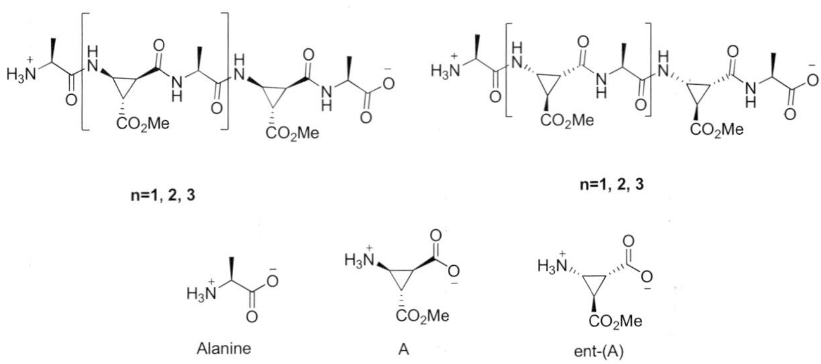

Figure 5: Structures of the α/β peptides with ratio 1:1 containing *cis*-AC^{3}C. The monomeric amino acid units are shown below.

The α/β peptides with *cis*-β-AC^{3}C are biologically important because of their selective binding with high affinity with Y1 receptor of the neuropeptide Y[54,55].

3.2 *β-amino acids with four membered rings*

Cyclobutane β-amino acids were also incorporated in the peptides to investigate their structural propensities in the peptides. There are two types of cyclobutane β-amino acid, *cis*-AC^{4}C and *trans*-AC^{4}C (Figure 2). Beta-dipeptide sequences containing *cis*-AC^{4}C and β-alanine adopted a hairpin like conformation in the solid state[28]. A tetra peptide containing two such *cis*-AC^{4}C and two β-alanine residues that alternated each other in sequence adopted a 10-helical conformation[29]. Homo oligomers of *cis*-AC^{4}C were also reported. These homo oligomers containing 2-8 *cis*-AC^{4}C showed a conformational bias to form structures with intramolecular hydrogen bonding[30,31]. The peptides preferred strand type conformation by forming six-membered *intra*-residue hydrogen bonding and these molecules self assembled to form nanosized fibrils and gels. Short peptide sequence like dipeptide sequences containing *trans*-AC^{4}C formed an eight-membered hydrogen bond[32,33] but longer sequences(such as hexamer and octamer) of the *trans*-AC^{4}C were stabilized through a 12-membered hydrogen bond and adopted a 12-helical conformation both in the solid state and in the solution[34]. The aza analogue of the AC^{4}C was also synthesized on the assumption that it would behave similarly like AC^{4}C[56]. Further, it was hypothesized that the additional nitrogen atom in the ring could act as a hydrogen bond acceptor that could provide the extra stability in the peptides nucleating different folds.

There is another example of a four membered β-amino acid that contains an oxetane ring[57]. The peptide hexamer (Figure 6) containing this amino acid gave a 10-helix in solution as confirmed by 2D NMR studies.

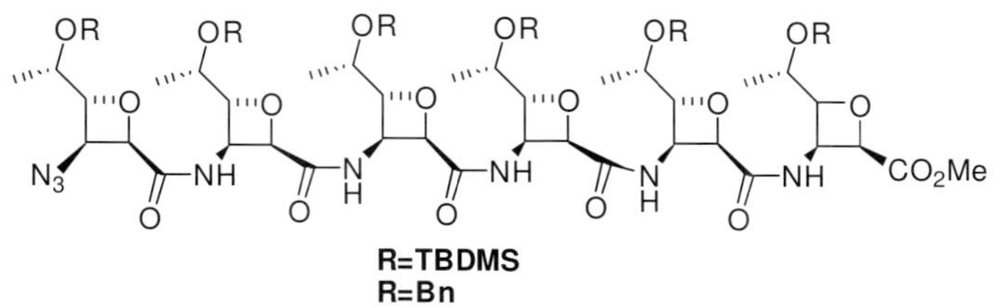

Figure 6: Structure of peptide hexamer containing oxetane ring.

3.3 *Cyclic β-amino acids with five membered rings*

β-peptide oligomer with another rigid cyclic β-amino acid that is trans-2-aminocyclopentane carboxylic acid (2-AC^{5}C) gives rise to different type helix form which is called as 12-helix as this helix is formed by forming 12-membered hydrogen bond in the backbone[35,36]. In 12-helix C=O group of ith residue forms hydrogen bonding with the NH group of the i+3 residue. Both the hexa and octa peptide oligomer containing 2-AC^{5}C gave 12-helical conformation in both solution and solid state[36]. In the CD peak values, the 12-helix obtained from β-peptide oligomer containing 2-AC^{5}C differs from the 14-helix formed in the from β-peptide oligomer containing 2-AC^{6}C[36]. The CD signature of 14-helix was a minimum at 215nm but in case of the helix obtained by AC^{5}C showed a minimum at 221 nm suggesting that the two helices are distinctively different from each other in solution.

The β-peptide oligomer with 2-AC^{5}C was not soluble in water. To make it water soluble a hydrophilic β-amino acid (trans-3-aminopyrrolidine-4-carboxylic acid (APC)) (Figure 7) residue was introduced along with 2-AC^{5}C in an alternative fashion[37]. After reverse phase HPLC purification the peptides were recovered as a trifluoroactetate salt. It has been shown that as the length of the oligomer increases the 12-helix stability also increases in these peptides. Further, a water soluble 12-helix was explored by introducing sulfonylated APC residue (S-APC) (Figure 7) and was used for side chain fictionalization via N-sulfonylation[58]. Such functionalized β-peptide 12-helices can mimic the conventional α-helices as they geometrically resemble each other.

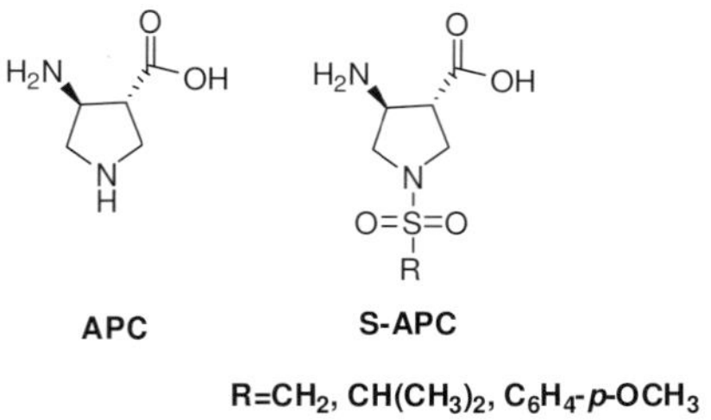

Figure 7: Structure of APC and N-sulfonylated APC (S-APC).

Five membered constrained β-amino acids are also incorporated in peptide along with the α-amino acids to form α-/β-peptides. In this type of α/β- peptides the side chains are manipulated by using α-amino acids and structural stability is provided by the use of β-amino acid in the peptide backbone. It has been shown in literature that an octapeptide (1)(Figure 8) containing 2-AC^{5}C and APC along with α-amino acids forms two types of helical structures, 11 helix and 14/15-helix[59]and these two structures interconvert rapidly in solutions like in peptide composed of only α-amino acids shows the interconversion of 3_{10}- and α-helical conformation[60]. The backbone hydrogen bonding pattern for a 11- helix is i to i+3 and for 14/15-helix it is i to i+4. It has been shown that in a 15 residue long α/β-peptide containing a five membered ring constraint favors 14/15-helical structure over 11-helix[61]. The presence of cyclic β-amino acid is necessary to get helical conformation in the α/β peptide and the substitution of cyclic β-amino acids with acyclic beta amino acids (2-5) (Figure 8) decreases the helix stability though in α/β-peptides the variation of α-amino acids also alter the helix stability.

1, AC-AC^{5}C-K-APC-Y-AC^{5}C-E-APC-A-NH$_2$

2, AC-β^{3}hLeu-K-APC-Y-AC^{5}C-E-APC-A-NH$_2$
3, AC-AC^{5}C-K-APC-Y-β^{3}hLeu-E-APC-A-NH$_2$
4, AC-AC^{5}C-K-APC-Y-AC^{5}C-E-β^{3}hLeu-A-NH$_2$
5, AC-β^{3}hLeu-K-β^{3}hLeu-Y-β^{3}hLeu-E-β^{3}hLeu-A-NH$_2$

Figure 8: Sequence of the octa peptides with varying amount of cyclic and acyclic amino acids.

GCN4-pLI is a 33 residue polypeptide which is composed of only α-amino acids[62-64], adopts very stable four helix bundle with heptad residue repeat (abcdefg) in the crystal state[64]. The α/β-peptide analog of GCN4-pLI with a heptad repeat residue containing β^3-amino acids (ααβαααβ) forms stable four helix bundle in the crystalline state[65]. Other types of residue repeats were also used to make analogs of the GCN4-pLI like αααβ, ααβ, ααβαβαβ and αβ and the four helix bundle stability decreases as the β-residue content increases in the sequence for example in ααβαβαβ and αβ residue repeat the helix stability is very poor[66]. But with the substitution of a rigid β-amino acid 2-AC^{5}C in place of the β^3 residue the helix bundle stability increases compared to their flexible analogs as the cyclic 2-AC^{5}C restricts the rotation θ along Cα-Cβ bond of the β-amino acid residue due to backbone cyclization[66].

As the five membered cyclic 2-AC^{5}C also increased the helix bundle stability in GCN4-pLI[66], it was also incorporated in the α/β/γ hybrid peptide where the peptide with a αγααβα hexad repeat could mimic a α-helix in solution[67]. This helix stability was assumed to be due to the presence of the β-residue though the presence of the γ-residue in this hybrid peptide could also contribute to the stability of the helix.

3.4 Cyclic beta amino acids with six membered rings

The case of six membered ring containing β-amino acids is a bit more confusing as this amino acid can have four types of stereoisomers as pointed out earlier in this review. We have attempted to synthesize all four stereoisomers wherein a tertiarybutyl group has been substituted on the ring to lock the conformation of the chair form. This has indeed enabled us to purify and study peptides with this amino acid and characterize the conformations preferred by them (unpublished).

It has been shown that short β-peptide oligomers with trans-2-aminocyclohexanecarboxylic acid adopt a β-peptide 14-helix conformation both in organic solvents and in the solid state[38,39]. In 14-helix the NH of the ith residue hydrogen bonds with the carbonyl group at the position i+2 to form 14-membered ring.14-helix is the type of helix where 14-membered ring is formed through hydrogen bond in the backbone. The conformational constraint is thus provided by the presence of the cyclohexane ring. In the series of peptides composed of varying amount of cyclic and acyclic residues the 14-helix (6-9) (Figure 9) stability decreases as number of acyclic residues increases in the peptide sequences[39]. (R,R,R)-*trans*-2,5 diaminocyclohexane carboxylic acid(2,5-DC^6C) was used as one of the cyclic beta amino acid to make the peptide water soluble . The acyclic amino acid homo-β-ornithine plays the same role as 2,5-DC^6C in this peptide. The helix stability was monitored by circular dichroism (CD) spectroscopy.

6, Cbz-DC^6C-AC^6C- DC^6C-AC^6C- DC^6C-AC^6C-OMe

7, Cbz-β^3hOrn- AC^6C- DC^6C-β^3hPhe- DC^6C-AC^6C-OMe

8, Cbz-β^3hOrn- AC^6C- β^3hOrn- AC^6C- β^3hOrn- AC^6C-OMe

9, Cbz-β^3hOrn- β^3hVal-β^3hOrn- β^3hPhe-β^3hOrn- β^3hVal-OMe

Figure 9: Sequences of the peptides containing varying amount of cyclic and acyclic β-amino acid residue.

3.5 Other cyclic β-amino acids

The cyclic amino acid, pyridine 3-carboxylic acid (PCA) (Figure 2) and nipecotic acid (NIP) (Figure 2) were used to form β-peptide oligomers. These cyclic amino acids don't have an amide hydrogen to facilitate hydrogen bonding in the peptides. In most of the regular secondary structures like α-helix, 3_{10}-helix and β-sheet, these structures are stabilized by hydrogen bonding between donor and acceptor in the backbone. But the β-peptide oligomers consisting of PCA and NIP adopts well-defined secondary structures[40] (for example- peptides with proline forms polyprolineI (PPI) and polyprolineII (PPII) helices though no backbone hydrogen bonding is present in these helices)[41]. The PCA and NIP oligomer structure analysis by CD spectroscopy revealed that the secondary structure propensity increases as the number of residues increase in the oligomer and the extent was nearly maximum when total four residues were placed in the oligomer sequences for both the cases[40].

Nipecoticacid (NIP) can form β-hairpin type of structures with other amino acids and the dinipecotic acid adopts the reverse turn conformation and forms the loop segment in the hairpin structure[42-44]. The homochiral dipnipecotic acid (R,R or S, S) (12,13) (Figure 10) does not induce the formation of the hairpin

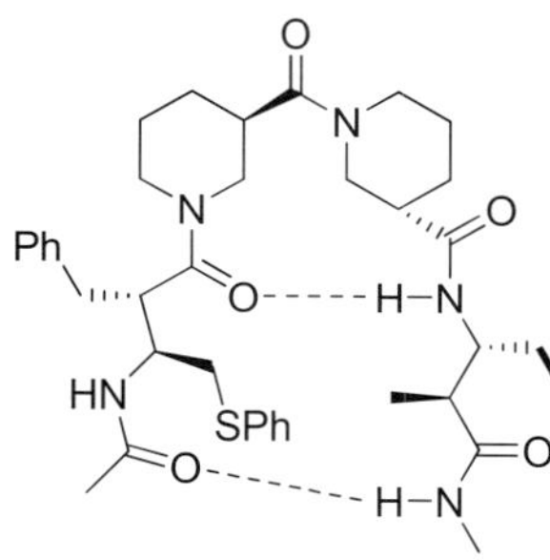

10

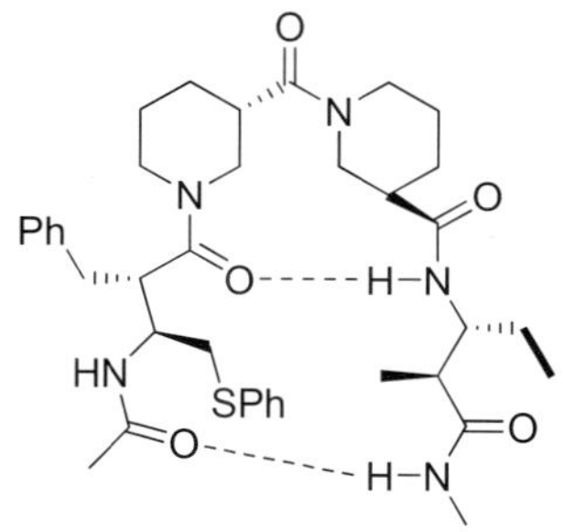

11

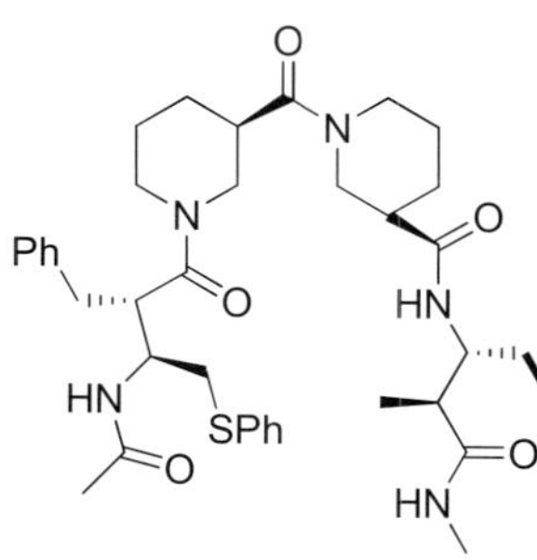

12

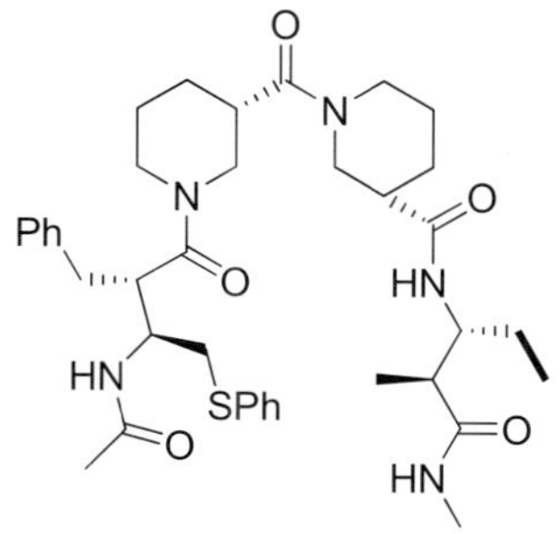

13

Figure 10: Structure of the peptides with different conformation of NIP. 10 and 11 shows hydrogen bonding to form hairpin structures.

structure as it does not form the required hydrogen bonding but both the heterochiral dinepocotic acid (R,S or S,R) (10,11) (Figure 10) promotes the formation of hairpin structure[44].

4. Conclusion

In conclusion it can be seen from the above discussion that cyclic beta amino acids are confomationally more versatile than their alpha amino acid analogs. Hence a detailed study is mandated before one starts to use them in constraining peptide conformations.

Acknowledgements

We thank Prof. C. Ramakrishnan and Mr. Vaibhav Sharma for allowing us to include results of their analysis in this review. This work was partially supported by IIT Kanpur and an senior research fellowship from the council of scientific and industrial research (CSIR) INDIA to Anju Duley.

References

1. J. S. Richardson, *Adv. Protein Chem.*, **34**, 167 (1981).
2. C. Branden and J. Tooze, *Introduction to Protein Structure*, Garland: New York, 1991.
3. J. M. Thornton, C. A. Orengo, A. E. Todd and F. M. Pearl, *J. Mol. Biol.,* **293**, 333 (1999).
4. G. A. Petsko and D. Ringe, *Protein Structure and Function (Primersin Biology)*; New Science Press Ltd.: London, 2004.
5. P. Balaram, *Curr. Opin. Struct. Biol.*, **2**, 845 (1992).
6. S. Aravinda, N. Shamala, R. S. Roy and P. Balaram, *Proc. Indian Acad. Sci.* (*Chem Sci.*) **115**, 373 (2003).
7. O. W. Griffith, *Annu. Rev. Biochem.*, **55**, 855 (1986).
8. D. Seebach, M. Overhand, F. N. M. Kühnle, B. Martinoni, L. Oberer, U. Hommel and H. Widmer, *Helv. Chim. Acta.*, **79**, 913 (1996).
9. D. Appella, L. A. Christianson, I. L. Karle, D. R. Powell and S. H. Gellman, *J. Am. Chem. Soc.*, **118**, 13071 (1996).
10. W. F. De Grado, J. P. Schneider and Y. J. Hamuro, *Peptide Res.*, **54**, 206 (1999).
11. H. Bestian, *Angew. Chem., Int. Ed. Engl.,* **7**, 278 (1968).
12. J. Masamoto, K. Sasaguri, C. Ohizumi and H. Kobayashi, *J. Polym. Sci., A-2,* **8**, 1703 (1970).
13. D. Seebach, S. Abele, K. Gademann, G. Guichard, T. Hintermann, B. Jaun, J. L. Matthews, J. V. Schreiber, L. Oberer, U. Hommel and H. Widmer, *Helv. Chim. Acta*, **81**, 932 (1998).

14. E. Juaristi, *Enantioselective Synthesis of â-Amino Acids*; Wiley-VCH: New York, 1997.
15. E. Juaristi, D. Quintana and J. Escalante, *Aldrichim. Acta.*, **27**, 3 (1994).
16. J. Podlech and D. Seebach, *Angew. Chem., Int. Ed. Engl.*, **35**, 471 (1995).
17. N. N. Romanova, A. G. Gravis and Y. G. Bundel, *Russ. Chem. Rev.*, **65**, 1083 (1996).
18. J. Venkatraman, S. C. Shankaramma and P. Balaram, *Chem. Rev.* **101**, 3131 (2001).
19. B. V. V. Prasad and P. Balaram, *CRC Crit. Rev. Biochem.*, **16**, 307 (1984).
20. I. L Karle and P. Balaram, *Biochemistry*, **29**, 6747 (1990).
21. C. Toniolo and E. Benedetti, *Macromolecules*, **24**, 4004 (1991).
22. S. Abele and D. Seebach, *Eur. J. Org. Chem.*, 1 (2000).
23. D. Seebach, S. Abele, S. Thierry, M. Hanggi, S. Gruner and P. Seiler, *Helv. Chim. Acta.*, **81**, 2218(1998).
24. G. V. M Sharma, P.S. Reddy, D. Chatterjee and A. C. Kunwar, *J. Org. Chem.*, **76**, 1562 (2011).
25. M. Andreini, C. Taillefumier, F. Cretien, V. Thery and Y. Chapleur, *J. Org. Chem.*, **74**, 7651 (2009).
26. S. Abele, P. Seiler, D. Seebach, *Helv. Chim. Acta.*, **82**, 1559 (1999).
27. F. Gnad and O. Reiser, *Chem. Rev.*, **103**, 1603 (2003).
28. M. Martín-Vila, E. Muray, G. P. Aguado, A. Alvarez-Larena, V. Branchadell, C. Minguillón, E. Giraltdand R. M. Ortuñoa, *Tetrahedron: Asymmetry,* **11**, 3569 (2000).
29. S. Izquierdo, M. J. Kogan, T. Parella, A. G. Moglioni, V. Branchadell, E. Giralt and R. M. Ortuño, *J. Org. Chem.,* **69**, 5093 (2004).
30. F. Rúa, S. Boussert, T. Parella, I. Díez-Pérez, V. Branchadell, E. Giralt and R. M. Ortuño, *Org. Lett.,* **9**, 3643 (2007).
31. E. Torres, E. Gorrea, K. K. Burusco, E. Da Silva, P. Nolis, F. Rúa, S. Boussert, I. Díez-Pérez, S. Dannenberg, S. Izquierdo, E. Giralt, C. Jaime, V. Branchadell and R. M. Ortuño, *Org. Biomol. Chem.,* **8**, 564 (2008).
32. E. Torres, E. Gorrea, E. Da Silva, P. Nolis, V. Branchadell and R. M. Ortuño, *Org. Lett.,* **11**, 2301 (2009).
33. S. Izquierdo, F. Rúa, A. Sbai, T. Parella, Á. Álvarez-Larena, V. Branchadell and R. M. Ortuño, *J. Org. Chem.,***70**, 7963 (2005).
34. C. Fernandes, S. Faure, E. Pereira, V. Thery, V. Declerck, R. Guillot and D. J. Aitken, *Org. Lett.*, **12**, 3606 (2010).
35. D. H. Appella, L. A. Christianson, D. A. Klein, D. R. Powell, X. Huang, J. J. Barchi and S. H. Gellman, *Nature*, **387**, 381 (1997).
36. D. H. Appella, L. A. Christianson, D. A. Klein, M. R. Richards, D. R. Powell and S. H. Gellman, *J. Am. Chem. Soc.,* **121**, 7574 (1999).
37. J. J. Barchi, X. L. Huang, D. H. Appella, L. A. Christianson, A. R. Durell and S. H. Gellman, *J. Am. Chem. Soc.*, **122**, 2711 (2000).

38. D. H. Appella, L. A. Christianson, I. L. Karle, D. R. Powell and S. H. Gellman, *J. Am. Chem. Soc.*, **118**, 13071 (1996).
39. D. H. Appella, J. J. Barchi, S. R. Durell and S. H. Gellman, *J. Am. Chem. Soc.*, **121**, 2309 (1999).
40. B. R. Huck, J. M. Langenhan and S. H. Gellman, *Org. Lett.,* **1**, 1717 (1999).
41. R. Rabanal, M. D. Ludevid, M. Pons and E. Giralt, *Biopolymers*, **33**, 1019 (1993).
42. Y. J. Chung, L. A. Christianson, H. E. Stanger, D. R. Powell and S. H. Gellman, *J. Am. Chem. Soc.,***120**, 10555 (1998).
43. B. R. Huck, J. D. Fisk and S. H. Gellman, *Org. Lett.,* **2**, 2607 (2000).
44. Y. J. Chung, B. R. Huck, L. A. Christianson, H. E. Stanger, S. Krauthauser, D. R. Powell and S. H. Gellman, *J. Am. Chem. Soc.,* **122**, 3995 (2000).
45. R. P. Cheng, S. H. Gellman and W. F. DeGrado, *Chem. Rev*, **101**, 3219 (2001).
46. D. Seebach, A. K. Beck and D. J. Bierbaum, *Chemistry and Biodiversity*, **1**, 1111 (2004).
47. J. A. Miller and S. T. Nguyen, *Mini-Review in Organic Chemistry*, **2**, 39 (2005).
48. P. G. Vasudev, S. Chatterjee, N. Shamala and P. Balaram, *Chem, Rev.* **111**, 657 (2011).
49. F. H. Allen, *ActaCryst.*, **B58**, 380 (2002).
50. I. J. Bruno, J. C. Cole, P. R. Edgington, M. Kessler, C. F. Macrae, P. McCabe, J. Pearson and R. Taylor, *ActaCryst.,* **B58**, 389 (2002).
51. C. Ramakrishnan, G. N. Ramachandran, *Biophysical Journal*, **5**, 910 (1965)
52. G. N. Ramachandran, C. Ramakrishnan, V. Sasishekharan, *J. Molecular Biol.* **7**, 95 (1963).
53. S. D. De Pol, C. Zorn, C. D. Klein, O. Zerbe and O. Reiser, *Angew. Chem. Int. Ed.,* **43**, 511 (2004).
54. N. Koglin, C. Zorn, R. Beumer, C. Cabrele, C. Bubert, N. Sewald, O. Reiser and A. G. Beck-Sickinger, *Angew. Chem.* **115**, 212 (2003).
55. N. Koglin, C. Zorn, R. Beumer, C. Cabrele, C. Bubert, N. Sewald, O. Reiser and A. G. Beck-Sickinger, *Angew. Chem. Int. Ed.* **42**, 202 (2003).
56. V. Declerck and D. J. Aitken, *J. Org. Chem.*, **76**, 708 (2011).
57. T. D. W. Claridge, J. M. Goodman, A. Moreno, D. Angus, S. F. Barker, C. Taillefumier, M. P. Watterson and G. W. J. Fleet, *Tetrahedron Lett.,* **42**, 4251 (2001).
58. H.-S. Lee, F. A. Syud, X. Wang and S. H. Gellman, *J. Am. Chem. Soc.*, **123**, 7721 (2001).
59. A. Hayen, M. A. Schmitt, F. N. Ngassa, K. A. Thomasson and S. H. Gellman, *Angew. Chem., Int. Ed.,* **43**, 505 (2004).
60. A. K. Bolin and G. L. Millhauser, *Acc. Chem. Res.,* **32**, 1027 (1999).
61. M. A. Schmitt, B. Weisblumand and S. H. Gellman, J. *Am. Chem. Soc.*, **126**, 6848 (2004).
62. E. K. O'Shea, R. Rutkowski and P. S. Kim, *Science,* **243**, 538 (1989).
63. E. K. O'Shea, J. D. Klemm, P. S. Kim and T. Alber, *Science,* **254**, 539 (1991).

64. P. B. Harbury, T. Zhang, P. S. Kim and T. Alber, *Science*, **262**, 1401 (1993).
65. W. S. Horne, J. L. Price, J. L. Keck and S. H. Gellman, *J. Am. Chem. Soc.,* **129**, 4178 (2007).
66. W. S. Horne, J. L. Price and S. H. Gellman, *PNAS,* **105**, 9151 (2008).
67. T. Sawada and S.H. Gellman, *J. Am. Chem. Soc.*, **133**, 7339 (2011).

PROTEIN DISULFIDE ANALYSIS AND DESIGN

SIDDHARTH PATEL, S. INDU*, C. RAMAKRISHNAN, RAGHAVAN VARADARAJAN

Molecular Biophysics Unit, Indian Institute of Science
Bangalore, Karnataka 560 012 India

The significant contribution of naturally occurring disulfide bonds to protein stability has encouraged development of methods to engineer non-native disulfides in proteins. These have yielded mixed results. We summarize applications of the program MODIP for disulfide engineering. The program predicts sites in proteins where disulfides can be stably introduced. The program has also been used as an aid in conformational analysis of naturally occurring disulfides in α-helices, antiparallel and parallel β-strands. Disulfides in α-helices occur only at N-termini, where the first cysteine residue is the N-cap residue of the helix. The disulfide occurs as a CXXC motif and can possess redox activity. In antiparallel β-strands, disulfides occur exclusively at non-hydrogen bonded (NHB) registered pairs of antiparallel β-sheets with only 1 known natural example occurring at a hydrogen bonded (HB) registered pair. Conformational analysis suggests that disulfides between HB residue pairs are under torsional strain. A similar analysis to characterize disulfides in parallel β-strands was carried out. We observed that only 9 instances of cross-strand disulfides exist in a non-redundant dataset. Stereochemical analysis shows that while the χ^{ss} angles are similar to those of other disulfides, the χ^1 and χ^2 angles show more variation and that one of the strands is generally an edge strand.

1. Introduction

Multiple forces of differing magnitude act concurrently on a protein to stabilize its folded, native state. Of these, non-covalent interactions include primarily hydrophobic interactions, hydrogen bonds, salt bridges and van der Waals interactions. Disulfide bonds are the most common covalent interaction which crosslink sequentially distant residues of the protein. Removal or reduction of a naturally occurring disulfide bond typically results in destabilization of the protein by about 2-5 kcal/mol[1]. This stability is conferred due to reduction of main chain entropy by the disulfide in the unfolded state of the protein[2].

Although naturally occurring disulfides are typically stabilizing, engineering new disulfide bridges in proteins have yielded mixed results with a few cases showing increased stabilization and several others leading to destabilization[3, 4].

*Current Address: Department of Biochemistry & Molecular Biology, Lederle Graduate Research Center 1021J, 710 N, Pleasant St. University of Massachusetts, Amherst, MA 01003, USA.

The destabilizing effect of such disulfides has been attributed to multiple reasons. The rearrangement of residues surrounding the introduced cysteines might carry an energetic penalty[2]. Loss of favorable native contacts would also be destabilizing[2, 5]. The stereochemistry of the engineered disulfides may be non-ideal. Hence the strain energy in the native state may offset the decreased entropy of the denatured state[6]. Finally, some disulfide bonds may stabilize the unfolded state due to favorable enthalpic interactions[7].

Understanding and characterizing disulfide bonds in the general context of protein structure and the specific context of different types of secondary structure is important as disulfide conformation may be linked to main-chain conformation. In addition to stabilizing proteins, engineered disulfides can act as conformational constraints and probes to validate protein secondary and tertiary structure[8].

2. MODIP – MOdeling of DIsulfide bridges in Proteins

MODIP[9] was developed as a computational tool to aid disulfide bond engineering. The program identifies residue pairs in proteins which are in the appropriate geometric orientation such that when mutated to cysteines, they would likely form a disulfide bridge. The procedure was further updated and improved by including information present in the structures available as the size of Protein Data Bank[10] (PDB) increased and by adding additional criteria to detect steric clashes at the identified sites[11].

The 1997 release of PDB containing 6017 entries was selected and a non-redundant set was made by retaining only those entries which had a sequence homology of less than 40% with any other entry and whose structures were at a resolution of 2.0Å or less. This resulted in 571 unique polypeptides out of 538 total proteins. These were further divided into a set of 172 entries containing 730 disulfide bridges.

At each pair of residues, i and j, MODIP evaluates different stereochemical parameters such as the non-bonded distances, $C^{\alpha}{}_{i}\ldots C^{\alpha}{}_{j}$ (r^{α}), $C^{\beta}{}_{i}\ldots C^{\beta}{}_{j}$ (r^{β}). For sites where a modeled disulfide appears feasible, it also calculates the torsion angles χ^{ss}, $\chi^{1}{}_{i}$, $\chi^{1}{}_{j}$. The r^{α} and r^{β} are required by MODIP to screen the initial selection of residue pairs whereas the torsion angles are required to assign grades to modeled disulfides. It is interesting to note that even after taking the larger size of the data set into account (as opposed to the 22 crystal structures from which the previous values were derived from) the values of the above parameters remained the same and thus were largely retained. The r^{α} and r^{β} were slightly relaxed to be more inclusive of borderline residue pairs and the values were set to upper limits of 7.0 and 4.7 Å and lower limits of 3.8 and 3.4 Å for r^{α} and r^{β} respectively. The values of χ^{ss}, χ^{1}_{i}, χ^{1}_{j} were retained with limits of 60° and 120° for $|\chi^{ss}|$, and 30° to 90° and 120° to 150° for $|\chi^{1}_{i}|$and $|\chi^{1}_{j}|$ angles respectively. The consistency of the parameter values between the older, smaller data set and the newer, larger data set only justifies their use for computational modeling.

The non-bonded energies involved in interaction with new sulfur atoms were calculated using an energy function of the form $Ae^{-\mu r} - Br^{-6}$ where r is the distance between two non-bonded atoms. Parameter values used were as given previously[12, 13]. Calculating the energies enabled classification of predicted sites based on not only geometric criteria, but also steric criteria and this lead to development of a 'grade factor' which reflected the suitability of a predicted site to accept a disulfide bond based on both geometrical and energetic criteria. MODIP predictions were also compared with experimental measurements of protein stability changes upon introduction of disulfides.

The cysteines involved in disulfide bonds were analyzed further with respect to their solvent accessibility, burial, side chain B factor values, residue separation and change in volume upon mutation. No clear distinctions could be made between locations of stabilizing and destabilizing disulfides. However stabilizing disulfides showed a general trend of occurring at regions of intermediate accessibility, mobility and depth. This suggests that stable disulfide bridges need to be in a region of the protein flexible enough that small destabilizing effects due to steric clashes can be tolerated. Further, the majority of stabilizing disulfides were between cysteines separated by a loop length of 25-75 residues and where the volume change upon mutation to cysteines was between +25Å^3 to -75Å^3. At small loop lengths there will be minimal effects of the disulfide on the conformational entropy of the denatured state. Also, large volume changes upon mutation tend to be destabilizing since positive volume changes often result in steric overlap whereas negative volume changes result in creation of a cavity and loss of van der Waals interaction[14, 15].

3. Disulfide Conformations in Different Secondary Structures

3.1 α-Helix N-Termini

Intrahelical disulfides have been known to occur in active sites of a number of oxidoreductase enzymes. These disulfides exist in the form of a CXXC motif (CYS-X-X-CYS). The redox potential and activity can be changed by mutating the XX residues. The CXXC motif has also been observed to play a functional role in metal binding proteins. Is this the only motif which forms a disulfide bond intrahelically? To answer this question, a comprehensive analysis of the stereochemical characteristics of disulfide bridges occurring in α-helices was carried out[4].

From the Protein Data Bank, structures with a resolution of 2Å or better and a sequence homology cut-off of 40% were chosen. This data set consisted of 5025 polypeptide chains with 2311 disulfide bonds. The α-helices were identified based on backbone torsion angles in the range of -120° to -10° and -120° to 20° for Φ and Ψ respectively. Residues with non-helical torsion angles immediately preceding the N-terminus of the helix and succeeding the C-terminus of the helix were labeled as N-cap and C-cap respectively. The helical segment was defined as extending from N-cap to C-cap with at least 4 residues occurring between the N and C-caps having α-helical torsion angles. The numbering for such helical segments started at 1 for the residue immediately following N-cap. A total of 29597 helical segments were identified from the above mentioned non-redundant data set[4].

The disulfides were modeled onto α-helices using MODIP. As an alternative to MODIP, *in-silico* mutations to CYS were also performed using the Swiss-PDB[16] software and the structures formed were minimized using the l-bgfs minimizer and OPLS-AA/L force field in the GROMACS[17] 3.3.3 package.

Upon scanning helical segments identified in the non-redundant data set by examining the SSBOND records in the PDB files, 35 naturally occurring disulfides were found, all formed between the N-cap and 3rd residue of the helix (N-cap-3). Thus naturally occurring intrahelical disulfides exist exclusively in the form of a CXXC motif. This indicates that the two CYS residues are in optimal geometric proximity and are present on the same face of the helix since an α-helix has 3.6 residues per turn.

In the above data set, there were a total of 642 CXXC motifs. Of these 102 occurred in α-helices such that the N-terminal CYS was an N-cap residue. Only 35 of these formed a disulfide bond, whereas the remaining 67 were non disulfide bonded. Looking at the functional importance of the CXXC motifs in these proteins, it was seen that 24 of 35 naturally occurring motifs were present in oxidoreductases, 1 was functionally active in metal ion transfer and the remaining 10 did not have any annotated functional roles. In the 67 instances of CXXC motifs which were non disulfide bonded, 45 were involved in metal ion coordination, 9 proteins had redox function indicating that the non disulfide bonded CXXC motifs can also probably be oxidized to form a disulfide bond. Indeed MODIP predicted 20 proteins from the non disulfide bonded set as potential sites for disulfide formation.

Apart from the above 102 cases where the first residue is an N-cap residue, 115 instances were identified in which all four residues of the CXXC motif have helical torsion angles. An additional 17 cases consisted of a CXXC motif in which the C-terminal CYS was the C-cap Cys. None of the above cases contained a disulfide bond. In addition, 408 non-helical CXXC motifs were found of which only 14 formed a disulfide bond, none of which had any annotated redox activity. Thus it can be concluded that the probability for a CXXC motif to form a disulfide bridge is extremely low for motifs occurring outside of the intrahelical N-cap-3 positions.

To rationalize this tendency, residue identities, main chain conformational preferences and side chain conformational preferences in intrahelical CXXC motifs were determined. It was seen that CXXC motifs occur primarily as CGPC where both Glycine and Proline occur at ~35% frequency. This can be attributed to the large number of proteins belonging to the thioredoxin family, other residues also occur between the two CYS residues.

The backbone torsion angle preferences of the three residues (XXC) following the N-cap belong to typical right handed α-helices, having mean (Φ,Ψ) values of ($-58\pm6.0^{\circ}$,$-45\pm9.5^{\circ}$),($-69\pm15.5^{\circ}$,$-27\pm15.2^{\circ}$) and ($-64\pm7.2^{\circ}$, $-40\pm6.8^{\circ}$) respectively. The N-cap CYS has a non-helical conformation with mean (Φ,Ψ) values of ($-112\pm25.2^{\circ}$,$106\pm25.4^{\circ}$). ω angle analysis revealed that there were no cis-peptides present in intrahelical disulfide bonded CXXC motifs. The stringent requirement of the N-cap CYS to be non-helical explains why intrahelical disulfide bonds occur exclusively at N-termini of helices. The favored C^{β}-C^{β} distance for residues i and i+3 within a helix is 5.87±0.55 Å whereas the favored C^{β}-C^{β} distance for disulfide formation is in the range of 3.4 Å – 4.6 Å. These distances can only be achieved if the N-cap CYS is non-helical.

The side chain torsion angle preferences of the CXXC motifs also differ from typical preferences of general disulfide bonds. The χ^{ss} for all disulfides shows two peaks at $+90^{\circ}$ and -90° whereas the χ^{ss} of intrahelical disulfides shows a single peak around $+90^{\circ}$. There are marked differences in the $\chi1$ and $\chi2$, χ^1_i and χ^1_j and χ^2_i and χ^2_j angles between intrahelical and other disulfide bonds. These are shown in Figure 1.

Noting the specific torsion angle preferences of intrahelical disulfide bonds, can we presume that the non-helical N-cap of every helix is a potential site for introduction of a disulfide bond? To this end, MODIP was run on the complete set of 29597 helical segments identified from the non-redundant database and could successfully predict 8779 sites with an N-cap-3 disulfide linkage out of a total of 20157 non-GLY residues at N-cap. Disulfides were computationally modeled and energy minimized at N-cap and N-cap-3 positions, both predicted by MODIP as well as not predicted by MODIP. In all cases disulfides could be modeled with reasonable bond lengths and angles, though in most cases dihedral angles were far from ideal. Hence while it appears possible that disulfides can be introduced at most α-helices with exposed termini, this remains to be experimentally confirmed.

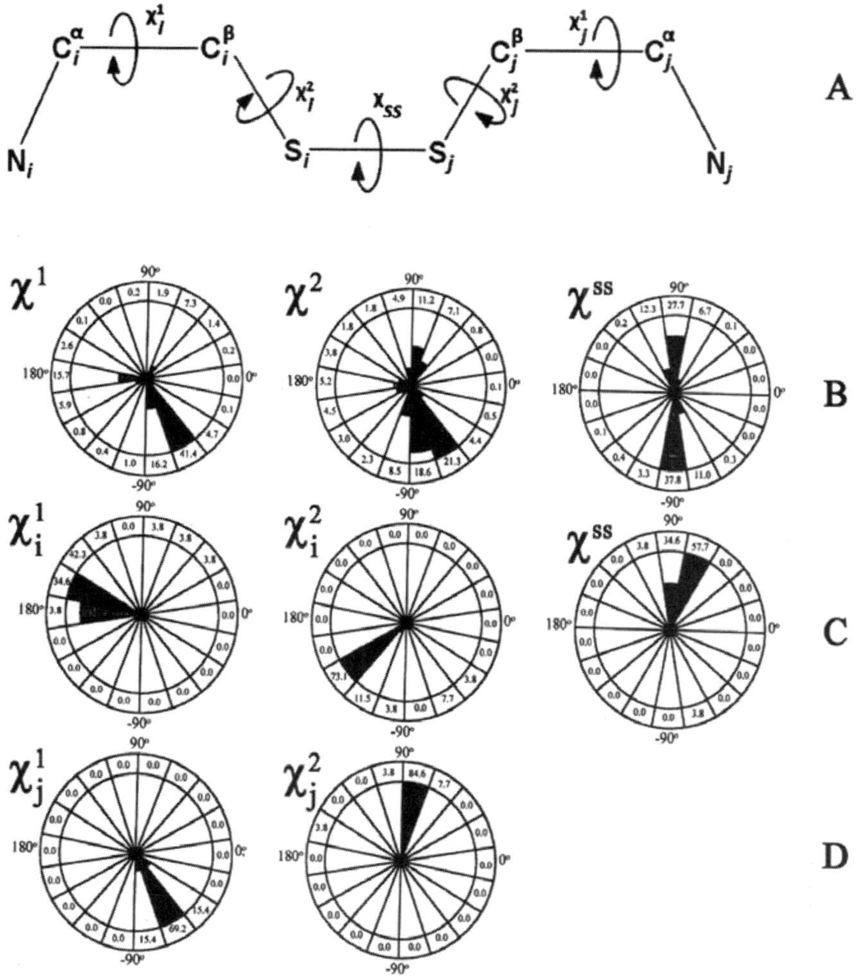

Figure 1: Disulfide torsion angles and their distributions. A: Torsion angle definitions. B–D: Circular histograms showing the percentage distributions for CYS side chain torsion angles for (B) all naturally-occurring disulfides. C, D: Naturally-occurring intrahelical disulfides. The percentage values for various angle ranges are indicated on the circumference of the histograms.

Further, when we observe that most non-intrahelical disulfide bonds are functionally inactive whereas intrahelical CYS residues of the N-cap-3 type often possess redox or metal binding activity, would it be reasonable to postulate that a helical CXXC motif is involved in redox or metal binding activity? To verify this, CYS pairs were experimentally introduced at multiple locations in two α-helices of an *E. coli* Thioredoxin protein lacking the wild type disulfide between residues 32 and 35. The CYS pairs were introduced at positions N-cap-3, 1-4 and 7-10. Of these, only CYS introduced at N-cap-3 could form disulfide bonds spontaneously during purification. The stabilities were characterized using chemical denaturation and DSC. All mutants were destabilized and showed redox activity, albeit lower than wild type thioredoxin. The decrease in stability can be attributed to the fact that there was a definite loss of favorable native interactions when WT residues were substituted by cysteines.

Thus, intrahelical disulfides only occur between the N-cap and N-cap-3 residues. This fact can be exploited to experimentally probe helix start sites in proteins whose structure is unknown. Replacing cysteine with selenocysteine and engineering diselenide derivatives can potentially be used to obtain phase information by anomalous scattering. Such diselenides may be better ordered than more conventionally used selenomethionine derivatives. In addition, (N-cap, N-cap-3) CYS residues may also be potential metal binding sites.

3.2 Cross-Strand Disulfides in Antiparallel β-Sheets

With limited protein structure data available, initial surveys[18, 19] found that there were no examples of disulfides linking adjacent β-strands. Thus it was previously thought that disulfides cannot be accommodated between residues of adjacent β-strands[18, 19]. With increase in size of protein structure databases, instances of disulfide bonds across adjacent β-strands were revealed[20, 21]. They typically occurred between registered pairs of residues in antiparallel β-strands. In antiparallel strands, there are two ways for residues to be in register (Figure 2), they can either be non-hydrogen-bonded (NHB) or hydrogen bonded (HB) with respect to a given partner strand, thus two kinds of disulfides can bridge adjacent β-strands, disulfides occurring on an NHB registered pair or a HB registered pair. Furthermore, it was thought that cross strand disulfides are energetically strained[22] and the frequency of occurrence of HB and NHB pairs was unclear. A recent study examined the nature of cross-strand disulfides occurring in registered pairs of antiparallel β-strands[23].

The non-redundant data set described above[4], containing 5025 polypeptides and 2311 disulfides was used. Antiparallel β-strands and residue pairs which are in register were identified using the "SHEET" records of the PDB files. These pairs were divided into NHB and HB pairs based on the N...O distance. If this distance was between 2.4 – 3.6 Å then the pair was considered as HB. A total of 54,207 NHB and 40,886 HB pairs were identified from the non-redundant data set.

From the complete data set, 90 disulfides were found which bridged residues occurring in adjacent strands of antiparallel β-sheets. Of these 90, 14 disulfides were between non registered cysteines. The remaining 76 disulfides occurred on adjacent strands between registered pairs and are referred to as cross-strand disulfides. Of these 76, 1 was a HB registered pair, the rest were NHB registered pairs.

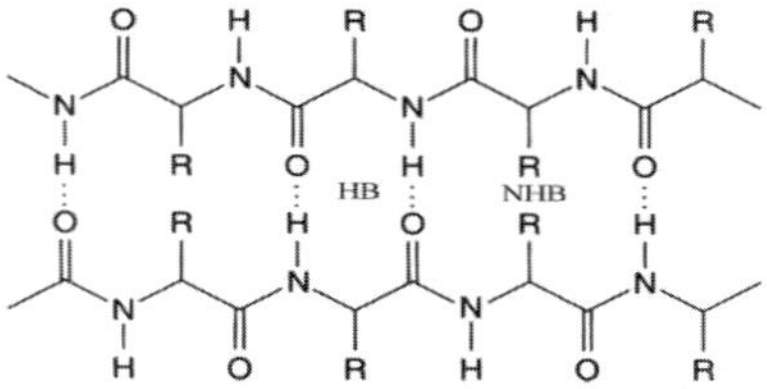

Figure 2: NHB (non-hydrogen-bonded) and HB (hydrogen-bonded) registered pairs in antiparallel β-strands.

On computing and comparing the Cα...Cα & Cβ...Cβ distances for all residues occurring in registered pairs, it was observed that these distance distributions for cross-strand disulfides significantly overlap with that of NHB pairs but not with HB pairs (Figure 3). This distance distribution may be one of the major contributing factors towards the highly preferential occurrence of cross-strand disulfides in NHB pairs.

Calculating the side chain torsion angles revealed more differences between disulfides in NHB pairs and all naturally occurring disulfides (Figure 4). While the χ^{ss} value ranges from +90° to -90° in all disulfides, the NHB pair disulfides only have a positive χ^{ss} value for all 75 pairs. A clear trend followed in the NHB disulfide pairs was that the values of $\chi^1_i \chi^2_i \chi^{ss} \chi^2_j \chi^1_j$ had the torsion angle signs as --+-- where "-" indicates angle of $0° > \chi > -180°$ and "+" indicates a value of $0° \leq \chi \leq 180°$.

For further insight into conformational constraints in cross-strand disulfides, MODIP was used to identify disulfide sites for disulfide introduction at both

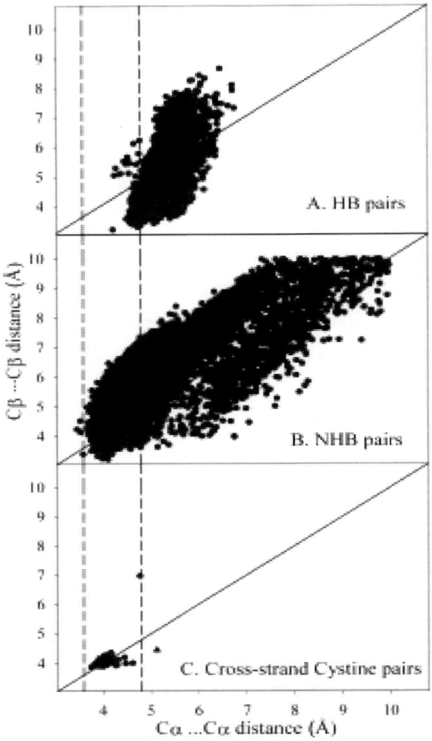

Figure 3: Cα...Cα and Cβ...Cβ distances for (A) HB registered pairs, (B) NHB registered pairs, and (C) cystines in registered pairs occurring in the non-redundant set. The point indicated by a triangle in (C) is for the sole HB cross-strand disulfide in PDB ID 1b56. The dashed lines in the plots enclose the observed range for NHB cystines.

NHB and HB residue pairs. Only A and B grade predictions were included for the analysis. While the total of 710 predictions made for NHB pairs was less than the 1507 predictions made for HB pairs, 37.2% of predictions for NHB pairs were of A grade as compared to 3.8% for HB pairs. This clearly indicates that NHB pairs are better sites for disulfide introduction. The most preferred combination of torsion angles for NHB pair predictions was "--+--" whereas for HB pairs it was "++--+". Thus the torsion angles of modeled disulfides are closer to the observed values for NHB pairs. This suggests that while disulfides can be introduced into NHB pairs, disulfides formed at HB pairs of residues might be strained.

Examples belonging to both groups were minimized in the GROMACS 3.3.3 package. NHB pairs which started out with the "--+--" pattern showed only small changes in torsion angles of <40° whereas HB registered pairs showed significant changes to their torsion angle values of >40° on minimization. This indicates that HB pair disulfides are likely to be more strained than disulfides at NHB pairs.

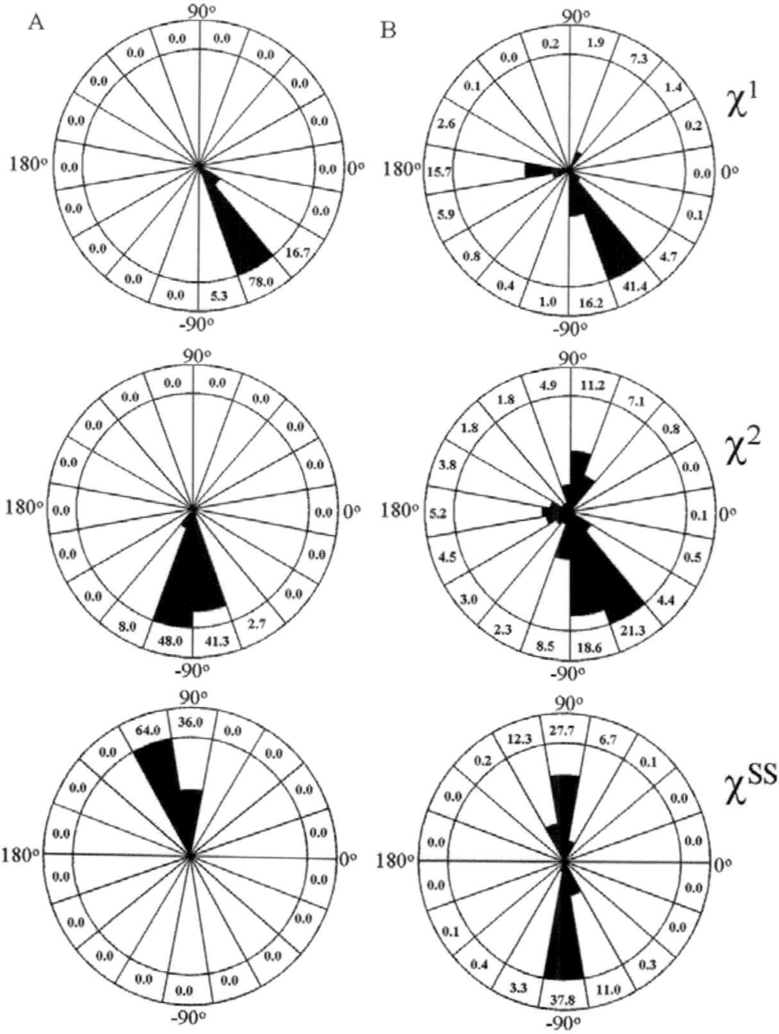

Figure 4: Circular histograms showing the percentage distributions for CYS side chain torsion angles for (A) NHB cross-strand disulfides and (B) all naturally occurring disulfides. The percentage values for various angle ranges are indicated on the circumference of the histograms.

Finally, 13 pairs of disulfides were experimentally introduced in NHB and HB pairs, either buried or exposed, in four model protein systems, leucine binding protein (LBP), leucine, isoleucine, valine binding protein (LIVBP), maltose binding protein (MBP) and Top7. In all four proteins, it was seen that disulfides introduced at exposed NHB pairs stabilized the protein whereas all of the HB pair disulfides were destabilizing.

The above studies, apart from rationalizing the preference of cross-strand disulfides towards NHB pairs and refuting claims of such bonds being strained[22], show that engineering disulfides in exposed NHB registered pairs of antiparallel strands is a robust method for stabilizing proteins.

3.3 Cross-Strand Disulfides between Parallel β-Strands

Studies on parallel β-sheets have previously reported that disulfide bridges between adjacent strands of parallel β-sheets occur rarely[24]. We report an analysis of existing structures in the Protein Data Bank to verify this fact.

To this end, the Nov 2011 release of the PDB was analyzed and 40099 entries containing parallel β-strands were identified using the SHEET record of the PDB files. MODIP calculations were carried out on all the above entries containing parallel β-strands. Disulfides were identified using the SSBOND record of the PDB files. Predicted disulfide sites in which at least one residue of the pair was a cysteine were 764 in number. Of these, 274 were identified to be real disulfides from the SSBOND records. These belonged to 255 unique chains and 135 unique entries. It is important to note that many of these disulfides were not true cross-strand disulfides because they occurred on collinear β-strands rather than adjacent β-strands. To determine the sequence homology within this set containing disulfide bonds in parallel β-strands, the sequences were aligned in a multiple sequence alignment (using ClustalW2[25]) and a dendrogram was generated. It was evident from the clustering in the dendrogram, that the majority of the proteins are in fact very similar in their sequence. Finally, the 255 unique chains containing 274 disulfides were culled using the PISCES[26] server to retain structures with a resolution of better than 2.0 Å and a sequence homology cut-off of 30%. Only 9 entries remained with a total of 10 disulfides, 9 of which were cross-strand disulfides.

Examination of the distribution of grades assigned by MODIP in its predictions hints at the reason behind the low frequency of cross-strand disulfides for parallel β-sheets. Of the total of 535795 predictions made for the complete data set, an overwhelming 67.5% predictions belong to grade C,

26.8% belong to grade B and just 5.6% predictions belong to grade A. MODIP penalizes predictions when geometric constraints for disulfide formation are not optimal and when there are steric clashes with neighboring atoms. Indeed it has been reported that cysteines in such cross-strand disulfides adopt less favorable *gauche-* $\chi 1$ conformation forming strained disulfide bonds[27] and this may be the reason why the frequency of such disulfides is so low. Recent studies on a model peptide where 2 peptides linked to D-prolyl-1,1-dimethyl-1,2-diaminoethane autonomously fold to form a two stranded parallel β-sheet show that a disulfide bond at the C-termini stabilizes the sheet but terminates the secondary structure. It was suggested that, to stably form the disulfide bond, one of the CYS residues has to adopt a non-β-strand Φ Ψ angle conformation[28]. Our analysis does not agree with this finding as has been elaborated below.

The 8 entries containing the 9 parallel β-strand, cross-strand disulfides were examined in detail with respect to their dihedral angles, residue and structural information from the PDB structure files and protein information such as SCOP classification, catalytic activity and cellular location. As indicated in Table 1, all disulfide bonds have the Φ, Ψ dihedral angles in the allowed region of the Ramachandran Map typically found in β-strands, thus refuting the recent claim that a stable cross-strand disulfide bond forms when one of the CYS residue adopts a non-sheet Φ-Ψ angle conformation[28]. The χ^{ss} angle values are largely similar to those seen for all disulfides with values ranging between $+80^{\circ}$ to $+110^{\circ}$ and -80° to -110°. However, the χ^1 angle values are more varied, with about half lying outside the typical range for disulfides. Similarly the χ^2 values show a fair amount of variation and do not show a definite trend. On observing the structural position of these cross-strand disulfides (Table 2), it is evident that one of the β-strands is generally an edge strand as seen in 5 out of 6 cases, the other 2 being beta barrels. Further, four structures have alternate conformations for the CYS pairs. Protein 1V0L adopts two separate conformations where a disulfide bond forms. Protein 3NGJ also adopts two separate conformations, one of which forms a cross-strand disulfide and the other binds a metal ion. Two proteins bind a metal ion and one has redox activity. Six of the eight proteins belong to the α/β class with three belonging to the TIM beta/alpha-barrel fold and the rest belonging to other different folds (Table 2).

One common feature observed in all cross-strand disulfides was the loss of co-planarity between the adjacent β-strands where the disulfide bond formed. This is clearly evident in Figure 5 which shows two examples of such strands twisted with respect to each other at the site of the disulfide bond.

Table 1. Dihedral angles for parallel strand disulfides.

PDB	Cys1	Φ^1	Ψ^1	Cys2	Φ^2	Ψ^2	χ^1_i	χ^2_i	χ^{ss}	χ^1_j	χ^2_j
1FO8	239	-164	167	305	-123	156	101	-76	114	-73	-111
1V0L	168	-114	143	201	-156	151	-54[a]	-27[a]	-90[a]	40[a]	115[a]
							-145[a]	83[a]	-60[a]	149[a]	-68[a]
1X6Z	129	-123	123	142	-105	150	177	-138	84	-57	-93
2DTX	173	-116	85	238	-119	125	-162	-165	116	-72	75
2JFG	208	-108	131	227	-106	93	33	149	-112	-87	5
2RAG	279	-93	123	356	-150	-170	-58	-17	-91	38	112
3KU3	52	-132	167	277	-147	161	-59	-58	-92	53	104
3NGJ	40	-123	118	61	-114	139	-85	-56	88	157	-129

[a]1V0L has 3 alternate conformations, 2 of which form a disulfide bond, thus there are two sets of side chain dihedral angles. All angles are in units of degrees.

Table 2. Residue information for parallel strand disulfides.

No.[a]	PDB	Position[b]	Occupancy		Fold
			Cys1	Cys2	
1	1FO8	Middle-Middle	1	1	Nucleotide-diphospho-sugar transferases
2	1V0L	End-Middle	0.33,0.33,0.33[c]	0.33,0.33,0.33[c]	TIM beta/alpha-barrel
3	1X6Z	Middle-End	0.54,0.38[c]	0.54,0.29[c]	Pili subunits
4	2DTX	Middle-End	1	1	NAD(P)-binding Rossmann-fold domains
5	2JFG	Middle-End	0.5,0.5[c]	0.5,0.5[c]	Ribokinase-like
6	2RAG	Middle-Middle	1	1	TIM beta/alpha-barrel
7	3KU3	Middle-Middle	1	1	Stalk segment of viral fusion proteins
8	3NGJ	Middle-Middle	0.6,0.4[c]	0.5,0.5[c]	TIM beta/alpha-barrel

[a] No. 6, 8 are metal binding, 4 shows redox activity. All proteins belong to class α/β except 3 which belongs to class α+β and 7 which belongs to class coiled coil proteins.

[b]Position with respect to start and end of the strand(s).

[c] In 1V0L the CYS residues have three alternate conformations, of which two form a disulfide. In 1X6Z, 2JFG and 3NGJ the CYS residues have two alternate conformations, of which one forms a disulfide.

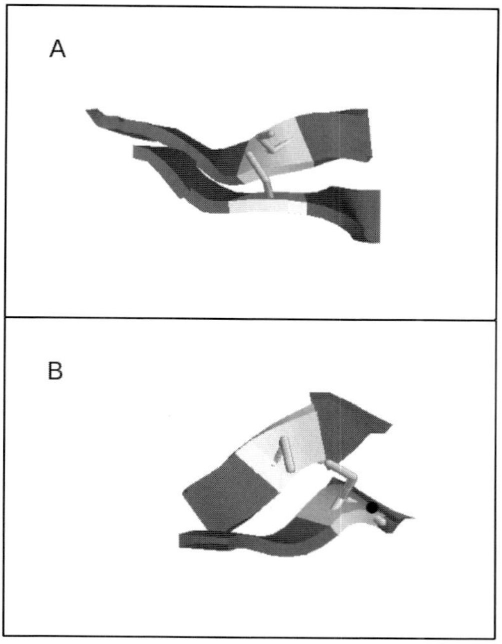

Figure 5: Residue strand orientation at the site of a disulfide bond between parallel strands (A) 1FO8, the β-strands are approximately coplanar beyond the disulfide but angled at the disulfide (B) 2RAG, the β-strands are tilted towards each other.

The above study aimed to characterize cross-strand disulfides occurring in parallel β-strands. Due to the low frequency of occurrence, no definite trend can be determined with respect to the side chain dihedral angles of the disulfide. However main chain dihedral angles all fall in the allowed region of the Ramachandran Plot. Most of the disulfides have at least one dihedral angle that lies outside the ranges typically observed for disulfides, suggesting this class of disulfides is strained.

4. Conclusions

Engineering stable disulfide bonds in proteins remains challenging as it is difficult to predict the energetic cost of protein reorganization to accommodate a disulfide bond. Characterizing disulfides in different secondary structures helps in refining criteria for disulfide engineering. Whereas disulfides at N-termini of α-helices and cross-strand disulfides in antiparallel β-strands occur in the structure database at a frequency sufficient to identify trends in dihedral angles and positional preferences, cross-strand disulfides in parallel β-strands are too rare to provide definite trends.

The propensity of disulfide bonds in different secondary structures to adopt specific χ angles which differ from those seen for disulfides occurring in other secondary structures has been observed but not always explained. Commonly used energy functions for proteins do not properly replicate observed disulfide dihedral angle distributions. Hence further work to understand disulfide dihedral angle preferences and their relation to protein backbone conformation is necessary.

Acknowledgments

Siddharth Patel acknowledges Council of Scientific and Industrial Research, Government of India for a Senior Research Fellowship. Figure 1 has been adapted from (4) and Figures 2, 3 and 4 have been adapted from (23).

References

1. B. Tidor and M. Karplus, *Proteins* **15**, 71 (1993).
2. T. E. Creighton, *Bioessays* **8**, 57 (1988).
3. K. Chakraborty, S. Thakurela, R. S. Prajapati, S. Indu, P. S. Ali, C. Ramakrishnan and R. Varadarajan, *Biochemistry* **44**, 14638 (2005).
4. S. Indu, S. T. Kumar, S. Thakurela, M. Gupta, R. M. Bhaskara, C. Ramakrishnan and R. Varadarajan, *Proteins* **78**, 1228 (2010).
5. C. Mitchinson and J. A. Wells, *Biochemistry* **28**, 4807 (1989).
6. B. A. Katz and A. Kossiakoff, *J. Biol. Chem.* **261**, 15480 (1986).
7. A. Doig and D. H. Williams, *J. Mol. Biol.* **217**, 389 (1991).
8. X. Ye, P. K. O'Neil, A. N. Foster, M. J. Gajda, J. Kosinski, M. A. Kurowski, J. M. Bujnicki, A. M. Friedman and C. Bailey-Kellogg, *Protein Sci.* **13**, 3298 (2004).
9. R. Sowdhamini, N. Srinivasan, B. Shoichet, D. V. Santi, C. Ramakrishnan and P. Balaram, *Protein Eng.* **3**, 95 (1989).
10. H.M. Berman, J. Westbrook, Z. Feng, G. Gilliland, T.N. Bhat, H. Weissig, I.N. Shindyalov and P.E. Bourne, *Nucleic Acids Res.* **28**, 235 (2000).
11. V. S. Dani, C. Ramakrishnan and R. Varadarajan, *Protein Eng.* **16**, 187 (2003).
12. R. Chandrasekaran and R. Balasubramanian, *Biochem. Biophys. Acta.* **188**, 1 (1969).
13. G. N. Ramachandran and V. Sasisekharan, *Adv. Protein Chem.* **23**, 283 (1968).
14. S. Chakravarty, A. Bhinge and R. Varadarajan, *J. Biol. Chem.* **227**, 31345 (2002).

15. R. Varadarajan and F. M. Richards, *Biochemistry* **31**, 12315 (1992).
16. N. Guex and M. C. Peitsch, *Electrophoresis* **18**, 2714 (1997).
17. E. H. B. Lindahl and D. van der Spoel, *J. Mol. Model* **7**, 306 (2001).
18. J. M. Thornton, *J. Mol. Biol.* **151**, 261 (1981).
19. J. S. Richardson, *Adv. Protein Chem.* **34**, 167 (1981).
20. P. M. Harrison and M. J. Sternberg, *J. Mol. Biol.* **264**, 603 (1996).
21. M. A. Wouters and P. M. Curmi, *Proteins* **22**, 119 (1995).
22. N. L. Haworth, L. L. Feng and M. A Wouters, *J Bioinform. Comput. Biol.* **4**, 155 (2006).
23. S. Indu, V. Kochat, S. Thakurela, C. Ramakrishnan and R. Varadarajan, *Proteins* **79**, 244 (2011).
24. N. Srinivasan, R. Sowdhamini, C. Ramakrishnan and P. Balaram, *Int. J. Pept. Protein Res.* **36**, 147 (1990).
25. M. A. Larkin, G. Blackshields, N. P. Brown, R. Chenna, P. A. McGettigan, H. McWilliam, F. Valentin, I. M. Wallace, A. Wilm, R. Lopez, J. D. Thompson, T. J. Gibson and D. G. Higgins, *Bioinformatics* **23**, 2947 (2007).
26. G. Wang and R. L. Dunbrack, Jr., *Bioinformatics*, **19**, 1589 (2003).
27. H. M. Fooks, A. C. Martin, D. N. Woolfson, R. B. Sessions and E. G. Hutchinson, *J. Mol. Biol.* **356**, 32 (2006).
28. A. M. Almeida, R. Li, and S. H. Gellman, *J. Am. Chem. Soc.* **134**, 75 (2012).
29. R. A. Laskowski, M. W. MacArthur, D. S. Moss and J. M. Thornton, *J. App. Cryst.* **26**, 283 (1993).

STRONGLY TWISTED AND COILED β-HAIRPINS AND THEIR ROLE IN PROTEIN FOLDING

ALEXANDER V. EFIMOV
Institute of Protein Research, Russian Academy of Sciences,
142290, Pushchino, Moscow Region, Russia
E-mail: efimov@protres.ru

A strongly twisted and coiled β-hairpin can be represented as a double-helical structure in which the strands are twisted and coiled in a right-handed sense. The double superhelix has both a concave and a convex surface. A distinctive feature of this structure is that it is always formed by the right-turned β-hairpin when viewed from the concave side of the superhelix. Another feature of this structure is that the φ, ψ values alternate along the polypeptide chain and φ, ψ values of the "inside" residues located on the concave surface have a strong tendency to fall in the β_E-region (extended conformations) and those of the "outside" residues in the β_P-region (polyproline region) of the Ramachandran map. A stereochemical analysis shows that glycines in "inside" and prolines in "outside" positions have to facilitate formation of coiled β-hairpins. Statistical data confirm these conclusions. Thus, the strongly twisted and coiled β-hairpin has the unique fold itself and can be taken as the starting structure in modeling of protein folds and folding pathways. The larger protein folds are obtained by stepwise addition of secondary structural elements to the root β-hairpin in accordance with a set of simple rules. A structural tree constructed using this approach includes 112 non-homologous proteins classified as a subclass of (α + β)-proteins and referred to as wrap-proteins.

1. Introduction

In globular proteins, the polypeptide chain folds over upon itself many times and the elements of secondary structure that are close together along the sequence have a strong tendency to be in close contact in the final three-dimensional structure.[1] Very often, α-helices and/or β-strands that are adjacent along the chain form a few well-defined types of super-secondary structures or structural motifs. While many different structural motifs have been observed to recur within globular proteins, only some of the motifs exhibit unique handedness and a unique overall fold.[2] The high frequency of occurrence of the structural motifs in unrelated proteins appears to be due to the relative stability of these kinds of folds. On the other hand, they can be formed rapidly because are composed of secondary structural elements adjacent along the chain. Such structural motifs have unique structures themselves and some of them can be a core around which the remainder of the molecule or domain is folded. It can be suggested that the structural motifs having unique folds act as nuclei or "ready-made" building blocks in protein folding. Alternatively, the motifs may be regarded as the starting structures in protein modeling.[3] In this paper, structure of the strongly twisted and coiled β-hairpin is considered and its role in protein modeling and folding is analyzed and discussed.

2. Two forms of β-hairpins

β-Hairpins can be right- or left-turned depending on whether the second β-strand runs on the right or the left relative to the first one when viewed from the same side (e.g. as viewed from the hydrophobic core). It is rather difficult to find a preference for one form of the β-hairpins if they are considered as isolated flat structures. Uniqueness appears at the level of higher order structures that include the β-hairpins or if the β-hairpin folds into a three-dimensional structure itself. Figure 1 represents examples of such higher order structures. A stereochemical analysis of non-homologous proteins shows that most β-hairpins in which SS-bridges are formed by cysteins located opposite to each other in the neighboring β-strands, are left-turned when viewed from the side where SS-bridges are situated.[4] Structures of the loop-hairpin type closed into cycles by SS-bridges

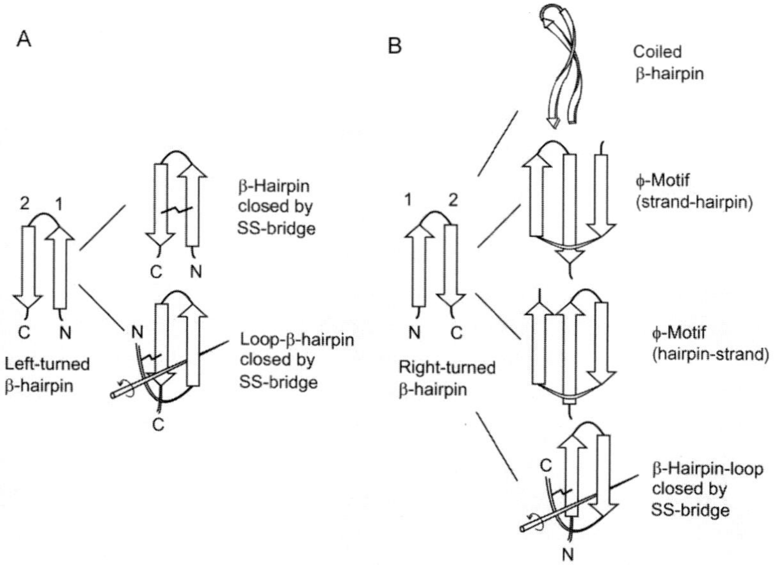

Fig. 1. Examples of higher order structures that include the left-turned (A) and the right-turned (B) β-hairpins. β-Strands are shown as arrows directed from the N- to C-ends and imaginary axes of the superhelices as straight bars.

also predominantly include the left-turned β-hairpins. In contrast, the hairpin-loop motif closed into a cycle by an SS-bridge includes the right-turned β-hairpin. Note, that both the motifs, the loop-β-hairpin and β-hairpin-loop, closed by SS-bridges form left-handed superhelices in three dimensions.[4, 5] Both types of the φ-motif, the strand-hairpin and hairpin-strand, include the right-turned β-hairpins when viewed from the crossover loop connecting the two edge β-strands.[6] If a long β-hairpin folds on to itself so that the β-strands of the two halves are packed orthogonally in the two different layers, it is also right-turned in most cases when viewed from the concave side.[7] If a β-hairpin is strongly twisted and coiled into a right-handed double-stranded superhelix, it is always right-turned when viewed from the concave side of the superhelix (Fig. 1B).

3. Structure of strongly twisted and coiled β-hairpins

β-Hairpins are widespread in proteins and occur both as isolated double-stranded antiparallel β-sheets and parts of multiple-stranded β-sheets. When viewed along the polypeptide chain direction, the β-sheets in proteins are almost invariably twisted in a right-handed sense.[8] An extended polypeptide chain with identical φ, ψ values is straight and has a structure corresponding to a β-strand in a flat or moderately twisted β-sheet. To maintain a large contact area without disturbing the hydrogen bonds, the strands must be coiled as well as twisted in the strongly twisted β-sheets.[9-11]

A strongly twisted and coiled β-hairpin can be represented as a double-helical structure in which the strands are twisted and coiled in a right-handed sense (Fig. 2a). Such a double superhelix has both a concave and a convex surface. A distinctive feature of this structure is that it is always formed by the right-turned β-hairpin when viewed from the concave side of the superhelix.[7] A similar superhelix can not be formed by a left-turned β-hairpin (Fig. 2d) because it is energetically unfavourable. Formation of a left-handed superhelix (Fig. 2c) is also unfavourable and these structures do not occur in proteins.[7]

Figure 3 shows a more detailed representation of a strongly twisted and coiled β-hairpin. One more feature of this structure is that the φ, ψ values alternate along the polypeptide chains.[7, 9-11] The φ, ψ values of the "inside" residues located on the concave surface of the superhelix have a strong tendency to fall in the β_E-region and those of the "outside" residues located on the convex surface fall in the β_P-region of the Ramachandran map (Fig. 4).[12-13] An analysis shows that "inside" residues can have φ, ψ values up to φ = -180°, ψ = 180° and even ε-conformation if these residues are glycines. It means that glycines in

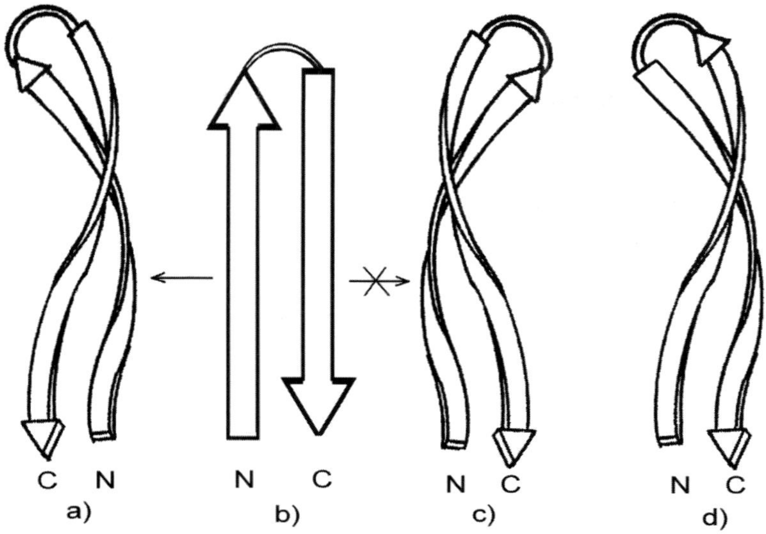

Fig. 2. A β-hairpin can be twisted and coiled to form the double superhelix shown in (a), but not those in (c) and (d).

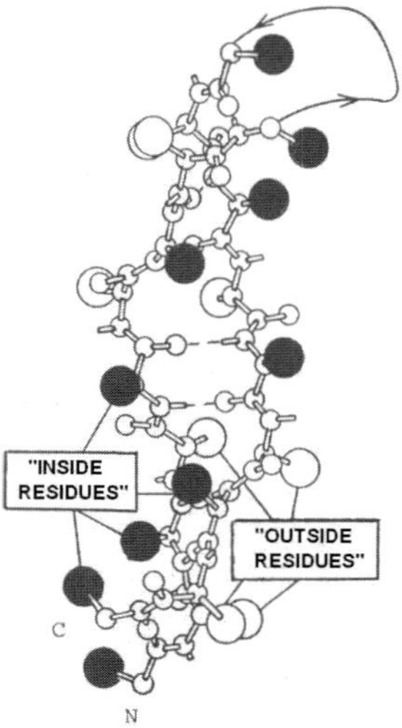

Fig. 3. A strongly twisted and coiled β-hairpin shown as a ball-and-stick model. Large black balls are "inside" side chains located on the concave surface of the double superhelix. "Outside" side chains are shown as large white balls.

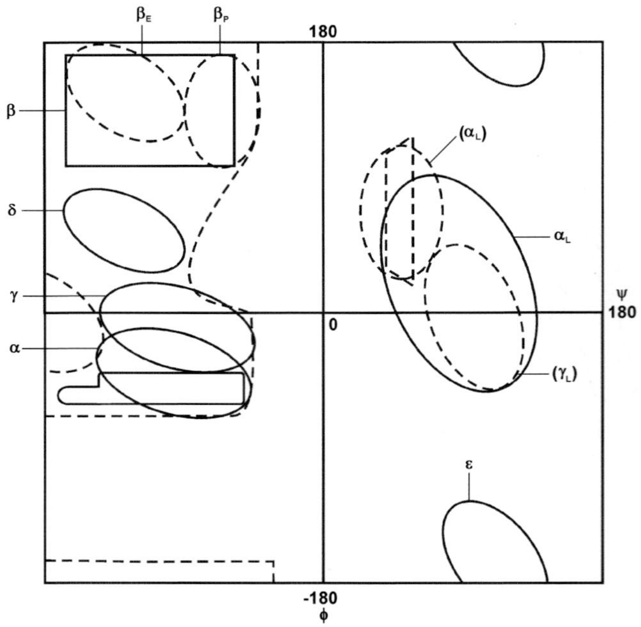

Fig. 4. Ramachandran map and six regions, α, β, γ, δ, ε and α_L, containing the overwhelming majority of experimental φ, ψ values.[12, 13] The large regions, β and α_L, can be subdivided into smaller subregions, β_E, β_P, α_L and γ_L, as shown with the dashed lines.

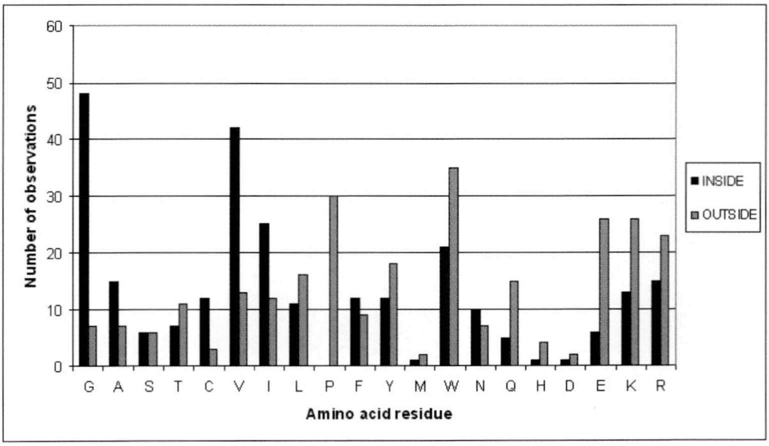

Fig. 5. Frequency of occurrence of amino acid residues in "inside" and "outside" positions of strongly twisted and coiled β-hairpins.

"inside" positions facilitate the coiling of β-hairpins.[7] On the other hand, prolines in "outside" positions have also to facilitate the formation of a coiled β-hairpin as they prefer the β_P-conformations. "Inside" positions cannot be occupied by prolines since very extended conformations are forbidden for them and they prevent from hydrogen bonding in these positions.[7] A statistical analysis confirms these conclusions.

The histogram presented in Fig. 5 shows frequency of occurrence of each amino acid residue in both "inside" and "outside" positions of strongly twisted and coiled β-hairpins found in 35 nonhomologous SH3-like domains.[14] As seen, glycines have a strong tendency to occupy the "inside" positions and prolines the "outside" positions. Moreover, prolines are not found in "inside" positions at all. It should also be noted that hydrophobic residues prefer to occupy the "inside" positions and hydrophilics the "outside" positions. In total, 386 strongly twisted and coiled β-hairpins have been found in non-homologous proteins and an analysis of them showed similar results.[14]

Thus, strongly twisted and coiled β-hairpins have the unique overall fold and handedness and are widespread in proteins. One may suggest that these structures can fold independently of the remaining part of the molecule and can act as nuclei and/or "ready building blocks" in protein folding.

4. Modeling of protein folds and folding pathways using the coiled β-hairpin as the starting structure

The problem of protein folding remains one of the central problems in structural biology. However, until now there are no experimental methods to observe protein folding in real time. Thus, theoretical modeling of protein folds and folding pathways is of particular value in solving the problem. The approach is based on the hypothesis that at the first step of protein folding a nucleus is formed and then the remaining part of the molecule or domain is folded around it.[15-17] The structural motif having a unique overall fold and handedness is taken as the starting structure in modeling. The larger structures are obtained by stepwise addition of α-helices and/or β-strands to the growing structure taking into account a restricted set of the rules inferred from the known principles of protein structure. Among these rules, prohibition of crossing connections, attention to handedness and a requirement for α-helices to be packed in α-helical layers and β-strands in β-layers are the most important. A general scheme that represents the starting structural motif (root motif), all the intermediate and completed structures connected by lines showing allowed pathways of structure growth, is referred to as the structural tree.[3]

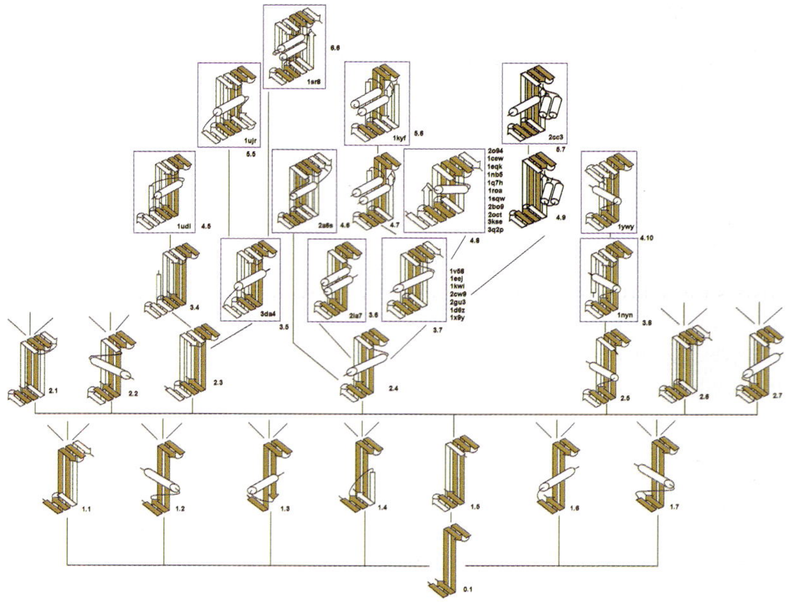

Fig. 6. A fragment of the structural tree for a subclass of (α + β)-proteins containing coiled β-hairpins. All structures are similarly oriented and are viewed from the concave surface of the root β-hairpin. The folds actually found in proteins are framed.

Figure 6 represents a fragment of the structural tree constructed for a subclass of (α + β)-proteins.[18] The strongly twisted and coiled β-hairpin is taken as the starting structure in modeling or as the root structural motif of the tree. The number of possible overall folds that can be obtained from one root structural motif is limited since the rules drastically reduce the number of allowed pathways of growth of starting and intermediate structures.[3] As seen, addition of an α-helix or a β-strand to the root β-hairpin at the first step can be done in different ways and results in formation of the folds shown in the bottom row (level) of Fig. 6. At the second step, addition of an α-helix or a β-strand to the structures of the bottom level can also be done in different ways etc. The complete structural tree includes 221 folds of which 67 occur in known proteins. The complete version of the tree is available at http://strees.protres.ru/ as an extension of structural classification of proteins PCBOST.[19] In total, our database includes 1242 PDB entries of (α + β)-proteins of this subclass and among them 112 are non-homologous. We suggest to call them wrap-proteins since most their structures can be represented as strongly twisted and coiled β-sheets wrapped around one or two α-helices.

5. Conclusions and Outlook

Thus, the strongly twisted and coiled β-hairpin has the unique fold itself and can act as a core around which the remainder of a protein molecule or a domain is folded. Overall folds of 112 non-homologous wrap-proteins can be obtained by stepwise addition of secondary structural elements to the root β-hairpin taking into account the same set of simple rules. The wrap-proteins can be classified as a subclass of (α + β)-proteins and levels of structural similarity between them can easily be observed by visual inspection of the tree (Fig. 6). The structural tree also shows possible folding pathways of wrap-proteins that lead to both known and yet unknown protein structures of this subclass. In our opinion, this information should help us in modeling of protein structures and in researching them.

Acknowledgements

This work was supported by Russian Foundation for Basic Research (Project No 10-04-00727). I would like to thank Ms E.A. Boshkova for help in drawing the Figures and Dr. A.M. Kargatov for help in preparing the manuscript.

References

1. M. Levitt and C. Chothia, *Nature* **261**, 552 (1976).
2. A.V. Efimov, *Structure* **2**, 999 (1994).
3. A.V. Efimov, *Proteins* **28**, 241 (1997).
4. E.V. Brazhnikov and A.V. Efimov, *Molecular Biology (Moscow)* **44**, 466 (2010).
5. P.M. Harrison and M.J.E. Sternberg, *J. Mol. Biol.* **264**, 603 (1996).
6. A.V. Efimov, *Biochemistry (Moscow)* **73**, 23 (2008).
7. A.V. Efimov, *FEBS Lett.* **284**, 288 (1991).
8. C. Chothia, *J. Mol. Biol.* **75**, 295 (1973).
9. K. Nishikawa and H.A. Scheraga, *Macromolecules* **9**, 395 (1976).
10. F.R. Salemme and D.W. Weatherford, *J. Mol. Biol.* **146**, 119 (1981).
11. C. Chothia, *J. Mol. Biol.***163**, 107 (1983).
12. A.V. Efimov, *Molecular Biology (Moscow)*, **20**, 208 (1986).
13. A.V. Efimov, *Prog. Biophys. Molec. Biol.* **60**, 201 (1993).
14. E.V. Brazhnikov and A.V. Efimov, in *Proteins and Peptides; V Russian Symposium*, Proceedings (Petrozavodsk, 2011) p. 190.
15. C. Levinthal, *Journal de Chimie Physique et de Physico-Chimie Biologique* **65**, 44 (1968).
16. O.B. Ptitsyn, *Dokl. Akad. Nauk SSSR* **210**, 1213 (1973).
17. D. B. Wertlaufer, *Proc. Natl. Acad. Sci. USA* **70**, 697 (1973).
18. E.A. Boshkova and A.V. Efimov, *Biochem. Biophys. Res. Commun.* (2012), in preparation.
19. A.B. Gordeev, A.M. Kargatov and A.V. Efimov, *Biochem. Biophys. Res. Commun.* **397**, 470 (2010).

IDENTIFICATION AND CONFORMATIONAL ANALYSIS OF HINGE REGIONS IN PROTEINS THAT UNDERGO DOMAIN SWAPPING

PRASHANT SHINGATE, ATUL K. UPADHYAY, U. S. RAGHAVENDER AND R. SOWDHAMINI

National Centre for Biological Sciences
Tata Institute of Fundamental Research
Bangalore, India
E-mail: mini@ncbs.res.in

The phenomenon of protein domain swapping, observed originally amongst a handful of protein examples, is acquiring attention not only due to statistically significant number of examples, but their relevance in neurodegenerative and other diseases. We describe an algorithm which could automatically detect ‘hinge regions’ in 87% of domain swapped examples, using structural properties such as higher inter-chain interactions, solvent accessibility and unstructured “coil” backbone conformation. This has enabled us to update our 3DSwap database which now contains data for 2057 protein structural entries where we observe swapping. Conformational analysis of four-residue hinge regions, in comparison with the equivalent region in the non-swapped domain counterpart, reveals fairly extended conformations are adopted by short hinges. Starting from one such hinge region, using metadynamics simulations, we show that there is a strong tendency to adopt to the more popular β-turn conformation as observed in the non-swapped domain structure. More detailed investigations of the inherently flexible, yet stable multiple conformations of the hinge regions that underlie the domain swapping structural transitions will be valuable.

1. Introduction

3D-domain swapping is one of the mechanisms in which a structural segment or complete domain of a monomeric unit of the protein interacts with other monomeric unit and *vice versa*, which result in intertwined dimeric or oligomeric structures[1]. Studies have shown the importance of 3D-domain swapping, such as mechanism for regulating function, as evolutionary strategy to create protein complexes and for promoting misfolding and aggregation resulting in numerous diseases, such as deposition diseases or conformational diseases, amyloidosis[2], serpinopathies and proteinopathies. Understanding the process of domain swapping by which most prions are formed would be a great help in identifying potential drug targets and in the therapeutic control of such diseases at an early age, if possible.

Swapped domain is a part of the protein which gets exchanged between monomeric subunits (see Figure 1 for a schematic). This could be a whole domain or a supersecondary structure in some proteins. The counterpart of the swapped domain in monomeric subunits is referred to as "non-swapped domain". The interface formed between two non-swapped domains is known as "non-swapped domain interface". The intermolecular interface formed between swapped domain of one subunit and non-swapped domain of another subunit is called as "swapped domain interface". A flexible linker region, that connects swapped domain with non-swapped domain is known as "hinge region" and is an important component since it permits movement of swapped domains acting as "conformational switches". Hinge region usually adopts loop conformation, but rarely adopts secondary structures based on the presence of specific residues. In this chapter, we show that the hinge regions could play a pivotal role in domain swapping by adopting distinct conformations. At least the short four-residue hinges, a reverse loop conformation is preferred in the monomeric form and S-shaped extended conformation in the domain-swapped form. Indeed, such changes could be induced by change in pH. We show, using molecular dynamics simulations that such excursions could happen.

In 1962, 3D-domain swapping was proposed as the mechanism of dimerization of RNase A[3]. In 1990, Tom Blundell and Christine Slingsby's groups solved and analyzed crystal structures of βB2-crystallin[4] dimer at 2.1 angstrom resolution to show that the linker region between two domains is extended, but domain interactions are analogous to those within monomeric gamma-crystallin[5] – thus providing the crystal structure report of domain swapping. The term "3D-domain swapping" was introduced by Eisenberg and group in their paper in 1994, while reporting the crystal structure of Diphtheria toxin[6]. Since then, a large number of protein structures are reported with the mechanism of 3D-domain swapping and deposited in Protein Data Bank (PDB[7]). The availability of large number of structural entries and the biomedical relevance of this phenomenon, had prompted us to ask the question that what may be the important criteria of protein to undergo this mechanism?

Detailed analysis of 3D-domain structures was performed and it was concluded that a protein fold must contain an exchangeable element, which we described as swapped domain, and these exchangeable elements must be linked to the rest of the protein by a flexible linker known as "hinge region". The hinge region must be flexible enough to adopt multiple backbone conformations and enable the swapped region to be stable in both as an intramolecular and intermolecular fashion[8,9,10].

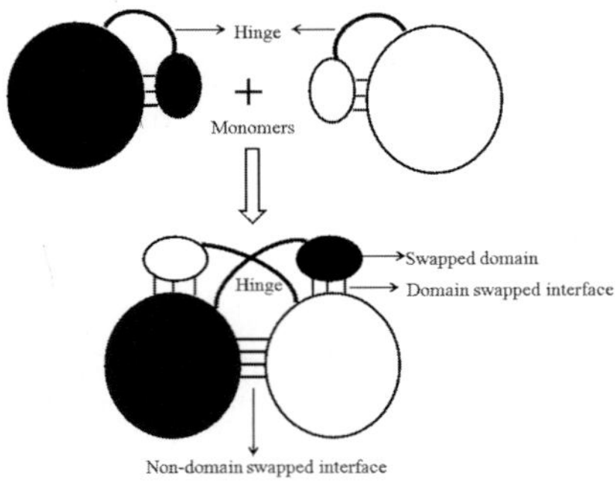

Figure 1: Schematic representation of domain swapped and non-swapped domains.

There are several examples of 3D-domain swapping as a mechanism of functional regulation *e.g.*, In glucose-fructose oxidoreductase, aspartate transcarbamylose 3D-domain swapping result in enhancement of functional properties of these enzymes. In case of p13suc1[11], which is a cyclin-dependent kinase subunit, it is proposed that 3D-domain swapping is a means to mask protein-protein recognition sites, and regulates cyclin-dependent kinase binding activity[12].

Identification of precise hinge boundaries enables us to point out critical hinge residues responsible for domain swapping. In majority of the domain-swapped proteins, such information is not easily available in the literature or PDB files and identification of hinge region using mere visualization software is not possible. Through substitution mutations in the hinge region, it is possible to identify critical hinge residues, but these experiments are time-consuming and difficult to perform on all linker residues. If structure of non-swapped domain of given domain swapped protein is available, termed as 'quasi-domain swapping' (Eisenberg paper reference), the hinge identification process is relatively easy. Our earlier hinge identification algorithm (HIDE)[13] depends on the availability of non-swapped domain homologue and therefore assigns domain swapped/ non-swapped domain status to proteins of interest. But in our study, we found that only 34% of total domain-swapped proteins have structure of their respective non-swapped domain. We now report a new hinge identification algorithm, HEIDI, which is independent of the availability of crystal structure of a homologue that is in non-swapped domain crystal state. The availability of crystal structures of domain-swapped proteins is increasing at a dramatic rate, hence it is important to identify hinge region using an automated algorithm.

2. Hinge Identification algorithm (HEIDI)

Hinge regions exhibit distinguishing structural features. Hinge region in domain-swapped proteins acquires extended conformation. Hence, characteristically, hinge region contains fewer intramolecular interactions, moderate to high intermolecular interactions, is positioned far from the core of the parent subunit, is solvent-accessible and retains extended backbone conformation. HEIDI exploits these structural features to pinpoint hinge boundaries.

The number of interactions is first normalized as follow:

$$n_i = \left(\frac{x_i - x_{avg}}{x_{\max} - x_{avg}} \right)$$

$$N_i = \left(\frac{X_i - X_{avg}}{X_{\max} - X_{avg}} \right)$$

n_i =Normalized intramolecular interactions

x_i =Number of intramolecular interactions at ith position

x_{avg} =Average number of intramolecular interactions for given proteins

$x_{\max}$ = Maximum number of intramolecular interactions present within given protein

N_i =Normalized intermolecular interactions

X_i =Number of intermolecular interactions at ith position

X_{avg} =Average number of intermolecular interactions for given proteins

$X_{\max}$ = Maximum number of intermolecular interactions present within given protein

The fragments identified to retain higher intermolecular interactions are further tested for their position relative to the core of respective subunit and whole protein. Hinge region lies away from the core of respective subunit and appears close to the centroid of whole protein. Top five regions are selected, on the basis of these critieria, for further evaluation. These candidates then are checked for the extended nature of backbone conformation and the most extended region is deemed as hinge region.

Further, solvent accessible surface area and secondary structural states are considered, including five flanking residues of the putative 'hinge', to identify precise boundaries of hinge region.

2.1. *Performance of HEIDI*

An in-house database of 3D-domain swapping proteins known as 3DSwap[14] (knowledgebase of proteins involved in 3D domain swapping), which have 293 PDB structures of proteins with cases of 3D domain swapping. Entries having large number of subunits or central loop swapping were excluded giving rise to 173 entries which were used to evaluate the performance of HEIDI. The algorithm was evaluated at two levels: At the first level, coverage of HEIDI was checked using number of cases where it could identify hinge region using following formula:

$$Coverage = \frac{R}{(R+W)} \times 100$$

R=Number of proteins in which whole or part of identified hinge region overlaps with reference hinge
W =Number of proteins in which, identified hinge and reference hinge are totally different

At this coarse level, hinge region could be identified using HEIDI for 83% domain swapped molecules with complete or partial overlapping literature hinge. At a finer level, the extent of overlapping was evaluated using following formula:

$$Accuracy = \left(\frac{2 \times R_c}{N_r + N_i} \right) \times 100$$

R_c =Number of residues common in reference hinge and identified hinge regions.
N_r =Total number of residues in reference hinge
N_i =Total number of residues in identified hinge

The accuracy at the finer level was around 67%, if strict literature-curated hinge boundaries were strictly considered. However, if the boundaries were relaxed upto five flanking positions, then the accuracy rose to 89%.

2.2. *Challenges in applying HEIDI*

Domain swapping mechanism is an oligomerisation mechanism which is not restricted to any class of the proteins. Besides this generalization, number of subunits, position of swapped domain, type of swapped domain, etc. varies case to case. Altogether, these variations affect the performance of the HEIDI. For instance, incomplete data (*e.g.* absence of coordinates for large portion in protein structure), central loop swapped proteins giving rise to false positives, intertwined domain swapped proteins and multiple hinge regions continue to pose challenges to the automatic identification of hinge regions using HEIDI.

3. Application of HEIDI on entire PDB to identify hinge regions

Starting from all the oligomeric structures of PDB (40,732 structures), subsequent to the generation of dimers using the transformation matrix where required, 5035 entries were potential domain-swapped oligomers as observed through our in-house DIAL server[15]. DIAL is an in-house webserver for the automatic identification of structural domains given the three-dimensional coordinates of a protein on the basis of clustering secondary structures and loops using their average distances. DIAL is an in-house webserver for the automatic identification of structural domains given the three-dimensional coordinates of a protein on the basis of clustering secondary structures and loops using their average distances. Domain-swapped entries can be identified since they would retain clusters with secondary structures arising from more than one chain.

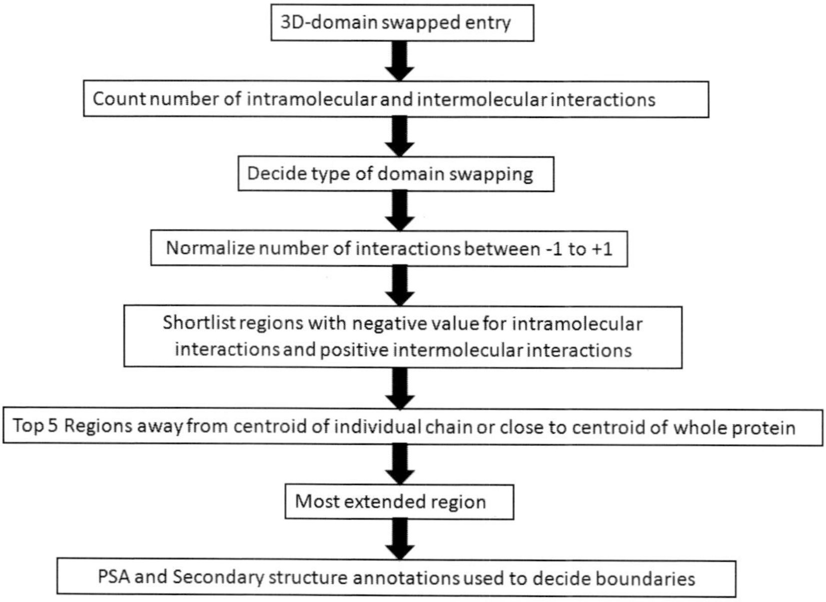

Figure 2: Flow of HEIDI algorithm (HingE Identification Independent of non-swapped domain).

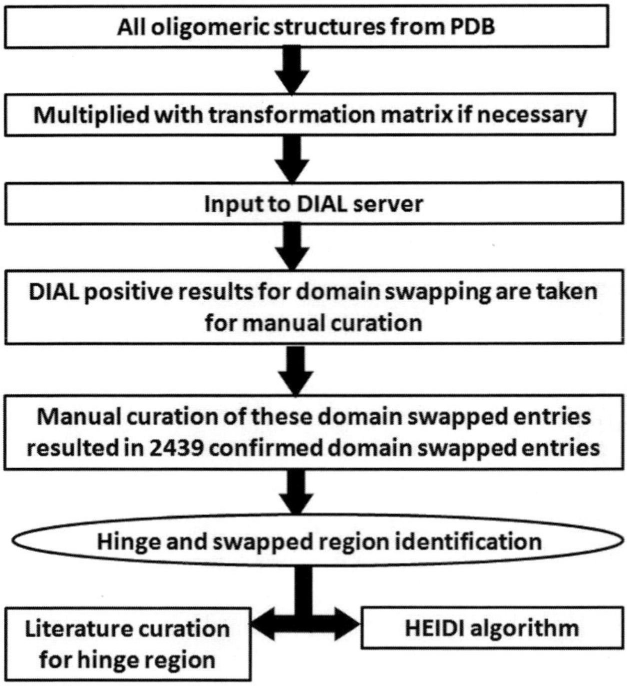

Figure 3: Schematic representation of collection of domain swapped and non-swapped domain.

Further, hinge boundaries could be identified for 2198 entries using HEIDI (workflow as described in Figures 2 and 3 and the data organised as 3DSWAP+ database and can be accessed at http://caps.ncbs.res.in/3dswapplus/).

4. Features of 3DSwap+ database

As in the previous version, 3DSwap+ database contains downloadable data at various levels: sequence (interface and hinge region), structure (for graphics visualization, interface regions, objective measures for extent of swapping, features such as hydrogen bonding, solvent accessibility and backbone conformation) and function (annotations from GO[16] and Pfam[17] databases).

In this update along with the above information we added few additional features such as effect of mutations in hinge region, swapped interface and non-swapped interfaces from experimental data. We also included information about homologues of domain swapped proteins by HIDE, effect of environmental changes as temperature, pH etc., on domain swapping.

5. Conformational analysis of hinge regions in domain-swapped form and comparison with hinge-equivalent region in non-swapped domain form

5.1 Analysis of backbone conformations (φ, ψ) of hinge regions

In our earlier study[18], analysis of (ϕ, ψ) conformations of hinge residues, it was observed that approximately 63% of (φ, ψ) conformations of domain swapped hinge residues lie in the first quadrant, while 31% of those in non-swapped domains lie in the first quadrant. A similar trend is observed with our latest dataset (Figures 4a, b). The current (ϕ, ψ) plot is generated using protein crystal structures having resolution better than 2Å from enriched dataset of domain-swapped proteins. In comparison to the residues at the hinge region, larger portion of residues in loops, equivalent to hinge region but in the non-swapped domain, resides in energetically less favourable partially allowed regions like α_L region and bridge region.

Structural superposition of all hinge regions from domain-swapped oligomers and non-swapped domain homologues was performed. Both structural superposition of domain-swapped hinge regions and non-swapped domain hinge region show high conformational preference for an overall extended conformation and reverse turn, respectively (Figure 5a and 5b). The extended conformation favors positioning of swapped domain such that native swapped domain interface can be restored. While in non-swapped domain to maintain globular structure of the protein, hinge region adopts reverse turn conformation while enduring some strain. This strain in hinge region might act as driving force for partial unfolding prior to domain swapping process.

5.2 Molecular Dynamics Simulations of a representative peptide fragment from hinge region

The observation of S-shaped extended peptide fragments in the hinge regions of the domain-swapped proteins prompted the investigation of the structural and conformational flexibility of a representative hinge region from a protein crystal structure. We focused on investigating the existence of conformationally stable canonical β-turn or structurally stable non-β-turn conformations[20] of the peptide fragments using molecular dynamics simulations. The peptide fragment GNVA (PDB: 1K46) was chosen for studying the conformational transitions of the dipeptide segment Asn-Val using vacuum MD simulations[21]. The peptide

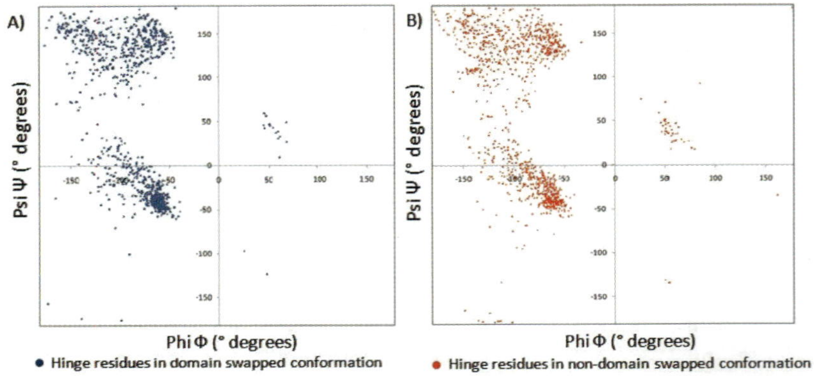

Figure 4: ϕ-ψ plots of four-residue Hinge region
a) ϕ-ψ plot of non-glycine hinge residues from domain swapped protein
b) ϕ-ψ plot of non-glycine hinge residues from non-swapped domains

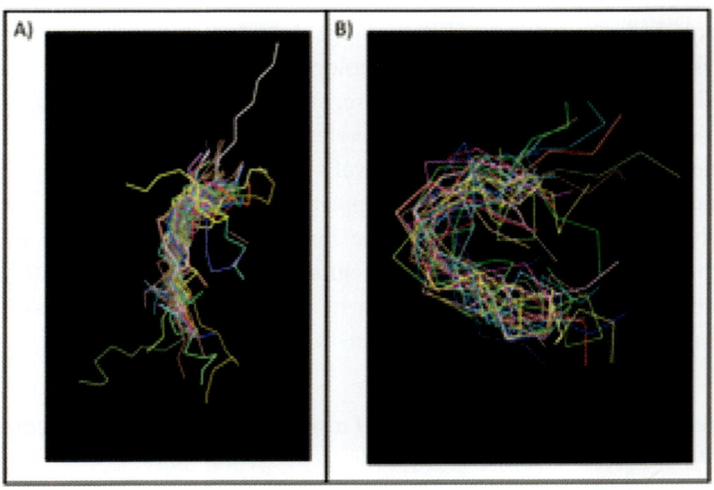

Figure 5: Superposition of 4-residue hinge regions in:
a) Extended conformation as in domain swapped form
b) Reverse loop as in non-swapped domain

GNVA, protected with acetyl and amide groups at N- and C-termini, and in a starting conformation identical to the crystal structure was chosen to investigate the region of (ϕ, ψ) space accessible to the tetra-peptide fragment, in a 10 ns vacuum MD simulation employing NVT ensemble (T = 300K, P = 1 bar). The results are represented in a Ramachandran map of the Asn(i+1)-Val(i+2) fragment (Figure 6a). The entire trajectory analysis of the MD simulation revealed that the Asn(i+1) adopts conformations which correspond to extended regions of the Ramachandran map, whereas Val(i+2) was seen to adopt mostly conformations in helical α_R ((ϕ, ψ) = (-60, -30)) and the bridge region (ψ~0°). This corresponds closely to a mixed Type I and Type II β-turn. The unique feature of this conformation is a 10-atom hydrogen bonded ring (C_{10}) involving the side-chain carboxamide group of Asn and the backbone amide NH group of Ala (Figure 6b).

It has been well established that Asn residue readily adopts conformations in α_L and partially allowed regions of the Ramachandran map, apart from regular conformations corresponding to L-amino acids[22]. This prompted us to investigate the structural stability of canonical β-turns in the GNVA peptide fragment using metadynamics simulations.

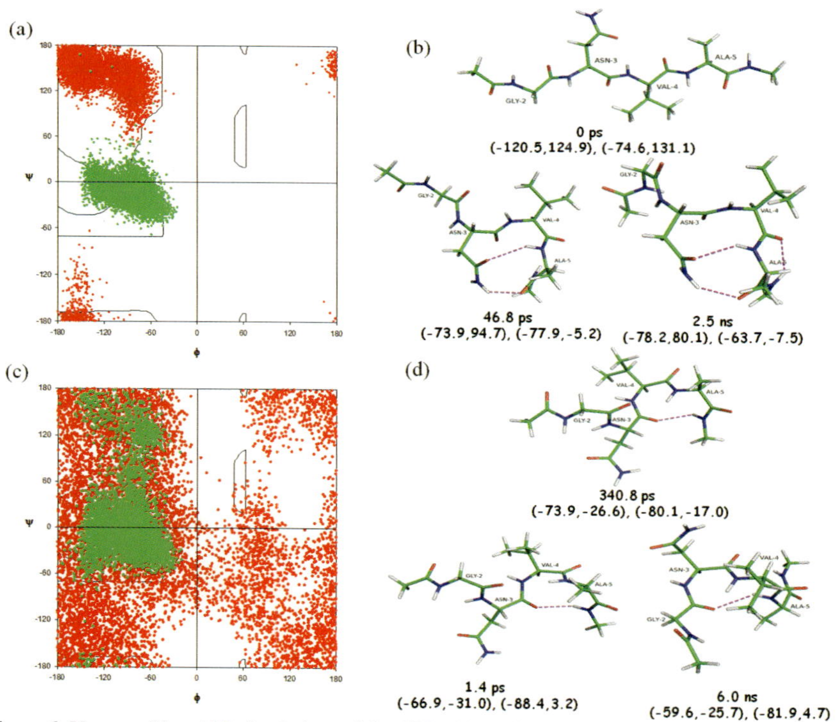

Figure 6: Vacuum 10 ns MD simulations of the GN(i+1)V(i+2)A peptide fragment from PDB:1K46. (a) plain MD, Ramachandran map of Asn-Val dipeptide, (b) representative structures at t = 0ps, 46.8 ps and 2.5 ns, (c) Metadynamics, Ramachandran map of Asn-Val dipeptide, (d) structural snapshots at t = 340.8 ps, 1.4 ns, and 6.0 ns. The backbone torsions are indicated in brackets and the hydrogen bonds are shown as dashed lines (magenta).

In a metadynamics simulation, the potential for one or more chosen variables ("CVs or collective variables") is modified by periodically adding a positive Gaussian potential at the location given by particular values of the variables. These repulsive Gaussians eventually fill up the well that is being sampled and force the calculation to sample at a different point, thus allowing the system to overcome large energy barriers. Metadynamics simulation (vacuum, 10 ns, NVT ensemble, T = 300 K and P = 1 bar) was performed on GNVA peptide fragment with backbone torsions (ϕ, ψ) of Asn as collective variables (parameters: width and height of the Gaussians were set to 5.0 Å and 0.03 Å for both ϕ and ψ). The peptide is seen to explore diverse backbone conformations, with Asn(i+1) adopting experimentally observed α_R, α_L as well as extended conformations. The Val(i+2) residue is also observed to sample a wider area of backbone conformational space than plain MD exploring both helical as well as extended conformations, though the helical and bridge conformations are relatively more populated (Figure 6c). Surprisingly, the peptide was seen to adopt a perfectly stable backbone-backbone hydrogen-bond mediated Type I β-turn in the course of simulations. Representative structures at a few time points along the trajectory are presented (Figure 6d).

The GNVA peptide fragment implicated as a hinge region in the domain swapped protein (PDB: 1K46) has been shown to adopt canonical Type I β-turn as well as mixed β-turn conformations in vacuum Metadynamics and plain MD simulations. Work towards unraveling the role of protein unfolding, mediated by the conformational switching of flexible yet structurally stable peptide fragments in the hinge regions, to decipher the molecular underpinnings of the forces responsible for domain swapping by performing MD simulations in the cellular context is being pursued presently.

6. Conclusion

Domain swapping is an important phenomenon since it is associated with neurodegenerative and other diseases. In comparison to 1990s, where the first few entries of domain swapped proteins were documented, there is an ever-increasing number of reports of such cases. However, there is less awareness to provide information about the presence of domain swap in structures and to provide exact domain swap boundaries and hinge regions, thus demanding automatic methods to identify them. We describe an objective method, called HEIDI, for the identification of hinge regions and use this approach to update our 3Dswap database to consider all entries in the Protein Data Bank. This method exploits structural features such as reduced intramolecular interactions, extended nature and increased solvent exposure of hinge regions to identify them given three-dimensional coordinates of a protein. The method was found to be more than 80% accurate in recognizing hinge regions and hence applied on close to 40,000 PDB entries. The updated version of this repository, 3DSwap+ database, can be accessed now for thousands of domain swapped PDB entries. Structural and functional details of these proteins have been curated and provided in a downloadable form.

Detailed analysis of backbone conformations of hinge regions and limited amount of molecular dynamics simulations of such regions reveal the presence of peculiar non-β-turn conformations[20] of hinge regions to retain extended structures. These, however, seem amenable to reach a more ordered β-turn conformation, as evident from molecular dynamics simulations of one such hinge region. If there is an inherent tendency of short hinge regions to retain regular β-turn conformation, it is unclear what drives the proteins to let hinge region stay extended in their domain-swapped form. This could be mere entropic stability of the relatively "open" system of the polypeptide, whilst engaged in intermolecular interactions. More detailed investigations and analyses will be valuable to understand the mechanism of domain swapping.

References

1. M. J. Bennetts, *Proc. Natl. Acad. Sci.*, **91**, 3127 (1994).
2. Y. Liu, G. Gotte, M. Libonati, and D. Eisenberg, *Protein Sci.*, **11**, 371 (2002).
3. A. M. Crestfield, W. H. Stein, and S. Moore., *Arch. Biochem. Biophys.*, **1**, 257 (1962).
4. B. Bax, R. Lapatto, V. Nalini, H. Driessen, P. F. Lindley, D. Mahadevan, T. L. Blundell, and C. Slingsby, *Nature.*, **347**, 776 (1990).
5. S. Najmudin, V. Nalini, H. P. Driessen, C. Slingsby, T. L. Blundell, D. S. Moss, and P. F. Lindley, *Acta Crystallogr. D Biol Crystallogr*, **49**, 223 (1993).
6. M. J. Bennett, S. Choe, and D. Eisenberg, *Proc. Natl. Acad. Sci.*, **91**, 3127 (1994)
7. H. Berman, K. Henrick, H. Nakamura, and J. L. Markley, *Nucleic Acids Res.* **35**, D301 (2007).
8. F. Rousseau, J. W. Schymkowitz, and L. S. Itzhaki, *Structure*, **11**, 243 (2003).
9. S. Yang, H. Levine, and J. N. Onuchic, *J. Mol. Biol.*, **352**, 202 (2005).
10. N. Koon, C. J. Squire, E. N. Baker, *Proc. Natl. Acad. Sci.*,**101**, 8295 (2004)
11. N. K. Nagradova., *Biochemistry-Moskow.*,**67**, 839 (2002).
12. K. H. Miller, J. R. Karr, and S. Marqusee, *J. Mol. Biol.*, **400**, 567 (2010).
13. K. Shameer, P. N. Shingate, S. C. Manjunath, G. Pugalenthi, and R. Sowdhamini, *Database (Oxford)*, **2011**, 42 (2011).

14. K. Shameer, G. Pugalenthi, K. K. Kundaswamy, P. N. Suganthan, G. Archunan, and R. Sowdhamini., *Bioinformatics and Biology Insights.,***4**, 33 (2010).
15. G. Pugalenthi, G. Archunan, and R. Sowdhamini, *Nucleic Acid Research,* **33**, 130 (2005).
16. M. Ashburner, C. A. Ball, J. A. Blake, D. Botstein, *et al., Nat Genet.* **25**, 25 (2000).
17. R. D. Finn, J. Mistry, B. Schuster-Bockler, S. Griffiths-Jones, *et al., Nucleic Acids Res.*, **34**, D247 (2006).
18. P. N. Shingate and R. Sowdhamini, *PLoS One,* **7**: e39305 (2012).
19. N. Srinivasan (1991). Conformational Studies on Globular Proteins: Data Analysis Ph.D. Thesis submitted in Indian Institute of Science, Bangalore, India.
20. C. Ramakrishnan, R. Sowdhamini, and P. Balaram *Coll. Czec. Chem. Commun.*, **53**, 2801 (1988).
21. Desmond Molecular Dynamics System, version 2.4, D.E. Shaw Research, New York, NY, 2010. Maestro-Desmond Interoperability Tools, version 2.4, Schrödinger, New York, NY, 2010.
22. N. Srinivasan, V. S. Anuradha, C. Ramakrishnan, R. Sowdhamini, and P. Balaram, *Int. J. Pept. Prot. Res.,* **44**, 112 (1994).

INTRINSICALLY DISORDERED PROTEINS: REVISITING THE STRUCTURE-FUNCTION PARADIGM

MEENAKSHI ANURAG

CSIR-Institute of Genomics & Integrative Biology
Mall Road, Delhi- 110007, India
E-mail: meenakshi.anurag@igib.in

GAJINDER PAL SINGH

Institute of Biochemistry, Biological Research Centre
Szeged, Hungary
E-mail: gajinder.pal.singh@gmail.com

DEBASIS DASH

CSIR-Institute of Genomics & Integrative Biology
Mall Road, Delhi- 110007, India
E-mail: ddash@igib.in

Almost a decade ago, the widely accepted and acknowledged sequence-structure-function paradigm faced a major challenge. Proteins were believed to be functional only when in structured/folded state with a well-defined and stable three dimensional conformation. Denatured proteins were found to loose function upon losing their structure. Though there were some indications about flexibility linked functionality of the proteins, it largely remained unnoticed till the beginning of this millennium. The discovery of intrinsically disordered proteins (also known as intrinsically unstructured, natively disordered, natively unfolded proteins) has provided new insights and explanations into functionality of many proteins. Intrinsically disordered proteins (IDPs) evolve rapidly and are particularly abundant in eukaryotes. They are associated with signaling, regulation and control pathways. Bioinformatics approaches are now widely used for quick and efficient identification of such proteins even at the proteome levels. This chapter highlights evolutionary trends and functional significance of IDPs, along with the methods to study intrinsically disordered regions.

1. Introduction

Mirsky and Pauling, in 1936, were the first to report denatured proteins[1], which were characterized by lack of defined conformation and hence were not crystallizable. Such evidences strengthened the structure-function paradigm. The first deviation, from well accepted trend of the role of stable conformation in function, was reported by Karush[2] for protein serum albumin that can bind to multiple molecules of varying shapes and sizes. It was inferred that the binding site exhibited large conformational modulations which provided it with substantial binding energy for an array of molecules. Karush pronounced it "Configurational adaptability". Independently, Koshland in 1958[3,4] floated the concept of Induced-fit binding that bore similarity to the concept of earlier proposed configurational adaptability (Fig 1). Later on, many studies showed domain conformational change upon substrate binding. These studies laid niche for the research focused on intrinsically disordered proteins. Such proteins have regions with missing electron density in X-ray diffraction pattern because of ill-defined structure.

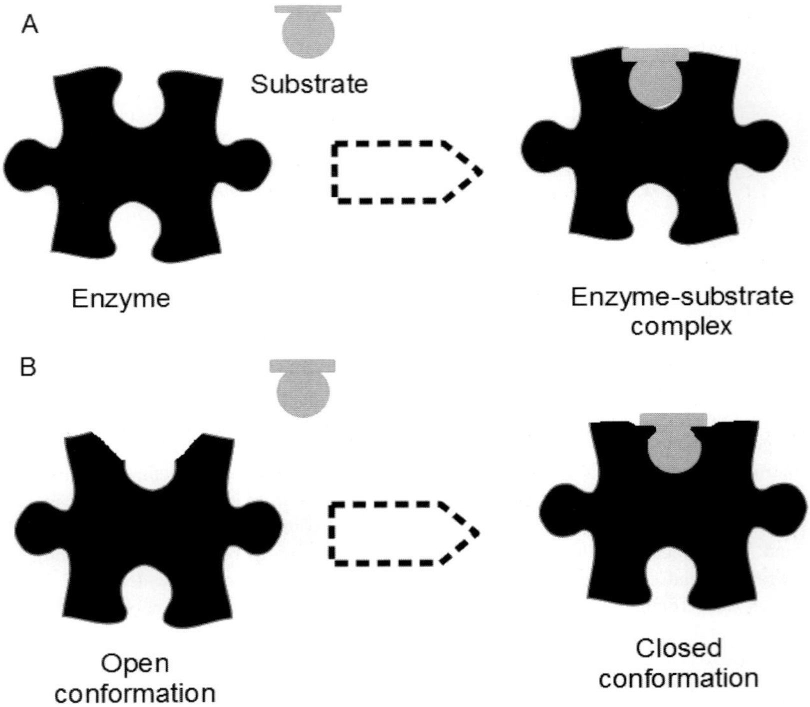

Figure 1. Proposed modes of substrate-protein binding. (A) The lock and key model first postulated in 1894 by Emil Fischer, according to which the binding site of the enzyme (enzyme is shown in black) has a unique geometric shape that is complementary to substrate (gray), similar to the complementarity of puzzle blocks. (B) Koshland in 1958 proposed the "induced-fit" mode of binding of substrate to the receptor, where the binding site of the receptor is conformationally flexible and there are structural transitions in the binding site (open to closed conformation).

Establishment of the relevance and functionality of intrinsically disordered regions (IDRs) has been a major milestone in the understanding of protein structure. Many proteins are now known to have IDRs and some are completely disordered. What guides a protein to remain disordered even in native state? It is well known that sequence dictates what kind of structure the peptide will attain; likewise the sequence also dictates if the peptide will be natively disordered. Studies have shown that amino acid composition of IDPs is skewed towards charged and hydrophilic amino acid residues[5,6]. IDRs have high solvent accessible surface as compared to structured proteins of similar lengths[7]. The presence of relatively large exposed surface facilitates its binding to multiple partners that may be either micromolecules like chemical entities or macromolecules like protein or DNA. Hence, intrinsically disordered proteins (IDPs) are crucial in interactions of many proteins.

2. Evolution - Disorder across Domains of Life

Evolutionary analysis helps in understanding the functional significance and mode of action of proteins. Genome wide bioinformatics predictions have shown that eukaryotes have much higher disorder as compared to prokaryotes (Fig 2). Wholly disordered proteins are much less prevalent than partially disordered ones[8,9]. The proteins that are wholly disordered on average constitute 2% of the bacterial proteome while in case of eukaryotes it is about 17% of the proteome. Disordered proteins with at least one long disordered region (consecutive >=30 disordered residues) on average constitute 3% of the archeal, 4.2% of the eubacterial and 33% of the eukaryotic proteome[9]. Disordered regions evolve significantly more rapidly than their ordered counterparts[10] making it less conserved across taxa. These regions have higher rates of insertions and deletions[11,12]. Hence, IDPs have higher capacity to rewire their interaction neighborhood through evolution[13]. Recently, Bellay et al.[14] classified disordered regions into highly evolving, conserved and non-conserved disorder. While highly evolving disorder is associated with signaling pathways, conserved disorder is associated with RNA binding and protein chaperone functions. Non-conserved disorder is likely much less functionally important than conserved ones.

Though IDRs are generally highly evolving, there are conserved short linear motifs (SLiMs are conserved short motifs which contribute in recognition or targeting) within IDRs[15] Higher incidence of positively selected residues in disordered regions of proteins (compared to structural regions) points to their importance in adaptive evolution of organisms where they occur [16].

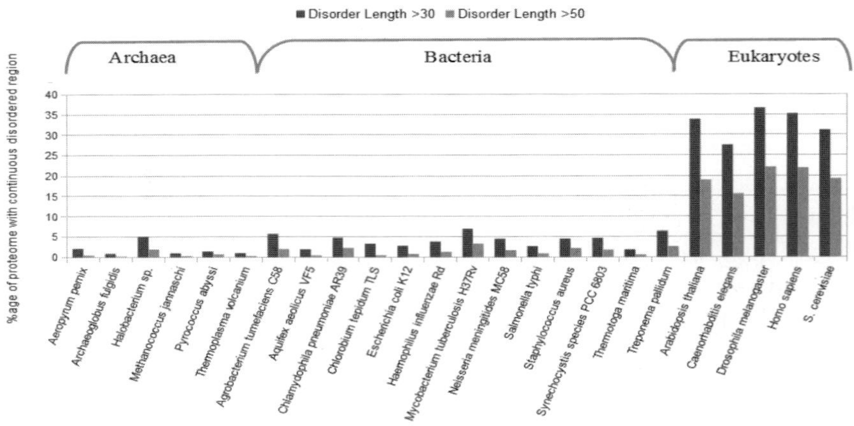

Figure 2. Percentage of proteome that comprises of proteins bearing atleast one long intrinsically disordered region (>30 – black bars and >50 – gray bars) across 24 genomes. The genomes include six archaea, thirteen bacteria and five eukaryotes . The predictions were made using DISOPRED2. (Adapted from Ward et al, 2004[9])

3. Functional Roles of Disordered Regions

For many of the functions, structure does seem to be necessary, particularly for the catalytic activity of proteins, where proper orientation of catalytically active residues is important during catalysis (though molten globule enzymes are also known[17]). But for many cellular functions of proteins, a compact 3D structure may not be a necessary requirement. In fact, disordered conformation may be advantageous for many functions. Sometimes, disorder in or around structured motifs, eg., coiled coils is essential for specific function and cellular compartmentalization of IDRs in biological systems[18]. Careful analysis of literature and bioinformatics analysis has identified functional classes overrepresented in IDPs. The most common function associated with IDPs is "binding" to macromolecules (Fig 3) like DNA, RNA and proteins and are particularly enriched in transcription and signal transduction categories[9]. IDRs may also be functional while remaining disordered as in linkers and entropic chains[19]. IDRs involved in binding may bind to their partners in a transient or a more stable manner.

Protein and RNA chaperones have also been found to be highly enriched in IDPs[20]. It has been proposed that ability of disordered regions to bind multiple partners, their transient binding capacity and their solubilization effect on the substrate upon binding might explain the prevalence of IDPs in chaperones.

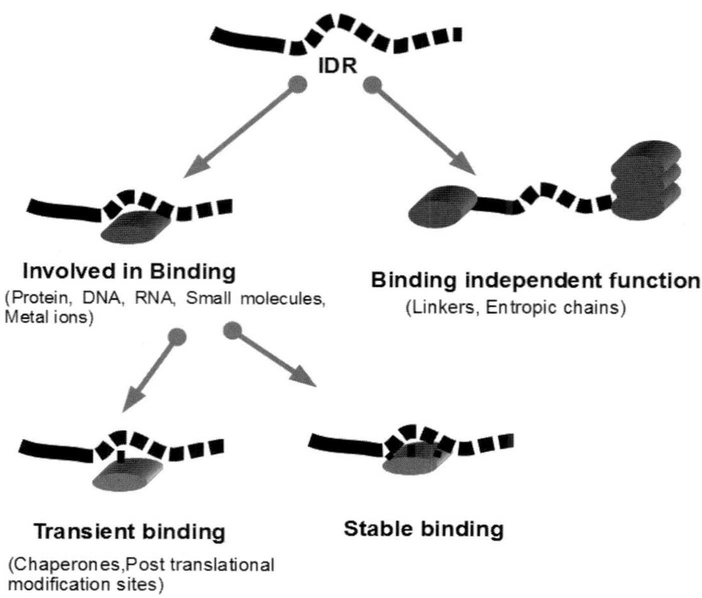

Figure 3. IDRs may function while remaining disordered as in flexible linkers and entropic chains, or more importantly they may function by binding to their interacting partner(s).

Studies on conformational flexibility in the protein degradation sites have reported that the denatured and flexible proteins degrade more rapidly than the compact ones[21].

4. Binding to Interaction Partners

4.1 Coupled Folding and Binding

Disordered regions generally fold upon binding to their partners (however many remain significantly disorder even after binding[22]). In some cases, both disordered partners may attain structure upon binding, as in binding between CBP and ACTR proteins[23]. Most commonly, binding is initiated by a residual structure in the disordered region, also known as Preformed Structural Elements (PSEs) or Molecular Recognition Features (MoRFs)[24]. Mechanism of coupled folding and binding is particularly well studied for interaction between phosphorylated KID domain (pKID) of CREB and KIX domain of CBP[25]. pKID is intrinsically disordered but folds upon binding to KIX domain. Interestingly this interaction was stabilized by hydrophobic residues in the binding interface[26]. The involvement of hydrophobic interactions in the interface was also seen in a large scale analysis of complexes involving IDRs. So even though, IDRs are depleted in hydrophobic residues, their interfaces are significantly more hydrophobic compared to the rest of the IDR and also compared to interfaces of globular proteins. The disorderness of pKID was also found to increase the rate of binding.

4.2 Transient Binding

During the binding of a disordered region to its binding partner, the entropy of the disordered region is reduced, reducing the binding affinity of the proteins. Thus it has been proposed that interactions involving disordered regions can be adjusted by evolution to be weak yet specific. In contrast, in interactions involving ordered regions, reducing the binding affinity necessarily reduces binding specificity. Thus there is an uncoupling of binding specificity and binding affinity in interactions involving disordered regions. In support of this hypothesis, proteins involved in signal transduction, which bind to their partners weakly, yet specifically, are significantly overrepresented in proteins with predicted disordered regions[27]. Also many examples are known from the literature in which disordered regions are involved in transient but specific interactions. Further many short binding motifs, which are frequently involved in transient binding, tend to occur in disordered regions. It should be noted however that not all interactions involving coupled folding and binding are transient, as IDRs may also be involved in stable complexes[28].

4.3 Higher Rate of Binding

It has been proposed that disordered regions may be able to quicken the search for their binding partners by virtue of their "increased radius of capture"[29]. This increased speed of interaction may be particularly important for proteins which are present in low numbers in cell e.g. transcription factors are typically present in cell only in few copies and are highly disordered. In particular, electrostatic interactions of disordered regions has been shown to increase their encounter rate[30].

4.4 Binding to Multiple Partners

The structural malleability of disordered regions has been proposed to allow the same or overlapping region(s) in the protein to bind to multiple partners e.g. GSK3beta protein interacts with both axin and FRAT through a disordered loop with significant overlap of the binding sites[31]. In many bioinformatics studies, it has been found that hub proteins (proteins interacting with large number of partners) were significantly more disordered than non-hubs in many organisms[33,34]. Intrinsic disorder was found to be a distinctive feature of viral proteins that interact with multiple human proteins[34]. There was a significant correlation between the number of target genes a transcription factor regulates and its intrinsic disorder[35]. For some structured hubs like 14-3-3 proteins and cdk2, it was found that their binding partners were highly disordered. A particularly interesting class of proteins was identified in which the same or overlapping disordered region mediates both inhibitory and activating activities on different or even same binding partner[36].

4.5 Disordered Regions as Sites for Post-Translational Modification

Disordered regions being more exposed on the surface are easily accessible to enzymes involved in post translational modifications. Phoshorylation sites have been found to be highly enriched in disordered regions in number of phosphoproteomic studies[37]. This trend was also noted previously and this observation was used to identify phoshorylation sites with increased confidence[38]. It has been proposed that multiple phoshorylations within disordered regions generate highly cooperative binding[39].

5. Identification of Disordered Regions and Proteins

5.1 Experimental Methods

X-ray crystallography, which has been the dominant technique for structure determination, does not work very well with IDRs. Disordered regions interfere with the crystallization of proteins from their saturated solution, which is a prerequisite for structure determination. Thus predicted disordered regions are often deleted by recombinant DNA technology prior to their expression, purification and crystallization. However proteins with short IDRs can crystallize and can be identified as regions with missing electron density. Though, it should be noted that not all regions with missing electron density are disordered, e.g. wobbly domains. Also IDRs may acquire structure as an artifact of crystallization process.

NMR, which determines structure of macromolecules in solution, is well suited to identify disordered regions and is the only direct technique by which disorderness can be studied at residue level resolution. But as NMR requires high protein concentrations, it has limitations in studying IDPs which aggregate at high concentration.

Far UV-Circular Dichroism (CD) spectroscopy is a very popular method to identify and quantify secondary structure in protein and lack thereof. Fourier-Transformed Infrared Spectroscopy (FTIR) also provides information on the secondary structure content of proteins. Fluorescence correlation spectroscopy (FCS) or FRET can also be used to quantify the hydrodynamic size of proteins at low concentration[40]. Non-cooperative unfolding upon addition of denaturation agents like urea or by increasing the temperature also indicate disorderness in proteins and can be measured using Differential Scanning Calorimetry (DSC). Small angle X-ray scattering gives information on the shape of the protein and IDPs can be identified by their lack of rigid core. Other techniques commonly used to study protein disorder include, site directed spin labeling, ultracentrifugation and gel-filtration, protease sensitivity, high speed atomic force microscopy and SDS–PAGE mobility.

Some proteome wide techniques for identification of IDPs have also been developed based on the observation that disordered proteins generally are resistant to protein unfolding and denaturing agents like heat, acid and urea, that lead to unfolding, aggregation and precipitation of globular proteins. Disprot database (http://www.disport.org) maintains the list of many experimentally characterized IDPs.

5.2 Computational Methods

The first disorder prediction method developed was based on the neural networks trained mostly on residues with missing electron density in X-ray crystal structure[41]. Since then, a large number of prediction programs[42-44] have been developed based on the different characteristics of disordered regions.

These prediction programs not only are useful in understanding how lack of structure relates to the function of the proteins, but also have practical applications in

1) X-ray crystallography – As IDRs interfere with crystallization process, it is important to correctly identify and delete these regions.

2) Protein annotation – Size and position of IDRs has been found to correlate with the functional classes of IDPs and algorithms trained on these characteristics can be used to annotate unannotated proteins[44].

3) Facilitating identification of short linear motifs – Short linear binding motifs tend to occur in IDRs. This information has been utilized to identify novel linear binding motifs and filter for functionally relevant binding sites[45].

Disorder prediction methods have been assessed from CASP5 in 2002. These assessments show that top predictors have fairly high accuracy (>0.8 based on ROC-AUC and ~0.75 based on average of sensitivity and specificity[46]). But CASP assessment is based mostly on prediction of residues missing in X-ray structure, which may not be the best definition of disorder. To identify potentially disordered regions in individual proteins, it is important to consider more than one method, based on different definitions of disorder. Disorder Metapredictors like MetaDisorder combine many existing disorder predictors to make more accurate disorder prediction[47].

6. Molecular Dynamics Studies of IDR

Intrinsically disordered regions are known to exhibit ensemble of structures which aid in rapid and promiscuous interactions with partners[48,49]. This mechanism is difficult to be captured experimentally because of short life span and conformational changes associated with interaction. Molecular dynamics simulations are being implemented in order to gain an insight into the mode of interaction and conformational change[50,51]. Coarse-grained simulations are helpful in getting a broader perspective of the coupled-binding and folding at a relatively low computational cost[52,53]. On the other hand, all-atom molecular dynamics simulation studies play a vital role in understanding the role of atoms of residues along with solvent (which are treated as explicit entities) in binding and conformational switching[51,52]. But these are computational intensive exercises which would require time as well as high computing power to run simulation for even smaller protein-ligand complexes. Also, efforts are being made to improve force-fields for simulation of IDRs[54,55]. The field is evolving at a fast pace and there are several exemplary studies which have shown folding mechanism of IDRs when interacting with partner[54].

7. Role in Diseases – Human Disorders, Pathogenic Diseases

Given the prevalence of IDPs in human proteome, it is perhaps not surprising that they are involved in many diseases[55]. Proteins associated with cancer, cardiovascular and neurodegenerative diseases are particularly enriched in IDPs[27,56,57].

It has been suggested that because of generally low affinity of interactions involving IDRs, these are better suited for inhibition by small molecules and thus may represent potentially new types of drug targets[58]. Some drug targets have an intrinsically disordered region (IDR) at their binding interface. These regions exhibit disordered-to-ordered transitions during the interaction with their substrates. One of the widely studied cases is the p53–MDM2 interaction[59,60]. p53 is an intrinsically disordered oncoprotein (Fig 4), which, when induced, results in apoptosis of tumor cells. There are several factors that regulate the activity of p53, one of them being the interaction with MDM2 (Fig 4). p53 binds with MDM2 via the intrinsically disordered interface and undergoes a structural transition. A small molecule, Nutlin, has been found to mimic the binding of p53 to MDM2 and thus competes with p53 to bind the MDM2 pocket[59,61,62].

With very little attained so far with structure based rational drug discovery approaches, disrupting protein-protein interactions involving disordered proteins is being considered vital for drug targeting.

Id	Interacting Partner	Interaction residue range
1	MDM2	18-26
2	HRMT1L2	1-83
3	WWOX	66-110
4	AXIN1	116-292
5	HIPK1	100-370
6	EF41	256-294
7	DNA	273-280
8	HIPK2	319-360
9	USP7	359-363
10	CARM1	300-393

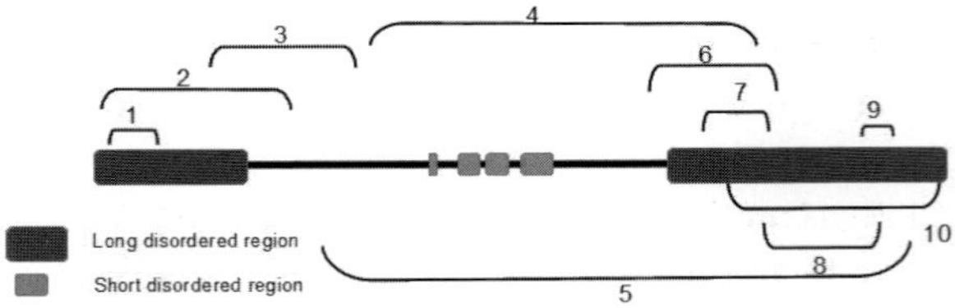

Figure 4. p53 is an oncoprotein which harbors two Long disordered regions (black) and short disordered region (light gray), The IDRs form the interacting interface with many partners. Regions which form interactive interface with ten partners (listed in the left hand side table) has been shown. FoldIndex was used for predicting IDRs.

Recent studies have aimed at identifying IDPs as potential drug target. A systematic selection protocol implemented to identify a set of crucial and targetable proteins in *M. tuberculosis*[61] lead to 13 IDPs which are essential for the survival of the organism. These proteins can be targeted to disrupt vital biological interactions and pathways. Using *in silico* approaches (docking and machine learning) inhibitors were proposed[62] against one of the prioritized protein named GlmU which is an essential bifunctional enzyme[63]. Hence, role of IDPs in pathogenesis and diseases needs to be explored further.

8. Conclusions and Outlook

Disordered regions can have different functions. They can bind to proteins transiently or more stably. They can bind nucleic acids, DNA or RNA; they may serve as sites for post-translation modification. They may function as linkers or have no observable function. Classification of disordered regions to these classes can be very useful and important step forward. These functional roles are expected to be influenced by various parameters of disordered regions like length, position, composition, charge, conservation, their position in interaction networks, expression level of the protein etc. IDPs have been associated with a wide range of diseases and disorders and their role in pathogens is also being explored. These can be further pursued as drug targets.

Acknowledgments

This work is supported by CSIR-Institute of Genomics and Integrative Biology. Meenakshi Anurag acknowledges ICMR for the Senior Research Fellowship.

References

1. A.E. Mirsky and L. Pauling, *Proc Natl Acad Sci USA* **22**, 439 (1936).
2. F. Karush, J. *Am. Chem. Soc.* **72**, 2714 (1950).
3. D.E. Koshland, *Proc Natl Acad Sci USA* **44**, 98 (1958).
4. D.E. Koshland, *Angewandte Chemie International Edition in English* **33**, 2375 (1995).
5. P. Romero, Z. Obradovic, X. Li, E.C. Garner, C.J. Brown, and A.K. Dunker, *Proteins* **42**, 38 (2001).
6. S. Vucetic, C.J. Brown, A.K. Dunker, and Z. Obradovic, *Proteins* **52**, 573 (2003).
7. R. Linding, J. Schymkowitz, F. Rousseau, F. Diella, and L. Serrano, *J. Mol. Biol.* **342**, 345 (2004).
8. A.K. Dunker, Z. Obradovic, P. Romero, E.C. Garner, and C.J. Brown, *Genome Inform Ser Workshop Genome Inform* **11**, 161 (2000).
9. J.J. Ward, J.S. Sodhi, L.J. McGuffin, B.F. Buxton, and D.T. Jones, *J. Mol. Biol.* **337**, 635 (2004).
10. C.J. Brown, S. Takayama, A.M. Campen, P. Vise, T.W. Marshall, C.J. Oldfield, C.J. Williams, and A.K. Dunker, *J. Mol. Evol.* **55**, 104 (2002).
11. C.J. Brown, A.K. Johnson, and G.W. Daughdrill, *Mol Biol Evol* **27**, 609 (2010).
12. C.J. Brown, A.K. Johnson, A.K. Dunker, and G.W. Daughdrill, *Current Opinion in Structural Biology* **21**, 441 (2011).
13. R. Mosca, R.A. Pache, and P. Aloy, *Mol. Cell Proteomics* **11**, M111.014969 (2012).
14. J. Bellay, S. Han, M. Michaut, T. Kim, M. Costanzo, B.J. Andrews, C. Boone, G.D. Bader, C.L. Myers, and P.M. Kim, *Genome Biology* **12**, R14 (2011).
15. A.N. Nguyen Ba, B.J. Yeh, D. van Dyk, A.R. Davidson, B.J. Andrews, E.L. Weiss, and A.M. Moses, *Sci. Signal.* **5**, rs1 (2012).

16. J. Nilsson, M. Grahn, and A.P. Wright, *Genome Biol* **12**, R65 (2011).
17. M. Vendruscolo, *Curr Opin Chem Biol* **14**, 671 (2010).
18. M. Anurag, G.P. Singh, and D. Dash, *Mol Biosyst* **8**, 346 (2012).
19. P. Tompa, *FEBS Lett.* **579**, 3346 (2005).
20. P. Tompa and P. Csermely, *FASEB J.* **18**, 1169 (2004).
21. K.S. Sandhu and D. Dash, *Proteins* **63**, 727 (2006).
22. M. Fuxreiter and P. Tompa, *Adv. Exp. Med. Biol.* **725**, 1 (2012).
23. S.J. Demarest, M. Martinez-Yamout, J. Chung, H. Chen, W. Xu, H.J. Dyson, R.M. Evans, and P.E. Wright, *Nature* **415**, 549 (2002).
24. C.J. Oldfield, Y. Cheng, M.S. Cortese, P. Romero, V.N. Uversky, and A.K. Dunker, *Biochemistry* **44**, 12454 (2005).
25. D. Eliezer and A.G. Palmer, *Nature* 447, 920 (2007).
26. K. Sugase, H.J. Dyson, and P.E. Wright, *Nature* **447**, 1021 (2007).
27. L.M. Iakoucheva, C.J. Brown, J.D. Lawson, Z. Obradović, and A.K. Dunker, *J. Mol. Biol.* **323**, 573 (2002).
28. H. Hegyi, E. Schad, and P. Tompa, *BMC Struct. Biol.* **7**, 65 (2007).
29. B.A. Shoemaker, J.J. Portman, and P.G. Wolynes, *Proc Natl Acad Sci USA* **97**, 8868 (2000).
30. H.-X. Zhou, X. Pang, and C. Lu, *Phys. Chem.* Chem. Phys. **14**, 10466 (2012).
31. R. Dajani, E. Fraser, S.M. Roe, M. Yeo, V.M. Good, V. Thompson, T.C. Dale, and L.H. Pearl, *EMBO J.* **22**, 494 (2003).
32. C. Haynes, C.J. Oldfield, F. Ji, N. Klitgord, M.E. Cusick, P. Radivojac, V.N. Uversky, M. Vidal, and L.M. Iakoucheva, *PLoS Comput. Biol.* **2**, e100 (2006).
33. G.P. Singh, M. Ganapathi, and D. Dash, *Proteins* **66**, 761 (2007).
34. L. Meyniel-Schicklin, B. de Chassey, P. André, and V. Lotteau, *Mol. Cell Proteomics* **11**, M111.014738 (2012).
35. G.P. Singh and D. Dash, *Proteins* **68**, 602 (2007).
36. D.M. Bustos and A.A. Iglesias, *Proteins* **63**, 35 (2006).
37. G.D. Amoutzias, Y. He, K.S. Lilley, Y.V. de Peer, and S.G. Oliver, *Mol Cell Proteomics* (2012).
38. L.M. Iakoucheva, P. Radivojac, C.J. Brown, T.R. O'Connor, J.G. Sikes, Z. Obradovic, and A.K. Dunker, *Nucleic Acids Res.* **32**, 1037 (2004).
39. P. Lenz and P.S. Swain, *Curr. Biol.* **16**, 2150 (2006).
40. S. Mukhopadhyay, R. Krishnan, E.A. Lemke, S. Lindquist, and A.A. Deniz, *PNAS* **104**, 2649 (2007).
41. Romero, Obradovic, and Dunker, Genome Inform Ser Workshop Genome Inform **8**, 110 (1997).
42. B. He, K. Wang, Y. Liu, B. Xue, V.N. Uversky, and A.K. Dunker, *Cell Research* **19**, 929 (2009).
43. D. Sethi, A. Garg, and G.P.S. Raghava, *Amino Acids* **35**, 599 (2008).

44. A. Lobley, M.B. Swindells, C.A. Orengo, and D.T. Jones, *PLoS Comput Biol* **3**, e162 (2007).
45. V. Neduva, R. Linding, I. Su-Angrand, A. Stark, F. de Masi, T.J. Gibson, J. Lewis, L. Serrano, and R.B. Russell, *PLoS Biol.* **3**, e405 (2005).
46. B. Monastyrskyy, K. Fidelis, J. Moult, A. Tramontano, and A. Kryshtafovych, *Proteins: Structure, Function, and Bioinformatics* **79**, 107 (2011).
47. L.P. Kozlowski and J.M. Bujnicki, *BMC Bioinformatics* **13**, 111 (2012).
48. M.M. Dedmon, K. Lindorff-Larsen, J. Christodoulou, M. Vendruscolo, and C.M. Dobson, *J. Am. Chem. Soc.* **127**, 476 (2005).
49. O. Awile, A. Krisko, I.F. Sbalzarini, and B. Zagrovic, *PLoS Comput Biol* **6**, e1000854 (2010).
50. J. Wang, Y. Wang, X. Chu, S.J. Hagen, W. Han, and E. Wang, *PLoS Comput Biol* 7, e1001118 (2011).
51. W. Zhang, D. Ganguly, and J. Chen, *PLoS Comput Biol* **8**, e1002353 (2012).
52. R.B. Best and G. Hummer, *J Phys Chem B* **113**, 9004 (2009).
53. M. Knott and R.B. Best, *PLoS Comput Biol* **8**, 7 (2012).
54. S. Rauscher and R. Pomès, *Biochemistry and Cell Biology* **88**, 269 (2010).
55. V.N. Uversky, C.J. Oldfield, and A.K. Dunker, *Annual Review of Biophysics* **37**, 215 (2008).
56. Y. Cheng, T. LeGall, C.J. Oldfield, A.K. Dunker, and V.N. Uversky, *Biochemistry* **45**, 10448 (2006).
57. S. Raychaudhuri, S. Dey, N.P. Bhattacharyya, and D. Mukhopadhyay, *PLoS ONE* **4**, e5566 (2009).
58. V.N. Uversky, *Expert Opinion on Drug Discovery* **7**, 475 (2012).
59. N.C. Arva, K.E. Talbott, D.R. Okoro, A. Brekman, W.G. Qiu, and J. Bargonetti, *Ethn Dis* **18**, 2 (2008).
60. B. Shan, D.-W. Li, L. Bruschweiler-Li, and R. Bruschweiler, *The Journal of Biological Chemistry* (2012).
61. M. Anurag and D. Dash, *Molecular BioSystems* **5**, 1752 (2009).
62. D. Singla, M. Anurag, D. Dash, and G.P.S. Raghava, *BMC Pharmacol.* **11**, 5 (2011).
63. W. Zhang, V.C. Jones, M.S. Scherman, S. Mahapatra, D. Crick, S. Bhamidi, Y. Xin, M.R. McNeil, and Y. Ma, Int. *J. Biochem. Cell Biol.* **40**, 2560 (2008).

INTRINSICALLY DISORDERED PROTEINS: REGULATION AND DISEASE

M. MADAN BABU

MRC Laboratory of Molecular Biology, Hills Road
Cambridge CB2 0QH, United Kingdom
E-mail: madanm@mrc-lmb.cam.ac.uk

ROBIN VAN DER LEE

MRC Laboratory of Molecular Biology, Hills Road
Cambridge CB2 0QH, United Kingdom
Centre for Molecular and Biomolecular Informatics, NCMLS
Radboud University Medical Centre, 6525 GA Nijmegen, The Netherlands

NATALIA SANCHEZ DE GROOT

MRC Laboratory of Molecular Biology, Hills Road
Cambridge CB2 0QH, United Kingdom

JÖRG GSPONER

Centre for High-Throughput Biology, Department of Biochemistry & Molecular Biology
The University of British Columbia, Vancouver, Canada
E-mail: gsponer@chibi.ubc.ca

Intrinsically disordered proteins (IDPs) are enriched in signaling and regulatory functions because disordered segments permit interaction with several proteins and hence the re-use of the same protein in multiple pathways. Understanding IDP regulation is important because altered expression of IDPs is associated with many diseases. Recent studies show that IDPs are tightly regulated and that dosage-sensitive genes encode proteins with disordered segments. The tight regulation of IDPs may contribute to signaling fidelity by ensuring that IDPs are available in appropriate amounts and not present longer than needed. The altered availability of IDPs may result in sequestration of proteins through non-functional interactions involving disordered segments (i.e., molecular titration), thereby causing an imbalance in signaling pathways. We discuss the regulation of IDPs, address implications for signaling, disease and drug development, and outline directions for future research.

1. Introduction

Most proteins need to adopt a defined three-dimensional structure to carry out their function. This is exemplified by the structures of numerous enzymes that highlight the need to precisely position functional groups of amino acids in spatial proximity to facilitate catalysis of chemical reactions. However, the last decade has witnessed a growing recognition that a large fraction of the genome of any organism encodes proteins that do not adopt a defined three-dimensional structure but are nevertheless important for cellular function. This group of proteins is referred to as intrinsically or natively disordered proteins (IDPs)[1-3]. Importantly, more than a third of the eukaryotic proteins have been shown to contain intrinsically disordered regions of over 30 residues in length[4], thereby establishing its prevalence.

IDPs lack a unique 3D structure, either entirely or in parts, when alone in solution[1-3]. This lack of structure in a protein can have several advantages[1-3,5-8]. For instance (i) IDPs provide a larger interaction surface area than globular proteins of a similar length, (ii) conformational flexibility, and the exposure of short linear peptide motifs and interaction-prone structural motifs (Molecular Recognition Features, MoRFs) allow IDPs to scaffold and interact with numerous other proteins, and (iii) diverse post-translational modifications facilitate regulation of their function and stability in a cell. Moreover, as many IDPs fold upon binding, they frequently interact with their targets with relatively high specificity and low affinity. Such a binding mechanism has been suggested to allow them to associate rapidly and initiate a signaling process and at the same time dissociate easily when the task is accomplished[9]. The structural and biochemical properties of disordered regions are ideal for proteins that mediate specific recognition of interaction partners or co-ordinate regulatory events in space and time. Accordingly, proteins that participate in binding, regulatory and signaling functions such as transcription factors and signaling proteins, are enriched for intrinsically disordered regions[1-4,7,8,10-13].

2. The need for regulating IDPs

Due to their unusual structural features and important functional properties, the presence of IDPs in a cell needs to be carefully monitored. Linear motif mediated interactions are highly abundant in nature and motif-binding protein domains are present in large families of related proteins with similar binding sequence preferences[7,14-16]. Therefore, they are likely to bind to each others' physiological targets if their concentration is increased[7,15]. The latter has been shown to be true in a number of large-scale studies involving the SH2, PTB and PDZ domain containing proteins[15,17]. In addition, scaffolding proteins have to be present in the cell in a defined concentration range to provide signaling flexibility[18-20] and to avoid molecular titration (also known as combinatorial inhibition or the pro-zone effect). Given the concentration restrictions of proteins that have these features, it is not surprising that altered availability (i.e., abundance and half-life) of IDPs is associated with several disease conditions such as cancer and neurodegeneration[13,21,22]. For instance, over-expression of the IDP Stathmin and under-expression of p27 have been linked with various types of cancer[23,24]. Similarly, over-expression and mutations in the IDPs tau, ataxin-1, synuclein and huntingtin are known to increase the risk of fibrillar aggregate formation and have been linked to pathological conditions in humans[25]. Importantly, several studies have shown that perturbation of microRNA based regulation of levels of the IDPs Atrophin and Pten, leads to neurodegeneration[26] and cancer[27], respectively. These examples suggest that maintaining the right endogenous levels of IDPs and the integrity of the associated regulatory mechanisms is crucial for normal cellular function.

3. General principles of IDP regulation

Several recent studies have investigated how IDPs are regulated in a cell. Such studies have analysed multiple large-scale datasets describing various aspects of regulation from a number of organisms, including human[28-36]. The general principles of regulation of IDPs emerging from these studies are described below.

3.1. IDPs are tightly regulated from transcript synthesis to protein degradation

In a recent study[28], we investigated how cellular systems have evolved to regulate the availability of IDPs. We integrated information about protein structure with multiple large-scale datasets describing aspects of gene regulation at the transcriptional and translational level (i.e., abundance, rate of synthesis and half-life of mRNA and proteins). We revealed that rather than a single regulatory mechanism, multiple mechanisms during transcription and translation controlled the availability of IDPs, many of which are present in low levels and for short periods of time (Figure 1A). In particular, we demonstrated that the short residence time is primarily due to post-transcriptional regulation via increased transcript clearance and proteolytic degradation. A similar analysis in *S. pombe* and human cells suggested that the tight regulation of IDPs is likely to be an evolutionary conserved strategy[28]. Based on these observations, we proposed that the tight control of IDPs might provide fidelity in signaling, regulation and recognition by minimizing the likelihood of (i) ectopic, non-functional interactions and (ii) inappropriately sequestering proteins into non-physiological protein complexes (i.e., molecular titration; see Figure 1B and 1C and their legend).

Consistent with these findings, Edwards *et al.*[30] showed that transcripts encoding disordered proteins in humans tend to have higher proportions of predicted miRNA target sites and higher mRNA decay rates, both of which are indicative of the lower gene expression levels of IDPs. The low mRNA abundance and IDP regulation by miRNA was earlier noted by Chen *et al.*[36]. In addition, Edwards *et al.*[30] demonstrated that disordered proteins have a greater proportion of predicted ubiquitination sites compared to structured proteins, an observation also noted recently by Radivojac *et al.*[37]. Collectively, these observations suggest that most disordered proteins and their transcripts are present in relatively low levels and for short periods of time compared to structured proteins. One should, however, note that (i) there are outliers to the observed global trends and (ii) the disordered proteins show a broad range of values for the various regulatory features (e.g., half-life, abundance, etc)[28,30,31]. Investigating the individual variations from the general trends might provide novel insights into the underlying biology.

One reason for the observed shorter half-life of IDPs could be attributed to their increased susceptibility for degradation by the proteasome and proteases[31,38,39]. For instance, several studies have revealed the importance of disordered segments for protein turnover by the ubiquitin dependent and independent proteasome degradation pathways[38-42]. Nevertheless, the observation that transcripts encoding disordered proteins are likely to be more tightly controlled and that this phenomenon is evolutionarily conserved in *S. cerevisiae*, *S. pombe* and humans[28] suggests another possibility: selective forces operate on both IDPs and their transcripts to precisely orchestrate their availability (i.e., abundance and time spent) in a cell (Figure 1A).

3.2. Dosage-sensitive genes are enriched in IDPs

Vavouri *et al.*[29] investigated why certain genes are harmful when over-expressed in a cell (i.e., dosage-sensitive genes). By studying various molecular causes through the analysis of large-scale datasets in yeast, they identified intrinsic protein disorder as an important determinant of dosage sensitivity. Specifically, they observed that proteins containing linear peptide motifs were enriched to be dosage sensitive. Moreover, they demonstrated that dosage-sensitive genes are tightly regulated at both mRNA and protein levels, and suggested that this control acts to prevent potentially harmful increases in protein concentration

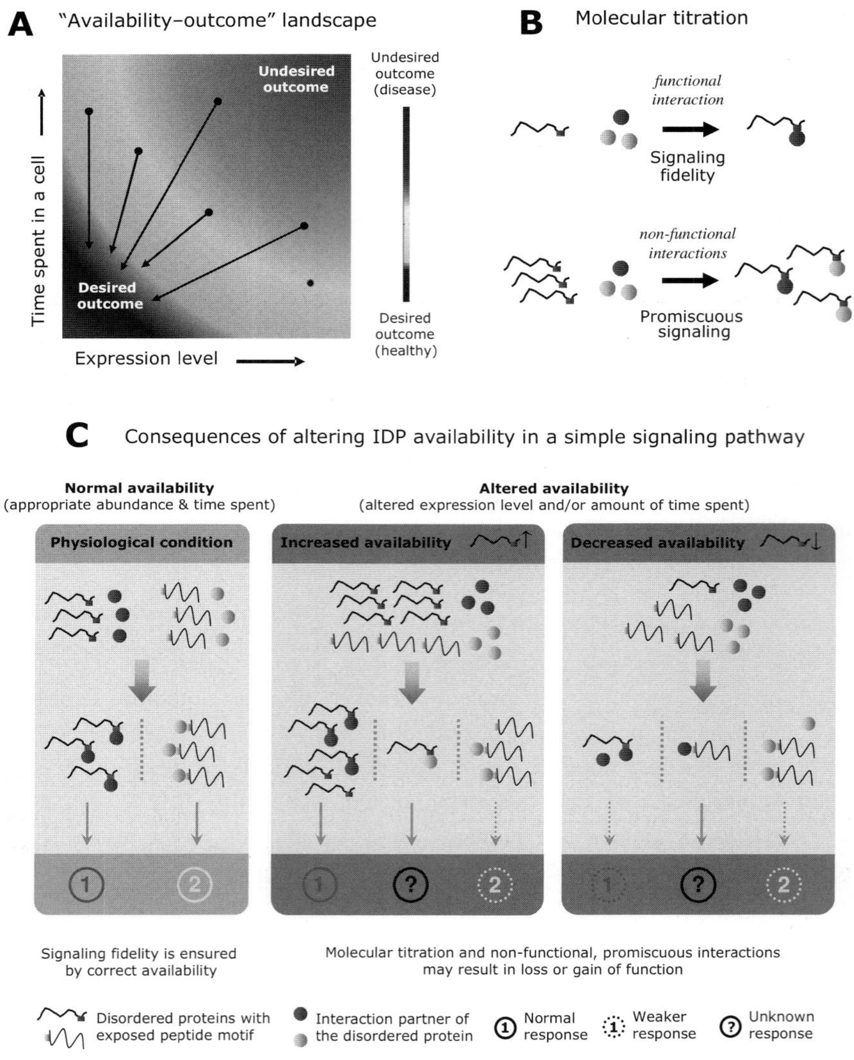

Figure 1: (A) The hypothetical "availability–outcome" landscape and the regulation of IDPs[28]. The presence of an IDP at appropriate levels and for the right amount of time results in the desired outcome (signaling fidelity; blue region of the landscape). Altered availability (e.g., presence at very high (or extremely low) levels or for long (or extremely short) periods of time) results in undesired outcomes (signaling promiscuity; red-yellow region). Within a cell, a coordinated, tight regulation (arrows) of IDPs (black dots) at several stages during transcription and translation ensures that they are present not only at appropriate levels (*x*-axis) but also for the right amount of time (*y*-axis). (B). Tight regulation of IDPs may provide fidelity in signaling and regulation by minimizing non-functional protein-peptide interactions. Wavy lines denote disordered proteins exposing a peptide

motif (small dark green box). At the right expression level and time period (top part), functional interaction is mediated with the relevant partner (dark green circle). In this condition, proteins that are structurally similar to the cognate interaction partner (light green circles, e.g., homologs) are unlikely to interact with the IDP as their affinity does permit interaction at low levels. However, at high concentrations of the IDP (bottom part), structurally similar proteins may interact and be sequestered due to non-functional interaction with the IDP, known as molecular titration. This may result in a perturbed cellular state and may cause cellular toxicity. (C) Potential consequences of altered availability of an IDP that is employed in a simple signaling pathway. Under physiological condition (left panel), disordered proteins 1 and 2 (wavy lines) both bind their relevant binding partners (dark and light green circles, respectively) using distinct peptide motifs (small dark and light green boxes), giving rise to correct signaling responses. In contrast, increased or decreased availability (i.e., abundance and time spent) of IDP 1 causes aberrant signaling responses (right panels). In the former situation, molecular titration results in a weakened or may even abolish response 2 (see also Figure 1B). In the latter situation, decreased availability of IDP 1 means that more of its binding partner will be unbound and available for non-physiological binding to IDP 2. As a result, both response 1 and 2 may be affected. Sequestration of protein into non-physiological complexes may result in altered cellular state due to the loss of function of an interacting protein or a gain of function of the non-physiological complex (shown as unknown response). The principles described above are also likely to be applicable to protein-nucleic acid interaction mediated regulation.

under physiological conditions. They showed that in worm, fly, mice and humans, the same properties are associated with dosage-sensitive genes and oncogenes. In order to explain their observations, the authors propose that disordered regions are prone to make promiscuous molecular interactions via linear motifs when their concentration is increased (Figure 1B) and that this is the likely cause of toxicity when genes are over-expressed. One should, however, note that not all IDPs are harmful when over-expressed as increased expression of many TFs that contain disordered segments have been documented to cause a non-lethal phenotype in yeast[43]. Though the authors[29] report that the number of aggregation prone residues do not correlate with dosage sensitivity, it remains to be investigated whether dosage-sensitive genes contribute to toxicity because of non-functional binding mediated by linear motif or sequestration of other proteins through aggregation prone segments (possibly via stretches enriched in glutamine or hydrophobic amino-acids)[34,35,44,45].

3.3. IDPs can be present in high levels and their availability can be finely tuned

Despite the reported tight regulation that results in low basal availability of a majority of IDPs, certain IDPs may be needed in high levels or for long periods of time in response to specific stimuli, under conditions of stress or during specific phases of the cell-cycle[28]. In fact, in some cases (e.g., titin), high levels may be required throughout the life-time of a cell. In light of the finding that IDPs are susceptible for proteasomal degradation[31,38-41], so-called nanny proteins have been proposed to stabilize an IDP by binding to the disordered segments that facilitate degradation[46].The principle mechanism of this model was shown in the tumor suppressor protein p53 and its paralog p73[39,46]. In other words, disordered proteins can have long lifetimes, simply because they become part of stable protein complexes (e.g., ribosomal proteins) or become involved in an interaction with other proteins (e.g., p53).As many IDPs have the capability to interact with several partners, often non-concurrently[47], it is likely that the stability of IDPs is modulated significantly by their affinity for the different interaction partners.

The balance between bound and unbound state of proteins can be influenced by post-translational modifications, particularly phosphorylation[24,48]. The polyelectrostatic model has been proposed recently to explain how the dynamics and the number of phosphorylation events can change the net charge on an IDP and thereby the binding affinity or the kinetics of interaction for a partner[49-51]. Consistent with previous observations[6], we reported that disordered proteins are on average targets of twice as many kinases as structured proteins in yeast[28] and observed that the targets of several kinases show a significant enrichment to be disordered. ~85% of the kinases that predominantly phosphorylate IDPs are either regulated in a cell-cycle dependent manner (e.g., Cdc28), or get activated upon exposure to particular stimuli (e.g., Fus3) or stress (e.g., Atg1). These results suggest that apart from basal transcriptional and translational regulation, post-translational modification of IDPs via phosphorylation is an important mechanism to fine-tune their availability in different conditions[28].

By investigating the absolute mRNA levels from over 70 different healthy human tissues, Edwards *et al.*[30] defined five different expression patterns of IDPs. Importantly, in contrast to the reported global trends, they identified a group of transcripts encoding IDPs such as ribosomal proteins that are highly expressed. The authors also observed that for a certain fraction of IDPs, there is a decrease in mRNA decay rates, lower proportions of miRNAs targeting their transcripts and lower proportions of ubiquitination sites. These properties were suggested to play a role in the fraction of IDPs with high expression levels. Interestingly, a recent study in yeast has shown that only proteins which are enriched in "evolutionarily flexible disorder" (i.e., disorder is conserved but not the underlying amino-acids) predominantly display dosage sensitivity and are expressed in low levels[52]. This is also the class of IDPs that is often involved in signaling and regulation. Taken together, these studies suggest that fine-tuning of IDP availability can be achieved by post-translational modifications and through other regulatory strategies (e.g., less miRNA targeting or formation of stable protein complexes). Such mechanisms can result in increased abundance and longer half-life via change in cellular localization or by protection from the degradation machinery as in certain phosphorylated forms of $p27^{kip1}$ [48]. Thus, not all IDPs need to be present in low levels and for short periods of time. Under appropriate conditions, various factors may tune the availability of IDPs and may permit them to be present at higher levels as dictated by cellular requirements.

4. Implications of the tight regulation

The observed patterns in the regulation of IDPs have several consequences. In the section below, we will discuss the implications for protein-protein interaction mediated signaling. The principles described are also likely to be applicable to regulation mediated by protein-nucleic acid interactions. We also discuss implications for disease and address strategies for development of drugs that target IDPs.

4.1. Signaling fidelity and flexibility

IDPs with signaling function are frequently re-used in multiple pathways that result in different, and often antagonistic, outcomes. Some examples include GSK3, which is used in the Wnt and Insulin signaling pathway, factors involved in cellular differentiation such as Oct4, Nanog, Sox2 and those involved in cell fate outcome such as components of the MAP kinase, RTK signaling, and the NFkB/P53/CBP-P300 pathway. In order to function properly, cells require that these IDPs are employed in the correct pathways at the right time to ensure an appropriate response. How do cells re-use the same components in multiple pathways, avoid cross-talk or undesirable outcomes and yet have the flexibility to react effectively to cellular conditions? Two major solutions have been defined in the literature (Figure 2; see legend for details): (i) spatial and temporal segregation of signaling proteins within a cell (Figure 2A) and (ii) increased signaling complexity (Figure 2B). The results from recent studies suggest a third major solution to this problem: the tight control of IDPs provides fidelity in signaling, regulation and recognition by ensuring that IDPs are available in appropriate levels and for the right amount of time (Figure 1A and 2C). A tight control of the steady state level of disordered proteins might minimize unwanted ectopic interactions involving linear motifs or MoRFs

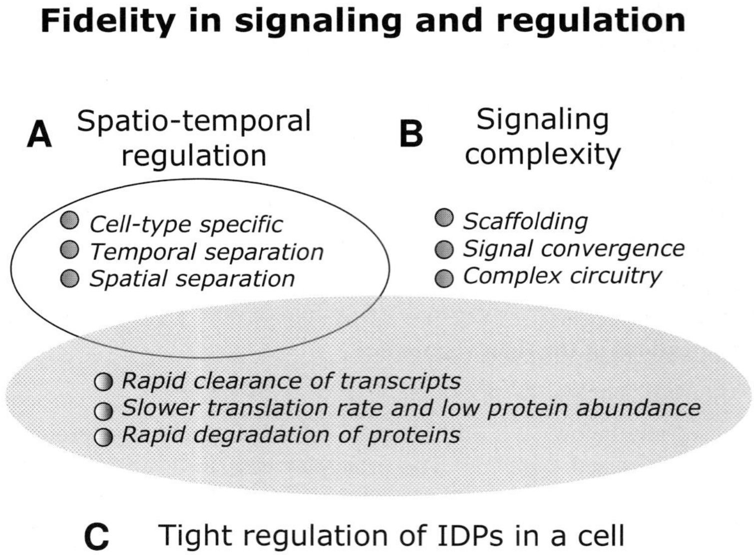

Figure 2: General mechanisms that facilitate re-use of the same component in multiple pathways and prevent erroneous cross-talk in signaling and regulation are (A) spatio-temporal segregation, (B) increased signaling complexity and (C) tight regulation of IDPs. Specific mechanisms of spatio-temporal segregation include (i) cell-type specific expression of the relevant components, (ii) temporal expression of specific components, and (iii) sub-cellular compartmentalization of components or setting up spatial gradients of signaling activity. Increased signaling complexity can be achieved by (i) scaffolding proteins that tether different signaling proteins and thereby isolate one pathway from another, (ii) convergence on a relatively small number of "interaction mechanisms" to provide context-dependent signaling or multi-site protein modification that dictate specific interactions and signaling outcomes and (iii) an intricate connectedness of a large number of interacting constituents that involves complex circuitry such as feed-back and bi-stable switches. In addition to spatio-temporal segregation and increased signaling complexity, recent findings suggest that global co-ordination of various regulatory steps ensures that IDPs are available in appropriate levels and for the right amount of time and may minimize erroneous cross-talk or ectopic interactions involving disordered segments in signaling and regulatory processes (see Figure 1). Please see the extended version at www .mrc-lmb.cam .ac.uk/ genomes /madanm /idp _ cosb _ extended .pdf for a referenced legend describing each of the aspects mentioned here.

(Figure 1B) and thereby reduce the likelihood of cross-talk between signaling pathways (Figure 1C). Furthermore, tight orchestration of protein abundance and half-life can help establish optimal scaffold-partner binding conditions, i.e. concentration ranges that maximize signaling flexibility[18-20,53]. The three strategies are not mutually exclusive and are likely to act together to achieve signaling fidelity and flexibility.

In addition, short half-lives at the transcript and protein level may facilitate quick adjustment of protein concentration when the environmental conditions or signaling inputs change[54]. Support for this concept comes from simple kinetic calculations, which show that mRNAs should have short half-lives to facilitate a rapid change in gene expression and ultimately protein level[55]. Moreover, recent studies demonstrate the importance of turnover of signaling proteins and their transcripts on signaling fidelity, flexibility and outcome. O'Dea *et al.*[56] investigated how degradation rates of IkB, which is an IDP and physically interacts and inhibits the NF-kB transcription factor, can affect signaling. The authors found that an intrinsic high degradation rate of free IkB was critical to maintain homeostasis between different signaling states and make the pathway sensitive for an input. Rutkowski *et al.*[57] investigated what determines a mammalian cell to selectively adopt an apoptotic or adaptive response upon mild ER stress. They discovered that survival is favored as a direct consequence of the intrinsic instability of mRNAs and proteins that promote apoptosis (e.g., Chop, an IDP) compared to those that facilitate adaptation (e.g., BiP, a structured protein). In this context, our observation[28] that most IDPs are less prone to stochastic variation in expression levels among individuals of a cell population (i.e., low expression noise), suggests that IDPs are available at precise levels and finely controlled. This would have a direct consequence on the kinetics and thermodynamics of a signaling event involving IDPs and may ensure minimal phenotypic variation in terms of signaling outcomes among individual cells in a population.

4.2. Drug development

Given the prevalence of IDPs and the important roles that they play in a cell, IDPs make potential drug targets [58]. Small molecules or short peptides have been designed to either bind to the flexible, disordered region or the interaction interface of their binding partners. For example, recent studies have found small molecules that shift the equilibrium of Myc-Max dimerization by binding to a disordered region of the Myc protein, thereby opening the possibility of counteracting the effects of Myc over-expression in many cancers[59]. Another study identified a cell-permeable compound that blocks the interaction between the bromodomain protein BRD4 and the disordered histone tail by binding specifically to the groove on the bromodomain [60]. This molecule was further shown to displace the BRD4 fusion oncoprotein from chromatin and establish an anti-proliferative effect in human squamous carcinoma cells. Moreover, designed peptides that target the PDZ domain of Dishevelled (a protein that is up-regulated in several cancers)have been shown to inhibit the Wnt signaling pathway[61]. These studies demonstrate that clefts within domains, which recognize peptide motifs, are indeed good targets for small-molecule based therapeutics.

The notion that IDPs are tightly regulated and several disease conditions primarily arise due to their altered availability, suggests an alternative strategy for drug development. In addition to identifying molecules that directly influence the interaction or stability of the protein itself, one can target regulatory enzymes that post-translationally modify IDPs, consequently affecting their availability. Indeed, Puca *et al.*[62] have shown that natural agents which inhibit Sirt1-deacetylase or siRNAs that reduce the abundance of the kinase HIPK2 lead to the increased stability of p53 and drive cancer cells towards apoptosis. Another study by Yuzwa *et al.*[63] rationally identified an inhibitor of tau deglycosylation to counter Alzheimer's disease condition. Since phosphorylation and glycosylation happen on the same residues, and because hyper-phosphorylation of tau is linked to Alzheimer's disease, the addition of an inhibitor of deglycosylase was shown to reduce hyper-phosphorylation of tau, thereby offering a viable route to treating the disease condition. Thus, the general principles of IDP regulation can be translated into possible intervention strategies by manipulating the regulatory mechanisms or enzymes that maintain their homeostasis: the appropriate concentration, post-translational modification status, localization and residence time of proteins in a cell (i.e., proteostasis network)[64].

5. Conclusions and future directions

In conclusion, the findings from recent studies highlight that the availability of IDPs is generally tightly regulated at multiple levels. This suggests that in addition to mutations that affect the IDPs themselves, mutations in the regulatory processes that influence the availability of specific IDPs may directly contribute to the disease phenotype. The discovery of such a broad level of control is not only important for understanding how the residence time of IDPs is regulated but should also serve to identify molecular strategies that nature has designed to precisely control availability of specific IDPs in a cell. These findings open up several important questions. Particularly, how the availability of IDPs is controlled in different tissues and during development and how certain IDPs are tolerated at high levels. Uncovering the impact of additional control mechanisms (e.g., specific sequence elements in transcripts, regulated unproductive splicing, etc.) on the availability of IDPs are fruitful lines for further research.

Another important direction of research emerges when one considers these findings from the point of view of disease causing pathogens such as viruses that cause cancer and bacterial pathogens. Given that IDPs are tightly regulated, viruses and bacterial pathogens may exploit this tight regulation for mediating successful infection. Indeed, several instances of pathogens that encode host-like peptide motifs, thereby titrating endogenous host proteins and altering cellular conditions, are now being discovered and documented[65-67]. Finally, the general principles discussed here have direct implications for (i) synthetic biology experiments that aim to rationally rewire signal transduction pathways to dictate specific cellular outcomes, minimize cross-talk between pathways, and in engineering gene regulatory circuits[19,68-70] and (ii) whole genome association studies aimed at identifying causal mutations that contribute to the diseases involving IDPs, since mutations in the modulators of IDP availability that can contribute to the disease phenotype may manifest as low-frequency alleles in human populations.

In short, this is an exciting time for researchers studying various aspects of IDP regulation. We believe that a combination of systems approach and careful experimental dissection of the molecular mechanisms involving IDPs will provide us with a better understanding of the roles of intrinsically disordered proteins and how mutations that perturb their levels cause human diseases.

Acknowledgments

Reprinted from Current Opinion in Structural Biology, 21 (3), M Madan Babu, Robin van der Lee, Natalia Sanchez de Groot, Jorg Gsponer, Intrinsically disordered proteins: regulation and disease, 432-440, Copyright (2011), with permission from Elsevier. We acknowledge MRC for funding and thank our colleagues for stimulating discussions. We thank Marija Buljan, Melis Kayikci, Charles Ravarani and Sreenivas Chavali for their helpful comments. MMB acknowledges the EMBO YI Programme and HFSP (RGY0073/2010) for support. JG is funded by PrioNet Canada and The University of British Columbia. RvdL acknowledges support from HFSP (RGY0073/2010). NSdG is a recipient of the FEBS Long-Term fellowship.We apologise for not being able to cite several relevant references due to lack of space. For an extended and more extensively referenced version of this manuscript, please see:
http://www.mrc-mb.cam.ac.uk/genomes/madanm/idp_cosb_extended.pdf

References

1. V. N. Uversky and A. K. Dunker, *Biochim. Biophys. Acta* **1804**, 1231 (2010).
2. H. J. Dyson and P. E. Wright, *Nat. Rev. Mol. Cell Biol* **6**, 197 (2005).
3. P. Tompa, *Structure and Function of Intrinsically Disordered Proteins*, 1 ed. (Chapman and Hall/CRC, 2009).
4. J. J. Ward, J. S. Sodhi, L. J. McGuffin, B. F. Buxton and D. T. Jones, *J. Mol. Biol* **337**, 635 (2004).
5. K. Gunasekaran, C. J. Tsai, S. Kumar, D. Zanuy and R. Nussinov, *Trends Biochem. Sci* **28**, 81 (2003).
6. L. M. Iakoucheva, P. Radivojac, C. J. Brown, T. R. O'Connor, J. G. Sikes, Z. Obradovic and A. K. Dunker, *Nucleic Acids Res* **32**, 1037 (2004).
7. F. Diella, N. Haslam, C. Chica, A. Budd, S. Michael, N. P. Brown, G. Trave and T. J. Gibson, *Front. Biosci* **13**, 6580 (2008).
8. J. Gsponer and M. M. Babu, *Prog. Biophys. Mol. Biol* **99**, 94 (2009).
9. P. E. Wright and H. J. Dyson, *Curr. Opin. Struct. Biol* **19**, 31 (2009).
10. A. Lobley, M. B. Swindells, C. A. Orengo and D. T. Jones, *PLoS Comput. Biol* **3**, e162 (2007).
11. M. Fuxreiter, P. Tompa, I. Simon, V. N. Uversky, J. C. Hansen and F. J. Asturias, *Nat. Chem. Biol* **4**, 728 (2008).
12. C. A. Galea, Y. Wang, S. G. Sivakolundu and R. W. Kriwacki, *Biochemistry* **47**, 7598 (2008).
13. L. M. Iakoucheva, C. J. Brown, J. D. Lawson, Z. Obradovic and A. K. Dunker, *J. Mol. Biol* **323**, 573 (2002).
14. V. Neduva and R. B. Russell, *Curr. Opin. Biotechnol* **17**, 465 (2006).
15. R. B. Jones, A. Gordus, J. A. Krall and G. MacBeath, *Nature* **439**, 168 (2006).
16. A. Stein and P. Aloy, *PLoS One* **3**, e2524 (2008).
17. A. Ernst, D. Gfeller, Z. Kan, S. Seshagiri, P. M. Kim, G. D. Bader and S. S. Sidhu, *Mol. Biosyst* **6**, 1782 (2010).
18. J. E. Ferrell, Jr., *Sci. STKE* **2000**, pe1 (2000).
19. S. A. Chapman and A. R. Asthagiri, *Mol. Syst. Biol* **5**, 313 (2009).
20. J. W. Locasale, A. S. Shaw and A. K. Chakraborty, *Proc. Natl. Acad. Sci. USA* **104**, 13307 (2007).
21. V. N. Uversky, C. J. Oldfield and A. K. Dunker, *Annu. Rev. Biophys* **37**, 215 (2008).
22. H. Hegyi, L. Buday and P. Tompa, *PLoS Comput. Biol* **5**, e1000552 (2009).
23. G. Baldassarre, B. Belletti, M. S. Nicoloso, M. Schiappacassi, A. Vecchione, P. Spessotto, A. Morrione, V. Canzonieri and A. Colombatti, *Cancer Cell* **7**, 51 (2005).

24. M. Grimmler, Y. Wang, T. Mund, Z. Cilensek, E. M. Keidel, M. B. Waddell, H. Jakel, M. Kullmann, R. W. Kriwacki and L. Hengst, *Cell* **128**, 269 (2007).
25. F. Chiti and C. M. Dobson, *Annu. Rev. Biochem* **75**, 333 (2006).
26. J. S. Karres, V. Hilgers, I. Carrera, J. Treisman and S. M. Cohen, *Cell* **131**, 136 (2007).
27. P. Mu, Y. C. Han, D. Betel, E. Yao, M. Squatrito, P. Ogrodowski, E. de Stanchina, A. D'Andrea, C. Sander and A. Ventura, *Genes Dev* **23**, 2806 (2009).
28. J. Gsponer, M. E. Futschik, S. A. Teichmann and M. M. Babu, *Science* **322**, 1365 (2008).
29. T. Vavouri, J. I. Semple, R. Garcia-Verdugo and B. Lehner, *Cell* **138**, 198 (2009).
30. Y. J. Edwards, A. E. Lobley, M. M. Pentony and D. T. Jones, *Genome Biol* **10**, R50 (2009).
31. P. Tompa, J. Prilusky, I. Silman and J. L. Sussman, *Proteins* **71**, 903 (2008).
32. O. Paliy, S. M. Gargac, Y. Cheng, V. N. Uversky and A. K. Dunker, *J. Proteome Res* **7**, 2234 (2008).
33. L. Ma, C. N. Pang, S. S. Li and M. R. Wilkins, *J. Proteome Res* **9**, 1218 (2010).
34. E. M. Marcotte and M. Tsechansky, *Cell* **138**, 16 (2009).
35. B. Ma and R. Nussinov, *Genome Biol* **10**, 204 (2009).
36. J. Chen, H. Liang and A. Fernandez, *Genome Biol* **9**, R107 (2008).
37. P. Radivojac, V. Vacic, C. Haynes, R. R. Cocklin, A. Mohan, J. W. Heyen, M. G. Goebl and L. M. Iakoucheva, *Proteins* **78**, 365 (2010).
38. G. Asher, N. Reuven and Y. Shaul, *Bioessays* **28**, 844 (2006).
39. P. Tsvetkov, N. Reuven, C. Prives and Y. Shaul, *J. Biol. Chem* **284**, 26234 (2009).
40. S. Prakash, L. Tian, K. S. Ratliff, R. E. Lehotzky and A. Matouschek, *Nat. Struct. Mol. Biol* **11**, 830 (2004).
41. T. Inobe, S. Fishbain, S. Prakash and A. Matouschek, *Nat. Chem. Biol* **7**, 161 (2011).
42. E. K. Schrader, K. G. Harstad and A. Matouschek, *Nat. Chem. Biol* **5**, 815 (2009).
43. G. Chua, Q. D. Morris, R. Sopko, M. D. Robinson, O. Ryan, E. T. Chan, B. J. Frey, B. J. Andrews, C. Boone and T. R. Hughes, *Proc. Natl. Acad. Sci. USA* **103**, 12045 (2006).
44. H. Olzscha, S. M. Schermann, A. C. Woerner, S. Pinkert, M. H. Hecht, G. G. Tartaglia, M. Vendruscolo, M. Hayer-Hartl, F. U. Hartl and R. M. Vabulas, *Cell* **144**, 67 (2011).
45. F. Fiumara, L. Fioriti, E. R. Kandel and W. A. Hendrickson, *Cell* **143**, 1121 (2010).

46. P. Tsvetkov, N. Reuven and Y. Shaul, *Nat. Chem. Biol* **5**, 778 (2009).
47. P. M. Kim, A. Sboner, Y. Xia and M. Gerstein, *Mol. Syst. Biol* **4**, 179 (2008).
48. I. M. Chu, L. Hengst and J. M. Slingerland, *Nat. Rev. Cancer* **8**, 253 (2008).
49. M. Borg, T. Mittag, T. Pawson, M. Tyers, J. D. Forman-Kay and H. S. Chan, *Proc. Natl. Acad. Sci. USA* **104**, 9650 (2007).
50. Z. Serber and J. E. Ferrell, Jr., *Cell* **128**, 441 (2007).
51. M. K. Malleshaiah, V. Shahrezaei, P. S. Swain and S. W. Michnick, *Nature* **465**, 101 (2010).
52. J. Bellay, S. Han, M. Michaut, T. Kim, M. Costanzo, B. J. Andrews, C. Boone, G. D. Bader, C. L. Myers and P. M. Kim, *Genome Biol* **12**, R14 (2011).
53. C. Kiel and L. Serrano, *Sci. Signal* **2**, ra38 (2009).
54. S. Legewie, H. Herzel, H. V. Westerhoff and N. Bluthgen, *Mol. Syst. Biol* **4**, 190 (2008).
55. J. E. Perez-Ortin, P. M. Alepuz and J. Moreno, *Trends Genet* **23**, 250 (2007).
56. E. L. O'Dea, D. Barken, R. Q. Peralta, K. T. Tran, S. L. Werner, J. D. Kearns, A. Levchenko and A. Hoffmann, *Mol. Syst. Biol* **3**, 111 (2007).
57. D. T. Rutkowski, S. M. Arnold, C. N. Miller, J. Wu, J. Li, K. M. Gunnison, K. Mori, A. A. Sadighi Akha, D. Raden and R. J. Kaufman, *PLoS Biol* **4**, e374 (2006).
58. S. J. Metallo, *Curr. Opin. Chem. Biol* **14**, 481 (2010).
59. A. V. Follis, D. I. Hammoudeh, H. Wang, E. V. Prochownik and S. J. Metallo, *Chem. Biol* **15**, 1149 (2008).
60. P. Filippakopoulos, J. Qi, S. Picaud, Y. Shen, W. B. Smith, O. Fedorov, E. M. Morse, T. Keates, T. T. Hickman, I. Felletar, M. Philpott, S. Munro, M. R. McKeown, Y. Wang, A. L. Christie, N. West, M. J. Cameron, B. Schwartz, T. D. Heightman, N. La Thangue, C. A. French, O. Wiest, A. L. Kung, S. Knapp and J. E. Bradner, *Nature* **468**, 1067 (2010).
61. Y. Zhang, B. A. Appleton, C. Wiesmann, T. Lau, M. Costa, R. N. Hannoush and S. S. Sidhu, *Nat. Chem. Biol* **5**, 217 (2009).
62. R. Puca, L. Nardinocchi, D. Givol and G. D'Orazi, *Oncogene* **29**, 4378 (2010).
63. S. A. Yuzwa, M. S. Macauley, J. E. Heinonen, X. Shan, R. J. Dennis, Y. He, G. E. Whitworth, K. A. Stubbs, E. J. McEachern, G. J. Davies and D. J. Vocadlo, *Nat. Chem. Biol* **4**, 483 (2008).
64. E. T. Powers, R. I. Morimoto, A. Dillin, J. W. Kelly and W. E. Balch, *Annu. Rev. Biochem* **78**, 959 (2009).
65. O. Aitio, M. Hellman, A. Kazlauskas, D. F. Vingadassalom, J. M. Leong, K. Saksela and P. Permi, *Proc. Natl. Acad. Sci. USA* (2010).

66. N. A. Sallee, G. M. Rivera, J. E. Dueber, D. Vasilescu, R. D. Mullins, B. J. Mayer and W. A. Lim, *Nature* **454**, 1005 (2008).
67. N. E. Davey, G. Trave and T. J. Gibson, *Trends Biochem. Sci* (2010).
68. C. Kiel, E. Yus and L. Serrano, *Cell* **140**, 33 (2010).
69. I. Nobeli, A. D. Favia and J. M. Thornton, *Nat. Biotechnol* **27**, 157 (2009).
70. E. C. O'Shaughnessy, S. Palani, J. J. Collins and C. A. Sarkar, *Cell* **144**, 119 (2011).

STRUCTURAL BASIS FOR THE REGULATION OF THE T-CELL TYROSINE KINASE ZAP-70

JONATHAN A. WINGER, PATRICK R. VISPERAS AND JOHN KURIYAN

Department of Molecular and Cell Biology, Howard Hughes Medical Institute
University of California, Berkeley, CA94720, USA
E-mail: kuriyan@berkeley.edu

ZAP-70 is a tyrosine kinase that functions as a key intermediary in the tyrosine phosphorylation cascade triggered by the activated T cell receptor. Using two SH2 domains arranged in tandem, ZAP-70 detects upstream phosphorylation signals and propagates them to downstream protein targets. Here we review the mechanism of ZAP-70 activation and regulation, with an emphasis on recent structural studies.

1. Introduction

This review article was inspired by the fact that one of us (JK) met Professor G.N. Ramachandran when he (JK) was in high school in India. Professor Ramachandran was gracious enough to spend about forty-five minutes talking to JK in his office at the Indian Institute of Science, Bangalore, and explained to JK the principles behind the phi-psi plot, or the Ramachandran map as we now know it. Much of this was incomprehensible to JK at that time, but this inspiring interaction seeded in JK a lifelong interest in protein conformation, and how changes in protein conformation underlie some of the most fundamental mechanisms in living systems. Some years ago JK had the privilege of contributing a review article on peptide recognition modules (such as the SH2 and SH3 domains) to a volume honoring Professor Ramachandran, which was published some time before his passing away. In this review article we continue a discussion of peptide binding modules, focusing on one particular enzyme that plays a key role in the response of cells in the immune system to activation by antigens.

1.1 T-cell signaling requires the tyrosine kinase ZAP-70

During the first steps of activation of a T-cell, the T-cell receptor (TCR) recognizes foreign peptide fragments bound to the Major Histocompatability Complex (MHC) on an Antigen Presenting Cell (APC). This binding event triggers calcium release and results in the activation of an intracellular signaling cascade known as the MAP kinase pathway, ultimately leading to an immune response. Early experiments demonstrated that portions of the TCR exhibited two separate activities: the α/β chains of the TCR functioned as the ligand binding portion and the γ, δ, ε and ζ chains of a TCR-associated membrane protein complex known as CD3 functioned as the signaling portion[1,2]. While it was known that following TCR stimulation there is robust and rapid phosphorylation of tyrosine residues on both the TCR and downstream effector proteins, neither the TCR nor the CD3 coreceptor were found to contain an intracellular catalytic domain like those found in membrane-spanning receptor tyrosine kinases.

Studies of T-cells defective in TCR signaling showed that these cells failed to exhibit tyrosine phosphorylation following TCR stimulation. Further experiments found that the cells lacked a protein called Lck, a Src-family tyrosine kinase[3]. Lck associates with the TCR upon T-cell activation and was found to be responsible for phosphorylation of the CD3 ζ-chain[4]. A tyrosine-phosphorylated protein of 70-kDa was found to be associated with the phosphorylated ζ-chain, and was later discovered through sequence analysis to contain a kinase domain[5,6]. Biochemical studies confirmed that this zeta-associated protein of 70-kDa (ZAP-70) has tyrosine kinase activity, is specifically recruited to the TCR upon TCR phosphorylation, is activated by Lck-mediated tyrosine phosphorylation[4,6,7] and is indispensable for downstream T-cell signaling[8-10]. A protein that functions similarly in B-cells was found to be a close relative of ZAP-70, known as Syk[11,12]. Subsequent studies revealed that ZAP-70 specifically phosphorylates the proteins Linker for Activation of T-cells (LAT)[13] and SH2-containing Leukocyte Protein of 76-kDa (SLP-76)[14]. LAT and SLP-76 act as adaptors that recruit various signaling molecules to propagate T-cell signaling downstream (Figure 1).

As a key intermediary in the tyrosine phosphorylation cascade triggered by the TCR, ZAP-70 recognizes and is activated by phosphorylation signals upstream, and also carries out phosphorylation reactions to further propagate the TCR signal downstream. ZAP-70 is well suited for this role, containing a pair of SH2 domains for phosphopeptide recognition and a kinase catalytic domain for phosphotransfer to downstream targets. Here, we review the current structural and mechanistic understanding of ZAP-70 activation and autoinhibition.

1.2 ZAP-70 is a non-receptor tyrosine kinase

The 500 or so protein kinases in the human genome share at their core a highly conserved catalytic kinase domain[15]. The kinase domain is an independently folding domain of approximately 300 residues that catalyzes the transfer of the γ-phosphate from ATP to serine, threonine, and tyrosine residues on substrate proteins. This phosphorylation event often acts as a switch to toggle a substrate protein between "ON" and "OFF" states.

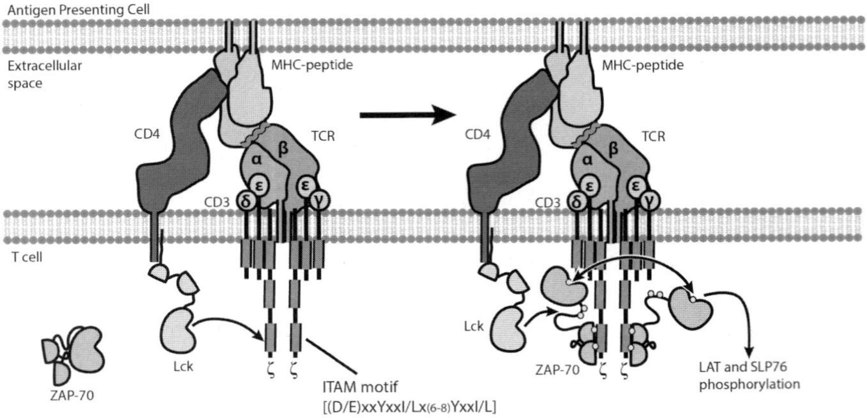

Figure 1: A model for T-cell activation. MHC-peptide complex binds to the TCR, leading to recruitment of CD4-associated Lck. Lck phosphorylates intracellular ITAMs that then interact with the tandem SH2 domains of ZAP-70 to recruit ZAP-70 to the TCR. Lck phosphorylates and activates TCR-bound ZAP-70, which subsequently phosphorylates the adaptor proteins LAT and SLP-76, leading to T-cell activation, proliferation, and differentiation.

Kinases that phosphorylate tyrosine residues can be grouped into two broad categories. The first category includes the receptor tyrosine kinases, which are often involved in the transfer of signals across the cell membrane in response to binding of an extracellular ligand such as a growth factor. The second category is comprised of the non-receptor tyrosine kinases, which are generally found in the cytoplasm and are involved in propagation of signals throughout the cell. ZAP-70 is a member of this second group.

Kinase domains exhibit a bi-lobed architecture composed of a primarily β-sheet N-terminal lobe (N-lobe) and a primarily α-helical C-terminal lobe (C-lobe)[16]. A linker connects the two lobes and acts as a hinge about which the two lobes of the kinase swivel according to the domain's activation state. The active site (ATP-binding pocket) is located at the interface of the N-lobe and C-lobe, and both lobes contribute catalytic residues to the active site.

Kinase domains assume a common active conformation that is optimal for phosphotransfer[17] (Figure 2A). A glutamate residue from a structurally conserved helix (helix αC) in the N-lobe forms an absolutely conserved salt bridge with a lysine that interacts with the α- and β-phosphates of ATP. An aspartate residue from the C-lobe coordinates magnesium, which is required for ATP binding and the phosphotransfer reaction. The N- and C-lobes assume a closed conformation that aligns all the catalytic residues in the active site. A glycine-rich loop formed between strands β1 and β2 folds down to complete the ATP-binding pocket.

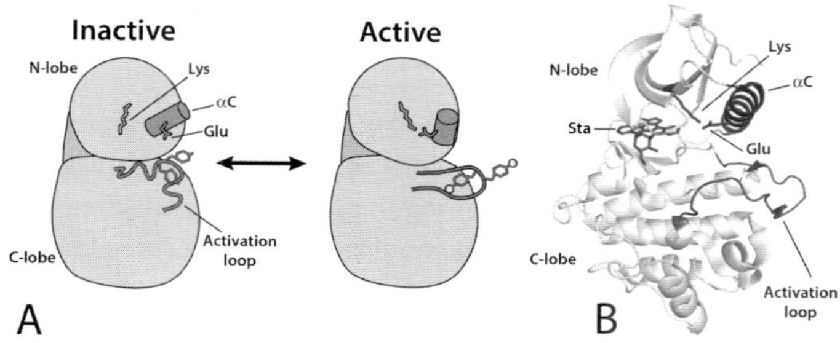

Figure 2: The isolated ZAP-70 kinase domain adopts an active conformation. (A) Cartoon comparison of active and inactive kinase domains. In an active kinase domain, helix αC is rotated and swung inward relative to the inactive conformation. This allows a salt bridge required for catalysis to form between a conserved N-lobe lysine and glutamate. The activation loop unfurls from the active site cleft, and the N- and C-lobes adopt a closed conformation that aligns catalytic residues in the active site. (B) Ribbon diagram of the ZAP-70 kinase domain bound to staurosporine (Sta). Key structural elements are shown in dark grey. Active ZAP-70: PDB ID 1U59.

Tyrosine kinases are often regulated by phosphorylation on a region of the catalytic domain called the activation loop. In inactive kinase domains, the activation loop is often folded inwards to occlude the active site. Phosphorylation of tyrosine residues in the activation loop causes it to unfurl, exposing the active site and acting as a docking site for substrate. The activation loop of ZAP-70 contains two tyrosine residues, Tyr 492 and Tyr 493, and biochemical and mutation studies have shown that phosphorylation of Tyr 493 is required for activation, while the effect of phosphorylation of Tyr 492 on kinase activity is more complex[18,19].

The X-ray crystal structure of the kinase domain of ZAP-70 bound to staurosporine, an ATP-competitive inhibitor, was determined in 2004[20]. The ZAP-70 kinase domain was found to be in an active conformation, likely due to the absence of the regulatory N-terminal domains. Indeed, biochemical studies indicate that the isolated kinase domain has a much higher activity than does full-length ZAP-70[21]. The overall fold of the active kinase domain of ZAP-70 strongly resembles that of other active kinases (Figure 2B). The N-lobe and

C-lobe are in a conformation that allows the completion of the active site by placing all key residues in position for catalysis. The activation loop adopts an unfurled, active conformation, and the structurally conserved helix αC is rotated inward, enabling formation of the catalytically critical salt bridge between Glu 386 and Lys 369. The side chain of Asp 479, part of a Asp-Phe-Gly motif conserved in all kinases, is positioned to coordinate a magnesium ion important for ATP binding and catalysis.

1.3 ZAP-70 contains a pair of accessory SH2 domains

Because of the central involvement of tyrosine phosphorylation in cell signaling, tyrosine kinases are subject to multiple levels of regulation. In addition to an activation loop that requires phosphorylation for kinase activation, non-receptor tyrosine kinases often contain accessory domains that maintain the kinase domain in an inactive state. For example, members of the Src-family kinases, the canonical non-receptor tyrosine kinase family, contain two N-terminal accessory domains, a phosphotyrosine-binding SH2 domain and a poly-proline type II (PPII) helix-binding SH3 domain. Src-family kinases are maintained in an autoinhibited state by two intramolecular interactions: an intramolecular interaction between the SH2 domain and a phosphorylated tyrosine in the C-terminal tail of the kinase domain, and an intramolecular interaction between the SH3 domain and a PPII-type helical conformation assumed by the SH2-kinase domain linker. Together, these interactions stabilize a conformation in which helix αC of the N-lobe is rotated outward, disrupting the arrangement of active site residues and holding the kinase domain in an inactive conformation.

Like Src-family kinases, ZAP-70 contains a pair of accessory domains N-terminal to the kinase domain. The accessory segment of ZAP-70, however, contains two SH2 domains in tandem, instead of the SH2-SH3 module found in Src-family kinases (Figure 3). The two SH2 domains are connected by an inter-SH2 linker (interdomain A) to form the tandem SH2 unit, which is joined to the C-terminal kinase domain by an SH2-kinase linker (interdomain B). The SH2-kinase linker contains a pair of tyrosine residues, Tyr 315 and Tyr 319, whose phosphorylation by Lck is a critical step in ZAP-70 activation[22,23].

1.4 Phosphorylation of the SH2-kinase linker is required for ZAP-70 activation

While phosphorylation of the activation loop of ZAP-70 is required for catalytic domain activation, additional phosphorylation is required for activation of the

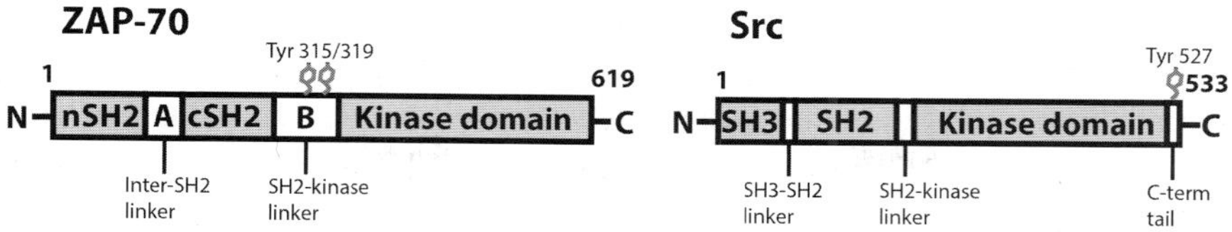

Figure 3: Comparison of the domain organization of Src and ZAP-70, both non-receptor tyrosine kinases. While the two proteins contain similar kinase domains, each has a different set of accessory domains. Regulatory tyrosine residues discussed in the text are shown.

full-length enzyme. Two tyrosine residues in the SH2-kinase linker, Tyr 315 and Tyr 319, are phosphorylated upon T-cell activation[23]. Using stably transfected Jurkat T cells, in which stimulation of TCR signaling leads to ZAP-70-mediated activation of the transcription factor NFAT, studies with a series of ZAP-70 constructs demonstrated that the Y319F mutation results in a dramatic decrease in NFAT activity, while the Y315F mutation causes a modest decrease in NFAT activity. Furthermore, murine T-cell hybridomas stably transfected with the Y319F mutant showed greatly reduced phosphorylation of the ZAP-70 substrates SLP-76 and LAT, compared to hybridomas stably transfected with wild-type ZAP-70[23]. These data indicate that phosphorylation of Tyr 315 and Tyr 319 is a key step in the activation of ZAP-70.

The requirement for phosphorylation of Tyr 315 and Tyr 319 in ZAP-70 activation suggests two possible roles for the SH2-kinase linker in regulation of the kinase domain: phosphorylation of these residues might be required for recruitment of downstream effector molecules, or the SH2-kinase linker might be involved in stabilizing the inactive state of ZAP-70. While the two tyrosine residues can simply serve a scaffolding function when phosphorylated[24-26], a more direct role for the SH2-kinase linker in ZAP-70 autoinhibition was supported by subsequent studies. In a ZAP-70-deficient Jurkat cell line, a ZAP-70 construct featuring a deleted SH2-kinase linker restored NFAT expression in response to stimulation of the TCR. The activity of this ZAP-70 deletion construct was even greater than the activity of wild-type ZAP-70. Crucially, recruitment of ZAP-70 to phosphorylated CD3 ζ-chains was unaffected by deleting the SH2-kinase linker[27]. Together, the data demonstrate that, while the SH2-kinase linker plays a role in regulating ZAP-70 activity, it is not required for recruitment to the TCR, nor is it required for downstream T-cell signaling.

One possible mechanism by which the residues Tyr 315 and Tyr 319 in the SH2-kinase linker might regulate ZAP-70 activation is by formation of an intramolecular autoinhibitory interaction, in a manner analogous to that observed for the receptor tyrosine kinase EPH receptor B2 (EPHB2). Structural and biochemical studies of EPHB2 kinase revealed that a pair of regulatory tyrosine residues in the juxtamembrane segment forms an intramolecular autoinhibitory interaction that is relieved upon phosphorylation. Using ZAP-70-deficient Jurkat cells, Brdicka and coworkers demonstrated that when Tyr 315 and Tyr 319 were mutated to alanine, or when the SH2-kinase linker was removed completely, ZAP-70 remained competent to rescue NFAT-driven luciferase activity. Furthermore, phosphorylation of transiently transfected ZAP-70 on Tyr 319 required co-transfection of Lck, as did phosphorylation of the ZAP-70 substrate LAT. However, the ZAP-70 Y315/319A double mutant was able to phosphorylate LAT in the absence of Lck[22]. Together, these data indicated that phosphorylation of Tyr 315 and Tyr 319 by Lck is required for activation of ZAP-70, and suggested that these residues participate in an autoinhibitory interaction that is relieved by mutation of the tyrosine residues to alanine.

1.5 The tandem SH2 unit undergoes a conformational change upon ITAM binding

A great deal of work has demonstrated that the tandem SH2 unit is instrumental in the high-affinity recruitment of ZAP-70 to specific motifs in the activated TCR[28-32]. These Immuno-receptor Tyrosine-based Activation Motifs (ITAMs) have the sequence YXX(L/I)X(6–8)YXX(L/I), in which the pair of tyrosine residues is phosphorylated by the Src-family kinase Lck upon TCR activation, and X can be any amino acid. While we focus on the ZAP-70 tandem SH2 domains here, much has also been learnt about the Syk tandem SH2 domains, and that information complements the information on ZAP-70[33-38].

The mechanism of high-affinity ITAM binding and recognition by the tandem SH2 domain unit of ZAP-70 was revealed in 1995, when Hatada and coworkers solved the structure of the tandem SH2 unit of ZAP-70 bound to a doubly phosphorylated ITAM peptide[39]. In the ITAM-bound structure (Figure 4A), the tandem SH2 unit resembles the letter Y, with the two SH2 domains adjacent to each other, connected by a coiled-coil-like stalk formed by the 65-amino acid inter-SH2 linker. The doubly phosphorylated ITAM peptide binds to both SH2 domains, which also interact with each other. The coiled coil formed by the inter-SH2 linker stabilizes the SH2 domains in the proper orientation for ITAM binding, and may constrain the distance between the phosphotyrosine binding sites, limiting the tolerated number of residues between phosphotyrosines in an ITAM motif. The C-terminal SH2 domain interacts with the N-terminal phospho-YXXL sequence of the ITAM in a canonical phosphopeptide-SH2 binding mode. However, the phosphotyrosine residue of the C-terminal phospho-YXXL ITAM sequence binds at the junction of the two SH2 domains and interacts with residues from both, leading to speculation that the N-terminal SH2 domain has an incomplete phosphotyrosine binding site[39].

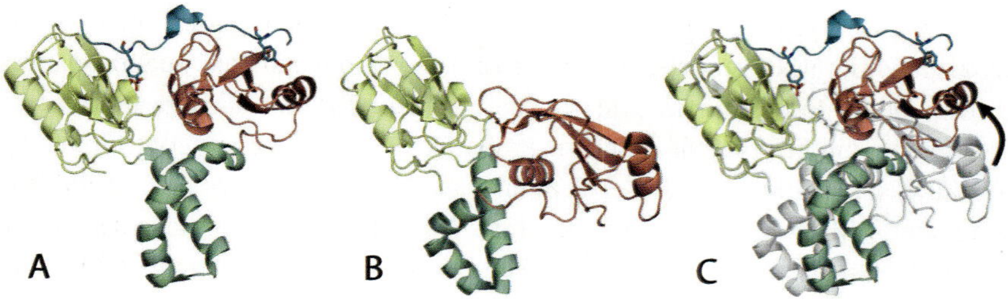

Figure 4: Structures of the apo and ITAM-bound tandem SH2 domains. (A) Ribbon diagram of the ITAM-bound ZAP-70 tandem SH2 domain, PDB ID 2OQ1. (B) Ribbon diagram of the apo ZAP-70 tandem SH2 domain, PDB ID 1M61. The inter-SH2 linker region is depicted in green, the N-terminal SH2 domain in yellow, and the C-terminal SH2 domain in red. The ITAM peptide is shown in teal. (C) Comparison of the structures of the apo (gray) and ITAM-bound tandem SH2 domains of ZAP-70. The molecules were superimposed by aligning the N-terminal SH2 domains. The arrow represents the direction of the large structural rearrangement that occurs upon ITAM binding.

A key conclusion drawn from the structure of the ITAM-bound tandem SH2 unit was that the interaction between the SH2 domains is likely quite weak, being mediated by a relatively small number of contacts, and both the inter-SH2 coiled coil and the ITAM itself likely provide much of the stabilization seen in the crystal structure. These observations suggested that the tandem SH2 domain unit in the absence of ITAM might exist in an entirely different conformation, and that a conformational switch in the tandem SH2 unit upon ITAM binding might drive allosteric regulation of the ZAP-70 kinase domain.

Evidence that such a conformational switch was possible was obtained in 2002 when the crystal structure of the unliganded (apo) tandem SH2 domains of ZAP-70 was published, along with an NMR study describing the conformational dynamics of the apo tandem SH2 unit[40]. Compared to the ITAM-bound structure, the C-terminal SH2 domain of the apo tandem SH2 unit is shifted away from the N-terminal SH2 domain so that the Y-shape of the ITAM-bound complex has now become more like an "L", and the tandem SH2 domains are not in a conformation competent to bind an ITAM (Figure 4B). The inter-SH2 coiled coil is in a similar conformation in the apo structure as it is in the ITAM-bound structure, except for one notable change. Part of the C-terminal helix of the coiled coil linker has melted to become an unstructured region, and the original position vacated by the melted helix is now occupied by one of the helixes of the C-terminal SH2 domain. The net result of this change is an increase in the buried surface between the inter-SH2 linker and the C-terminal SH2 domain, and it was postulated that this increase helps to compensate for the lack of stabilization provided by the binding of an ITAM peptide[40].

The instability of the inter-SH2 linker was further demonstrated by the accompanying NMR studies, which showed that in the apo state the resonances corresponding to the amino acids in the coiled coil region disappeared due to broadening, indicative of conformational exchange[40]. The conformational flexibility of the inter-SH2 linker demonstrated by the studies described above is consistent with previous experiments that showed that an antibody against the inter-SH2 linker no longer bound to the tandem SH2 domains or to full length ZAP-70 in the presence of doubly phosphorylated ITAM peptide or ITAM-containing ζ-chains[41]. Together, these data suggested a possible activation mechanism in which ITAM binding would result in the disruption of any contacts between the tandem SH2 unit and the kinase domain of ZAP-70, possibly leading to release of the tandem SH2 unit and relief of autoinhibition.

1.6 Overview of the structure of autoinhibited ZAP-70

While it was clear from structural and biochemical studies that the ZAP-70 kinase domain can be fully active in the absence of the tandem SH2 unit, and that the tandem SH2 unit can undergo large conformational rearrangements upon ITAM binding, it was unclear how the tandem SH2 unit inhibited the catalytic activity of the kinase domain, and how conformational changes in the tandem SH2 unit upon ITAM binding might result in relief of autoinhibition. Progress towards understanding regulation of the ZAP-70 kinase domain by the tandem SH2 unit was made in 2007, when, in collaboration with Art Weiss and coworkers, our lab solved the structure of a nearly full-length ZAP-70 construct containing the tandem SH2 unit, interdomains A and B, the kinase domain, and most of the C-terminal tail[42]. Expression and crystallization of the construct was facilitated by the mutation of two tyrosine residues in the SH2-kinase linker to phenylalanine (Y315F and Y319F), and mutation of a catalytically important aspartate residue (Asp 479) to asparagine. Together, these mutations render this ZAP-70 construct, termed ZAP-70-FF, inactive. In the ZAP-70-FF structure, the tandem SH2 unit is observed to pack against the distal side of the kinase domain, with the SH2 domains facing away from the kinase domain and the

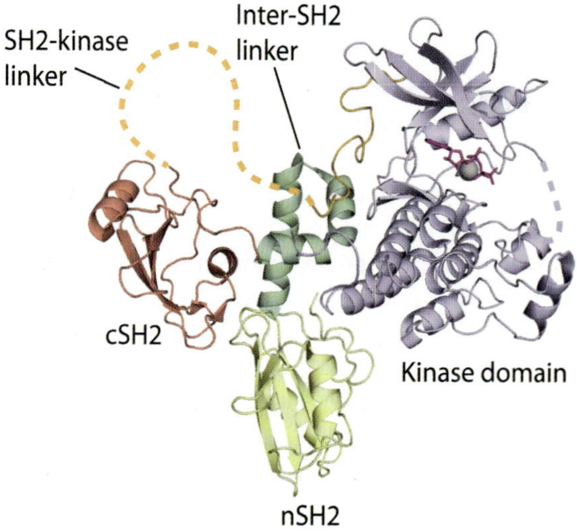

Figure 5: Crystal structure of autoinhibited ZAP-70.The N-terminal SH2 domain (yellow), inter-SH2 linker (green), C-terminal SH2 domain (red), SH2-kinase linker (orange), and kinase domain (blue) are shown in a ribbon diagram. The ATP analogue AMPPNP is depicted in magenta, and magnesium is depicted as a grey sphere. Disordered regions are depicted as dotted lines. Full-length ZAP-70: PDB ID 2OZO.

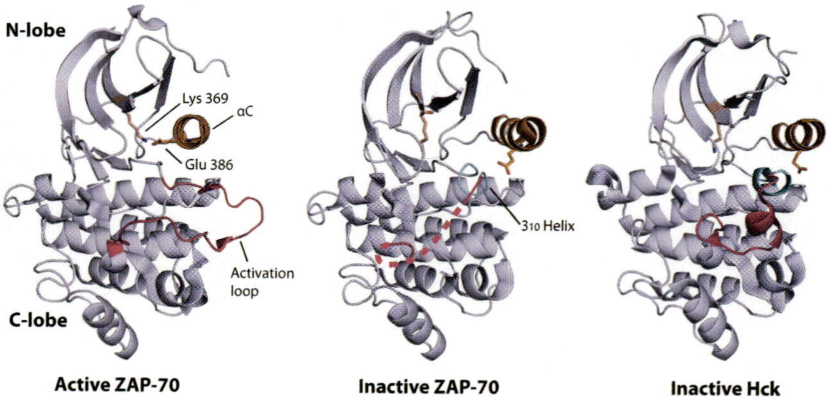

Figure 6: Comparison of active and inactive ZAP-70 kinase domains with the inactive kinase domain of Hck. The conformation of the kinase domain of autoinhibited full-length ZAP-70 resembles that of inactive Hck, a Src-family kinase. Helix αC is shown in orange, the activation loop is shown in magenta, and the short 3_{10}helix in the activation loop following the conserved Asp-Phe-Gly motif is shown in cyan. A disordered part of the activation loop is shown as a dotted line. Active ZAP-70: PDB ID 1U59. Inactive ZAP-70: PDB ID 2OZO. Inactive Hck: PDB ID 1QCF.

coiled coil of the inter-SH2 linker in contact with the C-lobe of the kinase domain (Figure 5). Part of the SH2-kinase linker is disordered, while the remainder is sandwiched between the inter-SH2 coiled coil and the N-lobe of the kinase. The tandem SH2 unit itself exhibits a conformation very similar to that seen in the structure of the isolated tandem SH2 unit solved previously[40].

Examination of the structure of ZAP-70 reveals that it is in an autoinhibited state: the kinase domain adopts an inactive conformation (Figure 6), similar to that observed for the structures of Src-family kinase catalytic domains in the presence of their regulatory domains[43,44]. Compared to its position in the active conformation, helix αC in the N-lobe is rotated outward, resulting in the removal of the conserved Glu 386 from the active site. As a consequence, the catalytically crucial residue Lys 369 is not in a position to interact with the phosphates of ATP. Instead, Lys 369 forms a salt bridge with Asp 479 of the conserved Asp-Phe-Gly motif, keeping the aspartate side chain from adopting a position competent to bind the catalytically required magnesium ion. The N-lobe is swung away from the C-lobe, further disrupting the active site. Part of the activation loop forms a short helix characteristic of the inactive conformation, which packs against and stabilizes the displaced conformation of helix αC. The remainder of the activation loop is unstructured.

1.7 The role of interdomain linkers in ZAP-70 autoinhibition

The inactive conformation of the kinase domain is maintained by interactions between the inter-SH2 linker, the SH2-kinase linker, and the N-lobe of the kinase domain. The SH2-kinase linker emerges from the N-terminus of the kinase domain, traces along the distal side of the N-lobe, and then interdigitates with the inter-SH2 linker. This packing of the linkers between the tandem SH2 unit and the kinase domain is referred to as the linker-kinase sandwich (Figure 7). The inter-SH2 linker and N-terminal SH2 domain also interact directly with the C-lobe of the kinase domain. Hydrophobic residues, contributed by the inter-SH2 linker, the SH2-kinase linker, and the kinase domain, form a hydrophobic core that stabilizes the linker-kinase sandwich. The importance of the hydrophobic core to the stability of the linker-kinase sandwich was confirmed by mutational analysis in vitro and in vivo[45]. However, because the conformation of the kinase domain where it contacts the linker-kinase sandwich does not differ significantly from the same region in the active kinase domain, it was proposed that the formation of the interface resulted in the rigidification of the kinase domain via the stabilization of an extensive hydrogen bonding network. It is reasonable to assume that phosphorylation of the regulatory tyrosine residues Tyr 315 and Tyr 319 in the SH2-kinase linker would result in disruption of the hydrophobic core that stabilizes the linker-kinase sandwich.

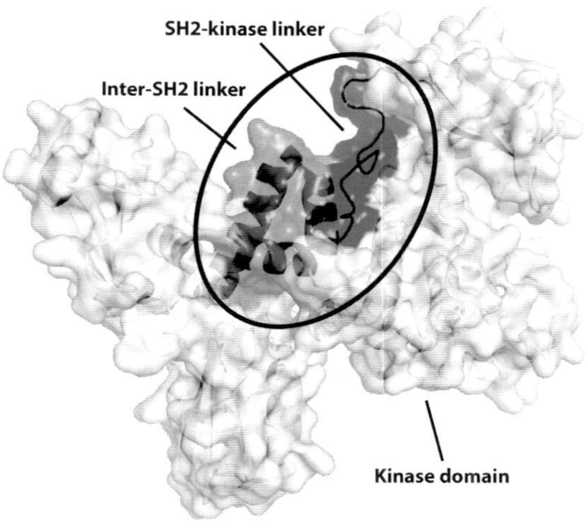

Figure 7: The linker-kinase sandwich of autoinhibited ZAP-70 (circled) is composed of the inter-SH2 linker (dark grey), the SH2-kinase linker (black), and the kinase domain (light grey). Autoinhibited ZAP-70: PDB ID 2OZO.

1.8 The mechanism of activation of ZAP-70

The arrangement of key regulatory features in the structure of ZAP-70 suggests a mechanism for activation of ZAP-70 via sequential disruption of the interactions that maintain the autoinhibited conformation (Figure 8). Upon activation of the TCR and phosphorylation of TCR-associated ζ-chains by Lck, ZAP-70 is recruited to the membrane by binding of the tandem SH2 unit to phosphorylated ITAMs. Engagement of the ITAM by the tandem SH2 unit reorganizes the SH2 domains, changing the conformation of the inter-SH2 coiled coil and resulting in the disengagement of the tandem SH2 unit from the kinase domain and disassembly of the linker-kinase sandwich. Upon disassembly of the linker-kinase sandwich, the SH2-kinase linker peels away from the N-lobe of the kinase domain, allowing the structural elements of the kinase domain to assume a catalytically competent conformation and exposing the regulatory tyrosine residues Tyr 315 and Tyr 319. The regulatory tyrosine

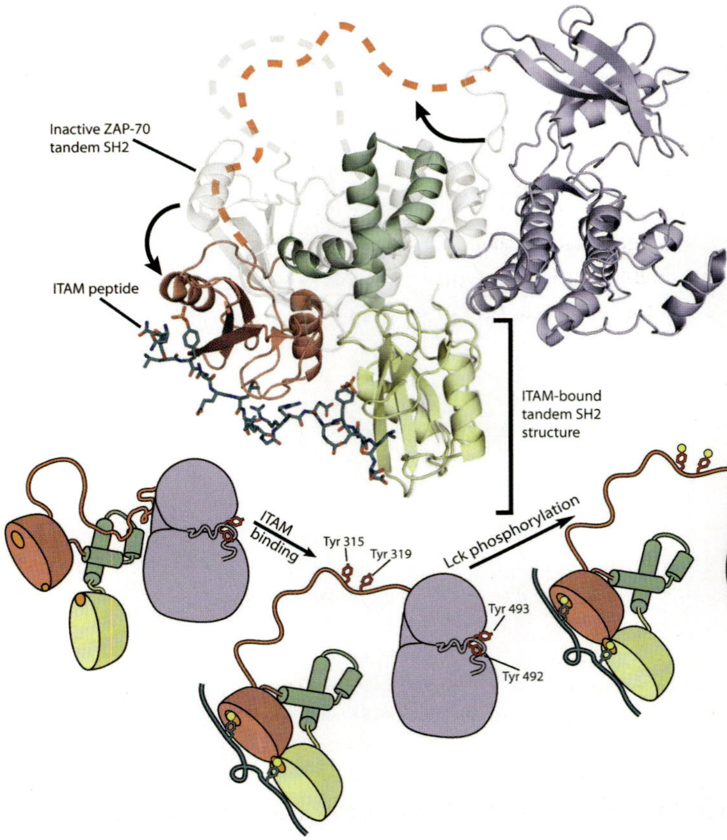

Figure 8: Structural insights reveal a model for the activation of ZAP-70. Top: The structure of the ITAM-bound tandem SH2 domains overlaid on the structure of full-length autoinhibited ZAP-70 by aligning the N-terminal SH2 domains of both structures. ITAM binding to the tandem SH2 domains of ZAP-70 (N-terminal SH2 domain in yellow, inter-SH2 linker in green, C-terminal SH2 domain in red, and ITAM peptide in teal) results in a large conformational shift relative to the tandem SH2 domains of inactive ZAP-70 (grey).The inactive ZAP-70 kinase domain (blue), and the SH2-kinase linker (orange) are also shown. Arrows depict structural rearrangements proposed to occur during ZAP-70 activation. Bottom: Schematic diagram of ZAP-70 conformational changes upon ITAM binding and Lck activation, as described in the text. Yellow circles indicate phosphorylation events on critical tyrosine residues. Inactive ZAP-70: PDB ID 2OZO. ITAM-bound tandem SH2 domains: PDB ID 2OQ1. Apo tandem SH2 domains: PDB ID 1M61.

residues are phosphorylated by Lck, preventing the re-formation of the inhibitory interaction between the SH2-kinase linker and the N-lobe of the kinase domain. The activation loop is also phosphorylated by Lck, resulting in the exposure of the active site of ZAP-70 for substrate binding and catalysis.

2. Conclusion

Structural and biochemical studies have revealed that the ZAP-70 kinase domain is stabilized in an inactive conformation by autoinhibitory interactions with its accessory domains. The inactive conformation of the ZAP-70 kinase domain is similar to the inactive conformation of Src-family kinases and many others. However, the arrangement of the SH2 domains and associated linkers, and the conformational changes that occur upon ITAM binding, provide ZAP-70 with a mechanism for maintaining the specificity of interaction and control of activation required for its central role in T-cell signaling.

Acknowledgments

We thank Art Weiss, Qingrong Yan, and members of the Kuriyan laboratory for valuable discussions.

References

1. M. B. Brenner, I. S. Trowbridge and J. L. Strominger, *Cell* **40**, 183 (1985).
2. P. S. Ohashi, T. W. Mak, P. Van den Elsen, Y. Yanagi, Y. Yoshikai, A. F. Calman, et al., *Nature* **316**, 606 (1985).
3. D. B. Straus and A. Weiss, *Cell* **70**, 585 (1992).
4. M. Iwashima, B. A. Irving, N. S. van Oers, A. C. Chan and A. Weiss, *Science* **263**, 1136 (1994).
5. A. C. Chan, B. A. Irving, J. D. Fraser and A. Weiss, *Proc. Natl. Acad. Sci.USA* **88**, 9166 (1991).
6. A. C. Chan, M. Iwashima, C. W. Turck and A. Weiss, *Cell* **71**, 649 (1992).
7. R. L. Wange, A. N. Kong and L. E. Samelson, *J. Biol. Chem.* **267**, 11685 (1992).
8. E. Arpaia, M. Shahar, H. Dadi, A. Cohen and C. M. Roifman, *Cell* **76**, 947 (1994).
9. A. C. Chan, T. A. Kadlecek, M. E. Elder, A. H. Filipovich, W. L. Kuo, M. Iwashima, et al., *Science* **264**, 1599 (1994).
10. M. E. Elder, D. Lin, J. Clever, A. C. Chan, T. J. Hope, A. Weiss, et al., *Science* **264**, 1596 (1994).
11. R. L. Geahlen, *Biochim. Biophys. Acta* **1793**, 1115 (2009).
12. A. Mocsai, J. Ruland and V. L. Tybulewicz, *Nat. Rev. Immunol.* **10**, 387 (2010).
13. W. Zhang, J. Sloan-Lancaster, J. Kitchen, R. P. Trible and L. E. Samelson, *Cell* **92**, 83 (1998).
14. J. Bubeck Wardenburg, C. Fu, J. K. Jackman, H. Flotow, S. E. Wilkinson, D. H. Williams, et al., *J. Biol. Chem.* **271**, 19641 (1996).

15. G. Manning, D. B. Whyte, R. Martinez, T. Hunter and S. Sudarsanam, *Science* **298**, 1912 (2002).
16. S. S. Taylor, D. R. Knighton, J. Zheng, L. F. Ten Eyck and J. M. Sowadski, *Annu. Rev. Cell Biol.* **8**, 429 (1992).
17. J. A. Adams, *Chem. Rev.* **101**, 2271 (2001).
18. A. C. Chan, M. Dalton, R. Johnson, G. H. Kong, T. Wang, R. Thoma, et al., *EMBO J.* **14**, 2499 (1995).
19. R. L. Wange, R. Guitian, N. Isakov, J. D. Watts, R. Aebersold and L. E. Samelson, *J. Biol. Chem.***270**, 18730 (1995).
20. L. Jin, S. Pluskey, E. C. Petrella, S. M. Cantin, J. C. Gorga, M. J. Rynkiewicz, et al., *J. Biol. Chem.* **279**, 42818 (2004).
21. C. Visco, G. Magistrelli, R. Bosotti, R. Perego, L. Rusconi, S. Toma, et al., *Biochemistry* **39**, 2784 (2000).
22. T. Brdicka, T. A. Kadlecek, J. P. Roose, A. W. Pastuszak and A. Weiss, *Mol. Cell. Biol.* **25**, 4924 (2005).
23. V. Di Bartolo, D. Mege, V. Germain, M. Pelosi, E. Dufour, F. Michel, et al., *J. Biol. Chem.* **274**, 6285 (1999).
24. B. L. Williams, B. J. Irvin, S. L. Sutor, C. C. Chini, E. Yacyshyn, J. Bubeck Wardenburg, et al., *EMBO J.* **18**, 1832 (1999).
25. J. Wu, Q. Zhao, T. Kurosaki and A. Weiss, *J. Exp. Med.***185**, 1877 (1997).
26. S. Gelkop and N. Isakov, *J. Biol. Chem.* **274**, 21519 (1999).
27. Q. Zhao, B. L. Williams, R. T. Abraham and A. Weiss, *Mol. Cell. Biol.* **19**, 948 (1999).
28. R. L. Wange, S. N. Malek, S. Desiderio and L. E. Samelson, *J. Biol. Chem.* **268**, 19797 (1993).
29. E. A. Ottinger, M. C. Botfield and S. E. Shoelson, *J. Biol. Chem.* **273**, 729 (1998).
30. W. Weissenhorn, M. J. Eck, S. C. Harrison and D. C. Wiley, *Eur. J. Biochem.* **238**, 440 (1996).
31. N. Isakov, R. L. Wange, W. H. Burgess, J. D. Watts, R. Aebersold and L. E. Samelson, *J. Exp. Med.* **181**, 375 (1995).
32. R. L. Wange, N. Isakov, T. R. Burke, Jr., A. Otaka, P. P. Roller, J. D. Watts, et al., *J. Biol. Chem.* **270**, 944 (1995).
33. K. Futterer, J. Wong, R. A. Grucza, A. C. Chan and G. Waksman, *J. Mol. Biol.* **281**, 523 (1998).
34. R. A. Grucza, K. Futterer, A. C. Chan and G. Waksman, *Biochemistry* **38**, 5024 (1999).
35. R. A. Grucza, J. M. Bradshaw, V. Mitaxov and G. Waksman, *Biochemistry* **39**, 10072 (2000).
36. S. Kumaran, R. A. Grucza and G. Waksman, *Proc. Natl. Acad. Sci.USA* **100**, 14828 (2003).
37. T. D. Groesch, F. Zhou, S. Mattila, R. L. Geahlen and C. B. Post, *J. Mol. Biol.* **356**, 1222 (2006).

38. Y. Zhang, H. Oh, R. A. Burton, J. W. Burgner, R. L. Geahlen and C. B. Post, *Proc. Natl. Acad. Sci.USA* **105**, 11760 (2008).
39. M. H. Hatada, X. Lu, E. R. Laird, J. Green, J. P. Morgenstern, M. Lou, et al., *Nature* **377**, 32 (1995).
40. R. H. Folmer, S. Geschwindner and Y. Xue, *Biochemistry* **41**, 14176 (2002).
41. L. Grazioli, V. Germain, A. Weiss and O. Acuto, *J. Biol. Chem.* **273**, 8916 (1998).
42. S. Deindl, T. A. Kadlecek, T. Brdicka, X. Cao, A. Weiss and J. Kuriyan, *Cell* **129**, 735 (2007).
43. F. Sicheri, I. Moarefi and J. Kuriyan, *Nature* **385**, 602 (1997).
44. W. Xu, S. C. Harrison and M. J. Eck, *Nature* **385**, 595 (1997).
45. S. Deindl, T. A. Kadlecek, X. Cao, J. Kuriyan and A. Weiss, *Proc. Natl. Acad. Sci.USA* **106**, 20699 (2009).

CONTROLLING CONFORMATIONAL DYNAMICS TO MODULATE PROTEIN FUNCTION

SHAHIR S. RIZK, MARCIN PADUCH AND ANTHONY A. KOSSIAKOFF
Biochemistry and Molecular Biology, The University of Chicago
Chicago, Illinois 60637, USA

In virtually all cases, protein function is governed by the protein's ability to adopt multiple conformations. Ligand binding events are a highly studied class of dynamic molecular adaptation. In a number of systems the adaptation is not driven simply by the ligand binding, but also by interactions with other molecules or ions. This type of allosteric modulation of protein structure, therefore, directly influences its function. Here we describe the development of synthetic antibodies from phage display libraries that are selected to specifically recognize different conformations of the same protein. We focus on maltose binding protein, as a model protein, which undergoes a large conformational change upon binding to its ligand, maltose. Using conformation-specific synthetic antibodies, the affinity of the protein was modulated by dynamically altering the equilibrium between the two conformations *in vitro* and *in vivo*. Further, the antibodies were used to rescue the function of a mutant with a drastically lower affinity. The implications for taking into account the conformational dynamics of proteins in designing affinity reagents expands their utility as biotechnology tools and will aid in the development of more sophisticated drugs that can target mechanisms of conformational regulation.

1. Introduction

1.1 The Relationship between Protein Structure and Function

It has become evident that the dynamic nature of proteins is an evolutionarily acquired trait that can play as significant a role in its function as its conformation and sequence. The proper function of proteins, in propagating specific molecular signals within the cell and maintaining homeostasis, relies on a programmed ability to alter conformational states triggered through environmental signals. The dynamic nature of protein conformation was hypothesized early on by studies of hemoglobin and other systems where the function of the protein is directly influenced by its conformation as a response to the changes in the concentrations of "allosteric effectors"[1,2]. This understanding provided an explanation of how basic protein functions, such as ligand binding and enzymatic activity, can be modulated by subtle changes in conformation. Clearly, dynamics is a major component linking protein conformation and protein function and is utilized in almost all aspects of biology, from receptor-ligand interactions, signal propagation, metabolic state regulation to gene expression and cell motility.

1.2 Periplasmic Binding Proteins

Protein scientists have recently begun to exploit the potential of engineering new functions through controlling changes in protein conformation. While initial attempts at "ligand design" focused on local interactions within a rigid backbone scaffold, more recent computational, rational and combinatorial approaches for protein engineering introduce additional backbone flexibility, conformational dynamics and interconversions between different structural forms.

A prime example has been the study of the bacterial periplasmic binding proteins. Members of this superfamily of proteins bind to a variety molecules (sugars, amino acids, anions and metals) using a conserved mechanism[3]. Each member undergoes a conformational change from an "open" to a "closed" form upon ligand binding characterized by a hinge-bending motion (Figure 1). Exploiting this conformational change has been particularly successful in developing biosensors for the detection of the diverse types of ligands by chemically or genetically attaching probes at regions that undergo a large change in conformation upon ligand binding, whereby the change in signal from the probe can be used to detect or quantify the concentration of the ligand[3-5].

1.3 Maltose Binding Protein

One of the most extensively studied members of the periplasmic binding proteins is maltose binding protein (MBP)[6-9]. NMR and fluorescence studies of MBP with mutations near the hinge region demonstrated the link between the conformation of MBP and its affinity for maltose[10]. This relationship formed the basis for a strategy to modulate the affinity of MBP for maltose. Indeed, it was shown that mutations that drive the equilibrium towards the "closed" form increase the affinity for maltose[11,12]. The variants that stabilize the "closed" form, accomplish this by "jamming" the hinge region with bulky side chains (Figure 2A). It was observed that the effective size of the side chain was directly related to the increase in affinity (Figure 2B). Ultimately, these types of mutational analyses were essential in determining the energetic differences between the "open" and "closed" forms in the presence and absence of ligands, as well as the energetics of a number of intermediate conformations[10].

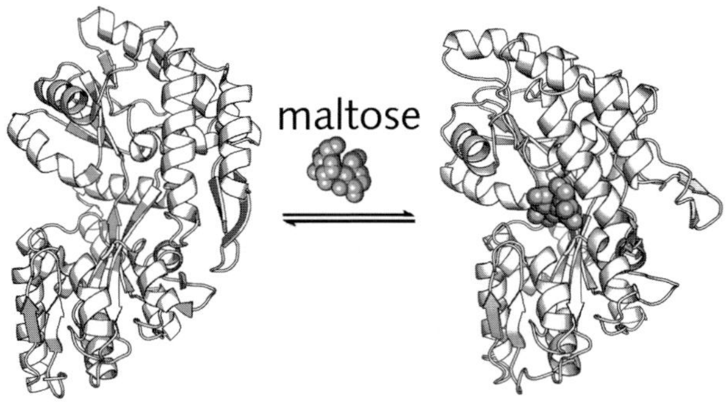

Figure 1: Ligand-induced conformational change in maltose binding protein (MBP). Binding of maltose (grey spheres) to MBP (ribbon) results in a large conformational change characterized by a hinge-bending motion, a mechanism which is conserved in all members of the periplasmic binding protein superfamily.

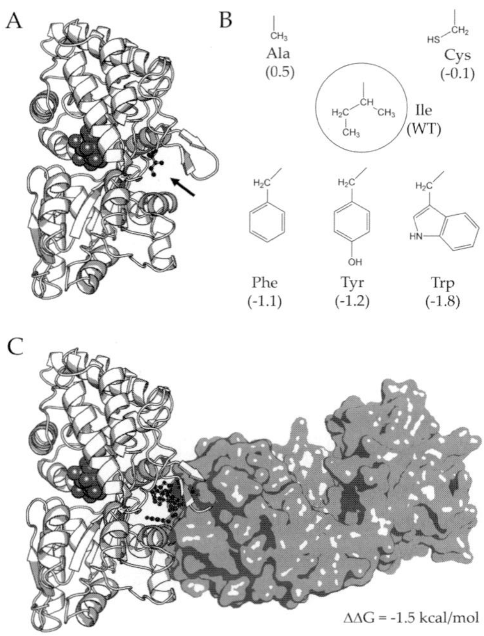

Figure 2: An isoleucine residue (ball and stick) at position 329 is located near the hinge region (arrow) away from the binding pocket. B. Mutations at position 329 which increase the size of the side chain, result in increased maltose affinity, displayed in parentheses as negative ΔΔG values (kcal/mol). C. Crystal structure of a closed-specific sAB (grey surface) in complex with maltose-bound MBP (ribbon). Side chains of the binding loops of the sAB (ball and stick) interact with MBP near at the hinge region near Ile329 (PDB code: 3PGF). The sAB forms a wedge that stabilizes the closed form of MBP.

2. Conformation-Specific Synthetic Antibodies

More recently, this conformational coupling has been exploited by generating a set of synthetic antibodies that were engineered through phage display to distinguish between the "open" and "closed" conformations of MBP[13]. The synthetic antibodies (or sABs) were selected from a large combinatorial library of an antibody Fab fragment scaffold displayed on bacteriophage[13]. The selections carried out in the presence of maltose resulted in sABs that generally have a higher affinity for the "closed" form of MBP (closed-specific sABs).

Conversely, sABs selected in the absence of maltose were found to have a higher affinity for the "open" form of MBP (open-specific sABs).

2.1 Using sABs to Modulate Protein-Ligand Interactions

The sABs showed the ability to distinguish between the different conformations of MBP with up to a 2.2 kcal/mol difference in binding energy between the two forms (Table 1). Based on these observations, the sABs were used as tools to modulate the conformation of MBP and in turn to dynamically modulate the affinity of MBP for maltose. Interestingly, the sABs where found to act as classical heterotropic allosteric effectors influencing both the affinity and cooperativity of ligand binding, where the closed-specific sABs (MCS) resulted in Hill coefficients with values > 1 and the open-specific sAB (MOS) with a Hill Coefficient < 1 (Table 1).

Table 1: Effect of MBP sABs on maltose affinity and binding cooperativity.

	Closed-specificity $\Delta\Delta G$ (kcal/mol)[a]	**Effect on maltose affinity $\Delta\Delta G$ (kcal/mol)[b]**	**Hill Coefficient**
MBP	N/A	N/A	1.0
MCS1	-2.2	-2.1	1.7
MCS4	-2.3	-2.0	1.3
MOS1	+1.4	+1.6	0.45

[a] Describes the difference in binding energy of a sAB for the apo and maltose bound forms of MBP ($\Delta G_{bound} - \Delta G_{apo}$)
[b] Describes the difference in MBP affinity for maltose in the presence and absence of sAB ($\Delta G_{MBP+sAB} - \Delta G_{MBP\ alone}$)

In contrast to previous work with MBP variants[10-12], engineering conformation-specific sABs demonstrated the ability to modulate the function of a protein without introducing mutations, but rather by adding an external force that could act as a "rheostat" of ligand binding affinity simply by modulating the energetics of the conformational equilibrium. Additionally, the specificity of each sAB for the closed form was equivalent to the enhancement of maltose binding affinity. For example, a 2 kCal/mol difference in binding affinity of a sAB between the "open" and "closed" conformations of MBP results in roughly 2 kCal/mol enhancement of the affinity of MBP for maltose[13] (Table 1).

A crystal structure of the complex between MBP and a closed-specific sAB revealed the mechanism through which the sAB enhances maltose binding affinity. The structure shows the sAB bound opposite to the maltose binding site near the hinge region (Figure 2C). One loop from the sAB extends into a cleft that is exposed only in the maltose-bound closed conformation of MBP and acts as a "wedge" that stabilizes the closed conformation. The structure supports the allosteric model for the effect of sABs on maltose binding and the positive cooperative binding between the sAB and maltose. Interestingly, the sAB binds near the site where mutations were introduced to generate higher affinity variants of MBP[10-12], reinforcing the direct link between the conformation at the hinge region and ligand binding affinity.

2.2 Using sABs to Rescue the Function of an Impaired Mutant

The observation that conformation-specific sABs can act as allosteric effectors presented a unique opportunity to explore the potential of modulating structure as a strategy to rescue functionally-impaired mutants. To test this hypothesis, a mutation was introduced at a critical tryptophan residue (to phenylalanine) within the binding pocket of MBP (Figure 3A) resulting in a large decrease in maltose binding affinity (~25-fold). In the presence of closed-specific sABs, the affinity of the mutant MBP was completely restored to physiological range (Figure 3B). In essence, the sABs act as rescue chaperones that restore the function of the mutant protein by shifting the equilibrium between the open and the closed conformation. This demonstrates that the function of a mutant protein can be rescued without re-introducing the natural amino acid at the site of mutation, but rather by the addition of an external factor that simply modulates the conformational landscape of the protein.

2.3 Modulating Protein-Ligand Interactions in vivo

This concept was further explored for the potential to modulate protein function in a living system by preferentially binding to a specific conformation of the protein. As with all members of the bacterial periplasmic binding proteins, MBP resides in the periplasm of *E. coli* as a soluble receptor in the absence of maltose. Upon ligand binding, the open-to-closed transition is detected by the membrane-associated multi-protein transporter complex, which binds to MBP

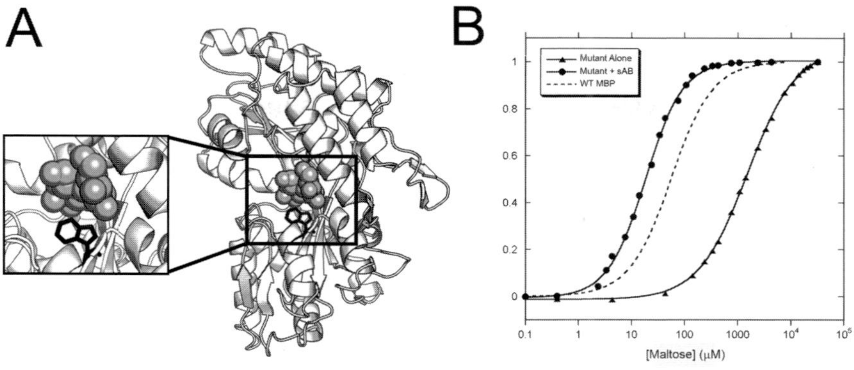

Figure 3: Rescuing protein function in a binding-deficient mutant. A. A close up of the interaction between Trp62 (black sticks) and maltose (grey spheres) within the binding pocket of MBP. B. A Trp62Phe mutant (triangles) results in a large decrease in affinity (~25-fold) relative to WT (dashed line). Addition of a close specific sAB to the mutant MBP restores the affinity for maltose to the physiological range (circles).

and uses the energy from ATP hydrolysis to force the opening and release of maltose to the transporter resulting in the transport of maltose into the cytoplasm[7,14,15]. Bacteria expressing closed-specific sABs where shown to thrive on low levels of maltose compared to bacteria expressing a control sAB, presumably due to the ability of the sAB to scavenge maltose by shifting the equilibrium of MBP to its closed state[13]. This experiment provided evidence that sABs engineered to preferentially recognize a specific conformation of MBP can influence its binding affinity for maltose *in vivo*.

3. Conclusions

The experiments describing the modulation of MBP ligand binding affinity underscored the importance of taking into consideration the roles conformational diversity and dynamics play in protein function. The strategy of targeting different conformations of the same protein can potentially be used to modulate enzymatic activity. Indeed, antibody fragments selected from a similar library were engineered to bind to different conformations of Caspase-1[16]. These antibodies could trap the enzyme in either an on-form or an off-form, resulting in activation or inhibition of the enzymatic activity, respectively. This strategy may also be useful in restoring the function of a mutant enzyme (or binding protein) where a mutation has disrupted the equilibrium between the functional and non-functional conformations. This may prove to be a transformative approach to tackle a large number of single point mutations that lead to whole classes of inherited disorders and enzyme deficiencies such as glycogen and lysosomal storage diseases[17,18], metabolic syndromes[19,20] and neurodegenerative disorders. Additionally, binding reagents that recognize the functional form of an enzyme would be a great asset in understanding how a particular mutation disrupts function by acting as conformational probes that can help to characterize energetic barriers induced by the mutation.

Acknowledgments

The authors thank A. and S. Koide for the sAB libraries, J. H. Heithaus and L. J. Bailey for insightful discussions. The work was partially funded by the following grants from the National Institutes of Health: U01GM09455, R01GM072688 and F32DK080619.

References

1. D. E. Koshland Jr., G. Nemethy and D. Filmer. *Biochemistry*, **5**, 365 (1966).
2. J. Monod, J. Wyman and J. P. Changeux. *J Mol Biol*, **12**, 88 (1965).
3. R. M. de Lorimier, J. J. Smith, M. A. Dwyer, L. L. Looger, K. M. Sali, C. D. Paavola, S. S. Rizk, S. Sadigov, D. W. Conrad, L. Loew and H. W. Hellinga. *Protein Sci*, **11**, 2655 (2002).
4. I. Alicea, J. S. Marvin, A. E. Miklos, A. D. Ellington, L. L. Looger and E. R. Schreiter. *J Mol Biol*, **414**, 356 (2011).
5. S. S. Rizk, M. J. Cuneo and H. W. Hellinga. *Protein Sci*, **15**, 1745 (2006).
6. J. Evenas, V. Tugarinov, N. R. Skrynnikov, N. K. Goto, R. Muhandiram and L. E. Kay. *J Mol Biol*, **309**, 961 (2001).
7. A. D. Gould and B. H. Shilton. *J Biol Chem*, **285**, 11290 (2010).
8. I. L. Medintz and J. R. Deschamps. *Curr Opin Biotechnol*, **17**, 17 (2006).
9. A. J. Sharff, L. E. Rodseth, J. C. Spurlino and F. A. Quiocho. *Biochemistry*, **31**, 10657 (1992).
10. O. Millet, R. P. Hudson and L. E. Kay. *Proc Natl Acad Sci USA*, **100**, 12700 (2003).
11. J. S. Marvin, E. E. Corcoran, N. A. Hattangadi, J. V. Zhang, S. A. Gere and H. W. Hellinga. *Proc Natl Acad Sci U S A*, **94**, 4366 (1997).
12. J. S. Marvin and H. W. Hellinga. *Nat Struct Biol*, **8**, 795 (2001).
13. S. S. Rizk, M. Paduch, J. H. Heithaus, E. M. Duguid, A. Sandstrom and A. A. Kossiakoff. *Nature structural & molecular biology*, **18**, 437 (2011).
14. M. L. Oldham, A. L. Davidson and J. Chen. *Curr Opin Struct Biol*, **18**, 726 (2008).
15. M. L. Oldham, D. Khare, F. A. Quiocho, A. L. Davidson and J. Chen. *Nature*, **450**, 515 (2007).
16. J. Gao, S. S. Sidhu and J. A. Wells. *Proc Natl Acad Sci U S A*, **106**, 3071 (2009).
17. T. M. Cox and M. B. Cachon-Gonzalez. *The Journal of pathology*, **226**, 241 (2012).
18. A. J. Reuser, M. A. Kroos, W. J. Visser and R. Willemsen. *Annales de biologie clinique*, **52**, 721 (1994).
19. K. Hyland. *The Journal of nutrition*, **137**, 1568S-72S; discussion 73S (2007).
20. P. J. Mason, J. M. Bautista and F. Gilsanz. *Blood reviews*, **21**, 267 (2007).

CONFORMATIONAL FEATURES OF SIGMA FACTOR/ANTI-SIGMA COMPLEXES

ANUJA JOSHI, JINAL K. SHUKLA, RAVI K. JAISWAL,
KAPIL GOUTAM AND B. GOPAL
Molecular Biophysics Unit,
Indian Institute of Science,
Bangalore 560 012

The association of σ factors with the RNA polymerase dictates the expression profile of a bacterial cell. Major changes to the transcription profile are achieved by the use of multiple σ factors that confer distinct promoter selectivity to the holoenzyme. The cellular concentration of a σ factor is regulated by diverse mechanisms involving transcription, translation and post-translational events. The number of σ factors varies substantially across bacteria. The diversity in the interactions between σ factors also vary- ranging from collaboration, competition or partial redundancy in some cellular or environmental contexts. These interactions can be rationalized by a mechanistic model referred to as the partitioning of σ space model of bacterial transcription. The structural similarity between different σ/anti-σ complexes despite poor sequence conservation and cellular localization reveals an elegant route to incorporate diverse regulatory mechanisms within a structurally conserved scaffold. These features are described here with a focus on σ/anti-σ complexes from *Mycobacterium tuberculosis*. In particular, we discuss recent data on the conditional regulation of σ/anti-σ factor interactions. Specific stages of *M. tuberculosis* infection, such as the latent phase, as well as the remarkable adaptability of this pathogen to diverse environmental conditions can be rationalized by the synchronized action of different σ factors.

1. Introduction

The core enzyme of bacterial transcription, the RNA polymerase (RNAP), is regulated by a complex network of multi-protein complexes. The RNAP is capable of nonspecific DNA binding and initiation of RNA synthesis from DNA ends or nicks, but requires a σ factor to initiate transcription. Transcription initiation is a multi-step process. The recognition of promoter DNA by the σ subunit of RNAP triggers conformational changes in both RNAP and promoter DNA. This step results in DNA unfolding and initiation of transcription. All bacteria possess a primary σ factor that is responsible for transcription of all housekeeping genes. Most bacterial genomes also encode alternative σ factors that can compete and replace the primary σ factor on the RNA polymerase under different environmental conditions[1,2].

The σ factors belong to two distinct protein families referred to as the σ^{70} and σ^{54} families. Structurally, σ factors can either form a stable closed complex that cannot transcribe (in the case of σ^{54}) or can spontaneously form an open complex that is competent for transcription (in the case of σ^{70}). σ factors of the σ^{70} family recognize promoters with conserved sequences at the -35 and -10 regions relative to the transcription start site whereas σ^{54} members recognize promoters having conserved elements near positions -12 and -24 from the transcription start site. In general, the number of σ factors in a bacterial species is proportional to the diversity of environmental conditions encountered by the bacterium. The distribution of the total number of σ factors among eighteen representative bacterial species is shown in Figures 1a and 1b. In this manuscript, we focus on the sequence and conformational features of σ factors belonging to the σ^{70} family. The subsequent conformational and mechanistic description is thus limited to this family of proteins.

The σ^{70} family is diverse and consists of four phylogenetically distinct clades[3,5]. Group I σ factors are responsible for basal level expression of most

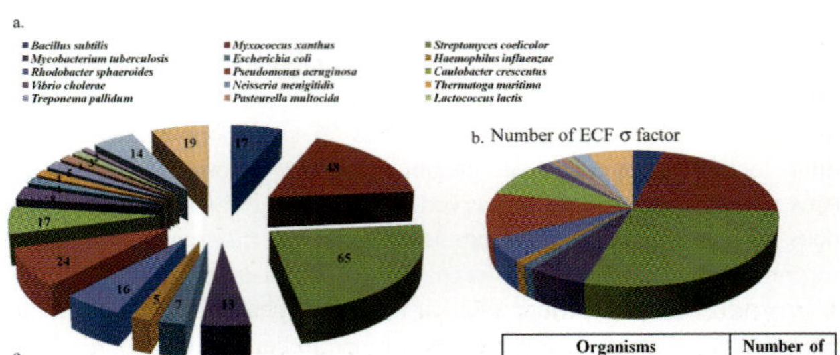

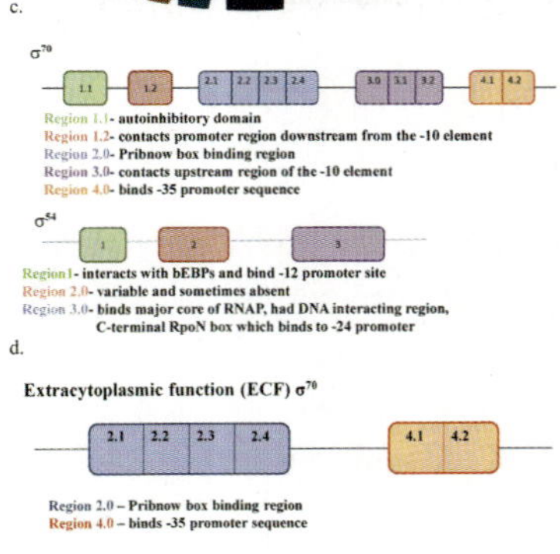

Organisms	Number of ECF σ factors
Bacillus subtilis	7
Myxococcus Xanthus	38
Streptomyces coelicolor	50
Mycobacterium tuberculosis	10
Escherichia coli	2
Haemophilus influenza	2
Rhodobacter sphaeroides	9
Pseudomonas aeruginosa	19
Caulobacter crescentus	14
Vibrio cholera	3
Neisseria meningitides	1
Thermatoga maritime	1
Trepanoma pallidum	1
Pasteurella multocida	2
Lactococcus lactis	1
Deinococcus radiodurans	1
Clostridium acetobutylicum	3
Bacillus halodurans	10

Figure 1: a. The distribution of the total number of σ factors in different bacterial species. b. The number of ECF (Extra-cytoplasmic function) σ factors in these species. c. The domain organization of the σ^{70} family of σ factors. d. The domain organization of the σ^{54} family of σ factors. e. The ECF σ factors have a minimalistic domain organization.

genes and are also known as primary σ factors. The group I σ factors comprise of four regions *viz.* region 1 (region 1.1, region 1.2), region 2 (region 2.1, 2.2, 2.3 and 2.4), region 3 (region 3.1 and 3.2) and region 4 (region 4.1 and 4.2) (Figure 1c). Group II σ factors also contain all four conserved regions but are dispensable for growth. Members of this group lack region 1.1 that functions as an auto-inhibitory module in the case of group I σ factors. Group III σ factors, also referred to as secondary σ factors, lack region 1 but retain regions 2, 3 and 4. The group IV σ factors retain a minimalistic architecture comprising of regions 2 and 4 and are usually more in number when compared to the others (Figure 1d). As group IV σ factors regulate stress responses related to the cell envelope, they are also referred to as ECF (Extracytoplasmic function) σ factors. The cellular concentration and activity of group IV σ factors are regulated by two prominent mechanisms- (i) Anti-σ factors that bind and inhibit their cognate σ factor, and (ii) appropriators that alter the activity of a specific RNAP holoenzyme. Adding to this complexity is the regulation of the anti-σ factors themselves. These vary from antagonistic action by anti–anti-σs, cooperatively by co-anti-σs, or post-translational mechanisms involving proteolysis and phosphorylation[4, 6-9].

2. Structural analyses of σ factor/anti-σ factor complexes

Over the past decade, several structures of σ/anti-σ factor complexes have been determined. These reveal a conserved fold for the inactive form of a σ factor bound to its cognate anti-σ factor (Table 1). In *Bacillus stearothermophilus*, the sporulation σ factor σ^F is regulated by its anti-σ factor SpoIIAB and an anti-anti-σ factor SpoIIAA. The structure of σ^F in complex with SpoIIAB revealed that SpoIIAB monomers form an asymmetric homodimer in complex with a single σ^F molecule. The RNAP-binding regions of σ^F are occluded in this asymmetric complex. This yields a model in which a σ factor is sandwiched between two anti-σ factors[10]. In the *Salmonella enterica* σ^{28}/FlgM complex, on the other hand, the FlgM wraps around σ^{28} in a way that it occludes the binding of RNAP to the σ factor[11]. The crystal structure of the cytoplasmic domain of *E. coli* RseA, an anti-σ factor in complex with its cognate σ factor σ^E, revealed that RseA functions as an anti-σ factor by sterically occluding the two major RNAP-binding elements of σ^E [12]. Another variant to this theme of inhibitory σ factor complexes are the σ/ZAS (Zinc associated anti-σ) complexes. The Zn^{2+} cofactor is released under oxidative stress conditions. This, in turn, leads to the dissociation of the σ/ZAS complex[13,14]. These four models of σ factor regulation are summarized in Figure 2.

Table 1: Structures of σ/anti- σ complexes.

		PDB ID	σ-factor	Reference
σ^E/ChrR	*R. sphaeroides*	2Z2S	Full length	Campbell EA. *et al.*, 2007
σ^E/RseA	*E. coli*	1OR7	Full length	Campbell EA. *et al.*, 2003
σ^{28}/FlgM	*Aquifex aeolicus*	1RP3	Domains 2,3 and 4	Sorenson MK *et al.*, 2004
σ^L_4/RslA	*M. tuberculosis*	3HUG	Domain 4	Thakur KG *et al.*, 2010
σ^F/SpoIIAB	*B. subtilis*	1L0O	Domain 3	Campbell EA *et al.*, 2002
σ^K/RskA	*M. tuberculosis*	3VDO	Full length	Shukla JK. *et al.*, communicated, 2012.
σ^D_4/RsdA	*M. tuberculosis*	3VEP	Domain 4	Jaiswal RK *et al.*, communicated, 2012.

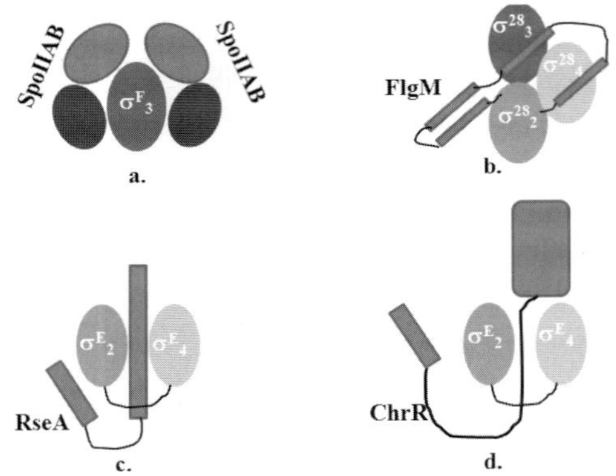

Figure 2: a. A sandwich model of σ/anti-σ interaction as seen from the σ^F/SpoIIAB crystal structure. b. The interaction between σ^{28} with its cognate anti-σ factor FlgM. FlgM wraps around the σ^{28} domains 2, 3 and 4. This occludes the RNAP binding regions. c. The crystal structure of the *Escherichia coli* σ^E/RseA complex shows RseA, the anti-σ factor, sandwiched between the σ^E_2 and σ^E_4. d. In *Rhodobacter sphaeroides* σ^E/ChrR, the anti-σ domain alone interacts with σ_2 and σ_4.

3. The anti-σ factors

An anti-σ factor prevents its cognate σ factor from binding the RNAP enzyme. The anti-σ's are broadly classified as (i) Cytoplasmic anti-σ factors and (ii) Membrane bound anti-σ factors.

Table 2: Secondary structural composition of cytosolic anti- σ factors.

	σ factor	Anti-σ factor	Secondary structure of anti-σ factor
Mycobacterium tuberculosis	σH	RshA	α helices
Mycobacterium tuberculosis	σF	RsbW	α helices, β sheets
Bacillus subtilis	σF	SpoIIAB	α helices, β sheets
Bacteriphage T4	σ^{70}	AsiA	α helices
Mycobacterium tuberculosis	σ^{28}	FlgM	α helices, β sheets

3.1 Cytosolic anti-σ factors

The characterized members of the cytoplasmic anti-σ factors include AsiA from bacteriophage T4, which inhibits *Escherichia coli* σ^{70}, FlgM, which inhibits the flagellar σ factor σ^{28}, *Bacillus subtilis* SpoIIAB, which inhibits the sporulation-specific σ factors σ^F and σ^G, *Bacillus subtilis* RsbW which inhibits the stress response σ factor σ^B and DnaK, a general regulator of the heat shock response, that inhibits the heat shock σ factor σ^{32} [5]. Cytosolic anti-σ factors are regulated by diverse mechanisms. Although these anti-σ factors differ in their regulatory mechanisms, all of them share common structural features (Table 2). The sequence diversity, perhaps, accounts for the differences in the regulatory mechanisms. An example in this context is SpoIIAB that regulates the activity of *B. subtilis* σ^F [15-17]. The crystal structure of σ^F in complex with SpoIIAB revealed that two SpoIIAB monomers associate with a single σ^F molecule giving rise to an asymmetric σ/anti-σ complex. This complex occludes the RNAP binding determinants of σ^F (Figure 2a)[10].

3.2 Membrane bound anti-σ factors

The membrane-associated anti-σ factors have three structural domains; (i) The cytosolic anti-σ domain (ASD) that directly interacts with a σ factor, (ii) a trans-membrane helix, and (iii) the extra-cytoplasmic domain (ECD) that serves as a receptor for environmental stimuli. In all ASDs structurally characterized thus far, helices H1 and H3 are required for interaction with σ_4 while H2 helps in positioning helices H1 and H3 to interact with the σ factor. The fourth helix (H4) binds σ_2 (Figure 4).

Multiple mechanistic routes have been demonstrated to regulate the release of a σ-factor from a σ/anti-σ factor complex (Figures 3b and 3c). The regulation

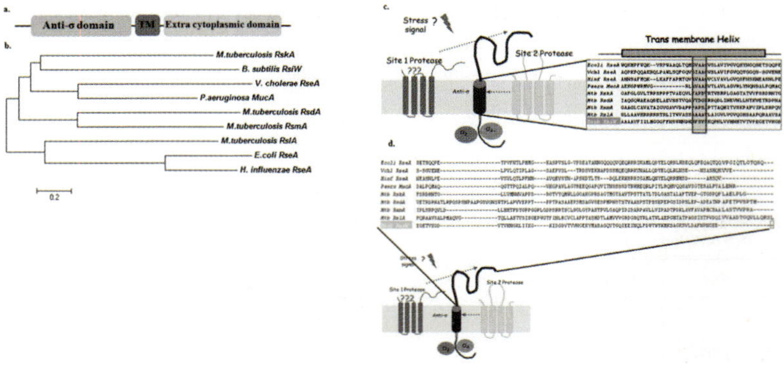

Figure 3: a. Domain organisation of the membrane bound anti-σ factors. b. Phylogenetic analysis of the membrane bound anti-σ factors. c. and d. A mechanistic model of activation of σ factors that are associated with membrane bound anti-σ factors. A sequence alignment of the trans-membrane helix and the extra-cytoplasmic domain suggests that this mechanism is likely to be broadly applicable in this class of anti-σ factors.

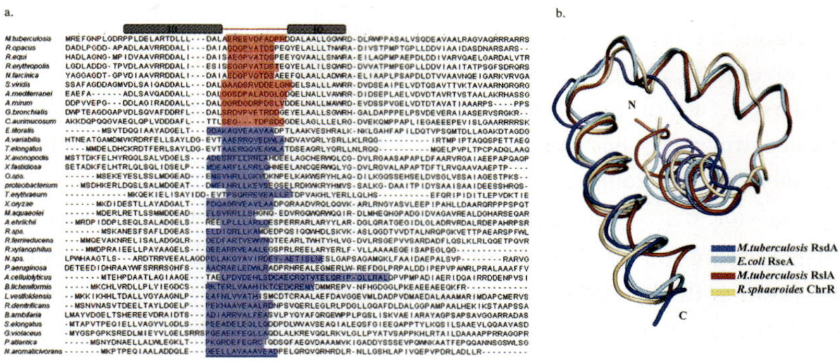

Figure 4: a. Multiple sequence alignment of RsdA homologs with canonical ASDs. This alignment of RsdA homologs with canonical ASD sequences was annotated using PSIPRED and trans-membrane helix predictions. Helix H2 is unstructured in the ASD of RsdA homologs (red colored segment) whereas this segment is predicted to have a helical conformation in canonical ASDs (blue in the alignment). b. The ASDs of *M. tuberculosis* RsdA (blue), *M. tuberculosis* RslA (sand), *R. sphaeroides* ChrR (red) and *E. coli* RseA (cyan) are superposed over three α helices (H1–H3).

of σ factor activity by a proteolytic cascade is best examined in the case of the *E. coli* ECF, σ^E. RseA, the anti-σ factor for σ^E, is a receptor protein. RseA is proteolyzed by a two-step mechanism- the proteolytic action of DegS triggers the activity of a membrane embedded protease RseP[18]. This results in the release of the cytosolic segment of RseA bound to σ^E. The anti-σ domain (ASD) of RseA is subsequently degraded by the ClpX-ClpP proteolytic complex. This regulatory mechanism has also been demonstrated for the *Bacillus subtilis* σ^W/RsiW and the *Pseudomonas aeruginosa* AlgU/MucA complexes. The existence of a similar proteolytic cascade in *M. tuberculosis* was demonstrated recently. *M. tuberculosis* Rip1 was identified as a site-2 protease for three membrane associated anti-σ factors RskA, RslA and RsmA[19].

3.3 The structurally conserved anti-σ domain

The interactions between an ASD and its cognate σ factor show substantial variations. This feature is seen in several characterized σ_4/ASD complexes. Thus while the conformation of σ_4 is distorted in both *E. coli* σ^{70}/AsiA and the *R. sphaeroides* σ^E/ChrR complexes, this is not in the case in either *E. coli* σ^E/RseA or the *M. tuberculosis* σ^L/RslA complexes. The distortion induced in σ_4 upon ASD binding is also different in these complexes. Indeed, a comparison between the ChrR and RseA ASDs suggested that while the secondary structure and tertiary arrangement of helices 1-3 are likely to be conserved, sequence diversity leads to differences in σ_4 interactions.

The structure of *M. tuberculosis* anti-σ^D (RsdA), determined recently, revealed a variant of the ASD in that it has only two helices. Helix H2 is a loop in RsdA. This is seen from the superposition of the structures of RslA (r.m.s.d. of 2.7 Å over 27 α carbon positions), RseA (r.m.s.d of 2.1 Å over 33 α carbons) or ChrR (r.m.s.d. of 2.0 Å over 34 α carbons) on RsdA (Figure 4). Superposition of the ASDs also suggests that helix H3 is the most conserved secondary structural element when compared to helices H1 and H2. A sequence search revealed other ASDs similar to RsdA (Figure 4).

The ASD can also serve as a receptor for environmental stimuli. In the case of *M. tuberculosis* σ^L/RslA, for example, conformational change upon Zn^{2+} release governs the dissociation of the σ^L/RslA complex.

The ASD thus demonstrates surprisingly diverse regulatory mechanisms embedded in a simple four helical structural scaffold. While structural conservation is essential for binding a cognate σ factor, incorporation of different, perhaps multiple, regulatory mechanisms in an ASD provides a route to achieve temporal changes in σ factor levels corresponding to diverse environmental stimuli[20] (Figure 4).

4. Promoter selectivity partially rationalizes the multiple σ factor model of bacterial transcription

As described earlier, the functions of several alternate σ factors are regulated by protein-protein interactions. The σ competition model suggests that competition amongst different σ factors to bind to the RNAP results in changes in the transcription pattern. Collaboration amongst σ factors is relatively less understood. Can this be rationalized better by the cellular concentration of different σ factors or differences in promoter recognition by ECF σ factors?

The hypothesis for this analysis involves two assumptions, (i). The σ factor is always in excess of the RNAP and (ii). σ factors compete for RNAP binding with the corollary that weak promoters are subject to competition amongst σ factors. The second assumption is based on studies in the *E. coli* model system[5]. Based on the intracellular concentration measurements of a σ factor, its affinity to bind the core RNAP and initiate transcription *in vitro*, Ishihama and colleagues were able to demonstrate a gradation of binding affinities for the alternate σ factors[21]. This followed the order $\sigma^A > \sigma^N > \sigma^F > \sigma^H / \sigma^{FecI} > \sigma^E > \sigma^S$. Based on these experiments, they were able to conclude that the total number of all σ factors combined (~1200 molecules/cell) was more than that of the RNAP (~700 molecules/cell) not engaged in the transcription cycle. A key conclusion of this study was that the majority of the RNAP in the cell must be associated with one of the σ factors.

The influence of promoter recognition features is best exemplified in the case of *B. subtilis* σ factors. There are ten σ factors in *B. subtilis*[22]. Based on an analysis of a subset of genes regulated by σ^X, σ^V and σ^W certain sequence features in the -35 promoter element seem poised for functional collaboration. For example, genes under the control of both σ^W and σ^X have identical -35 element sequences at the promoter. Even in the case where the sequences are not identical, common sequence features suggest a more relaxed promoter recognition criterion. This is seen in the case of genes regulated by σ^V, σ^X and σ^W (Figure 5). If the promoter recognition at the -35 element is less specific in the case of ECF σ factors, can affinity changes between the RNAP and σ factors provide a rationale for multi-σ partitioning of transcription? To better understand this process, we obtained a theoretical estimate of the binding affinity between the σ factors and the RNAP in both *E. coli* and *B. subtilis* (Figure 5). The free energy for the docking of these assemblies in the case of *E. coli* is mostly in agreement with the experimental values reported by the Ishihama group[21]. This enabled us to use the *B. subtilis* estimates for binding

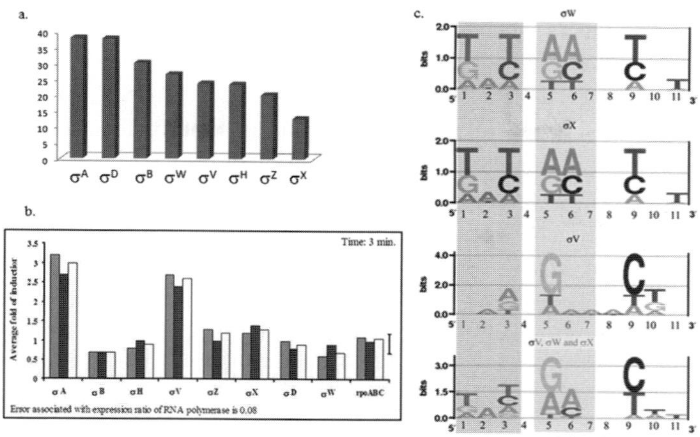

Figure 5: a. Free energy of dissociation (ΔG) for the interaction between the RNAP and σ factors from *B. subtilis*. b. Induction of various σ factors and the RNAP genes when treated with vancomycin.[22] c. Shared promoters between ECF σ factors show common features apart from the 'AAC' motif.

affinity in this analysis. The estimates of the binding affinity follow the order $\sigma^A > \sigma^D > \sigma^B > \sigma^W > \sigma^H > \sigma^V > \sigma^Z > \sigma^X$. To evaluate if these estimates rationalize σ factor competition, we examined the case of σ factor concentration as a function of antibiotic stress using Vancomycin[23]. The induction of the stress response regulator σ^B upon incubation with Vancomycin suggests that the transcription profile from this stage is regulated mostly by σ^B. σ^V levels also vary upon induction (Figure 5). The transcriptional response to this stress is thus controlled by the primary-like σ factors σ^A and σ^B acting in conjunction with σ^V. The main inference from this analysis is that while cellular concentrations of σ factors dictate the transcriptional profile, collaborative arrangements between ECF σ factors is governed at the level of promoter recognition.

5. The σ /anti-σ factor complexes in *Mycobacterium tuberculosis*

Mycobacterium tuberculosis has ten ECF σ factors. The large number of ECF σ factors in *M. tuberculosis* is consistent with the diverse microenvironments encountered by this bacillus inside the host. A phylogenetic analysis was performed on membrane associated anti-σ factors. We note that each of the four *M. tuberculosis* membrane-bound anti-σ factors lie in a different cluster. This is perhaps a reflection of diverse functions (Figure 3a)[25]. The predominant environmental stimulus in host-pathogen interactions is oxidative stress.

An activated macrophage produces reactive oxygen intermediates (ROI) and reactive nitrogen intermediates (RNI)[26,27]. About 264 genes have been shown to vary in their expression levels in response to hydrogen peroxide and nitric oxide, collectively. These genes fall into ten different functional categories[28]. Two ECF σ factors, σ^E and σ^H regulate transcription of atleast 15% of these genes under oxidative stress (Figure 6a, 6b)[29-33.]

The collaboration between σ factors, described in the earlier section, partially rationalizes the oxidative stress response in *Mycobacterium tuberculosis*. σ^H and σ^E play an important role in response to oxidative stress. The interaction of σ^H with RshA is also regulated by phosphorylation of RshA by the protein kinase PknB[31,32,34,35]. Like σ^H, σ^E is regulated by multiple regulatory processes. σ^E can be transcribed from three different promoters viz. P1, P2 and P3 depending on the environmental conditions (Figure 6c). Apart from these mechanisms of regulation, the σ^E/RseA complex is also regulated by cytosolic proteases[33,34,37,38]. A rapid response to oxidative stress is thus achieved by collaboration between σ^H and multiple isoforms of σ^E.

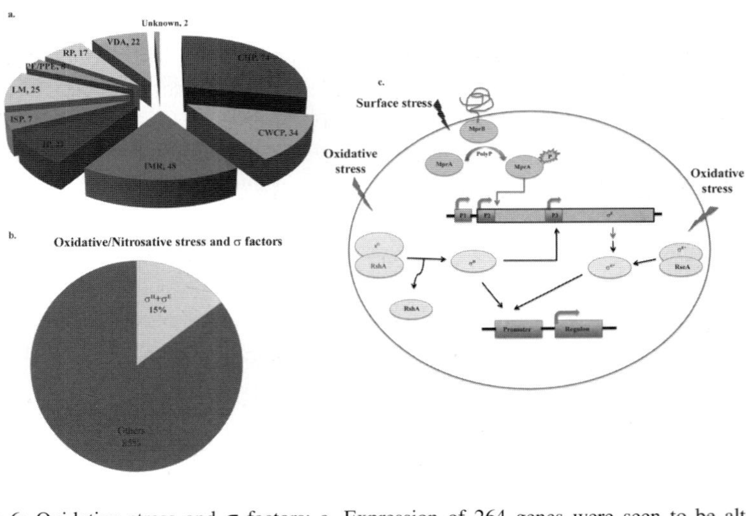

Figure 6: Oxidative stress and σ factors: a. Expression of 264 genes were seen to be altered in response to H_2O_2 and NO stress. These involve ten different functional categories: CHP- conserved hypothetical proteins, CWCP-cell wall and cell processes, IMR- Intermediary metabolism and respiration, IP-information pathways, ISP-insertion sequences and phages, LM-lipid metabolism, PE/PPE- PE/PPE gene family, RP- regulatory proteins, VDA- virulence, detoxification and adaptation, UNKNOWN- unknown function, b. σ^H and σ^E together regulate the expression of atleast 15% of these genes. c. Under oxidative stress conditions, RshA and RseA, the ZAS anti-σ factors release σ^H and σ^E. σ^E has multiple promoters (P1, P2 and P3) and alternative transcription and translational start sites[33, 36- 38].

In summary, structural and mechanistic studies on σ factors reveal different regulatory mechanisms that were hitherto poorly understood. Collaboration between different σ factors can be rationalized by accounting for both, the cellular levels of free, active σ factors as well as promoter fidelity. The emphasis in this manuscript was on the protein-protein interactions in a subset of alternative σ factors. In this context, the ECF σ/anti-σ interaction is a fascinating example of protein engineering. Functional limitations restrict the conformational space that can be sampled, resulting in the conserved fold of a σ factor and the ASD despite poor sequence similarity. However, multiple regulatory mechanisms are embedded in this common structural scaffold. While the ASD functions in some cases as a receptor for environmental stress, specific sequence motifs in other ASDs lead to the recruitment of proteases for selective degradation of the anti-σ factors. These studies thus reveal layers of complexity that are embedded into a deceptively simple protein fold. Further studies are clearly needed to understand these features that dictate prokaryotic transcription.

Acknowledgements

The work in this laboratory is supported by grants from the Wellcome Trust UK, and Department of Biotechnology, India. We gratefully acknowledge Archana Seetharaman and Krishan Gopal Thakur for their inputs and analysis of *B. subtilis* σ factors. Jinal K. Shukla and Kapil Goutam are research fellows of Council of Scientific and Industrial Research (CSIR), India.

References

1. M. S. Paget and J. D. Helmann, *Genome Biol.*, **4**, 203 (2003).
2. K. Murakami and S. A. Darst, *Curr. Opin. Struct. Biol.*, **13**, 31 (2003).
3. C. A. Gross, C. Chan, A. Dombroski, T. Gruber, M. Sharp, J. Tupy and B. Young, *Cold. Spring. Harb. Symp. Quant Biol.*, **63**, 141 (1998).
4. S. Sterberg, T. Peso-Santos, and V. Shingler, *Ann. Rev. Microbiol.*, **65**, 37 (2011).
5. T. M. Gruber and C. A. Gross, *Ann. Rev. Microbiol.*, **57**, 441 (2003).
6. E. A. Campbell, L. F. Westblade and S. A. Darst, *Curr. Opin. Microbiol.*, **11**, 121 (2008).
7. A. Feklistov and S. A. Darst, *Genes Dev.*, **23**, 2371 (2009).
8. M. Raffaelle, E. I. Kanin, J. Vogt, R. R. Burgess and A. Z. Ansari, *Mol. Cell*, **20**, 357 (2005).

9. G. Mittenhuber, *J. Mol. Microbiol. Biotechnol.*, **4**, 77 (2002).
10. E. A. Campbell, S. Masuda, J. L. Sun, O. Muzzin, C. A. Olson, S. Wang and S. A. Darst, *Cell*, **108**, 795 (2002).
11. M. K. Sorenson and S. A. Darst, *Mol. Cell*, **14**, 127 (2004).
12. E. A. Campbell, J. L. Tupy, T. M. Gruber, S. Wang, M. M. Sharp, C. A. Gross and S. A. Darst, *Mol. Cell*, **11**, 1067 (2003).
13. E. A. Campbell, R. Greenwell, J. R. Anthony, S. Wang, L. Lim, K. Das, H. J. Sofia, T. J. Donohue and S. A. Darst, *Mol. Cell*, **27**, 793 (2007).
14. K. G. Thakur, T. Praveena and B. Gopal, *J. Mol. Biol.*, **397**, 1199 (2010).
15. L. Duncan and R. Losick, *Proc. Natl. Acad. Sci. USA*, **90**, 2325 (1993).
16. P. Aldridge and K. Hughes, *Curr. Opin. Microbiol.*, **5**, 161 (2002).
17. S. Alper, L. Duncan and R. Losick, *Cell*, **77**, 195 (1994).
18. M. S. Brown, J. Ye, R. B. Rawson and J. L. Goldstein, *Cell,* **100**, 39 (2000).
19. J. G. Sklar, H. Makinoshima, J. S. Schneider, and M. S. Glickman, *Mol. Microbiol.*, **77**, 605 (2010).
20. L. N. Kinch, K. Ginalski and N. V. Grishin, *Protein Sci.*, **15**, 84 (2006).
21. H. Maeda, N. Fujita and A. Ishihama, *Nuc. Acids Res.*, **28**, 3497 (2000).
22. W. G. Haldenwang, *Microbiol. Rev.*, **59**, 1 (1995).
23. J. D. Helmann, *Adv. Microb. Physiol.*, **46**, 47 (2002).
24. K. Tamura, D. Peterson, N. Peterson, G. Stecher, M. Nei and S. Kumar, *Mol. Biol. Evol.*, **28**, 2731 (2011).
25. S. Ehrt and D. Schnappinger, *Cell. Microbiol.*, **11**, 1170 (2009).
26. G. Lamichhane, *Front. Microbiol.*, **2**, 176 (2011).
27. M. I. Voskuil, I. L. Bartek, K. Visconti and G. K. Schoolnik, *Front. Microbiol.*, **2**, 105 (2011).
28. J. M. Lew, *Tuberculosis* (Edinb.), **1991**, 1 (2011).
29. M. I. Voskuil , G. K. Schoolnik, E. Dubnau, M. Gomez and I. Smith, *J. Exp. Med.*, **198**, 705 (2003).
30. R. Manganelli, M. I. Voskuil, G. K. Schoolnik, E. Dubnau, M. Gomez and I. Smith, *Mol. Microbiol.*, **45**, 365 (2002).
31. S. Raman, T. Song, X. Puyang, S. Bardarov, W. R. Jacobs and R. N. Husson, *J. Bacteriol.*, **183**, 6119 (2001).
32. R. Manganelli, M. I. Voskuil, G. K. Schoolnik and I. Smith., *Mol. Microbiol.*, **41**, 423 (2001).
33. T. Song, S. L. Dove, K. H. Lee and R. N. Husson, *Mol. Microbiol.*, **50**, 949 (2003).
34. S. T. Park, C-M. Kang, and R. N. Husson, *Proc. Natl. Acad. Sci. USA*, **105**, 13105 (2008).

35. R. Manganelli, and R. Provvedi, *Mol Microbiol*, **75**, 538 (2010).
36. V. Dona, S. Rodrigue, E. Dainese, G. Palù, L. Gaudreau, R. Manganelli and R. Provvedi, *J Bacteriol.*, **190**, 5963 (2008).
37. H. He, R. Hovey, J. Kane, V. Singh, T. C. Zahrt, *J. Bacteriol.*, **188**, 2134 (2006).
38. S. Barik, K. Sureka, P. Mukherjee, J Basu and M Kundu., *Mol. Microbiol.*, **75**, 592 (2010).

FOLD TYPE II PYRIDOXAL 5′-PHOSPHATE DEPENDENT ENZYMES: STRUCTURE, SUBSTRATE RECOGNITION AND CATALYSIS

S. BISHT[1], S.R. BHARATH[1], H. S. SAVITHRI[2] AND M. R. N. MURTHY[1]
[1]*Molecular Biophysics Unit, Indian Institute of Science, Bangalore, India*
[2]*Department of Biochemistry, Indian Institute of Science, Bangalore, India*

Enzymes utilizing pyridoxal 5′-phosphate dependent mechanism for catalysis are observed in all cellular forms of living organisms. PLP-dependent enzymes catalyze a wide variety of reactions involving amino acid substrates and their analogs. Structurally, these ubiquitous enzymes have been classified into four major fold types. We have carried out investigations on the structure and function of fold type I enzymes serine hydroxymethyl transferase and acetylornithine amino transferase, fold type II enzymes catabolic threonine deaminase, D-serine deaminase, D-cysteine desulfhydrase and diaminopropionate ammonia lyase. This review summarizes the major findings of investigations on fold type II enzymes in the context of similar studies on other PLP-dependent enzymes. Fold type II enzymes participate in pathways of both degradation and synthesis of amino acids. Polypeptide folds of these enzymes, features of their active sites, nature of interactions between the cofactor and the polypeptide, oligomeric structure, catalytic activities with various ligands, origin of specificity and plausible regulation of activity are briefly described. Analysis of the available crystal structures of fold type II enzymes revealed five different classes. The dimeric interfaces found in these enzymes vary across the classes and probably have functional significance.

Abbreviations: DSD: D-serine deaminase; SR: serine racemase; LSD: L-serine dehydratase; TRPSβ: tryptophan synthase β; DCyD: D-cysteine desulfhydrase; ACC:; ACCD: 1-amino-1-carboxy cyclopropane deaminase; DAPAL: diaminopropionate ammonia lyase; PLP: pyridoxal 5′ phosphate; TS: threonine synthase; TD: threonine deaminase; OASS: O-acetylserine sulfhydrase; OPSS: phosphoserine sulfhydrase; CBS: cystathionine β synthase; SR: serine racemase

1. Introduction

Pyridoxal 5′-phosphate (PLP) was recognized as a cofactor of enzymes confined to amino acid transamination and decarboxylation reactions till the 1950s. Since then, the horizon of PLP-dependent activities has expanded to include a large number of functions that cover five of the six classes listed by the enzyme commission[1]. These activities include transfer of amino group, deamination, decarboxylation, racemization, elimination and replacement of groups at α, β and γ positions of amino acids. As of 2009, 121 unique genes with 2167 different amino acid sequences exhibiting 184 different enzyme activities associated with 161 EC numbers and belonging to 149 different families of proteins that depend on PLP for activity were identified[1]. This constitutes ~4% of all known enzyme activities.

PLP-dependent enzymes have been classified into four major fold types, I – IV, based on available crystal structures and sequence comparisons[2-4]. The four folds have been named after the most extensively studied enzyme belonging to that class. The majority of the structurally characterized PLP-dependent enzymes belong to fold type I or aspartate aminotransferase family. The less studied fold type II enzymes are represented by tryptophan synthase β-subunit. Fold type III and IV are represented by alanine racemase and D-amino acid aminotransferase, respectively. Functionally, PLP-dependent enzymes are also classified as α, β or γ family depending on the carbon atom at which the enzyme brings about covalency changes.

In most of the PLP-dependent enzymes, PLP is covalently linked to a lysine through a Schiff base linkage, referred to as the internal aldimine form. On approach of the substrate, a transaldimination reaction takes place and PLP gets covalently bound to the substrate, forming an external aldimine and releasing the lysine side chain (Figure 1a). The conjugated π-electron system of PLP plays a major role in the reactions catalyzed by PLP-dependent enzymes. Dunathan (1966) proposed that the bond cleaved during the reaction is oriented parallel to the π-electron system of PLP and is labialized by resonance and hyperconjugation effects (Figure 1b)[5]. PLP also serves as an electron sink and stabilizes reaction intermediates.

Fold type II PLP-dependent enzymes generally belong to the β family and catalyze diverse reactions in pathways that include degradation of D- and L-Ser, L-Thr, D-Cys, β substituted Ser, Cys and Ala derivatives, D- and L-diaminopropionate (DAP) and 1-amino, 1-carboxy cyclopropane (ACC); biosynthesis of tryptophan, threonine and cysteine; and serine racemization. Studies on fold type II PLP-dependent enzymes is of much interest as these enzymes catalyze reactions in key cellular metabolic pathways and hence have potential applications. Detailed reviews are available on structure and function of tryptophan synthase[6-10], O-acetylserine sulfhydrylase[11-14], and cystathionine β-synthase[15-17].

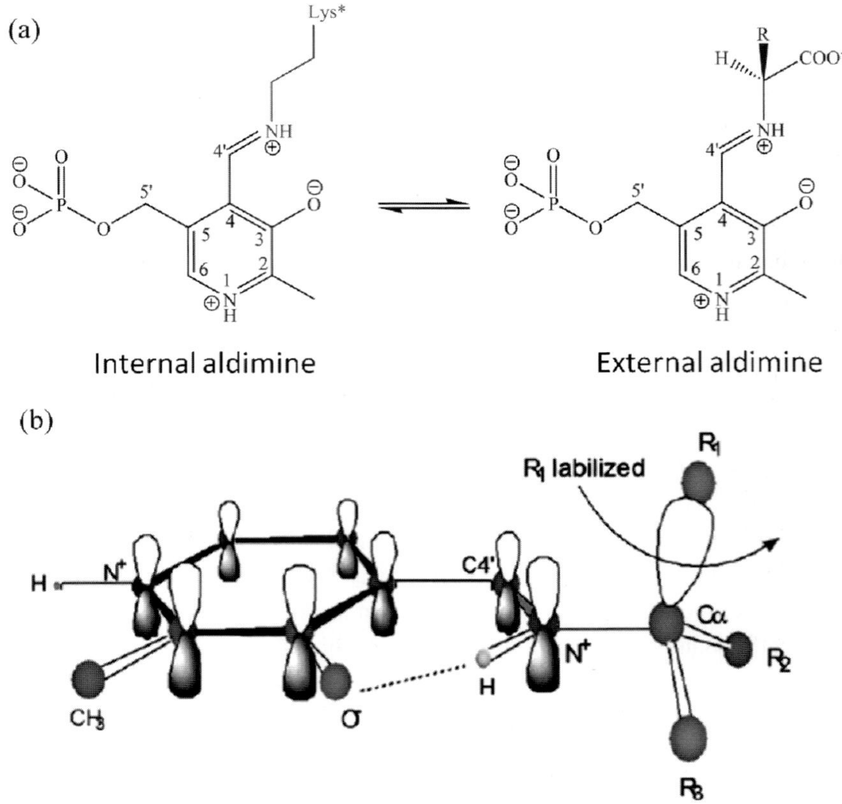

Figure 1. Chemistry of PLP based reactions (a) Transimination reaction with an incoming substrate amino acid leads to the formation of the external aldimine. This step is common to most of the PLP-dependent enzymes. (b) Schematic representation of Dunathan's hypothesis showing the bond destined to be cleaved by PLP-dependent reaction. The active sites of PLP-dependent enzymes specifically orient one bond of the substrate linked to the cofactor (external aldimine) such that it is normal to the plane of Schiff base/pyridine ring. (Reproduced from Toney, 2005.)

2. Classification of fold type II PLP-dependent enzymes

Based on flexible structural alignment using Vorolign[18], fold type II enzymes could further be divided into five classes: (i) cysteine synthases, (ii) DCyD and ACC deaminases (ACCD), (iii) threonine and tryptophan synthases (TS and

TrpS), (iv) monomeric threonine synthases and (v) a diverse set of enzymes including L-serine dehydratase (LSD), D-serine dehydratase (DSD), catabolic and anabolic threonine deaminases (TD), diaminopropionate ammonia lyase (DAPAL) and serine racemase (SR). Class I cysteine synthases O-acetylserine sulfhydrase (OASS), O-phosphoserine sulfhydrase (OPSS) and cystathionine β synthase (CBS) use different substrates to synthesize cysteine. DCyD from *Salmonella typhimurium* (*St*DCyD), the structure of which was determined in our laboratory, shows interesting differences in specificity from the related ACCD and a protein of unknown function referred to as ACCD homolog protein (AHP). Class III synthases involved in threonine and tryptophan synthesis are dimeric proteins. TrpS is a multisubunit enzyme, where the β subunit has the topology of fold type II enzymes. In contrast to class III, Class IV synthases are monomeric. Structural and functional studies on Class V enzymes DSD, catabolic TD from *S. typhimurium* (*St*DSD, *St*TD), DAPAL from *E. coli* (*Ec*DAPAL) have been carried out in our laboratory[19-21]. Salient features emerging from these studies on the subtle control of specificity and catalysis in PLP-dependent enzymes are summarized in the rest of the review.

3. Protomer structure of fold type II PLP-dependent enzymes

Protomers of fold type II PLP-dependent enzymes consist of two domains of unequal size. Both the domains consist of a β-sheet core surrounded by α-helices. The organization of the two domains is illustrated in Figure 2 using the structure of DCyD[22], which is one of the smallest fold type II PLP dependent enzymes. The secondary structural elements of DCyD, except for the N-terminal helix HI and the short helix H5, are fully conserved among all other structurally characterized fold type II enzymes. In other enzymes, additional segments or domains are found inserted at several places. In all fold type II enzymes, the active site lies in a cleft between the two domains. The lysine residue to which PLP is covalently attached usually occurs at the end of a loop that connects the two domains. Five strands (S2, S7, S8, S9 and S10 of DCyD) and two helices (H9 and H10 of DCyD) of the large domain are in spatially analogous positions in PLP-dependent enzymes of all fold types[23].

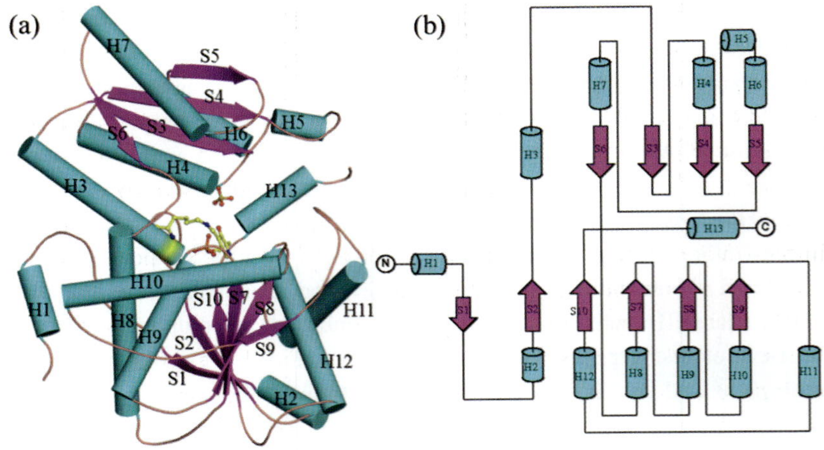

Figure 2. Structure and topology of DCyD (a) The secondary structural elements of the protomer are labeled. Helices are shown as blue cylinders and their labels have a prefix H. The strands are shown as magenta arrows and their labels are numbered sequentially with a prefix S. (b) Topology of DCyD: Helices are shown as cyan cylinders, while the strands are shown as magenta arrows. N and C termini of the protomer are also marked.

4. Oligomerization and subunit interfaces

Most fold type II PLP-dependent enzymes are functional as homodimers. However, a few are known to function as monomers or tetramers. Examination of the available crystal structures of dimeric fold type II enzymes revealed three distinct interfaces, designated here as A, B and C. The residues that interact at similar interfaces may not be structurally equivalent due to the low level of sequence identity among these. However, the interfaces are characterized by the way the two protomers relate to each other spatially and the disposition of the secondary structural elements with respect to the interface. Dimers of A-type such as *St*DCyD (Figure 3a) have an interface constituted by residues from both large and small domains and the buried surface area ranges from 1500-4000 Å^2. The wide range in the buried surface area is due to differences in the size of protomers and presence or absence of swapping segments between the two protomers. The interface in B-type dimers such as catabolic *St*TD (Figure 3b) is smaller with buried surface areas in the range of 800-1200 Å^2 and is mostly contributed by the large domain. The C-type dimerization, observed only in *Ec*DAPAL (Figure 3c), involves interactions from the large domains only and has a buried surface area of 1050 Å^2. Figures 3d-g illustrate the diversity of these interfaces. The polypeptide in gray represents the protomers and the segments involved in interfacial interactions are highlighted in colour. All the dimeric interfaces are symmetrical and the protomers are related by a rotation of ~180°. *St*DCyD shown in Figure 3d is one of the simplest examples of A-type dimers. Unlike a few other dimers of A-type, *St*DCyD does not possess swapped segments between the protomers. The regions that are involved in interface of *St*DCyD are common to all other A-type dimers. The B-type interface (Figure 3e) involves a subset of segments that constitute A-type interface. Similarly, the C-type interface (Figure 3f) is a subset of B-type interface. This suggests that the three interfaces A, B and C might have evolved from a common ancestral interface. LSD has an interesting interface (Figure 3g). It is similar to B-type interface but also involves extensive interaction of a helix from the small domain. As the interaction of the helix from the small domain is also found in A-type interface, LSD dimer could be viewed as an evolutionary link between A and B-type interfaces.

A class wise distribution of interfaces is also observed. In class I, II and III of fold type II enzymes, only A-type dimerization is observed. All enzymes belonging to the class IV are functional monomers. The interfaces in class V are variable. In these enzymes, holo-LSD, catabolic TD and SR are B-type dimers, DAPAL forms a C-type dimer, DSD is a monomer[20] while the two anabolic TDs exhibit different tetrameric organizations[24, 25]. Interestingly, apo-form of LSD from *Rattus norvegicus* forms a different type of dimer not observed in any other fold type II enzyme[26]. The transition from holo to apo form in LSD is accompanied by a large change in its quaternary structure. The relative abundance of A-type dimers indicates that the interface found in holo-form of LSD might be the ancestral dimeric structure, that has further evolved to B and C type organizations.

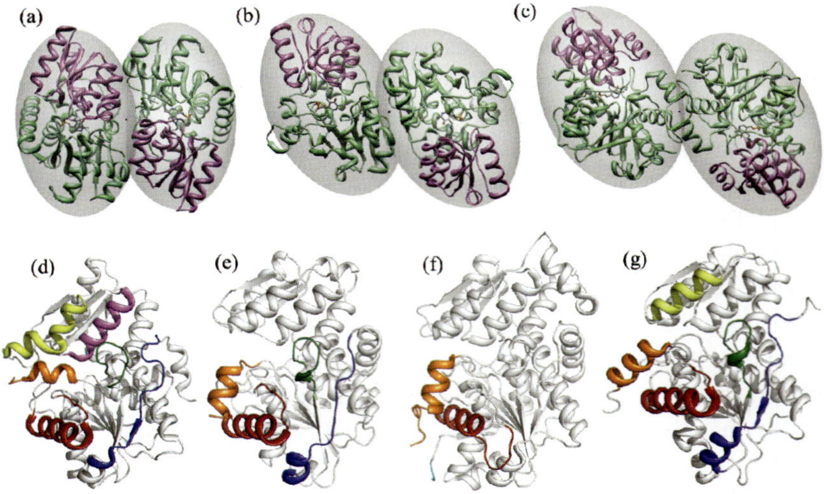

Figure 3. Three distinct dimerization modes observed in fold type II enzymes. Relative orientation of two protomers in A, B and C type dimers are represented using structures of (a) DCyD, (b) catabolic TD and (c) DAPAL. Two-fold symmetry axis that relates the two protomers is perpendicular to the plane of paper and is marked by a black dot. Secondary structure elements that are involved in interface are highlighted in red, orange, blue, pink, green and yellow for (d) DCyD, (e) catabolic TD, (f) DAPAL and (g) LSD. LSD may be considered as an evolutionary link between A and B type dimers (see text).

5. Active site geometry

The active site of fold type II enzymes occurs in a cleft formed between the two domains and is constituted by residues from a single polypeptide chain of the usually dimeric protein. In some enzymes, binding of PLP is associated with structural rearrangements, either in the form of disorder to order transitions or relative movement of the two domains or changes in the position or conformation of residues close to the active site. All the enzymes with the exception of class II have eight loops that define the active site pocket. Figure 4a highlights these loops (labeled 1-8) in DAPAL. Loops 2-4 and 5-8 are part of small and large domains, respectively, while loop 1 connects the two domains. Loop 6 is absent in class II enzymes and only seven loops define the active site. In enzymes of other classes, loop 6 is the most dynamic segment of the active site. In *Ec*DAPAL[21] and *A. thaliana* TS[27], this loop is disordered in apo-form and the active site is exposed to the solvent with a wide opening. In LSD from *Rattus norvegicus*, this loop is in different conformations in the apo and holo forms[26] and in the latter form, it protrudes into the active site. In the apo-form, the movement of the loop away from the active site along with inter-domain movement leads to an open structure in which active site of each protomer is blocked by the small domain of the other protomer.

6. Environment of PLP in fold type II enzymes

In all PLP-dependent enzymes, the conjugated π-electron system of PLP acts as an electron sink and stabilizes reaction intermediates and facilitates catalysis. PLP has four ionizable groups (pyridine nitrogen, the hydroxyl group at C3, Schiff base nitrogen and the phosphate group) in the physiological pH range. It is likely that the ionization state of the cofactor is a key factor in catalysis[28]. The polypeptide fold influences the microenvironment of the active site by modifying the pKa values of the ionizable groups.

PLP is involved in a large number of interactions at the active site such as covalent (Schiff base) linkage of C4′ with a Lys residue, interactions of the phosphate group, stacking of pyridine ring and hydrogen bonding interactions of the pyridine N1 and O3. In fold type II PLP-dependent enzymes, loop 1 provides the Lys residue for covalent attachment to PLP and main chain N atoms of loop 5 form extensive hydrogen bonding interactions with the oxygen atoms of phosphate group. Loop 5, also sometimes referred to as phosphate binding loop, is rich in glycines and is mainly comprised of hydrophobic residues in all fold type II enzymes. Figure 4b shows the PLP binding interactions in *Ec*DAPAL. The phosphate group of PLP is situated close to the amino terminus of a helix, H10 in *St*DCyD. The atoms interacting with the non-ester oxygen atoms of the phosphate group form two nearly parallel planer layers[29]. The first layer of interacting atoms coincides with the plane of non-ester oxygen atoms.

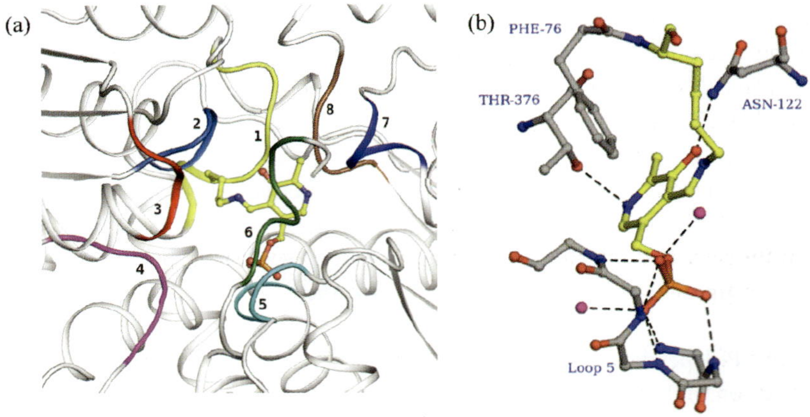

Figure 4. Active site and PLP binding environment in *Ec*DAPAL. (a) Eight loop regions that form the active site in *Ec*DAPAL are highlighted in color and are marked 1-8. PLP is shown in ball and stick representation. (b) Interactions between the polypeptide and PLP. PLP is covalently linked to the active site Lys forming an internal aldimine (yellow). Phosphate group is stabilized by interactions with main chain N atoms of loop 6 and water molecules (red balls). Pyridine ring forms a perpendicular (T shaped) stacking with Phe76 while pyridine N and OH are involved in H-bonding with Thr376 and Asn122, respectively.

7. Differences in PLP-binding geometry of fold type II enzymes

In fold type I enzymes, the pyridine ring stacks with an aromatic residue or histidine. In contrast, in fold type II enzymes, stacking interactions are not significant and with the exception of class II, the pyridine ring faces loop 6 on the substrate binding side and loop 1 on the opposite side. In fold type II class I, the pyridine ring is placed between a glycine (Gly228 in *St*OASS) from the N-terminal of loop 6 and a small hydrophobic residue, mostly valine (Val140 in *St*OASS), from loop 1. In classes III, IV and V, PLP lies between a similar glycine residue from loop 6 (Gly288 in *Ec*DAPAL) and a Phe from loop 1 (Phe76 in *Ec*DAPAL, Figure 4b). Aromatic ring of Phe is normal to the pyridine plane. Only exceptions to this are TrpS from class III and DSD from class V. In case of TrpS, Phe is substituted by a His residue. Corresponding residue is an Ile in DSD. In class II enzymes, which lack loop 6, phenolic side chain of a tyrosine from loop 7 forms a parallel stacking interaction with the pyridine ring. This class is unique in having a residue stacking the pyridine ring on the substrate binding side of PLP. On the opposite side, an asparagine residue faces PLP ring from a position corresponding to Phe76 of *Ec*DAPAL. Presence of a parallel stacking interaction similar to fold type I enzymes may represent convergent evolution as PLP enzymes of fold type I and II do not have significant sequence or structural similarities.

8. Importance of interactions of pyridine nitrogen and hydroxyl groups

Pyridine N1 interactions in fold type II enzymes is distinct from those of other PLP dependent enzymes. N1 is hydrogen bonded to a charged Asp or Glu in most fold type I enzymes. This is likely to enhance the electron withdrawing capacity of PLP. In contrast, an Arg is close to N1 in alanine racemase of fold type III leading to lower electron withdrawing potential of PLP. In fold type II enzymes, the residue hydrogen bonding with N1 is a Ser, Thr or Cys. Hence, the electron withdrawing propensity of PLP in fold type II enzymes may be intermediate of fold type I and III enzymes. In most of the fold type II enzymes, loop 8 contributes the interacting residue (Thr376 in *Ec*DAPAL). ACCD is an exception with a glutamate residue (from loop 7) hydrogen bonded to N1[30, 31]. It is the only enzyme in fold type II that acts on a cyclic substrate that lacks a Cα proton. In fold type I enzymes, where pyridine N1 is usually in protonated form, quinonoid intermediates are commonly observed. In fold type II enzymes, instead of a carbanion intermediate, a neutral α,β unsaturated aldimine is likely to be stable. An aminoacrylate intermediate has been observed in the crystal structures of *Ec*DAPAL[21], *M. tuberculosis* OASS[32], *Drosophila* CBS[33] and *S. typhimurium* TrpS[34].

In all fold type II enzymes, loop 2 provides a partner for hydrogen bonding with O3 of pyridine hydroxyl. With the exception of TrpS, the interacting residue is an asparagine (Asn122 in *Ec*DAPAL, Figure 4b). TrpS has a glutamine at the corresponding position[35].

9. Substrate binding and catalysis

Several PLP-dependent enzymes have been observed to undergo inter-domain movement leading to a closed form upon substrate binding. As interactions between the ligand and protein are enhanced upon domain closure, the structural change might be considered as an induced fit. Open and closed structures of

*St*DSD[20] are related by rotation of the small domain by 15° with respect to the large domain (Figure 5a). Similar ligand-induced movements have also been observed in DSD from *E. coli*[36], SR from *Schizosaccharomyces pombe*[37], *Rattus norvegicus* and *Homo sapiens*[38], and OASS from *S. typhimurium*[39]. Loops 2-4, which are part of the small domain, undergo displacement upon domain closure. Loop 3, the most distant of the three loops from the active site, shows the largest deviation. Domain movement leads to repositioning of catalytic residues such that the geometry is suitable for catalysis. Substrate induced domain movement is thought to be an important factor determining substrate specificity[40].

In all PLP-dependent enzymes, binding of substrate leads to the formation of a Schiff base between the amino group of substrate and C4′ of PLP (external aldimine). This leads to release of Lys side chain and rotation of PLP pyridine ring by 10-30° (Figure 5b). The carboxyl of the bound substrate in all the

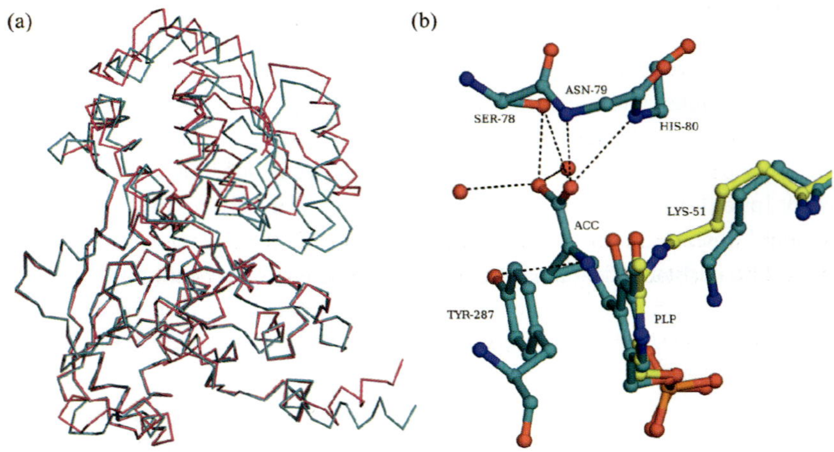

Figure 5. Substrate induced conformational changes and environment of substrate binding in DSD. (a) Open (magenta) and closed (blue) conformations. (b) Superposition of internal and external aldimine structures of DCyD showing the 15° flip in the orientation of PLP on binding ACC (a substrate analog). Parallel stacking interaction of pyridine with Tyr287, as observed in DCyD, is maintained in both internal and external aldimine forms. Residues from loop 5 and water molecules (red balls) form stabilizing interactions with the carboxyl group of PLP bound ligand. Side chains of Ser79 and His80 are masked for clarity.

enzymes is stabilized by interactions with main chain N atoms and a few side chain atoms of loop 2 (Figure 5b). In contrast, in fold type I enzymes, an arginine residue interacts with the substrate carboxyl making a salt bridge. After the formation of external aldimine, with the exception of ACCD, reaction is initiated by abstraction of Cα proton in all fold type II enzymes. The substrate, ACC lacks a Cα proton and therefore is deaminated by a different mechanism[41]. Lys residue from loop 1 that forms Schiff base with PLP in the internal aldimine form is free in the external aldimine form and could act as a base for abstraction of Cα proton from L-amino acid substrates[21, 39, 42, 43]. Similarly, the active site Lys might carry out nucleophilic attack or abstraction of Cβ proton in ACCD[30]. In DSD, DAPAL and SR, Thr, Asp and Ser, respectively, appear to abstract the Cα proton while in DCyD tyrosine residue from loop 7 may act as a general base for proton abstraction[20, 21, 37, 38]. In enzymes that catalyze D-amino acids, substrate Cα proton points in a nearly opposite direction to that of Cα proton of L-amino acids, as the carboxyl and amino groups are anchored at similar positions. Enzymes like DAPAL and SR that are active towards both isoforms of substrates are found to possess two suitably oriented catalytic residues that might act as general base for abstraction of Cα proton for each isoform. Residues at the active site that are not involved in carboxyl binding or proton abstraction but at hydrogen bonding distance to the substrate are from loops 2, 3, 4 and 6. Although these residues are not conserved in fold type II enzymes, a main chain oxygen atom, mostly from a Gly (Gly288 in *Ec*DAPAL) occupies an equivalent position in all the enzymes except for class II, where the side chain hydroxyl of Tyr stacked against PLP occupies an equivalent position.

10. Catalysis with non-cognate ligands

Most of the fold type II enzymes are capable of binding and catalyzing reactions with other analogous compounds at lower rates compared to their cognate substrate. In lyases such as LSD, DSD, TD, and DAPAL, the reaction rate with the cognate substrate is highest and side reactions with other substrate are much slower. DSD can also catalyze reaction with D-Thr and D-allo Thr, while D-DAP acts as an inhibitor[20, 44]. DAPAL can efficiently degrade both D- and L-DAP and cleave D- and L-Ser with lower efficiency[21, 45-47]. In most cases, examination of the active site geometry in high resolution X-ray crystal structures does not directly reveal structural rationale for differences in the rates of activity with different ligands. The protein residues that form stabilizing interactions with the amino acid substrate (with amino and carboxyl groups) are mostly conserved. Selectivity might be brought about by the nature of the external aldimine complex, orientation of the pyridine ring and the substrate, the ability of the substrate to induce domain movement leading to a catalytically active closed form and consequent interactions of the substrate with various functional groups of the active site. However, open and closed forms have not been demonstrated to occur in all PLP-dependent enzymes. In fold type II enzymes, the bound substrate has to be appropriately oriented to maximize the overlap between the bond to be cleaved and π-electron system[5]. The extent of overlap would depend on the angle between the C-H bond (to be cleaved) and π-orbital system. The substrates that bind in a favorable orientation are therefore likely to be more efficiently catalyzed. Thus, further extensive structural studies are needed to gain insights into the relation between ligand binding and efficiency of catalysis.

11. Catalytic function

Different PLP-dependent enzymes with similar active site geometries and acting on the same substrate do not always lead to the same final product. LSD, SR, and TrpS utilize the same substrate; L-Ser. LSD cleaves L-Ser to pyruvate, water and ammonia. SR catalyzes the interconvertion of L-Ser and D-Ser. It also exhibits a low rate of dehydratase activity. TrpS converts L-Ser to tryptophan by a β-replacement reaction. A general scheme of reaction that seems to be applicable to PLP-dependent enzymes is shown in Figure 6. In all these enzymes, the initial steps of the reaction are the same. The amino acid substrate gets covalently linked to PLP as an external aldimine and then the Cα hydrogen is abstracted by a general base, leading to the formation of a quinonoid intermediate. In SR, an additional hydrogen atom is attached to Cα resulting in inversion of the substrate enantiomeric form. In LSD, the β-group is eliminated with the release of a water molecule leaving PLP-aminoacrylate at the active site. Subsequently, aminoacrylate is released and non-enzymatically cleaved to pyruvate and ammonia. TrpSβ follows the same sequence of reactions till the release of the β-group followed by the addition of an indole moiety generated by the α-subunit of TrpS to the Cβ of PLP-aminoacrylate leading to the formation of a second quinonoid intermediate. This is followed by addition of a proton to the same face of Cα from which it was initially abstracted, leading to tryptophan with no change in chirality. Although the structures of all these enzymes are available, it has not been straightforward to understand the subtle differences in the enzymatic properties. This lack of deep insight is reflected in the limited success in interconverting the activities of these enzymes.

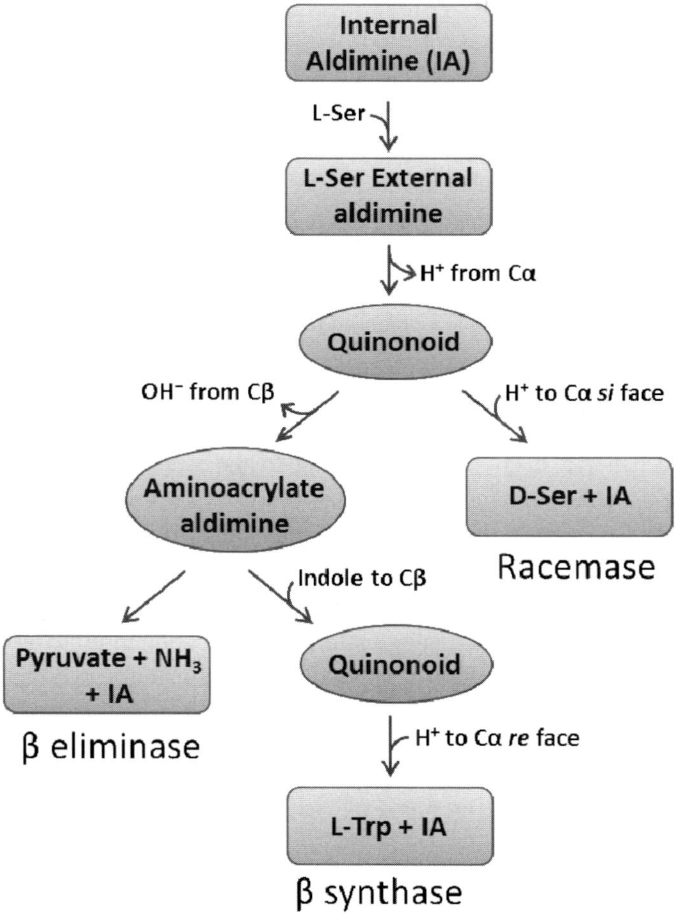

Figure 6. Scheme of reaction steps involved in catalysis by LSD, SR and TrpS.

12. Conclusions

PLP-dependent enzymes are involved in wide variety of chemical transformations of amino acids and their analogues required for maintenance of cellular functions in all known living organisms. Fold type II enzymes are specialized in catalyzing both synthesis and degradation of amino acids involving covalency changes at β or γ carbon of the substrates. Among the 17 known fold type II enzymes, 13 have a representative crystal structure. The crystal structures share many similarities in spite of differences in their specificities and functions. The fold type II enzymes are excellent examples of proteins with low sequence identity but with very similar three-dimensional topology. Although structures are available, understanding the structural bases of their specificity has been challenging. It is likely that subtle changes in their structures result in considerable changes in the substrate specificities and reactions catalyzed. Because of their key roles, these enzymes have industrial and clinical applications and are potential targets for anti-microbial drugs.

13. Legacy of Prof. G.N. Ramachandran

The Ramachandran diagram corresponding to the structural models of PLP-dependent enzymes determined in our laboratory is shown in Figure 7. These include Acetyl ornithine aminotransferase, catabolic TD, DSD and DCyD from *S. typhimurium*; DAPAL from *E. coli* and active site mutants of Serine hydroxymethyltransferase from *Bacillus stearothermophilus*. Ramachandran map remains one of the most important tools for validation of protein structures. It is appropriate that we end this review by citing the map for fold type II PLP-dependent enzymes determined at MBU, IISc. The map not only validates the structures but also reflects the mix of secondary structural elements of these proteins.

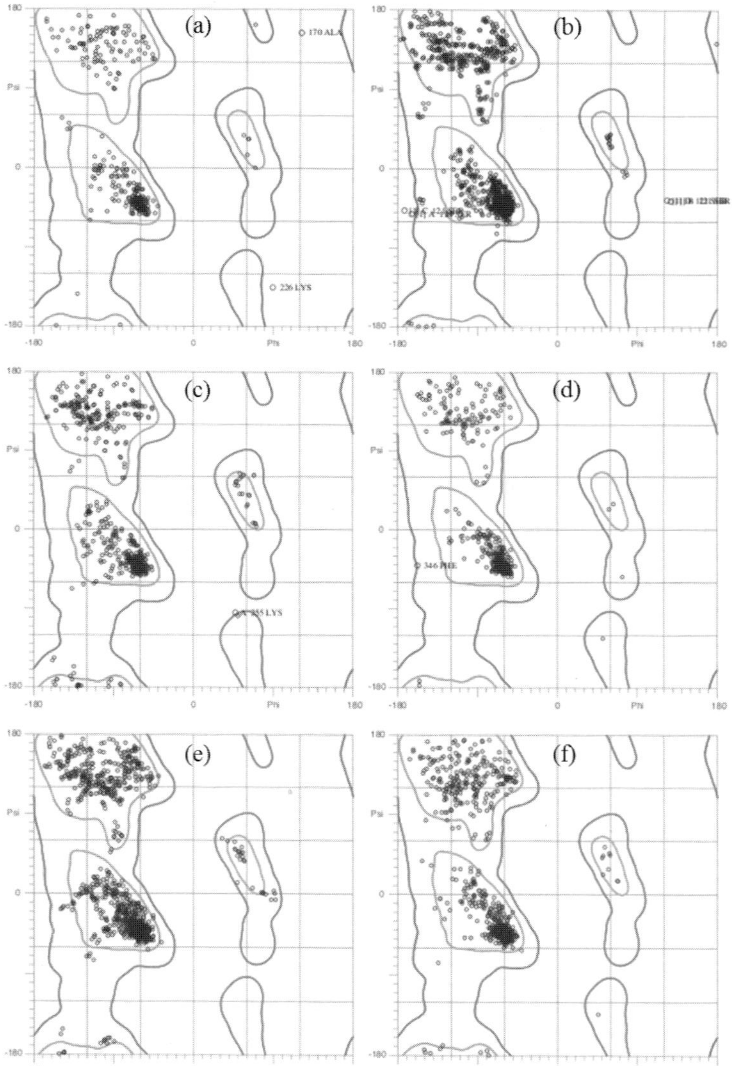

Figure 7. Ramachandran diagrams corresponding to highest resolution crystal structures of (a) N341A Serine hydroxymethyltransferase (PDB Id: 2VI8), (b) catabolic *St*TD (PDB Id: 2GN0), (c) Acetyl ornithine aminotransferase (PDB Id: 2PB2), (d) *St*DSD (PDB Id: 3R0X), (e) *St*DCyD (PDB Id: 4D9B) and *Ec*DAPAL (PDB Id: 4D9I).

References

1. R. Percudani and A. Peracchi, *EMBO Rep* **4**, 850 (2003).
2. N. V. Grishin, M. A. Phillips and E. J. Goldsmith, *Protein Sci* **4**, 1291 (1995).
3. J. N. Jansonius, *Curr Opin Struct Biol* **8**,759 (1998).
4. P. K. Mehta, P. Christen, *Adv Enzymol Relat Areas Mol Biol* **74**, 129 (2000).
5. H. C. Dunathan, *Proc Natl Acad Sci U S A* **55**, 712 (1966).
6. S. Raboni, S. Bettati and A. Mozzarelli, *Cell Mol Life Sci* **66**, 2391 (2009).
7. M. F. Dunn, D. Niks, H. Ngo, T. R. Barends and I. Schlichting, *Trends Biochem Sci* **33**, 254 (2008).
8. T. R. Barends, M.F. Dunn, M. F and I. Schlichting, *Curr Opin Chem Biol* **12,** 593 (2008).
9. E. W. Miles, S. Rhee and D. R. Davies, *J Biol Chem* **274**, 12193 (1999).
10. E. W. Miles, *Advances in Enzymology and Related Areas of Molecular Biology* **64**, 93 (1991).
11. M. Meier, M. Janosik, V. Kery, J. P. Kraus and P. Burkhard, *EMBO J* **20**, 3910 (2001).
12. A. Mozzarelli, S. Bettati, B. Campanini, E. Salsi, S. Raboni, R. Singh, F. Spyrakis, V. P. Kumar and P. F. Cook, *Biochim Biophys Acta* **1814**, 1497 (2011).
13. W. M. Rabeh, and P. F. Cook, *J Biol Chem* **279**, 26803 (2004).
14. P. F. Cook, *Biochim Biophys Acta* **1647**, 66 (2003).
15. S. M. Aitken, P. H. Lodha and D. J. K. Morneau, *Biochimica et Biophysica Acta (BBA) - Proteins & Proteomics* **1814**, 1511 (2011).
16. R. Banerjee, R. Evande, O. Kabil, S. Ojha and S. Taoka, *Biochim Biophys Acta* **1647**, 30 (2003).
17. R. Banerjee and C. G. Zou, *Arch Biochem Biophys* **433**, 144 (2005).
18. F. Birzele, J. E. Gewehr, G. Csaba and R. Zimmer, *Bioinformatics* **23**, e205 (2007).
19. D. K. Simanshu, H. S. Savithri and M.R. Murthy, *J Biol Chem* **281**, 39630 (2006).
20. S. R. Bharath, S. Bisht, H. S. Savithri and M. R. Murthy, *FEBS J* **278**, 2879 (2011).
21. S. Bisht, V. Rajaram, S. R. Bharath, J. N. Kalyani, F. Khan, A. N. Rao, H. S. Savithri and M. R. Murthy, *J Biol Chem* **287**, 20369 (2012).
22. S. R. Bharath, S. Bisht, R. K. Harijan, H. S. Savithri and M. R. Murthy, M. R. *PLoS One* **7**, e36267 (2012).
23. K. A. Denessiouk, A. I. Denesyuk, J. V. Lehtonen, T. Korpela and M. S. Johnson, *Proteins* **35**, 250 (1999).

24. D. T. Gallagher, G. L. Gilliland, G. Xiao, J. Zondlo, K. E. Fisher, D. Chinchilla and E. Eisenstein, *Structure* **6**, 465 (1998).
25. E. Gonzales-Vigil, C. M. Bianchetti, G. N. Phillips and G. A. Howe, *Proceedings of the National Academy of Sciences* **108**, 5897 (2011).
26. T. Yamada, J. Komoto, Y. Takata, H. Ogawa, H. C. Pitot and F. Takusagawa, *Biochemistry* **42**, 12854 (2003).
27. C. Mas-Droux, V. Biou and R. Dumas, *J Biol Chem* **281**, 5188 (2006).
28. O. Hur, D. Niks, P. Casino and M. F. Dunn, *Biochemistry* **41**, 9991 (2002).
29. A. I. Denesyuk, K. A. Denessiouk, T. Korpela, and M. S. Johnson, *J Mol Biol* **316**, 155 (2002).
30. M. Yao, T. Ose, H. Sugimoto, A. Horiuchi, A. Nakagawa, S. Wakatsuki, D. Yokoi, T. Murakami, M. Honma and I. Tanaka, *J Biol Chem* **275**, 34557 (2000).
31. S. Karthikeyan, Q. Zhou, Z. Zhao, C. L. Kao, Z. Tao, H. Robinson, H. W. Liu and H. Zhang, *Biochemistry* **43**, 13328 (2004).
32. R. Schnell, W. Oehlmann, M. Singh and G. Schneider, *J Biol Chem* **282**, 23473 (2007).
33. M. Koutmos, O. Kabil, J.L. Smith and R. Banerjee, *Proc Natl Acad Sci U S A* **107**, 20958 (2010).
34. H. Ngo, N. Kimmich, R. Harris, D. Niks, L. Blumenstein, V. Kulik, T. R. Barends, I. Schlichting and M. F. Dunn, *Biochemistry* **46**, 7740 (2007).
35. C. C. Hyde, S. A. Ahmed, E. A. Padlan, E. W. Miles and D. R. Davies, *J Biol Chem* **263**, 17857 (1988).
36. D. V. Urusova, M. N. Isupov, S. Antonyuk, G. S. Kachalova, G. Obmolova, A. A. Vagin, A. A. Lebedev, G. P. Burenkov, Z. Dauter, H. D. Bartunik, V. S. Lamzin, W. R. Melik-Adamyan, T. D. Mueller and K. D. Schnackerz, *Biochim Biophys Acta* **1824**, 422 (2011).
37. M. Goto, T. Yamauchi, N. Kamiya, I. Miyahara, T. Yoshimura, H. Mihara, T. Kurihara, K. Hirotsu and N. Esaki, *J Biol Chem* **284**, 25944 (2009).
38. M. A. Smith, V. Mack, A. Ebneth, I. Moraes, B. Felicetti, M. Wood, D. Schonfeld, O. Mather, A. Cesura a nd J. Barker, *J Biol Chem* **285**, 12873 (2010).
39. P. Burkhard, G. S. Rao, E. Hohenester, K. D. Schnackerz, P. F. Cook and J. N. Jansonius, *J Mol Biol* **283**, 121 (1998).
40. A. C. Eliot and J. F. Kirsch, *Annu Rev Biochem* **73**, 383 (2004).
41. T. Ose, A. Fujino, M. Yao, N. Watanabe, M. Honma and I. Tanaka, *J Biol Chem* **278**, 41069 (2003).
42. E. W. Miles, *Advances in Enzymology and Related Areas of Molecular Biology* 93 (2006).
43. R. Evande, S. Ojha and R. Banerjee, *Archives of Biochemistry and Biophysics* **427**, 188 (2004).
44. D. Dupourque, W. A. Newton and E. E. Snell, *J Biol Chem* **241**, 1233 (1966).

45. T. Nagasawa, K. Tanizawa, T. Satoda and H. Yamada, *J Biol Chem* **263**, 958 (1988).
46. T. Uo, T. Yoshimura, T. Nishiyama and N. Esaki, *Biosci Biotechnol Biochem* **66**, 2639 (2002).
47. F. Khan, V. R. Jala, N. A. Rao and H. S. Savithri, *Biochem Biophys Res Commun* **306**, 1083 (2003).

INTRAMOLECULAR ISOPEPTIDE BONDS: NOVEL POST-TRANSLATIONAL MODIFICATIONS IN BACTERIAL PILI AND CELL-SURFACE ADHESINS

NEIL G. PATERSON
School of Biological Sciences, University of Auckland
Private Bag 92019, Auckland, New Zealand

HAE JOO KANG
Division of Molecular Biosciences, Membrane Protein Crystallography Group
Imperial College London, London SW7 2AZ, United Kingdom

EDWARD N. BAKER
School of Biological Sciences, University of Auckland
Private Bag 92019, Auckland, New Zealand

Despite the presence of many and varied functional groups among the 20 amino acids commonly used in proteins, covalent bonds between these groups are rare. Only disulfide bonds, between pairs of Cys residues are relatively common. Nevertheless, protein structure analyses continue to reveal novel covalent linkages that result from intramolecular reactions in proteins. Here we describe one such example in which isopeptide bonds are formed between the side chains of lysine and asparagine (or aspartic acid) in an autocatalytic process during protein folding. Bonds of this kind were first discovered in the course of structural analyses of the component proteins of the pili expressed by *Streptococcus pyogenes*. Isopeptide bonds are now known to be widely present in other pili and in cell-surface adhesins, where they provide resistance to physical and chemical stress and may be an evolutionary alternative to disulfide bonds.

1. Introduction

The many beautiful pictures of proteins that adorn current biochemistry textbooks are a testament to the remarkable variety of three dimensional structures that these molecules can adopt. While the possible folds that can be adopted are governed by the geometric and energetic factors so elegantly captured by the Ramachandran plot, the stabilities and fine structural details of proteins depend on networks of interactions between their constituent amino acids. Most of these are non-covalent: hydrophobic interactions, hydrogen bonds and ion pairs (salt bridges). Despite the variety of functional groups carried by amino acid side chains, the only covalent cross-links commonly found within proteins are disulfide bonds, formed between pairs of cysteine residues. These bonds do not cause protein folding; rather they form spontaneously when two Cys residues are brought together during protein folding. Disulfide bonds are thus a classic example of a spontaneous intramolecular reaction, occurring as a post-translational modification.

In recent years, some unexpected post-translational modifications have been discovered, often associated with intramolecular reactions that generate novel functional elements, for example the Cys-Tyr bond in galactose oxidase[1], which creates a radical centre, and the chromophore in green fluorescent protein[2]. Here we describe a novel post-translational modification that was discovered in the course of structural studies of the pili expressed by the Gram-positive pathogen *Streptococcus pyogenes*[3]. In this case, an autocatalytic intramolecular reaction results in the formation of isopeptide bonds that cross-link lysine and asparagine (or aspartic acid) side chains in the protein interior. These bonds now appear to be relatively widespread among bacterial cell-surface proteins[4].

2. Isopeptide bonds in proteins

Isopeptide bonds are amide bonds that differ from conventional peptide bonds in that one or both of the two reacting groups is not an α-amino or α-carboxyl group. These bonds are well known as intermolecular linkages, for example in ubiquitylation[5], sumoylation[6] and sortase-mediated cell-surface protein anchoring[7], where the ε-amino group of a lysine side chain reacts with an α-carboxyl group of another polypeptide, and in transglutamination[8] where a lysine ε-amino group is linked to the carboxyamide group of a glutamine side chain. These reactions are all enzyme-mediated. In contrast, the isopeptide bonds discussed here are intramolecular, forming cross-links between amino acid side chains within proteins, and they are formed autocatalytically, without enzymatic intervention. Only one such example, of a self-generated isopeptide bond, had previously been seen, in the capsid of the bacteriophage HK97 where intermolecular Lys-Asn isopeptide bonds form during capsid maturation, linking the protein subunits together in "chain-mail" rings[9]. Emerging data from pili and cell-surface adhesins now suggests, however, that autocatalytically-formed isopeptide bonds are a common post-translational modification in such systems.

2.1. *Initial discovery in bacterial pili*

Bacterial pili are long, hair-like protein assemblies whose general function is to engage with host cells to enable colonization and infection[10]. In doing so, these assemblies are exposed to significant shear stresses and must thus be flexible but mechanically strong. The pili on Gram-negative organisms, which are of several types[11,12], are arranged as bundles of non-covalently associated subunits (pilins), gaining strength from the multiplicity of these interactions. The pili on Gram-positive organisms are very different, however. These have only begun to be characterized in the past 10 years, and the model that has emerged is of a linear polymer in which as many as 100-200 pilin subunits are arranged in tandem, like beads on a string. These pili are assembled by sortase enzymes, which cleave the polypeptide chain of one subunit at a specific sequence motif near its C-terminus, and then generate an isopeptide bond linkage joining the new C-terminal carboxyl group to a lysine side chain on the next subunit[13,14] (Figure 1). In this way a covalent polymer is formed. But how are the individual subunits stabilized against unfolding in the absence of any form of "bundling"?

The answer to this came from the structural analysis of the major pilin Spy0128 from the pili expressed by *Streptococcus pyogenes*, also known as Group A Streptococcus (GAS). This is the pilin subunit that forms the shaft of the GAS pilus, successive copies being polymerized by a specific sortase to create a fully covalent polymer, with isopeptide bonds linking the subunits. The Spy0128 structure[3] revealed two tandem immunoglobulin-like (Ig-like) domains. Unexpectedly, each domain contained an isopeptide bond cross-link (Figure 2), formed by a reaction between the side chains of a lysine residue and an asparagine residue contributed by the first and last β-strands of the domain, respectively. This assignment was strongly supported by the electron density and was confirmed by mass spectrometry[3,15]. Also associated with each isopeptide bond was a glutamic acid residue whose carboxyl group was hydrogen bonded to the C=O and N-H groups of the isopeptide bond moiety. Mutagenesis showed this residue to be essential for isopeptide bond formation[3].

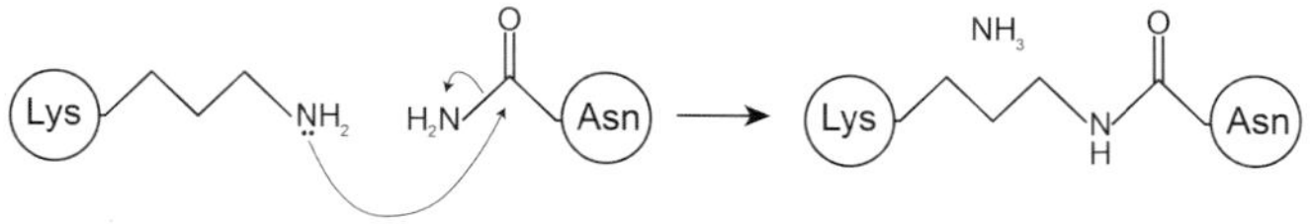

Figure 1. Isopeptide bond formation. The ε-amino group of a lysine residue attacks the γ-carbon atom of an asparagine side chain, with the liberation of ammonia.

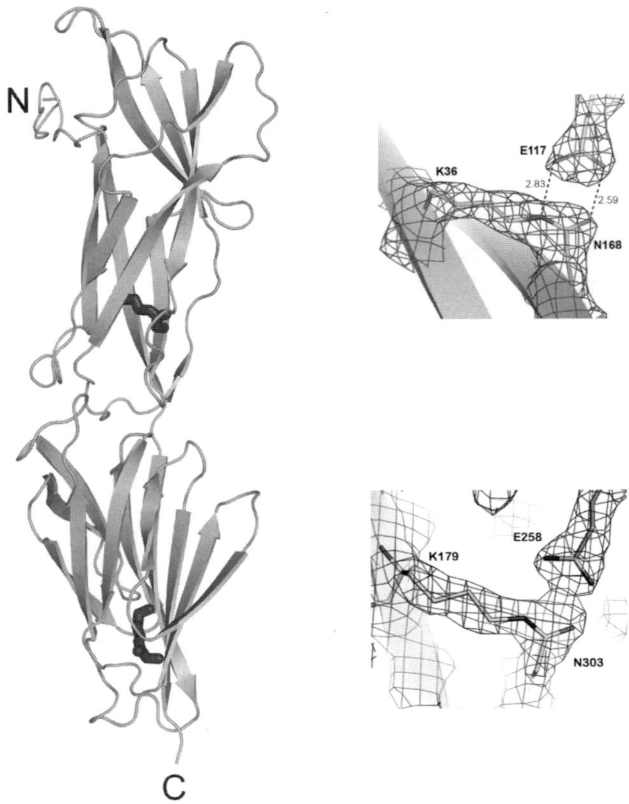

Figure 2. Structure of Spy0128, with its two Ig-like domains, is on the left, showing an isopeptide bond near the bottom of each domain. On the right are shown the Lys, Asn and Glu residues in each case. Note the continuous electron density for the isopeptide bond, extending from Lys to Asn.

2.2. *Conserved structural context*

Since their discovery in Spy0128, similar internal isopeptide bonds have been found in a number of other cell-surface proteins, both in the pilin proteins of other Gram-positive bacteria, and in a number of multi-domain cell-surface molecules[4]. In all cases identified to date, the proteins are involved in adhesion to host cells, or other surfaces, and are thus required to be robust and able to withstand strong shear forces. Most strikingly, the domains in which such bonds have so far been found fall exclusively into two sub-types of Ig-like domain, designated as CnaA and CnaB (Figure 3). These take their names from the cell-surface adhesin Cna, from *Staphylococcus aureus*, in which several A-type domains are followed by a series of tandem B-type domains that are presumed to form a stalk supporting the collagen-binding A domains[16]. The archetypal CnaA fold has a jelly roll-like topology, forming 4-stranded and 5-stranded antiparallel β-sheets that pack as a β-sandwich[17]. The CnaB fold comprises 7 β-strands organized as a 3- and a 4-stranded β-sheet, also packed as a β-sandwich, with all strands antiparallel except for the first and last, which are parallel[18].

An intriguing feature, suggestive of a common evolutionary origin, is that all internal isopeptide bonds identified so far are located at conserved positions within their CnaA or CnaB domains (Figure 3). In CnaA domains a lysine from the first β-strand is linked to an asparagine from the second-last β-strand, cross-linking the two β-sheets. In CnaB domains a lysine from the first β-strand forms an isopeptide bond with an asparagine (or aspartic acid) from the last β-strand, thus cross-linking these first and last strands. In each case an acidic residue, Asp for CnaA domains and Glu for CnaB domains, is contributed from another β-strand to provide the necessary catalytic function, and in each case the isopeptide bond is buried in the hydrophobic core of the domain as required for the bond-forming reaction (see section 3).

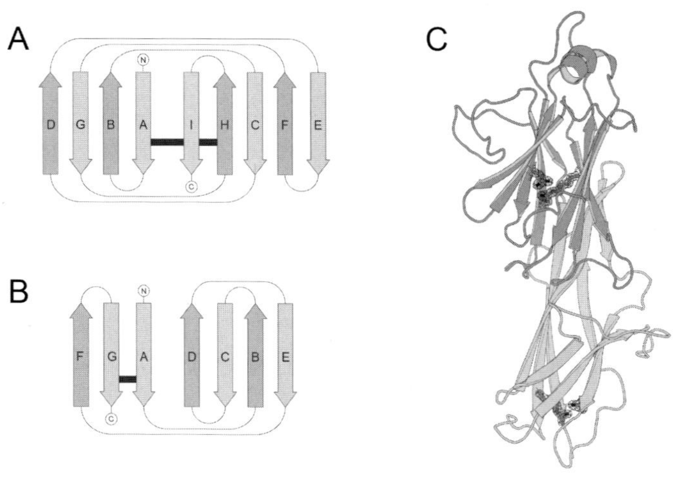

Figure 3. CnaA and CnaB domains. In (A) the archetypal CnaA fold is shown, with a black bar showing the typical location of an isopeptide bond, linking strands A and H. In (B) a similar topology diagram is shown for a CnaB domain, with its isopeptide bond linking strands A and G. In both cases the core fold is often embellished with inserted loops and additional secondary structure elements. (C) shows the middle (CnaA) and C-terminal (CnaB) domains of the major pilin SpaA, with the isopeptide bonds and catalytic acidic residues highlighted in their electron density in each.

2.3. *Occurrence in pili and other cell-surface adhesins*

Gram-positive pili consist of a polymeric shaft, in which some 100-200 copies of the major (backbone) pilin are joined end-on-end, with a specialized adhesin subunit at the tip for attachment to host cells and often, but not always, another specialized pilin at the base, joining the whole assembly to the cell wall[19,20].

Crystal structures are now available for the major pilin subunits from a number of different Gram-positive bacteria: Spy0128 from *S. pyogenes*[3], SpaA from *Corynebacterium diphtheriae*[21], GBS80 from *Streptococcus agalactiae*[22], FimA from *Actinomyces oris*[23], RrgB from *Streptococcus pneumoniae*[24-26] and BcpA from *Bacillus cereus*[27] (Figure 4). These proteins vary widely in size and sequence, but follow a common modular pattern based on tandem CnaA and CnaB domains (reviewed in [28]). The simplest, Spy0128, comprises just two tandem CnaB domains, each containing an isopeptide bond cross-link in the generic position. SpaA, GBS80 and FimA have three tandem domains, with a CnaA middle domain between two CnaB domains; BcpA has four tandem domains, a middle CnaA domain preceded by two CnaB domains and followed by a third CnaB domain; and RrgB also has four domains, one CnaA and three CnaB, but with one of the CnaB domains being an insertion in the CnaA domain and packed laterally against it.

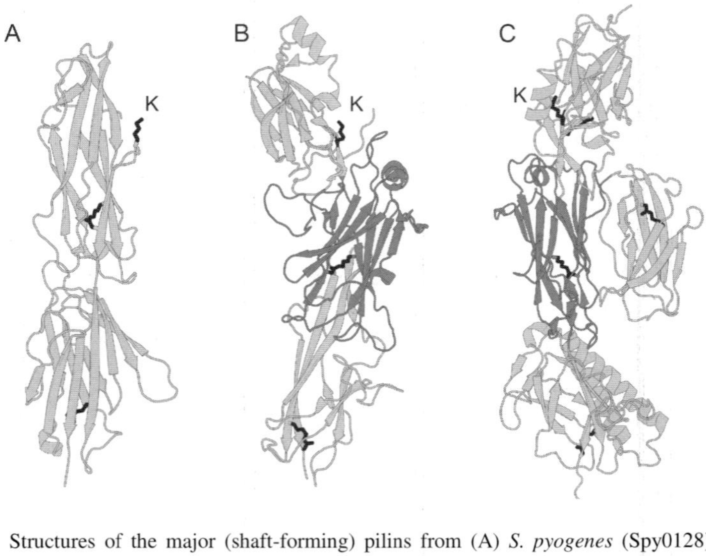

Figure 4. Structures of the major (shaft-forming) pilins from (A) *S. pyogenes* (Spy0128), (B) *C. diphtheriae* (SpaA) and (C) *S. pneumoniae* (RrgB). CnaA-type domains are in darker shades, CnaB-type domains in lighter shades. The N-terminal domain is at the top and C-terminal domain at the bottom in each case. The internal isopeptide bonds are in black and the lysine side chains used to connect to the preceding pilin in the assembly are also in black and identified with a K.

A defining feature of the pilus shaft is that it is stabilized throughout its length by isopeptide bonds, both between subunits and within subunits. Each subunit is attached to the next by a sortase-mediated intermolecular isopeptide bond, while within the subunits each domain, whether CnaA-type or CnaB-type contains an internal isopeptide bond cross-link. This means that despite its long, slender dimensions, typically only 30-40 Å thick, the assembly is maintained by a succession of covalent linkages throughout its length[28]. The only exception is that in most cases the N-terminal domain either has an isopeptide bond that only forms after the intermolecular linkage is made, or (as in SpaA) lacks one completely. The N-terminal domain in these cases differs, however, in that it is outside the load-bearing core of the pilus, as it precedes the intermolecular linkage; thus, an isopeptide bond here is of less significance to the stability of the shaft. Less is known about adhesin subunits at the tip of the pilus, but the two structures that are available, for RrgA from *S. pneumoniae*[29] and Cpa from *S. pyogenes*[30], show that both contain CnaB domains with internal isopeptide bond cross-links, thus extending the load-bearing structure of the shaft.

Finally, as more structures of cell-surface adhesins, or fragments of them, are solved, it becomes increasingly apparent that internal isopeptide bonds are common here, too, wherever CnaB-like or CnaA-like domains are used. The Staphylococcal adhesin Cna has isopeptide bonds in its collagen-binding A domain and in its repetitive B region[3,31]; the latter comprises three repeats, each made up of two isopeptide bond-containing CnaB domains[31]. Isopeptide bonds are also present in the C-terminal domains of cell-surface adhesins from *Streptococcus gordonii*[32] and *Streptococcus mutans*[33], oral bacteria involved in biofilm and dental plaque formation, and in a CnaB domain from the fibronectin-binding adhesin FbaB from *S. pyogenes*[34]. The latter has a well-characterized Lys-Asp isopeptide bond. Bioinformatic studies[3,34] suggest that these examples may be merely the tip of the iceberg, since thousands of CnaB and CnaA domains are predicted to occur in other cell-surface proteins. Sequence searches are difficult, however, as the Lys, Asn/Asp and catalytic Asp/Glu residues are widely separated, all being on separate β-strands. Moreover, 'split' domains in which two parts of a CnaB or CnaA domain are interrupted by a large insertion but are nevertheless held together by an isopeptide bond, as in RrgA[29] and Cpa[30], would be impossible to predict. Nevertheless, close examination of predicted CnaA/CnaB domains suggests that isopeptide bonds are common.

2.4. *Contributions to stability*

The prevalence of isopeptide bonds in pili and other adhesins suggests that their primary role is in imparting resistance to mechanical stress, since these long, thin molecules and assemblies – essentially only one molecule thick – will be subject to large tensile forces when engaged with host cells. A recent study of Spy0128 polymers by atomic force microscopy[35] confirms this view, showing that the molecules are virtually inextensible with their isopeptide bonds in place but are easily pulled apart without them. The strategic location of the bonds, cross-linking the first and last strands of CnaB domains and cross-linking the two β-sheets of CnaA domains, has the effect of creating a covalent core along the principal axis of pilus and adhesin structures built from such domains[4,36].

Other studies show that the isopeptide bonds also contribute substantially to thermal stability and resistance to proteolysis. Circular dichroism analysis shows that the melting temperature of wild type Spy0128 is about 85 °C, but that loss of either of its two isopeptide bonds reduces this by 25 °C and loss of both gives a highly unstable, mostly-unfolded protein[37]. Likewise, thermal shift studies of the *S. pneumoniae* pilins RrgA, RrgB and RrgC show that the isopeptide bonds are important for thermal stability[38], and in FbaB the Lys-Asp bond in its CnaB domain has been shown to contribute remarkable conformational, pH and thermal stability[34]. Effects on proteolysis are more equivocal, probably depending on where proteolytic sites are located, but the isopeptide bonds in both Spy0128[37] and BcpA[27] have been shown to give very strong resistance to proteolysis.

3. Mechanism of isopeptide bond formation

The isopeptide bond-forming Lys, Asn and Glu residues in Spy0128 are buried in the hydrophobic core of each domain, surrounded by aromatic and other hydrophobic residues[3]. This strongly suggests that bond formation cannot be enzyme-catalyzed and must be generated by an autocatalytic reaction that takes place when the residues are brought together in the protein interior during folding. It further implies that the pKa values for the key lysine and glutamic acid residues will be significantly perturbed by their hydrophobic environment, which strongly favours the neutral forms of these amino acids; unprotonated NH_2 for the lysine ε-amino group and protonated COOH for the glutamic acid carboxyl group. Indeed, the latter is clearly protonated in the Spy0128 crystal structure as in both domains it forms a short hydrogen bond with the carbonyl oxygen of the isopeptide moiety[3].

Several theoretical studies using quantum mechanical/molecular mechanical (QM/MM) calculations[34,39] have converged on a mechanism of isopeptide bond formation that is fully in accord with structural and mutagenesis data. The overall reaction for the formation of the Lys-Asn bonds found in Spy0128 and most other cases is shown in Figure 2, but examples of Lys-Asp isopeptide bonds also exist[30,34] and fit the same mechanism; the only difference is that Lys-Asn bonds result in the loss of a molecule of ammonia whereas for Lys-Asp bonds a water molecule is lost. Key features of the mechanism are that the Lys ε-amino group must be unprotonated to permit nucleophilic attack on the γ-carbon of the Asn carboxyamide group (or Asp carboxyl group), and that the adjacent Glu carboxyl group acts as a proton shuttle, taking a proton from the amino group for ultimate transfer to the ammonia or water leaving group. There is more to be learned yet, as mutagenesis shows that the environment of the isopeptide bond is subtly tuned to the use of either Asn or Asp[34].

4. Evolutionary questions

So far, all characterized intramolecular isopeptide bonds are in Ig-like domains of the CnaA or CnaB sub-types, located in conserved positions that are specific to the particular sub-type. Moreover, the catalytic acid seems to be consistently Asp in CnaA domains and Glu in CnaB domains. These observations suggest a common evolutionary origin and may also reflect a common folding mechanism. On the other hand, it is also true that many CnaA and CnaB domains do not have such bonds, and whether they have been gained or lost during evolution is a matter of speculation. Whether internal isopeptide bonds may be found in other domain types is unknown, but given the wide range of proteins that are yet to be characterized this is not unlikely.

Whatever their origin, internal isopeptide bonds can be seen as an evolutionary alternative to disulfide bonds. Both types of bonds are formed when the participating residues are brought together during protein folding. It is notable that all the examples of proteins with internal isopeptide bonds are from Gram-positive bacteria, many of which lack the full disulfide bond-forming machinery (the periplasmic dsb proteins) that is characteristic of Gram-negative bacteria[40,41]. Moreover the Type 1 and Type IV pili of Gram-negative organisms and other adhesive structures such as Dr fimbriae all depend on conserved disulfide bonds for their stable formation[42-44]. This raises the possibility that isopeptide bonds may have arisen in Gram-positive bacteria as an alternative strategy for stabilizing types of proteins (cell-surface pili and other adhesins) that are stabilized by disulfide bonds in other organisms.

Acknowledgments

We gratefully acknowledge research support from the Marsden Fund of New Zealand and the Health Research Council of New Zealand.

References

1. S. J. Firbank, M. S. Rogers, C. M. Wilmot, D. M. Dooley, M. A. Halcrow, P. F. Knowles, M. J. McPherson and S. E. V. Phillips, *Proc. Natl. Acad. Sci. USA* **98**, 12932 (2001).
2. R. Heim, D. C. Prasher, and R. Y. Tsien, *Proc. Natl. Acad. Sci. USA* **91**, 12501 (1994).
3. H. Kang, F. Coulibaly, F. Clow, T. Proft and E. N. Baker, *Science* **318**, 1625 (2007).
4. H. Kang and E. N. Baker, *Trends Biochem. Sci.* **36**, 229 (2011).
5. C. M. Pickart, *Annu. Rev. Biochem.* **70**, 503 (2001).
6. R. T. Hay, *Mol. Cell* **18**, 1 (2005).
7. L. A. Maraffini, A. C. Dedent and O. Schneewind, *Microbiol. Mol. Biol. Rev.* **70**, 192 (2006).
8. J. W. Weisel, *Adv. Protein Chem.* **70**, 247 (2005).
9. W. R. Wikoff, L. Liljas, R. L. Duda, H. Tsuruta, R. W. Hendrix and J. E. Johnson, *Science* **289**, 2129 (2000).
10. T. Proft and E. N. Baker, *Cell. Mol. Life Sci.* **66**, 613 (2009).
11. G. Waksman and S. J. Hultgren, *Nat. Rev. Microbiol.* **7**, 765 (2009).
12. L. Craig, M. E. Pique and J. A. Tainer, *Nat. Rev. Microbiol.* **2**, 363 (2004).
13. J. L. Telford, M. A. Barocchi, I. Margarit, R. Rappuoli and G. Grandi, *Nat. Rev. Microbiol.* **4**, 509 (2006).
14. A. P. Hendrickx, J. M. Budzik, S. Y. Oh and O. Schneewind, *Nat. Rev. Microbiol.* **9**, 166 (2011).
15. H. J. Kang, M. Middleditch, T. Proft and E. N. Baker, *Biopolymers* **91**, 1126 (2009).
16. J. M. Patti, J. O. Boles and M. Hook, *Biochemistry* **32**, 11428 (1993).
17. Y. Zong, Y. Xu, X. Liang, D. R. Keene, A. Hook, S. Gurusiddappa, M. Hook and S. V. Narayana, *EMBO J.* **24**, 4224 (2005).
18. C. C. Deivanayagam, R. L. Rich, M. Carson, R. T. Owens, S. Danthuluri, T. Bice, M. Hook and S. V. Narayana, *Structure* **8**, 67-78 (2000).
19. S-Y. Oh, J. M. Budzik and O. Schneewind, *Proc. Natl. Acad. Sci. USA* **105**, 13703 (2008).
20. M. Hilleringmann, P. Ringler, S. A. Muller, G. De Angelis, R. Rappuoli, I. Ferlenghi and A. Engel, *EMBO J.* **28**, 3921 (2009).
21. H. J. Kang, N. G. Paterson, A. H. Gaspar, H. Ton-That and E. N. Baker, *Proc. Natl. Acad. Sci. USA* **106**, 16967 (2009).
22. K. Vengadesan, X. Ma, P. Dwivedi, H. Ton-That and S. V. Narayana, *J. Mol. Biol.* **407**, 731 (2011).

23. A. Mishra, B. Devarajan, M. E. Reardon, P. Dwivedi, V. Krishnan, J. O. Cisar, A. Das, S. V. Narayana and H. Ton-That, *Mol. Microbiol.* **81**, 1205 (2011).
24. G. Spraggon, E. Koesema, M. Scarselli, E. Malito, M. Biagini, N. Norais, C. Emolo, M. A. Barocchi, F. Giusti, M. Hilleringmann, R. Rappuoli, S. Lesley, A. Covacci, V. Masignani and I. Ferlenghi, *PLoS One* **5**, e10919 (2010).
25. N. G. Paterson and E. N. Baker, *PLoS One* **6**, e22095 (2011).
26. L. Mortaji, C. Contreras-Martel, M. Moschioni, I. Ferlenghi, C. Manzano, T. Vernet, A. Dessen and A. M. Di Guilmi, *Biochem. J.* **441**, 833 (2012).
27. J. M. Budzik, C. B. Poor, K. F. Faull, J. P. Whitelegge, C. He and O. Schneewind, *Proc. Natl. Acad. Sci. USA* **106**, 19992 (2009).
28. H. J. Kang and E. N. Baker, *Curr. Opin. Struct. Biol.* **22**, 200 (2012).
29. T. Izore, C. Contreras-Martel, L. El Mortaji, C. Manzano, R. Terrasse, T. Vernet, A. M. Di Guilmi and A. Dessen, *Structure* **18**, 106 (2010).
30. J. A. Pointon, W. D. Smith, G. Saalbach, A. Crow, M. A. Kehoe and M. J. Banfield, *J. Biol. Chem.* **285**, 33858 (2010).
31. K. Vengadesan and S. V. Narayana, *Protein Sci.* **20**, 759 (2011).
32. C. Forsgren, R. J. Lamont and K. Persson, *J. Mol. Biol.* **397**, 740 (2010).
33. M. R. Larson, K. R. Rajashankar, P. J. Crowley, C. Kelly, T. J. Mitchell, L. J. Brady and C. Deivanayagam, *J. Biol. Chem.* **286**, 21657 (2011).
34. R. M. Hagan, R. Bjornsson, S. A. McMahon, B. Schomburg, V. Braithwaite, M. Bukl, J. H. Naismith and U. Schwarz-Linek, *Angew. Chemie* **49**, 8421 (2010).
35. J. Alegre-Cebollada, C. L. Badilla and J. M. Fernandez, *J. Biol. Chem.* **285**, 11235 (2010).
36. T. O. Yeates and R. T. Clubb, *Science* **318**, 1558 (2007).
37. H. J. Kang and E. N. Baker, *J. Biol. Chem.* **284**, 20729 (2009).
38. L. El Mortaji, R. Terrasse, A. Dessen, T. Vernet and A. M. Di Guilmi, *J. Biol. Chem.* **285**, 12405 (2010).
39. X. Hu, H. Hu, J. A. Melvin, K. W. Clancy, D. G. McCafferty and W. Yang, *J. Amer. Chem. Soc.* **133**, 478 (2011).
40. R. J. Dutton, D. Boyd, M. Berkmen and J. Beckwith, *Proc. Natl. Acad. Sci. USA* **105**, 11933 (2008).
41. B. Heras, S. R. Shouldice, M. Totsika, M. J. Scanlon, M. A. Schembri and J. L. Martin, *Nat. Rev. Microbiol.* **7**, 215 (2009).
42. M. D. Crespo, C. Puorger, M. A. Scharer, O. Eidam, M. G. Grutter, G. Capitani and R. Glockshuber, *Nat. Chem. Biol.* **8**, 707 (2012).
43. H. Harvey, M. Habash, F. Aidoo and L. L. Burrows, *J. Bacteriol.* **191**, 6513 (2009).
44. R. Piatek, P. Bruzdziak, M. Wojciechowski, B. Zalewska-Piatek and J. Kur, *Biochemistry* **49**, 1460 (2010).

STRUCTURAL BIOINFORMATICS APPROACHES FOR DECIPHERING BIOSYNTHETIC CODE OF SECONDARY METABOLITES

SHRADHA KHATER AND DEBASISA MOHANTY
Bioinformatics Center, National Institute of Immunology, New Delhi – 110067, India

Polyketides and non-ribosomal peptides constitute a major class of pharmaceutically important secondary metabolites with diverse biological functions. They are biosynthesized by polyketide synthase (PKS) and non-ribosomal peptide synthetase (NRPS) family of megasynthases. These enzymes can generate enormous diversity in the chemical structures of secondary metabolites by combinatorial use of a limited number of catalytic domains and subtle variations of amino acids in the active site pockets of these enzymatic domains. In this review, we discuss recently developed structural bioinformatics approaches which help in correlating sequence and structural features of the PKS and NRPS megasynthases to the chemical structures of their secondary metabolites biosynthetic products.

1. Introduction

From among the large number of metabolites biosynthesized by different organisms, the secondary metabolites biosynthesized by microbial, fungal and plant species have been of particular interest as they constitute a large repertoire of pharmaceutically important natural products[1-5]. Of these secondary metabolite natural products, polyketides and non-ribosomal peptides comprise the largest family. Polyketide synthase (PKS) and non-ribosomal peptide synthetase (NRPS) function as a template for their biosynthesis. Decoding the so called "biosynthetic code of secondary metabolite" by which these megasynthases generate diversity in their chemical structures is crucial for identifying and predicting new secondary metabolite natural products and engineering altered natural products[6].

Biosynthesis of polyketides by PKSs resembles assembly line mechanism of fatty acid biosynthesis[3, 4]. Short chain fatty acids activated as acyl CoA esters are used as starter/extender substrates by acyltransferase (AT) domain, which loads these units onto an acyl carrier protein (ACP). ACP carries the intermediates to different domains throughout the catalytic cycle. Ketosynthase (KS) accepts the starter/extender units from ACP and catalyze polyketide chain elongation by decarboxylative Claisen condensation reaction. The resulting ketide group is optionally reduced by ketoreductase (KR), dehydratase (DH) and enoylreductase (ER) domains. Final polyketide chain is released from ACP and cyclized in a variety of different ways by thioesterase (TE) domains. In absence of reductive steps highly reactive poly-β-ketones are formed that can be cyclized, aided by product template (PT) domain producing wide-ranged aromatic polyketides. Depending on the type of organization of these catalytic domains PKSs have been classified into three general classes: type I, type II and type III. Type I PKSs are multidomain megasynthases containing multiple catalytic domains on a single polypeptide chain. Type I PKSs are further subdivided into modular and iterative. Type I modular PKSs have separate set of catalytic domains, used only once in an "assembly-line" process, for each round of extension (Figure 1). Unlike bacterial type I PKSs, iterative fungal type I systems reuses a single set of catalytic domain repeatedly to yield the final product. Based on the number of reductive steps iterative type I PKSs are further categorized as non-reducing (PKS), partially reducing and highly reducing PKSs. NR-PKSs are responsible for synthesis of aromatic polyketides in fungi (Figure 1). In type II PKSs each domain is a free-standing protein that associates to form a multi-enzyme complex and is used iteratively to yield a final product. In contrast to type I and type II PKSs, type III PKSs are simple homodimeric proteins. Starter substrate is directly loaded onto the active site Cys, without the help of ACP, where it is decarboxylatively condensed with extender units iteratively. The enzyme bound product is then cyclized to produce structurally diverse scaffolds.

Similar to modular PKSs, NRPSs perform the condensation of proteogenic as well as non-proteogenic amino acid in an assembly line fashion to form diverse peptide natural products[3] (Figure 1). The adenylation (A) domain of NRPSs is a gatekeeper enzyme, which recognizes and activates starter and extender units. The adenylated monomer is then covalently attached to dedicated carrier protein, peptidyl carrier protein (PCP), which transfers intermediate from one active site to another. Condensation (C) domain carries out peptide bond formation between two PCP-bound intermediates. The release of mature product is catalyzed by thioesterase (TE) domain, which is present at the distal C-terminus of these enzymes. Upon reaching its full length, the peptide is either hydrolytically cleaved from the protein to produce linear product or an intramolecular nucleophilic attack releases a macrocyclic structure. PKS and NRPS can be optionally modified by tailoring enzymes to introduce new

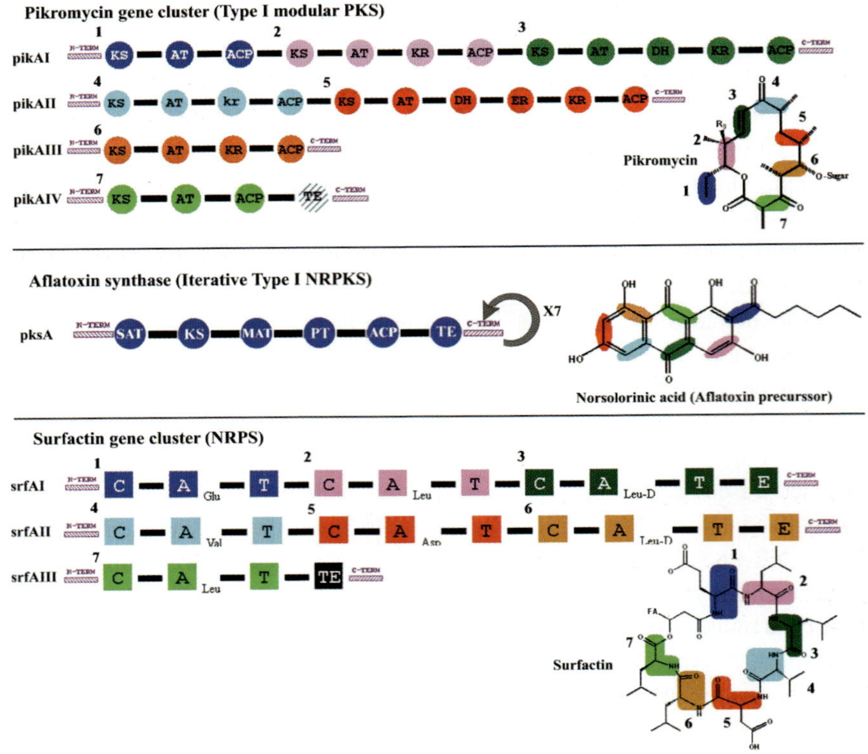

Figure 1: Typical examples of domain organization of type I PKS and NRPS gene clusters along with the chemical structures of their biosynthetic products. In case of modular PKS and NRPS clusters each module has been numbered and the functional group added by each module has respective number in the chemical structure. The middle panel depicts an example of type I iterative PKS, containing a single module that is utilized iteratively to yield norsolorinic acid.

functionalities. Depending on the organization of their catalytic domains, PKS and NRPS can generate different types of secondary metabolites. In recent years, significant advances have been made in experimental characterization of the secondary metabolite products of large number PKS/NRPS gene clusters from bacterial and fungal organisms. Analysis of these clusters have helped in development of a number of bioinformatics tools to identify number of modules and catalytic domains and hence predict the structure of their linear product[7-9]. These tools use a homology based approach to identify catalytic domain in a query sequence, by pairwise alignment to template sequences or profiles, from known PKS/NRPS clusters. However, the selection of starter/extender substrates, chain length and number of iterations and also various different cyclization modes is controlled by subtle variations in the amino acid sequences of key catalytic domains (Figure 2). As it is extremely difficult to predict these features from sequence analysis alone, structural bioinformatics approaches have been used to decipher substrate recognition mechanisms. These structural bioinformatics approaches have been aided by major developments in structural biology of these megasynthases. In this review, we briefly describe developments in structure based *in silico* methods which attempt to decipher chemical structures of secondary metabolites encoded by various PKSs and NRPSs gene clusters.

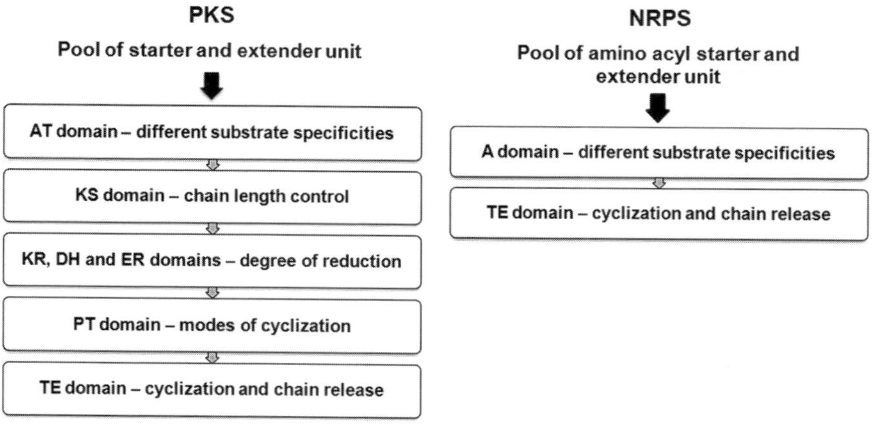

Figure 2: Schematic diagram depicting role of different catalytic domains in generation of metabolic diversity by controlling, selection of starter/extender substrates (AT of PKS and A of NRPS), chain length by number of condensation steps (KS of iterative PKS), regioselectivity of ketoreduction (KR in PKS) and different modes of cyclization (PT & TE in PKS and TE in NRPS).

2. Structure based *in silico* approaches for understanding structure-function relationship of PKS and NRPS megasynthases

Significant progress has been made in the field of structural biology of PKS and NRPS megasynthases in the last decade. Currently structures for different domains like KS, AT, KR, DH, ARO/CYC, ACP, PT, TE of PKSs and C, A, PCP and TE of NRPSs (Figure 3). These structures have been extensively discussed in several recent reviews[10-12]. Several structures for multidomain fragments like KS-AT, PCP-C and PCP-TE or complete termination module of the surfactin NRPS are also available. We describe here, how structures of some

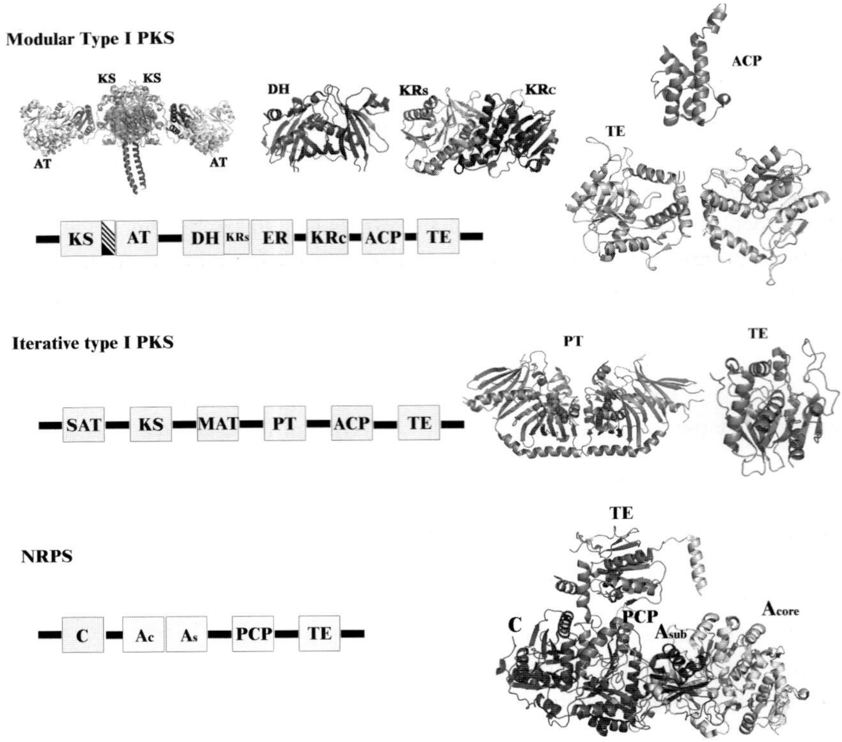

Figure 3: Currently available three dimensional structures for different catalytic domains or multidomain fragments of modular PKS, type I iterative PKS and NRPS. In case of NRPS, structure of the complete termination module is available, while for modular PKS structures largest structure corresponds to the KS-AT didomain along with the intervening linker region. The three dimensional structures correspond to the PDB entries 2HG4 (KS-AT), 3EL6 (DH), 2FR0 (KR), 2JU2 (ACP), 1MO2 (TE - modular), 3HRQ (PT), 3ILS (TE - iterative), 2VSQ (NRPS).

of these key catalytic domains of PKS and NRPS have revealed substrate recognition mechanism during biosynthesis of polyketides and nonribosomal peptides and helped in formulation of rules for deciphering biosynthetic products of PKS/NRPS gene clusters.

2.1. *Recognition of starter and extender units by A and AT domains*

Because of their gatekeeper role, AT and A domain are major determinants of specificity for starter and extender substrates during biosynthesis of polyketides and non-ribosomal peptides respectively. Availability of Phe bound crystal structure of A domain of gramicidin S synthase[13] revealed the specific amino acids of A domain are involved in substrate recognition. Challis *et al.* and Stachelhaus *et al.* successfully utilized this structural information to decipher the so called "specificity conferring code" of A domain by aligning sequences of different adenylation domains of known specificity to the gramicidin S crystal structure[13-15]. Based on this structural bioinformatics approach for identifying specificity determining residues (SDR), several automated tool for prediction of A domain specificity have been developed by different groups during the last decade. NRPS-PKS[16] predicts A domain specificity based on the simple comparison of putative substrate binding residue in the query sequence with SDRs of A domains derived from a training set of 152 A domains from 22 NRPS gene clusters. NRPSpredictor2[17, 18] uses Transductive Support Vector Machines (TSVM) approach for predicting substrate specificities of A domains using active site residues. In contrast to these active site residue based methods, NRPSsp[19] uses Hidden Markov Model (HMM) profiles based on entire sequence of A domains to predict their substrates. Table 1 shows results of a benchmarking study carried out by us to evaluate the accuracies of these A domain specificity prediction tools on a data set of 156 bacterial and 20 fungal A domains. As can be seen, structure based methods like NRPS-PKS and NRPSpredictor2 show better performance compared to sequence based methods like NRPSsp. Even though substrate specificities of bacterial A domains can be predicted with an accuracy above 60% by both NRPS-PKS and NRPSpredictor2, both the tools perform poorly for fungal A domains. This is not surprising because fungal A domains are poorly represented in the training set of A domains used by NRPS-PKS and NRPSsp and even though

Table 1: Results from benchmarking of different softwares for prediction of substrate specificity of adenylation domains on a dataset of 156 bacterial A domains and 20 fungal A domains with known substrates. The numbers indicated are percentage correct predictions, when the predicted amino acid is exactly the same as the known substrate (Exact) or when they are chemically similar (Similar).

	Total no. of A domains in test set	Percentage of correct predictions					
		NRPS-PKS		NRPS Predictor2		NRPSsp	
		Exact	Similar	Exact	Similar	Exact	Similar
BACTERIAL	156	60	73	65	79	46	62
FUNGAL	20	15	20	25	25	15	20

NRPSpredictor2 has specialized SVMs for fungal A domains, the structural template used for mapping SDRs is a bacterial A domain.

In fact recently available structure for eukaryotic A domain[20] that activates large hydroxamate amino acid indicate the binding pocket is constituted by 17 amino acids. Therefore, use of this new structure by different structural bioinformatics tools is likely to help in better prediction of specificity of fungal A domains. In order to identify the SDRs of PKS AT domains, Yadav *et al.*[21] carried out a structure based binding pocket analysis using crystal structure of AT domain from E. Coli FAS as template and sequences of 187 PKS AT domains with known specificity. This study could successfully identify distinct binding pocket signatures for methylmalonate, malonate and other starter substrates. This approach has been implemented in NRPS-PKS software and prediction results have been validated by experimental studies[22]. Similar approaches have been used in other softwares like ASMPKS[23], Clustscan[24], NP.searcher[25] to predict AT specificity.

2.2. *Substrate specificity of KS and CHS domains*

The KS domains of type I and type II PKS as well as FAS enzymes adopt a thiolase fold and catalyze chain elongation by decarboxylative Claisen condensation of ACP bound substrates. Even though KS domain sequences from modular PKSs do not show phylogenetic clustering as per their substrates[8], a very high degree of correlation with substrate specificity is seen in the phylogenetic clustering of KS domains from trans-AT PKSs as well as fungal iterative PKSs. Biochemical studies and mutagenesis experiments[26] guided by crystal structures of KS domains from type II PKS and FAS suggested that, size and shape of substrate-binding channel holds the code for pathway specific chain length control. The crystal structure of KS-CLF domain[27, 28] revealed distinct cavity for accommodating the cognate substrate. The acyl binding pocket has been shown to be narrow and more specific in FAS KS and type II PKS, while DEBS KS3 and KS5 have wider pockets consistent with their reported substrate tolerance[29-34]. Analysis of the substrate binding pockets in the structural models of iterative KS domains by Yadav *et al.* revealed that size and hydrophobicity of substrate-binding pocket shows distinct correlation with the number of iteration and degree of saturation of the resultant polyketide products[35]. Therefore, these studies have shown that active site residues of KS domains can give valuable clues for predicting number of iterations and hence the polyketide products of iterative PKSs. The chalcone synthase (CHS) superfamily of type III PKSs are homodimeric proteins which adopt thiolase fold similar to the KS domains. However, unlike KS domains which act on ACP bound substrates, these CHS like enzymes catalyze iterative decarboxylative Claisen-type condensation of CoA derivatives of starter and extender substrates to generate enzyme bound β-carboxy intermediates. They also catalyze chain release of final product and its cyclization to synthesize a variety of different chemical structures. The elucidation of the first crystal structure of plant CHS[36] revealed distinct compartments in the active site pocket of the enzyme for binding of starter and extender moieties and cyclized product. The structure of pks18[37, 38] from *Mycobacterium tuberculosis* revealed a novel substrate binding tunnel which facilitates utilization of long chain fatty acid starter substrates by bacterial type III PKSs[39]. These and other crystal structures of type III PKS elucidated in recent years have helped in understanding how type III PKSs utilize different starter/extender substrates, control chain length and cyclization by subtle variation of the active site pocket residues[40]. Based on analysis of these structures, bioinformatics tools like NRPS-PKS facilitate identification of the specificity determining residues in type III PKSs and their putative substrates/products.

2.3. *Stereo specificity of Ketoreductase (KR) domain*

KR domains catalyze stereospecific reduction of diketide group. Crystal structures for KR domains from type I (EryKR1 and TylKR1) and type II PKS have been reported[41-46] and both have been seen to adopt short-chain dehydrogenase/reductase fold. Type I KR contains two tightly packed subdomains, one being catalytic and the other being truncated non-catalytic subdomain. Even though catalytic and structural subdomains of type I KR have similar fold, they lack distinct similarity in sequence and in PKS modules containing ER domains, structural and catalytic subdomains of KR are separated in sequence by the intervening ER domain (Figure 2). Structure based phylogenetic analysis has revealed that, type I KR domains could be divided into 6 categories on the basis of α and β substituent chirality[42, 47, 48] and some of these predictions have been verified experimentally[49, 50]. Based on these studies, rules for prediction of type I KR substrate specificity have been incorporated in software like Clustscan[24] and SBSPKS[51]. In contrast to type I KRs, type II KRs contains a single domain. What differentiates the substrate specificities for type I and II KRs is a long insertion between helices 6 and 7, which forms the substrate binding cleft. The substrate specificity of type II KR is characterized by a blend of factors like enzyme conformation, interaction between substrate-binding and catalytic residue at molecular levels and dynamics of the binding cleft[45].

2.4. *Regioselectivity of ring cyclization by Product Template (PT) domain*

PT domains could not be identified by conventional domain identification tools because of their uniqueness to NRPKSs. The polypeptide stretch intervening the AT and ACP domain of NRPKS was considered a non-catalytic linker sequence. The development of the novel Udwary-Merski algorithm (UMA), that differentiates domains from linkers by using secondary structure, local hydrophobicity and family homology, helped in identification of this new domain in PksA, an iterative type I PKS[52]. Pioneering biochemical experiments[53] established that these PT domains indeed catalyze cyclization of highly reactive polyketide backbone produced by NRPKS. The availability of the crystal structure of PT domain[54] in complex with substrate palmitate and a product analog revealed mechanistic details of the enzymatic reaction catalyzed

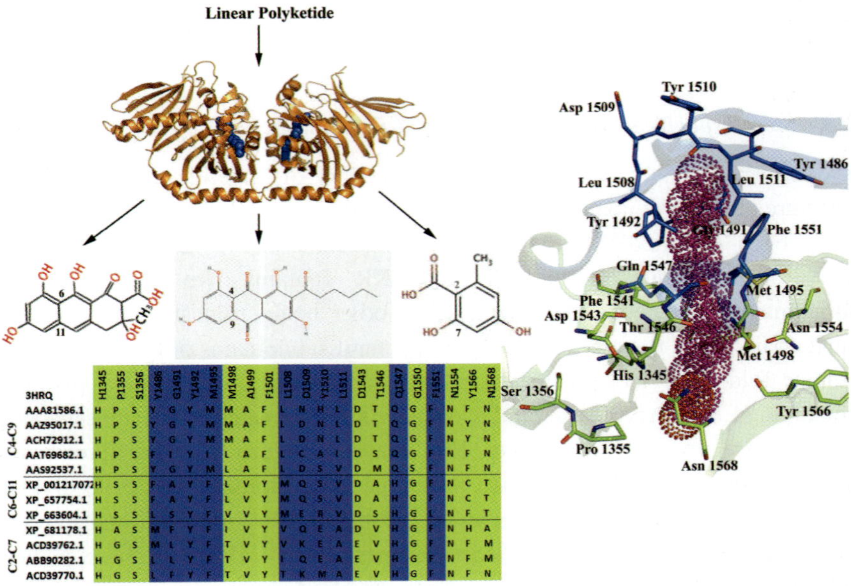

Figure 4: The structural fold of PT domain (PDB ID: 3HRQ) and different types of cyclization reactions (either 6-11 or 4-9 or 2-7 modes) it can catalyze to convert a non reduced linear polyketide into aromatic compounds (top left panel). Depiction of the hexyl binding pocket (blue sticks) and cyclization pocket (green sticks) in the three dimensional structure of PksA PT domain (right panel). List of residues lining the hexyl binding (blue) and cyclization (green) pockets in PT domains with known cyclization patterns (bottom left).

by this domain. The structure showed a modified 'double hot dog' fold and harbored a binding site consisting of phosphopantathiene (PPT) binding region, cyclization chamber and hexyl-binding region. While the cyclization chamber contains the catalytic dyad (Asp 1543/His 1345) to initiate region-specific cyclization and accomodate the cyclized product, the hexyl binding pocket set deep in the core of the protein accepts the hexanoyl starter substrate. Prior to cyclization, the polyketide substrate binds essentially in a linear manner with a slight kink towards the cyclization chamber. As can be seen from Figure 4, alignment of the sequences of various PT domains with distinct first ring cyclization patterns like C4-C9, C2-C7 and C6-C11, with the structure of PksA PT domain reveals interesting correlation between active site sequence profile and cyclization pattern[54, 55]. The residue lining the cyclization chamber show maximum variations to accommodate different substrates and different cyclization modes[56, 57]. Interestingly in case of NRPKS which do not utilize hexanoyl starter units, Gly 1491 at the base of the hexyl-binding region changes to a bulky residue closing the pocket. Li *et al.*[57] have recently proposed a sequence based methods for classification and prediction of regioselectivity of PT domains and validated the predictions by experimental studies. Therefore, combining binding pocket information from crystal structure of PT domains with phylogenetic analysis of experimentally characterized PT domains could indeed lead to development of automated tools for prediction of cyclization modes of uncharacterized NRPKS PT domains.

2.5. *Chain release & cyclization by Thioesterase (TE) domain*

Thioesterase (TE) domains catalyze the release of the final polyketide or nonribosomal peptide product. TE domains are not only capable of simple hydrolysis, but also macrolactamization (C-N), macrolactonization (C-O) and Claisen cyclization (C-C)[55]. Structures of TE domains from type I and II PKSs have been solved. All of them have classic α/β hydrolase fold. In spite of conserved fold and catalytic triad TE domains exhibit distinct substrate specificity and regioselectivity for different types of cyclization. As the substrate binding pocket consists of α/β hydrolase core on one side and lid on the other side, the variability in this lid region governs the complementarities between substrate and the binding pocket. Modular PKS TEs from erythromycin and pikromycin synthase catalyzing macrolactonization contain a long, open amphipathic channel that runs across the protein[58] and substrate-enzyme interaction is predominantly hydrophobic. In case of NRPS TEs, a flexible lid region is also shown to be covering the hydrophobic binding pocket in the closed form. However, in contrast to the open tunnel of TEs from modular PKS, NRPS TEs have a substrate binding channel that is closed at both ends and opens from the side[59, 60]. The substrate binding pocket of iterative type I PKS TEs (eg: PksA TE) resembles NRPS TEs[61]. Recently elucidated structure of TE domain from curacin A cluster defines a new family of TEs, with very low sequence similarity to other PKS TEs but with a conserved α/β hydrolase fold. In a departure from the mechanism that PKS and NRPS TEs employ to catalyze chain release, curacin A TE first transfers a sulfonate group onto the β-hydroxyl, which is then followed by TE mediated hydrolysis, decaroboxylation and sulfate elimination to form a terminal alkene group[62]. This structure depicts an active site intact for hydrolysis and a dedicated Arg residue to recognize β-sulfate. Additional thioesterase genes, called type II TEs are also found in several type I PKS and NRPS gene clusters as free-standing enzymes[63, 64]. These TEs function as editing enzymes with the capability of removing stalled reaction products. Structural comparison of trans- and cis-acting TEs[65] reveals that, though the mouth of the substrate binding pocket of type II TEs is wider, the active site cavity is much shallower. This makes type II TEs specific for short acyl groups. Even though the crystal structures of various PKS and NRPS TE domains have helped in understanding mechanism of substrate recognition by TE domains, identification of specific residues which govern regioselectivity of macrocyclization has been a difficult task because of the conformational flexibility of the binding pocket arising from movement of the lid region.

3. Summary and future prospects

Structure based *in silico* analysis of PKS and NRPS megasynthases from known secondary metabolite biosynthetic pathways has helped in understanding the mechanistic details of how these megasynthases control metabolic diversity. However, only in case of A and AT domains it has been possible to formulate automated predictive rules for correlating substrate specificities of these catalytic domains to their active site residue profiles. It is also necessary to benchmark the predictive ability of these tools on appropriate test sets. As discussed earlier our benchmarking analysis suggests that alternate structural template should be used for prediction of substrate specificity of fungal A domains. Even though automated computational tools are available to predict the starter/extender specificity of CHS domains and regioselectivity of KR domains, their performance has not been systematically benchmarked. Based on catalytic residue profiles cyclization modes of non-reducing polyketides by PT domain can be predicted. However, it will be necessary to develop automated computational tools for prediction of PT calayzed cyclization modes and validate the results on larger datasets. In contrast to PT domains, the types of macrocyclizations catalyzed by TE domains are much more complex. Secondly no structures are available for TE domains in complex with the substrate, making correct identification of binding modes of their substrates a challenging task. Since the flexible lid region is also believed to be a part of the binding pockets of TE domains, it will be necessary to incorporate ligand and receptor flexibilities while evaluating complementarities between TE domains and their substrates. Recent docking and MD simulations on enzyme-substrate complexes involving DH and KR domains of modular PKS has addressed such questions[66]. It is envisaged that, in future availability of more number of crystal structures and improvement in computational methods will further enhance predictive ability of tools for deciphering chemical structures of secondary metabolites and it might be possible to start CASP[67] type blind prediction contests for evaluating their true potential as proposed by Wong *et al.*[4].

Acknowledgements

Authors thank Director, NII for encouragement and support. SK thanks DBT, India for award of BINC fellowship. The work has been supported by grants to NII from Department of Biotechnology (DBT), India. DM also acknowledges financial support from DBT, India under BTIS project and National Bioscience Career Development award.

References

1. G. L. Challis, *J Med Chem* **51**, 2618 (2008).
2. C. T. Walsh and M. A. Fischbach, *J Am Chem Soc* **132**, 2469 (2010).
3. M. A. Fischbach and C. T. Walsh, *Chem Rev* **106**, 3468 (2006).
4. F. T. Wong and C. Khosla, *Curr Opin Chem Biol* **16**, 117 (2012).
5. J. W. Li and J. C. Vederas, *Science* **325**, 161 (2009).
6. D. E. Cane, C. T. Walsh and C. Khosla, *Science* **282**, 63 (1998).
7. S. Anand and D. Mohanty, Handbook of research on computational and systems biology: Interdisciplinary applications 380 (2011).
8. H. Jenke-Kodama and E. Dittmann, *Nat Prod Rep* **26**, 874 (2009).
9. B. O. Bachmann and J. Ravel, *Methods Enzymol* **458**, 181 (2009).
10. S. C. Tsai and B. D. Ames, *Methods Enzymol* **459**, 17 (2009).
11. M. Strieker, A. Tanovic and M. A. Marahiel, *Curr Opin Struct Biol* **20**, 234 (2010).
12. H. L. Condurso and S. D. Bruner, *Curr Opin Chem Biol* **16**, 162 (2012).

13. E. Conti, T. Stachelhaus, M. A. Marahiel and P. Brick, *Embo J* **16**, 4174 (1997).
14. G. L. Challis, J. Ravel and C. A. Townsend, *Chem Biol* **7**, 211 (2000).
15. T. Stachelhaus, H. D. Mootz and M. A. Marahiel, *Chem Biol* **6**, 493 (1999).
16. M. Z. Ansari, G. Yadav, R. S. Gokhale and D. Mohanty, *Nucleic Acids Res* **32**, W405 (2004).
17. C. Rausch, T. Weber, O. Kohlbacher, W. Wohlleben and D. H. Huson, *Nucleic Acids Res* **33**, 5799 (2005).
18. M. Rottig, M. H. Medema, K. Blin, T. Weber, C. Rausch and O. Kohlbacher, *Nucleic Acids Res* **39**, W362 (2011).
19. C. Prieto, C. Garcia-Estrada, D. Lorenzana and J. F. Martin, *Bioinformatics* **28**, 426 (2011).
20. T. V. Lee, L. J. Johnson, R. D. Johnson, A. Koulman, G. A. Lane, J. S. Lott and V. L. Arcus, *J Biol Chem* **285**, 2415 (2010).
21. G. Yadav, R. S. Gokhale and D. Mohanty, *J Mol Biol* **328**, 335 (2003).
22. O. A. Trivedi, P. Arora, A. Vats, M. Z. Ansari, R. Tickoo, V. Sridharan, D. Mohanty and R. S. Gokhale, *Mol Cell* **17**, 631 (2005).
23. H. Tae, E. B. Kong and K. Park, *BMC Bioinformatics* **8**, 327 (2007).
24. A. Starcevic, J. Zucko, J. Simunkovic, P. F. Long, J. Cullum and D. Hranueli, *Nucleic Acids Res* **36**, 6882 (2008).
25. M. H. Li, P. M. Ung, J. Zajkowski, S. Garneau-Tsodikova and D. H. Sherman, *BMC Bioinformatics* **10**, 185 (2009).
26. Y. Tang, S. C. Tsai and C. Khosla, *J Am Chem Soc* **125**, 12708 (2003).
27. X. Zhu, F. Yu, X. C. Li and L. Du, *J Am Chem Soc* **129**, 36 (2007).
28. A. T. Keatinge-Clay, D. A. Maltby, K. F. Medzihradszky, C. Khosla and R. M. Stroud, *Nat Struct Mol Biol* **11**, 888 (2004).
29. T. Maier, M. Leibundgut and N. Ban, *Science* **321**, 1315 (2008).
30. J. G. Olsen, A. Kadziola, P. von Wettstein-Knowles, M. Siggaard-Andersen, Y. Lindquist and S. Larsen, *FEBS Lett* **460**, 46 (1999).
31. A. C. Price, C. O. Rock and S. W. White, *J Bacteriol* **185**, 4136 (2003).
32. X. Qiu, A. E. Choudhry, C. A. Janson, M. Grooms, R. A. Daines, J. T. Lonsdale and S. S. Khandekar, *Protein Sci* **14**, 2087 (2005).
33. Y. Tang, C. Y. Kim, Mathews, II, D. E. Cane and C. Khosla, *Proc Natl Acad Sci U S A* **103**, 11124 (2006).
34. Y. Tang, A. Y. Chen, C. Y. Kim, D. E. Cane and C. Khosla, *Chem Biol* **14**, 931 (2007).
35. G. Yadav, R. S. Gokhale and D. Mohanty, *PLoS Comput Biol* **5**, e1000351 (2009).
36. J. L. Ferrer, J. M. Jez, M. E. Bowman, R. A. Dixon and J. P. Noel, *Nat Struct Biol* **6**, 775 (1999).
37. R. Rukmini, V. M. Shanmugam, P. Saxena, R. S. Gokhale and R. Sankaranarayanan, *Acta Crystallogr D Biol Crystallogr* **60**, 749 (2004).

38. R. Sankaranarayanan, P. Saxena, U. B. Marathe, R. S. Gokhale, V. M. Shanmugam and R. Rukmini, *Nat Struct Mol Biol* **11**, 894 (2004).
39. P. Saxena, G. Yadav, D. Mohanty and R. S. Gokhale, *J Biol Chem* **278**, 44780 (2003).
40. I. Abe, *Curr Opin Chem Biol* **16**, 179 (2012).
41. A. T. Keatinge-Clay and R. M. Stroud, *Structure* **14**, 737 (2006).
42. A. T. Keatinge-Clay, *Chem Biol* **14**, 898 (2007).
43. P. Javidpour, T. P. Korman, G. Shakya and S. C. Tsai, *Biochemistry* **50**, 4638 (2011).
44. P. Javidpour, A. Das, C. Khosla and S. C. Tsai, *Biochemistry* **50**, 7426 (2011).
45. T. P. Korman, Y. H. Tan, J. Wong, R. Luo and S. C. Tsai, *Biochemistry* **47**, 1837 (2008).
46. A. T. Hadfield, C. Limpkin, W. Teartasin, T. J. Simpson, J. Crosby and M. P. Crump, *Structure* **12**, 1865 (2004).
47. P. Caffrey, *Chembiochem* **4**, 654 (2003).
48. R. Reid, M. Piagentini, E. Rodriguez, G. Ashley, N. Viswanathan, J. Carney, D. V. Santi, C. R. Hutchinson and R. McDaniel, *Biochemistry* **42**, 72 (2003).
49. A. Baerga-Ortiz, B. Popovic, A. P. Siskos, H. M. O'Hare, D. Spiteller, M. G. Williams, N. Campillo, J. B. Spencer and P. F. Leadlay, *Chem Biol* **13**, 277 (2006).
50. H. M. O'Hare, A. Baerga-Ortiz, B. Popovic, J. B. Spencer and P. F. Leadlay, *Chem Biol* **13**, 287 (2006).
51. S. Anand, M. V. Prasad, G. Yadav, N. Kumar, J. Shehara, M. Z. Ansari and D. Mohanty, *Nucleic Acids Res* **38**, W487 (2010).
52. J. M. Crawford, A. L. Vagstad, K. C. Ehrlich and C. A. Townsend, *Bioorg Chem* **36**, 16 (2008).
53. J. M. Crawford, A. L. Vagstad, K. P. Whitworth, K. C. Ehrlich and C. A. Townsend, *Chembiochem* **9**, 1019 (2008).
54. J. M. Crawford, T. P. Korman, J. W. Labonte, A. L. Vagstad, E. A. Hill, O. Kamari-Bidkorpeh, S. C. Tsai and C. A. Townsend, *Nature* **461**, 1139 (2009).
55. J. M. Crawford and C. A. Townsend, *Nat Rev Microbiol* **8**, 879 (2010).
56. C. Hertweck, *Angew Chem Int Ed Engl* **48**, 4688 (2009).
57. Y. Li, Image, II, W. Xu, I. Image and Y. Tang, *J Biol Chem* **285**, 22764 (2010).
58. S. C. Tsai, H. Lu, D. E. Cane, C. Khosla and R. M. Stroud, *Biochemistry* **41**, 12598 (2002).
59. S. D. Bruner, T. Weber, R. M. Kohli, D. Schwarzer, M. A. Marahiel, C. T. Walsh and M. T. Stubbs, *Structure* **10**, 301 (2002).
60. S. A. Samel, B. Wagner, M. A. Marahiel and L. O. Essen, *J Mol Biol* **359**, 876 (2006).

61. T. P. Korman, J. M. Crawford, J. W. Labonte, A. G. Newman, J. Wong, C. A. Townsend and S. C. Tsai, *Proc Natl Acad Sci U S A* **107**, 6246 (2010).
62. J. J. Gehret, L. Gu, W. H. Gerwick, P. Wipf, D. H. Sherman and J. L. Smith, *J Biol Chem* **286**, 14445 (2011).
63. A. R. Butler, N. Bate and E. Cundliffe, *Chem Biol* **6**, 287 (1999).
64. Y. Doi-Katayama, Y. J. Yoon, C. Y. Choi, T. W. Yu, H. G. Floss and C. R. Hutchinson, *J Antibiot (Tokyo)* **53**, 484 (2000).
65. A. Koglin, F. Lohr, F. Bernhard, V. V. Rogov, D. P. Frueh, E. R. Strieter, M. R. Mofid, P. Guntert, G. Wagner, C. T. Walsh, M. A. Marahiel and V. Dotsch, *Nature* **454**, 907 (2008).
66. S. Anand and D. Mohanty, *BMC Struct Biol* **12**, 10 (2012).
67. J. Moult, J. T. Pedersen, R. Judson and K. Fidelis, *Proteins* **23**, ii (1995).

ACCELERATED MOLECULAR DYNAMICS: AN EFFICIENT ENHANCED SAMPLING METHOD FOR BIOMOLECULAR SIMULATIONS

LEVI C.T. PIERCE[1], WILLIAM SINKO[1] AND J. ANDREW McCAMMON[1,2,3]
[1]*Department of Pharmacology, University of California San Diego*
La Jolla, California 92093-0365
[2]*Department of Chemistry & Biochemistry, and NSF Center for Theoretical Biological Physics, University of California San Diego, La Jolla, California 92093-0365*
[3]*Howard Hughes Medical Institute, University of California San Diego*
La Jolla, CA 92093-0365

Currently, access to rare biomolecular events on long time scales is limited to hundreds of microseconds using conventional molecular dynamics (MD), but generally simulations are run for only a few hundred nanoseconds due to the computational resources needed for these simulations. In order to access relevant events occurring on long time sales, milliseconds and beyond, using conventional hardware, enhanced sampling methods are commonly used. Accelerated molecular dynamics (aMD) is increasingly used in biomolecular simulation because it efficiently allows access to events on long timescales without increasing the number of simulation steps required. Here we review the aMD method and discuss the achievements and limitations of this enhanced sampling simulation technique.

1. Introduction

Why do we need enhanced sampling methods to explore the conformational space of a protein? Current, standard hardware and software allow a scientist to conduct molecular dynamics simulations on the order of hundreds of nanoseconds in a reasonable amount of time. However, many biological processes of interest extend far beyond hundreds of nanoseconds into the microsecond and millisecond range[1-6]. Therefore, we either need faster hardware or more efficient software in order to achieve improved conformational space sampling.

Since 2007, the computational chemistry field has seen two significant hardware advances for simulating biomolecules. The first is Anton, a specialized super computer developed by D.E. Shaw Research whose only purpose is to run molecular dynamics simulations[7]. Not only has Anton pushed the limits of conventional molecular dynamics simulation beyond the millisecond mark, but it has also been a great tool for developing improved force fields, which are essential for any simulation[8,9]. Anton has made a significant advancement in what timescales are possible in MD simulation, but it is currently not widely available for computational chemists. The second advancement has been harnessing the power of graphics processor units (GPUs) for computational simulations of biomolecules[10-12]. Using a single graphics processor unit, simulations on the microsecond timescale can now be accessed by any research group that has a spare desktop computer in the lab. GPU computing is certainly much more widely available than Anton supercomputers, and provides a more reasonably priced alternative to more traditional central processing unit (CPU) clusters.

While hardware developments will continue to increase the speed of MD simulations, we now focus on how we can achieve the enhanced sampling from our simulations using methodology changes. The methods can be broken down into two basic groups, 1) those that require some *a priori* knowledge of the phase space to be explored and 2) those that require no preexisting knowledge. Some methods that fall into the first class are metadynamics[13,14], conformational flooding[15, 16], adaptive biasing force[17-19], and umbrella sampling. Examples of those that fall into the second class are replica exchange, hyperdynamics[20,21], parallel replica dynamics[22], temperature accelerated dynamics[23] and accelerated molecular dynamics (aMD)[24]. This review focuses on aMD which was inspired by Voter's early work [20,21]. aMD applies a bias to the underlying potential energy landscape, promoting escape from energetic traps, allowing the biomolecule to transition from one state to another quickly and efficiently. By varying the amount of bias applied, different levels of acceleration can be achieved, effectively varying the enhancement in conformational sampling[25]. Since the aMD simulation is non-canonical due to the flattened energy surface a weighting factor, based on the energetic modification, must be applied to each frame to recover the canonical ensemble. However, recovery of accurate statistics is an increasing challenge as the acceleration is increased[26].

2. Theory

In its original form, the aMD method modifies the potential energy landscape by raising energy minima that lie below a defined threshold level, while leaving those areas lying above the threshold unmodified. As a result, barriers separating adjacent energy basins are reduced, providing the simulation with more rapid access to conformational space that cannot be easily accessed in a cMD simulation. Historically, aMD was first implemented by Hamelberg *et al.* within the framework of the sander module in the AMBER 7 package and was used to study several small peptide and protein systems[24,27]. The method modifies the energy landscape by adding a boost potential, $\Delta V(r)$, to the original potential energy surface when $V(r)$ is below a pre-defined energy level E, as shown in equation 1.

$$\Delta V(r) = \begin{cases} 0, & V(r) \geq E \\ \frac{(E-V(r))^2}{\alpha + (E-V(r))} & V(r) < E \end{cases} \tag{1}$$

The parameter α modulates the depth and the local roughness of the energy basins on the modified potential (Figure 1).

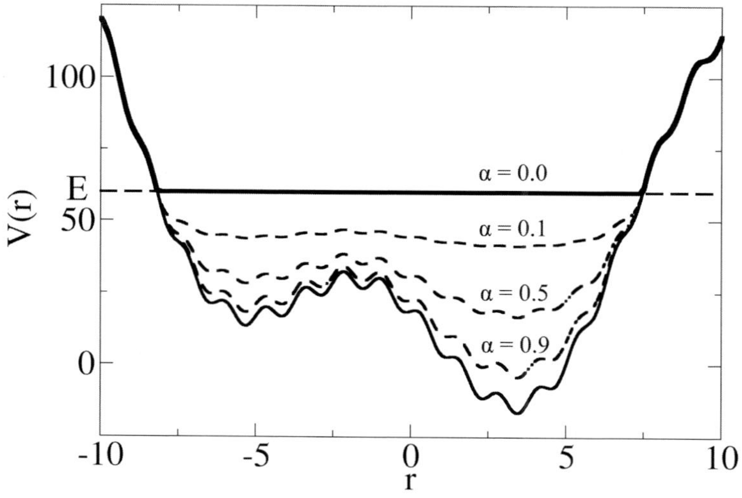

Figure 1: A 2D representation of the potential energy surface (solid line) modified and flattened by aMD using different values of α (dashed lines), E is represented by a horizontal dashed line. As α is decreased the modified energy surface flattens to a greater extent.

These equations may be applied to any potential energy term, but originally were applied only to torsional terms, which play a critical role in conformational determination in classical mechanical simulations. In principle, this approach also allows the correct canonical average of an observable, calculated from configurations sampled on the modified potential energy surface, to be fully recovered from the aMD simulations. In order to simultaneously enhance the sampling of internal and diffusive degrees of freedom a dual boosting approach was employed, based on separate torsional and total potential energies as shown in equation 2 and 3.[28]

$$V(r) = V_o(r) + V_t(r) \tag{2}$$

$$V^*(r) = \{V_o(r) + [V_t(r) + \Delta V_t(r)]\} + \Delta V_T(r) \tag{3}$$

Here $V_o(r)$ is the original potential, $V_t(r)$ is the total potential of the torsional terms, $\Delta V_t(r)$ and $\Delta V_T(r)$ are the boost potentials applied to the torsional terms $V_t(r)$ and the total potential energy $V_T(r)$, respectively.

The parameterization of aMD simulations is an active area of research, however significant conformational changes have been observed using the following formulas: The torsional boost parameter, E(dih), is set equal to the average dihedral energy, V_o(dih), obtained from a short cMD simulation plus 4 kcal/mol times the number of solute residues and the α(dih) parameter is then set to 1/5 of this value[28-30]. The following was used in an accelerated simulation of BPTI[29]:

E(dih) = 595 kcal mol^{-1} + (4kcal mol^{-1} residue^{-1} *58 solute residues)
= 827 kcal mol^{-1}

α (dih) = (1/5)*(4kcal mol^{-1} residues^{-1} * 58 solute residues) = 46.4 kcal mol^{-1}

For the total boost parameter, E(tot), the value should be equal to 0.16 * total number of atoms plus the average total potential energy, V_o(tot), obtained from a short cMD simulation and the α(tot) should simply be 0.16 * total number of atoms. The following was used in an accelerated simulation of BPTI[29]:

E(tot) = -47128 kcal mol^{-1} + (0.16kcal*mol^{-1} atom^{-1} * 18,226 atoms)
= -44212 kcal mol^{-1}

α (tot) = (0.16kcal mol^{-1} atom^{-1} * 18,226 atoms) = 2916 kcal mol^{-1}

de Oliveira *et al.* demonstrated that the mobility of water could be increased 35% while still recovering liquid water structural ensemble averages that matched those of a cMD simulation[31]. Selecting appropriate boost parameters for complex biomolecules was determined primarily by two publications exploring dynamics of Ras and Proline Racemase[28,30] In practice, the length of the cMD simulation used for parameterization can be ~10-50ns. These parameter guidelines have been found to work well for both small systems and those as large as ~200,000 atoms.

One of the current limitations of aMD is achieving accurate free energy estimates of the conformational space explored. This is due to large fluctuations in the values of the bias, $\Delta V(r)$, occurring within the minima along the potential energy surface. Frames with the largest values of $\Delta V(r)$ may dominate the final statistics after the reweighting procedure is carried out using the exponential factor $\exp(\beta\Delta V(r))$. Commonly, the problem is compounded further by the fact that the great majority of frames are not saved from the simulation since structural information is data intensive, so the sample size is smaller. High energy states are much more prevalent in aMD simulations than cMD simulations due to a greater number high energy intermediate states crossed, as well as the increase in fluctuation of isolated intermolecular interactions such as bond vibrations and rotations. The structures collected with high energy values will have a small value of $\Delta V(r)$ and will contribute very little to the statistics when reweighted. Reweighting issues tend to grow with the complexity of the accelerated system. Shen and Hamelberg have examined different approaches for dealing with the reweighting problem, such as approximating the exponential function, and we refer the reader to their paper for a more detailed discussion of the reweighting problem[26]. While a method for directly computing kinetics from a single aMD simulation has not been determined, Hamelberg *et al.* has demonstrated that, using several aMD simulations at various acceleration levels, one can compute kinetic rates for reactions[32].

3. Implementations

There are currently two molecular dynamics packages that officially support accelerated molecular dynamics (aMD), NAMD[33] and AMBER12[33,34]. We also note that a highly efficient GPU implementation has been released with AMBER12 in which the authors demonstrate access to events on the millisecond timescale. There have also been numerous unreleased variations of the method including selective aMD[35], replica exchange aMD[36], inverted aMD[37], and windowed aMD[38].

4. Applications

4.1 Alanine Dipeptide

The quintessential system for demonstrating improved conformational space sampling is alanine dipeptide. This simple system is often used to test new sampling algorithms due to the fact that near convergence can be achieved in a reasonable amount of time using conventional molecular dynamics (cMD). Convergence can be monitored using a description of phase space mapped out by the phi and psi angles, or a Ramachandran plot[39]. Experimental data has characterized several regions on the Ramachandran plot, which can be used to determine when convergence is ultimately achieved. Once an average dihedral energy from a short ~10ns cMD simulation has been determined, aMD may be parameterized. In this case we accelerate the dihedrals of the system and promote exploration of phase space as shown in Figure 2.

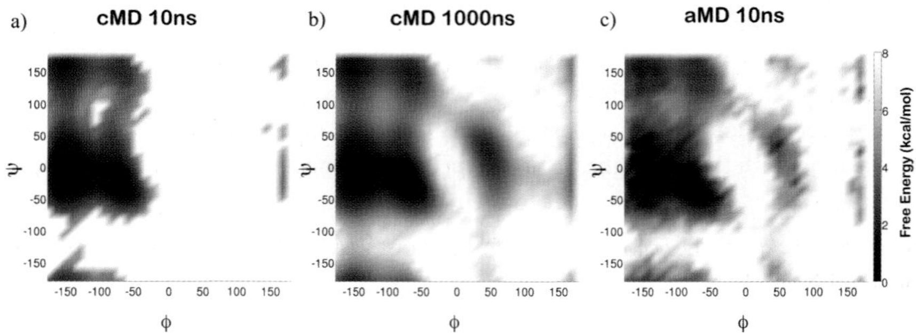

Figure 2: A Ramachandran plot of alanine dipeptide simulations in explicit water showing the increased sampling of a short aMD (10ns) simulation over a short cMD simulation (10ns). The short aMD simulation shows conformational space sampling, which is similar to that of a long cMD simulation (1000ns).

4.2 Ras and Proline Racemase

While alanine dipeptide is a great test system for many biomolecular simulation procedures the ultimate goal of aMD is to explore large conformational changes of complex biomolecules that cannot be seen with cMD simulations. One of the rate limiting factors of large conformational change is diffusional motion of protein domains. To overcome this limitation the acceleration is applied not only to the dihedrals but also to the total potential energy. Using the dual boost method effectively increases the diffusion rate of the water and promotes the occurrence of large diffusive motions[27]. Simulation of Ras using the dual boost method demonstrated that nucleotide-dependent conformational transitions could be observed using aMD where conventional molecular dynamics using the same computational effort failed to achieve the same conformational transition[28]. While the Ras protein contained ~190 residues the Proline Racemase system was more than twice the size of Ras containing ~414 residues. The application of aMD to this system produced a dramatic conformational change covering a span of more than 100Å in principal component space[30].

4.3 GET3

As mentioned earlier one of the limitations of aMD is determining an accurate description of the free energy surface explore by complex biomolecules. One of the common methods to describe the conformational space explored throughout the course of a biomolecular simulation is using principal component analysis (PCA)[28,30] or full correlation analysis (FCA)[40]. Wereszczynski and McCammon used two of the primary modes identified by FCA from an initial aMD simulation to describe a set of collective variables for a more rigorous free energy analysis using adaptive biasing force[41]. Even with this large GET3 protein, aMD combined with FCA could efficiently describe a set of collective variables for the use of the adaptive biasing force method allowing the authors to determine free energies of the different conformational states associated with nucleotide binding.

4.4 BPTI

Large conformational changes within biomolecules are certainly associated with long timescales inaccessible to conventional molecular dynamics yet there are subtle conformational changes, often associated with specific dihedral angles within biomolecules, which are also inaccessible using cMD. One such example is the rotation about the Chi-38 Chi-14 dihedral angle in bovine pancreatic trypsin inhibitor (BPTI), which is known experimentally to occur on the microsecond to millisecond timescale[42]. This conformational change was observed using brute force cMD by the D.E. Shaw Research group from a 1.031ms simulation performed using their Anton super computer. Recently, Pierce *et al.* also observed these millisecond timescale events using a highly

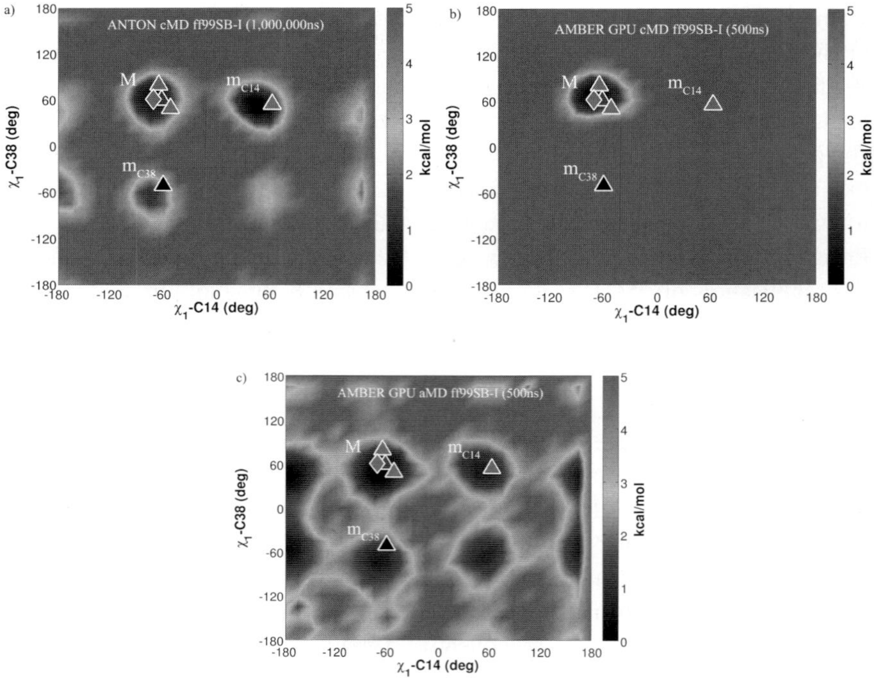

Figure 3: The χ1-C14 vs. χ1-C38 dihedral angle free energy surfaces of (a) 1ms cMD simulation (b) 500ns cMD ff99SB-I simulation and (c) 500ns aMD ff99SB-I simulation. The long-lived stable states are plotted onto the free energy surface and are labeled as triangles. The crystal structure, 5PTI, is demarked by the diamond. The major state M is labeled along with the minor states mC14 and mC38.

efficient aMD implementation running on a single graphics card, as shown in (Figure 3)[29]. The authors also demonstrate how aMD can be used to compute NMR chemical shift differences, which can be directly correlated to an effective timescale.

4.5 REXAMD

Using aMD for exploration of long timescale conformational changes of biomolecules has proven successful, but the problem of generating statistics for the canonical ensemble is still an active area of research. One approach to address the statistical problem of sampling in aMD is Replica-Exchange Accelerated Molecular Dynamics (REXAMD)[36]. In this method several replicas at increasing levels of acceleration are simulated and a Monte Carlo based exchange is used to swap between replicas. The idea being that the replica at the bottom is not accelerated and needs no re-weighting to recover the canonical distribution. The authors demonstrate the effectiveness of this method by computing free energy in the framework of thermodynamic integration on a small system. This is certainly a promising approach and it will be interesting to see how well this works for larger systems where the sampling problem is more prevalent. Additionally, we remind the readers that any multi-replica approach requires more computing resources than a single simulation.

4.6 Selective aMD

Another approach that has been developed to deal with the statistical sampling problem is selective aMD[35]. This method only accelerates selected torsions within the protein thereby reducing the overall statistical noise associated with bias for each frame in the simulation. The idea of selectively accelerating torsions was also demonstrated by only accelerating the torsions of the substrate within the binding pocket of pin1[43]. In this case the reweighting procedure was not hampered as much by the re-weighting issues.

4.7 Inverted aMD and Windowed aMD

Others have attempted to invert the aMD equations so that the low energy wells would be sampled on an unmodified potential surface, and only high energy transition states would be accelerated[37]. This alleviates the reweighting problem in simple systems but may cause exploration of extremely high energy states in more complex systems. A balance of increased sampling and the recovery of statistics has been a challenge in complex systems. For this reason windowed aMD has been proposed which accelerates the simulation between a window of high and low energy values to protect the exploration of very high energy barriers. This method was shown to be effective at describing the free energy surfaces of peptides and calculating the free energy difference between vancomycin and its bacterial glycopeptide ligands[38].

5. **Conclusion**

Achieving sufficient sampling of conformational space represents an ongoing challenge for most biologically relevant systems studied using conventional molecular dynamics simulations. The aMD method approaches this problem by modifying the system's potential energy landscape and reducing barriers separating different states. Here we have discussed aMD theory, major challenges for aMD, biomolecular simulations that employed the aMD method, and outlined several variants of aMD. While conformational states occurring on long timescales can be accessed with aMD, acquiring accurate statistics for the relevant conformational and energetic states is still a challenge.

Acknowledgments

This work was supported in part by NSF, NIH, HHMI, CTBP, NBCR, and the NSF Supercomputer Centers.

References

1. J. Kubelka, T.K. Chiu, D.R. Davies, W.A. Eaton and J. Hofrichter, *J. Mol. Biol.* **359**, 546 (2006).
2. R.D. Schaeffer, A. Fersht and V. Daggett, *Curr. Opin. Struct. Biol.* **18**, 4 (2008).
3. P.L. Freddolino and K. Schulten, *Biophys. J.* **97**, 2338 (2009).
4. M.K. Gilson and H.X. Zhou, *Annu. Rev. Biophys. Biomol. Struct.* **36**, 21 (2007).
5. E. Lindahl and M.S. Sansom, *Curr. Opin. Struct. Biol.* **18**, 425 (2008).
6. F. Khalili-Araghi, J. Gumbart, P.C. Wen, M. Sotomayor, E. Tajkhorshid and K. Schulten, *Curr. Opin. Struct. Biol.* **19**, 128 (2009).
7. D.E. Shaw, M.M. Deneroff, R.O. Dror, J.S. Kuskin, R.H. Larson and J.K. Salmon, et al. Anton, a special-purpose machine for molecular dynamics simulation. Communications of the ACM 2008:91.
8. D.E. Shaw, P. Maragakis, K. Lindorff-Larsen, S. Piana, R.O. Dror and M.P. Eastwood, et al., *Science* **330**, 341 (2010).
9. K. Lindorff-Larsen, S. Piana, K. Palmo, P. Maragakis, J.L. Klepeis and R.O. Dror, et al., *Proteins* **78**, 1950 (2010).
10. M.J. Harvey, G. Giupponi and G.D. Fabritiis, *J. Chem. Theory Comput.* **5**, 1632 (2009).
11. M.S. Friedrichs, P. Eastman, V. Vaidyanathan, M. Houston, S. Legrand and A.L. Beberg, et al., *J. Comput. Chem.* **30**, 864 (2009).
12. A.W. Gotz, M.J. Williamson, D. Xu, D. Poole, S. Le Grand and R.C. Walker, *J. Chem. Theory Comput.* **8**, 1542 (2012).
13. G. Bussi, A. Laio and M. Parrinello, *Phy. Rev. Lett.* **96**, 090601 (2006).
14. V. Leone, F. Marinelli, P. Carloni, M. Parrinello, *Curr. Opin. Struct. Biol.* **20**, 148 (2010).
15. H. Grubmüller, *Phys. Rev. E.* **52**, 2893 (1995).

16. O.F. Lange, L.V. Schäfer and H. Grubmüller, *J. Comput. Chem.* **27**, 1693 (2006).
17. E. Darve and A. Pohorille, *J. Chem. Phys.* **115**, 9169 (2001).
18. E. Darve, D. Rodriguez-Gomez and A. Pohorille, *J. Chem. Phys.* **128**, 144120 (2008).
19. J. Henin, G. Fiorin, C. Chipota and M.L. Klein, *J. Chem. Theory Comput.* **6**, 35 (2009).
20. A.F. Voter, *Phy. Rev. Lett.* **78**, 3908 (1997).
21. A.F. Voter, *J. Chem. Phys.* **106**, 4665 (1997).
22. A.F. Voter, *Phys. Rev. B.* **57**, R13985 (1998).
23. M.R. Sorensen and A.F. Voter, *J. Chem. Phys.* **112**, 9599 (2000).
24. D. Hamelberg, J. Mongan and J.A. McCammon, *J. Chem. Phys.* **120**, 11919 (2004).
25. U. Doshi and D. Hamelberg, *J. Chem. Theory Comput.* **7**, 575 (2011).
26. T. Shen and D. Hamelberg, *J. Chem. Phys.* **129**, 034103 (2008).
27. D. Hamelberg, C.A. de Oliveira and J.A. McCammon, *J. Chem. Phys.* **127**, 155102 (2007).
28. B.J. Grant, A.A. Gorfe and J.A. McCammon, *PLoS Comput. Biol.* **5**, e1000325 (2009).
29. L.C.T. Pierce, R. Salomon-Ferrer, C. Augusto F. de Oliveira, J.A. McCammon and R.C. Walker, *J. Chem. Theory Comput.* (2012).
30. C.A.F. de Oliveira, B.J. Grant, M. Zhou and J.A. McCammon, *PLoS Comput. Biol.* **7**, e1002178 (2011).
31. C.A.F. de Oliveira, D. Hamelberg and J.A. McCammon, *The Journal of Physical Chemistry B* **110**, 22695 (2006).
32. D. Hamelberg, T. Shen and J. Andrew McCammon, *J. Chem. Phys.* **122**, 241103 (2005).
33. Y. Wang, C.B. Harrison, K. Schulten and J.A. McCammon, *Comput. Sci. Discov.* **4**, (2011).
34. T.A.D. D.A. Case, T.E. Cheatham, III, C.L. Simmerling, J. Wang, R.E. Duke, R. Luo, R.C. Walker, W. Zhang, K.M. Merz, B. Roberts, S. Hayik, A. Roitberg, G. Seabra, J. Swails, A.W. Goetz, I. Kolossvai, K.F. Wong, F. Paesani, J. Vanicek, R.M. Wolf, J. Liu, X. Wu, S.R. Brozell, T. Steinbrecher, H. Gohlke, Q. Cai, X. Ye, J. Wang, M.-J. Hsieh, G. Cui, D.R. Roe, D.H. Mathews, M.G. Seetin, R. Salomon-Ferrer, C. Sagui, V. Babin, T. Luchko, S. Gusarov, A. Kovalenko, and P.A. Kollman., *University of California, San Francisco* (2012).
35. J. Wereszczynski and J.A. McCammon, *J. Chem. Theory Comput.* **6**, 3285 (2010).
36. M. Fajer, D. Hamelberg and J.A. McCammon, *J. Chem. Theory Comput.* **4**, 1565 (2008).
37. C.A.F. de Oliveira, D. Hamelberg and J.A. McCammon, *J. Chem. Theory Comput.* **4**, 1516 (2008).

38. W. Sinko, C.A.F. de Oliveira, L.C.T. Pierce and J.A. McCammon, *J. Chem. Theory Comput.* **8**, 17 (2012).
39. G.N. Ramachandran, C. Ramakrishnan and V. Sasisekharan, *J. Mol. Biol.* **7**, 95 (1963).
40. O.F. Lange and H. Grubmuller, *Proteins* **70**, 1294 (2008).
41. J. Wereszczynski and J.A. McCammon, *Proc. Natl. Acad. Sci. U. S. A.* (2012).
42. M.J. Grey, C. Wang and A.G. Palmer, 3rd, *J. Am. Chem. Soc.* **125**, 14324 (2003).
43. H.A. Velazquez and D. Hamelberg, *Biochemistry* **50**, 9605 (2011).

STRUCTURAL DYNAMICS OF *E.coli* PORPHOBILINOGEN DEAMINASE DURING THE TETRAPOLYMERISATION OF PORPHOBILINOGEN

MEENAKSHI PRADHAN*, HARINI SRINIVASAN*, NAVNEET BUNG* AND GOPALAKRISHNAN BULUSU

TCS Innovation Labs-Hyderabad (Life Sciences Division), Tata Consultancy Services Limited, Hyderabad 500081, India

E-mail: g.bulusu@tcs.com

Porphobilinogen deaminase catalyses the formation of 1-hydroxymethylbilane through a stepwise polymerisation of four units of porphobilinogen using a unique dipyrromethane cofactor as a primer. Structural and biochemical studies have suggested residues with catalytic importance, but their specific role in the mechanism and the dynamic behaviour of the protein remains unknown. Dynamics of the protein through three stages of chain elongation were studied using MD simulations to understand the concomitant structural changes. Observations suggest that domain 1 and domain 2 move apart while the cofactor turn region (240-243) moves into the active site and is inclined towards domain 2, thus creating space for the pyrrole moieties added at each stage. Results also suggest possible role of D50, K55 and R149 in active site loop modulation.

1. Introduction

Porphobilinogen deaminase (E.C 2.5.1.61, transferase), also known as 1-hydroxymethylbilane synthase or preuroporphyrinogen synthase, is present in most organisms and is the third enzyme in the heme biosynthesis pathway. It catalyses the stepwise polymerisation of four units of porphobilinogen (PBG) into a linear tetrapyrrole, preuroporphyrinogen[1] (Fig. 1).

Porphobilinogen deaminase (PBGD) has a unique cofactor, dipyrromethane (DPM), covalently attached to a cysteine through a thioether bond. DPM acts as a primer during the polymerisation reaction and remains attached to the protein permanently and irreversibly[1]. Assembly of the cofactor, and subsequent addition of four PBG molecules takes place through a sequence of steps: (1) deamination of the incoming PBG substrate, (2) nucleophilic attack by the terminal pyrrole ring of the enzyme bound cofactor on the substrate and (3) deprotonation[1] (Fig. 1).

Mutations in human PBGD (hPBGD) cause acute intermittent porphyria, with elevated levels of the heme precursors ALA (5-aminolevulinic acid) and

---

*Authors with equal contribution.

PBG in the urine[3]. PBGD is of specific importance in the *Plasmodium falciparum* heme biosynthesis pathway too. Studies have shown that the malarial parasite synthesises heme *de novo*, despite acquiring heme from host red cell hemoglobin in the intraerythrocytic stage[4]. Inhibition of this heme synthesis leads to death of the parasite. Moreover, *P. falciparum* PBGD (PfPBGD) is speculated to function as two enzymes – porphobilinogen deaminase and uroporphyrinogen III synthase, similar to that found in *Leptospira interrogans* PBGD (LiPBGD)[5]. Thus, PBGD can be a potential drug target. The mechanism of tetrapyrrole chain elongation in PBGD is still not clear. *Escherichia coli* PBGD (EcPBGD), being comparatively less complex than the human and *P. falciparum* system, can serve as a good model to enhance the understanding of the mechanism.

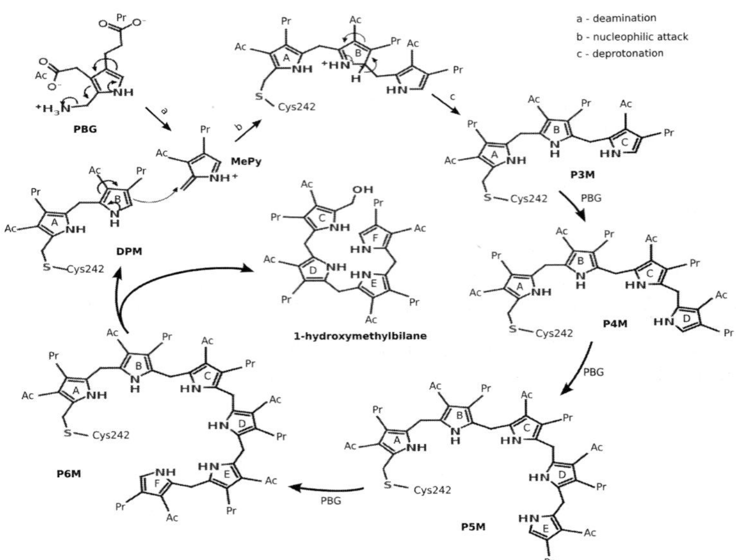

Figure 1: Schematic representation of tetrapyrrole chain elongation catalyzed by PBGD. The scheme shows the deamination of porphobilinogen (PBG) to methylene pyrrolenine (MePy), the nucleophilic attack by the B ring of dipyrromethane (DPM) on MePy forming an intermediate that undergoes deprotonation to form a tripyrrole moiety (P3M). Subsequent additions of PBG elongate the chain to form a tetrapyrrole moiety (P4M), pentapyrrole moiety (P5M) and hexapyrrole moiety (P6M). Hydrolysis of the tetrapyrrole from the protein gives 1-hydroxymethylbilane leaving behind the cofactor. The rings of the elongating pyrrole chain are labelled as A, B, C, D, E and F starting from the pyrrole ring covalently attached to Cys 242. The acetate and propionate side-groups of the pyrroles are denoted by -Ac and -Pr, respectively.

1.1. PBGD so far

The three-dimensional structures of EcPBGD and hPBGD reported so far have shown that the protein consists of three α/β domains (Fig. 3(a)) connected by hinge segments with domain 1 and 2 being very similar to tranferrins and periplasmic binding proteins[2]. DPM cofactor is covalently attached to sulfur of C242 (C261 in hPBGD) on a flexible turn (cofactor turn) of domain 3. The active site cleft occurs at the interface between domain 1 and 2. The coordinates for most of the residues in the active site loop region, 40-63 (55-78), were missing in all the available crystal structures due to high mobility of this segment[2, 3].

Detailed structural and biochemical studies on EcPBGD indicated its single catalytic site[1]. K55 and K59 were reported to be essential for catalytic activity of the enzyme[6]. Mutational studies have reported the importance of several arginines in enzyme substrate complex formation, substrate binding and chain elongation. R232H and R11H were seen to have a detrimental effect on the activity of the enzyme[7]. Residues R131, R132, R149 and R155 interact with acetate (-Ac) and propionate (-Pr) side groups of the DPM cofactor and stabilise it[8]. Interactions of the cofactor, in both oxidised and reduced forms, with residues of the enzyme have been studied and D84 has been suggested to play a key role in catalysis, by stabilizing the positive charges on the pyrrole rings during the chain elongation process[2].

Structural studies on hPBGD hint to a similar single catalytic site[3] as observed in EcPBGD[1]. Structure of hPBGD suggested that D99 is a critical catalytic residue[9]. The motion along the hinge regions predicted in hPBGD could facilitate the sequential entry of PBG molecules. The hinge residue, H120 was shown to be critical for enzyme activity and H120P mutation leads to inactivation[9]. Mutational studies on the hPBGD indicated that D99, R149, R167 and R173 were involved in cofactor assembly and chain elongation[10].

In contrast to the human and E. coli enzyme, PfPBGD has a leucine (L116) in place of a lysine (K55 in EcPBGD, K70 in hPBGD). Mutational studies showed that L116K has higher activity than the wild type PfPBGD producing urogen I (non-enzymatic product) along with urogen III, the product of wild type PBGD[5].

The catalytic mechanism has been hypothesised based on biochemical and mutational studies, but details of the dynamic behaviour of the protein and the accommodation of the growing tetrapyrrole chain in the active site cleft during elongation remains unknown. Louie et al., (1996) hypothesised two mechanisms for accommodating the elongating chain[2]. The sliding active site model suggests that the cofactor is accommodated in the cavity within the protein and domains adjust themselves in order to shift the binding site and catalytic site residues near the terminal pyrrole ring. The moving chain model states that the elongating chain is progressively pulled past the catalytic site placing the penultimate and terminal rings in the substrate binding site. Though biochemical studies recognised residues with catalytic importance, their specific role in the catalytic mechanism remains unclear. The exit mechanism of 1-hydroxymethylbilane, after the catalysis, is also unknown.

2. Methods

2.1. Preparation of starting structures

Structure of *E.coli* PBGD with the cofactor, DPM (in reduced form)[2], covalently attached to C242, was taken from PDB (2YPN; resolution: 2.3 Å)[11]. The coordinates of the missing residues, 43-59, were loop modelled using Modeller 9V8[12]. Stages of chain elongation studied in the current work include – DPM (PBGD with the DPM cofactor), P3M (PBGD after the first catalytic addition of PBG to DPM, i.e., with P3M moiety) and P4M (PBGD after addition of two molecules of PBG to DPM, i.e., with P4M moiety). The starting structure for each stage was prepared by covalently attaching a moiety of methylene pyrrolenine (MePy) close to the polypyrrole cofactor in the active site cavity of the protein and subjecting it to energy minimization. Parameters for the covalently attached cofactor DPM and the subsequent chain extensions, i.e. tripyrrole (P3M) and tetrapyrrole (P4M) were obtained from ATB server[13].

2.2. Molecular dynamics simulations

Explicit solvent united-atoms simulations of the different stages of PBGD were performed using Gromacs 4.5.5[14] with G53a6 forcefield[15] on *EKA* supercomputer (Computational Research Laboratories, India). The PBGD systems were solvated in an octahedron box with a 9 Å layer of SPCE water; protein charges were neutralised with addition of Na+ ions. Energy minimisation was carried out using Steepest Descent method for 7000 cycles or till convergence was achieved. NVT and NPT position restrained equilibration was done for 200ps and 1ns respectively with V-rescale temperature coupling[16] and Parrinello-Rahman pressure coupling[17] for the protein and non-protein parts separately. The temperature was gradually raised from 0K to 300K at 3K/ps. Bond lengths were constrained using LINCS algorithm[18]. Periodic boundary conditions were employed to minimise edge effects and Particle Mesh Ewald[19,20] was used for full electrostatic computations with interpolation order of 4, tolerance of 1e-5 and Fourier spacing of 1.6Å. MD simulation for DPM and P3M stages were carried out for 35ns each and the coordinates were saved after every 10ps. On addition of an extra pyrrole ring to P3M, it was observed that the domains 1 and 2 move apart. In order to prevent this, harmonic distance restraints were applied to centers of mass of the domains for 15ns, followed by unrestrained dynamics of 35ns in the P4M stage.

2.3. *Trajectory and structural analyses*

Trajectories were analysed using VMD 1.9.1[21] and Gromacs 4.5.5. The root mean square deviations (RMSDs) of the backbone atoms for all stages of the protein, the radius of gyration (Rgyr) and hydrogen bonds were calculated using VMD plugins. The loop modelled and energy minimised 2YPN structure was used as the reference for RMSD. The root mean square fluctuations (RMSFs) of the Cα atoms, the solvent accessible surface area (SASA) of the entire protein were calculated using Gromacs. The principal component analysis and dynamic cross correlation matrix (DCCM) calculation, of the Cα atoms, were done for the individual stages (DPM, P3M and P4M). Additionally, combined essential dynamics[22] was done on the concatenated trajectory of the three stages to analyse the cumulative effect. DCCM was calculated using a script. The interactions indicated between residues are represented as the minimum distance between the possible donor and acceptor atoms of the respective residues.

3. Results and Discussion

3.1. *RMSD, Radius of Gyration and SASA*

The overall RMSD of the protein as observed from Table 1 shows that it progressively increases from DPM to P4M. The domain-wise contribution to the overall RMSD shows that in the case of domain 1 the deviation in P3M and P4M stages is similar and is higher than that in DPM. For domain 2 and domain 3 a progressive increase is observed from DPM to P4M similar to that of overall RMSD. The RMSD of the residues in the active site loop region (44-60) is seen to increase going from DPM to P3M stage, followed by a slight decline in the RMSD for P4M stage.

The average Rgyr and SASA values in Table 1 show that the protein opens slightly in each stage to accommodate the increasing polypyrrole chain within the active site. The fluctuation in SASA values of the active site loop region along with the RMSD of the loop indicate that the loop is generally flexible in all stages of chain elongation, as expected. The slight dip in the SASA value of the loop region in P3M suggests it may be interacting with parts of the protein.

Table 1: RMSD, SASA and Rgyr values averaged over the trajectory with their standard deviation.

		RMSD, Å	SASA, Å^2	Rgyr, Å
DPM	Overall	2.7±0.4	16632.7±337.8	19.8±0.3
	Domain1	3.6±0.5		
	Domain2	1.5±0.3		
	Domain3 without terminal residues (308-313)	1.6±0.3		
	Loop	7.2±1.0	1345.1±92.1	
P3M	Overall	4.1±0.1	16852.7±281.0	20.1±0.2
	Domain1	5.6±0.2		
	Domain2	3.0±0.6		
	Domain3 without terminal residues (308-313)	1.9±0.2		
	Loop	10.6±0.3	1121.3±95.3	
P4M	Overall	4.8±0.2	17233.8±307.6	20.3±0.2
	Domain1	5.6±0.2		
	Domain2	5.3±0.3		
	Domain3 without terminal residues (308-313)	2.9±0.2		
	Loop	9.8±0.6	1236.9±98.3	

3.2. RMSF

The RMSF plots for the three stages of chain elongation are illustrated in Fig. 2(a). From the figure it can be observed that in the DPM and P4M stages the fluctuation in the active site loop region is high, while that in domain 2 is low. The reverse is observed in the P3M stage where the fluctuation is high in parts of domain 2 and low in the loop region. This indicates that in order to accommodate the growing pyrrole chain either the active site loop or domain 2 readjusts itself. The fluctuations in domain 3 are seen to increase continuously through the stages, which suggest that it dynamically readjusts to the growing pyrrole chain.

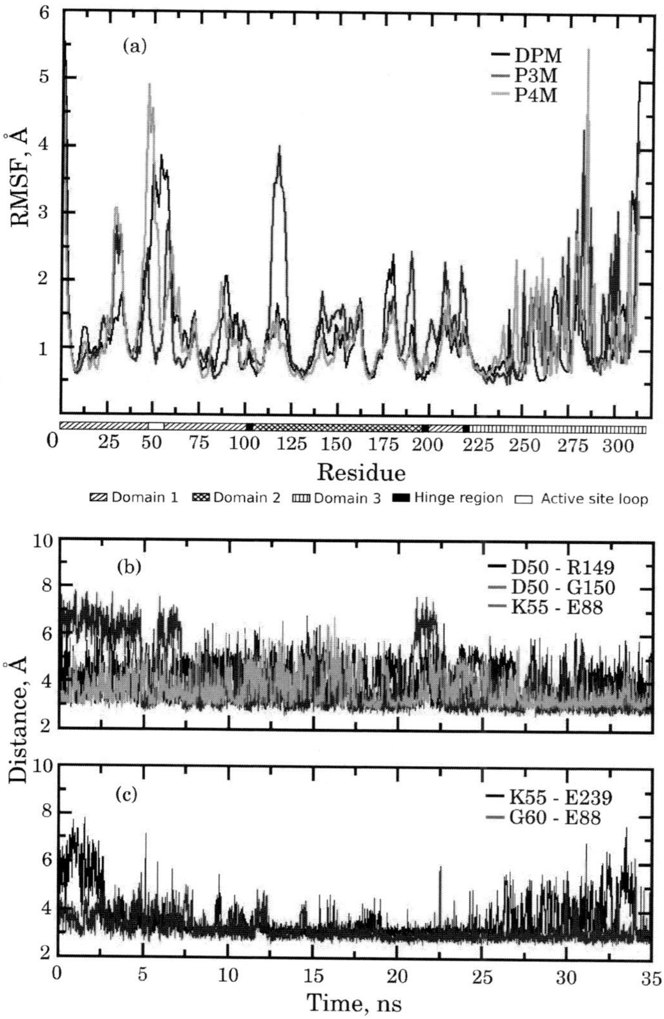

Figure 2: (a) Comparative RMSF of the protein Cα atoms during stages of chain elongation (DPM, P3M and P4M). (b) Interaction of D50 with R149, D50 with G150 and K55 with E88, during the P3M stage, (c) Interaction of K55 with E239 and G60 with E88 during the P4M stage.

3.3. Changing Dynamics of Active Site Loop

Dynamics of the flexible active site loop segment is observed to be different in each stage of the pyrrole chain elongation. Persistent interactions between the loop residues and residues in the rest of the protein are tracked along the length of the simulation in each stage. The flexibility of loop can be attributed to the residues D50 and K55. The interactions of these residues in the three stages are tracked to correlate it with the loop movement. In the DPM stage, the loop region is seen to move back and forth from the active site cleft. During this movement D50 interacts with R149 and K55 interacts with Q243, E305, V306 and Y307 for some time. At other times K55 interacts with N308 and E88, thus explaining the back and forth motion of the loop. During the P3M stage, the loop is seen to remain close to the active site; D50 interacts with both R149 and G150, while K55 interacts with E88 (Fig. 2(b)). In the P4M stage, the loop is seen to slide downwards, which leads to the loss of the interaction of D50 with R149 that is replaced by the interaction of G60 with E88. Also, K55 is seen to interact with E239 (Fig. 2(c)). These interactions, supported by RMSD and SASA analyses of the loop, probably govern the motion of the loop region during the catalytic mechanism. Mutations of R149 and K55 residues showed decrease in the enzyme activity, emphasizing their role in loop movement[6, 7].

3.4. Combined essential dynamics

Cumulative effect of the domain movements observed during the individual steps of the chain elongation process can be studied using combined essential dynamics (ED). The combined ED captured most of the protein motions in the first two principal components. The principal component 1 shown in Fig. 3(c) indicates domain 2 to be negatively correlated to domain 1, while parts of domain 1 are seen to be positively correlated to some regions of domain 3 and negatively correlated to other regions in domain 3. The movement of the cofactor turn is observed to be slightly towards domain 2 with the active site loop residues moving away synchronously. This indicates that with the growing pyrrole chain, domain 3, especially the cofactor turn region, moves into the active site cleft with an inclination towards domain 2. This movement is aligned with the motion of domain 1 and domain 2 away from each other, thus re-adjusting themselves for the growing pyrrole chain. Major fluctuations in regions of domain 1 and domain 2 can also be seen from the RMSF of principal component 1 (Fig. 3(e)). The principal component 2 (Fig. 3(d)) captures the back and forth movement of the active site loop region, which can be seen from the RMSF of principal component 2 (Fig. 3(f)).

The above motions are cross-verified by DCCM, calculated from the combined trajectories of DPM, P3M and P4M stages. DCCM presented in Fig. 3(b) shows that domain 1 (1-99, 200-217) is negatively correlated to domain 2 (105-193).

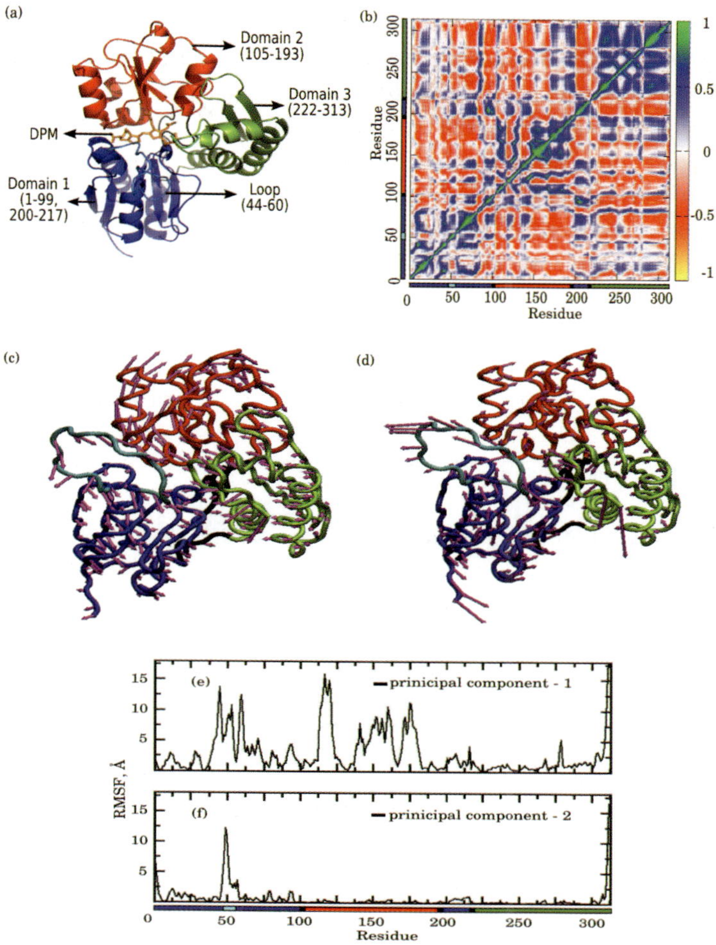

Figure 3: (a) Loop modelled structure of PBGD covalently bound to cofactor DPM (orange) (PDB ID: 1PDA) with the domains (1 in blue, 2 in red, and 3 in green), active site loop (cyan) and hinge regions between domains (black). (b) DCCM plot of the concatenated simulation trajectory showing overall correlation between domain residues of PBGD. Negative correlation is in red and yellow, whereas the colours blue and green represent positive correlation. Motion of domains in combined essential dynamics for (c) principal component-1 and (d) principal component-2. RMSF of Cα atoms corresponding to (e) principal component-1 showing maximum fluctuation in domains 1 and 2 and (f) principal component-2 showing maximum fluctuation in the active site loop region.

Mildly positive correlations can be observed in few regions of domain 1 with domain 2. Residues in domain 3 (222-313) are seen to be partly positively correlated to residues 1-45 and 80-90 of domain 1 and partly negatively correlated to residues 46-80, 90-100 and 200-217 of domain 1. Residues of domain 2 (105-169) are mostly negatively correlated to domain 3 (222-313) with positive correlation in region 170- 199 of domain 2. These observations map to the motions inferred from the combined ED.

The inter domain hydrogen bond analysis also supports the domain movement with decrease in hydrogen bonds between domain 1 and domain 2 (Fig. 4(a)), while there is an increase in interactions between domain 1 and domain 3 (Fig. 4(b)) from DPM to P4M stage. The interaction between domain 2 and domain 3 is more in case of P4M than in the DPM and P3M stages (Fig. 4(c)) because of movement of the cofactor turn as discussed earlier.

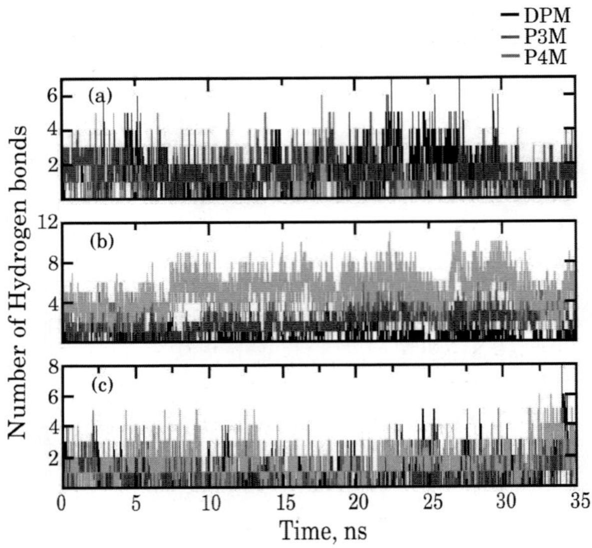

Figure 4: Number of hydrogen bonds between (a) domain 1 and domain 2, (b) domain 1 and domain 3, (c) domain 2 and domain 3 during the DPM, P3M and P4M stages in grey scale.

Negative correlation of domain 3 with the region 90-100 in domain 1 during the P3M and P4M stage simulations, seem to cause the helical residues 210-217 to be stretched, leading to the formation of a nascent β-strand through a hydrogen bond interaction between I98 and L219 backbones.

The movements in domain 1 are seen to reposition the catalytic residues, which are a part of this domain, near the terminal pyrrole ring. Increasing interaction between the growing cofactor and domain 2 seem to cause the upward movement in the cofactor turn towards domain 2. These domain movements are similar to those hypothesised by Louie et al, (1996) in their sliding active site model[2]. Also, the increasing distance between the domain 1 and domain 2 due to hinge motion upon pyrrole chain elongation are similar to those suggested by Song et al., (2009)[9]. Similar domain movements were also observed in structurally analogous binding proteins, such as lactoferrin[23].

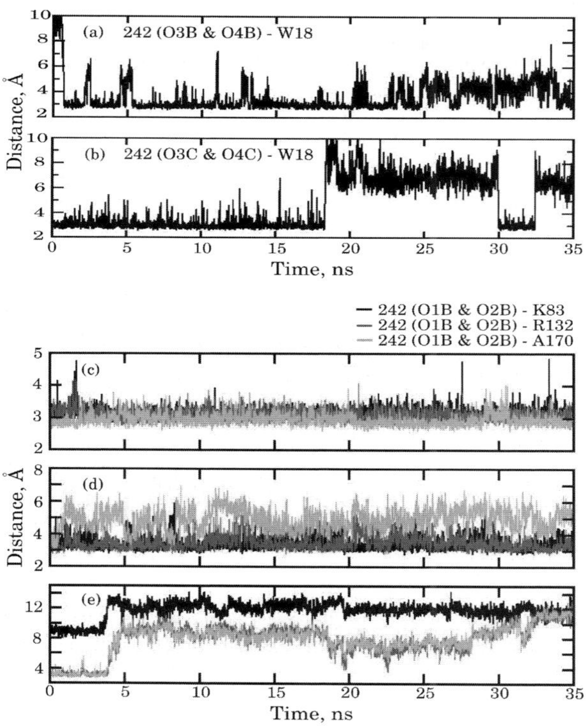

Figure 5: Minimum distance between W18 and (a) O3B and O4B of B-ring, carboxylate oxygens of -Pr chain, in P3M stage, (b) O3C and O4C of C-ring, carboxylate oxygens of -Pr chain, in P4M stage. Minimum distance of the carboxylate oxygens (O1B and O2B) of -Ac chain in residue 242 with active site residues K83, R132, A170 in (c) DPM stage, (d) P3M stage and (e) P4M stage in grey scale.

3.5. *Dynamics of the growing pyrrole chain*

The accommodation of the elongating pyrrole chain can be understood by studying the electrostatic interactions of the growing pyrrole chain linked to C242 in each of the three stages with W18, K83, R132 and A170. There is no interaction of residue 242 with W18 in the DPM stage, while the interaction is observed in P3M and P4M stages. In P3M stage, W18 is seen to interact with the carboxylate oxygens of the -Pr chain of ring B (Fig. 5(a)), while in P4M it is seen to interact with the carboxylate oxygens of the -Pr chain of ring C (Fig. 5(b)). This shift in interaction seen in the P4M stage is a result of the upward movement of the cofactor i.e., in the P4M stage the C ring occupies the position that is occupied by the B ring in P3M stage.

The carboxylate oxygens of the -Ac chain of ring B are seen to interact with residues K83, R132 and A170 in the DPM stage (Fig. 5(c)). In the P3M stage interaction with A170 is lost, while that with K83 and R132 persist (Fig. 5(d)). In the P4M stage all interactions are lost after the initial 4ns of the simulation (Fig. 5(e)). This supports the observation that the elongating pyrrole chain accommodates itself in the active site cleft with an inclination towards domain 2, giving the cofactor turn a slight upward motion, with the ring B rotating towards the hollow space between domain 1 and domain 2 as a result of which it loses the interactions with the protein residues in the P4M stage.

4. Conclusion

The first three stages of tetrapyrrole chain elongation in EcPBGD were investigated by MD simulations. The compactness of the overall protein was seen to decrease progressively from the DPM to P4M stage. The simulations support the role of D50, K55 and R149 in the active site loop modulation and catalytic mechanism as suggested by biochemical studies. The domain movements observed during each stage, suggest that domain 1 and domain 2 move away from each other while the cofactor turn moves into the active site with an inclination towards domain 2, thus creating space for the additional pyrrole moieties added at each stage. Analysis of simulations of two more stages, P5M and P6M are in progress and insights from these studies can help in understanding the mechanism of tetrapyrrole biosynthesis, not only in *E. coli*, but also in other organisms.

Acknowledgement

Authors wish to thank Dr. Alpesh Malde, University of Queensland for help with parameters of the cofactors and Dr. Monika Sharma, International Institute of Information Technology for providing the script to perform DCCM calculations.

References

1. G. V. Louie, P. D. Brownlie, R. Lambert, J. B. Cooper, T. L. Blundell, S. P. Wood, M. J. Warren, S. C. Woodcock and P. M. Jordan, *Nature*, 359, 33-39 (1992).

2. G. V. Louie, P. D. Brownlie, R. Lambert, J. B. Cooper, T. L. Blundell, S. P. Wood, V. N. Malashkevich, A. Hädener, M. J. Warren and P. M. Shoolingin-Jordan, *Proteins: Struct., Funct., and Genet.*, 25, 48-78 (1996).
3. R. Gill, S. E. Kolstoe, F. Mohammed, A. Al D-Bass, J. E. Mosley, M. Sarwar, J. B. Cooper, S. P. Wood and P. M. Shoolingin-Jordan, *Biochem. J.*, 420, 17-25 (2009).
4. Z. Q. Bonday, S. Taketani, P. D. Gupta and G. Padmanaban, *J. Biol. Chem.*, 272, 21839-21846 (1997).
5. V. A. Nagaraj, R. Arumugam, B. Gopalakrishnan, Y. S. Jyothsna, P. N. Rangarajan and G. Padmanaban, *J. Biol. Chem.*, 283, 437-444 (2008).
6. A. Hädener, P. R. Alefounder, G. J. Hart, C. Abell and A. R. Battersby, *Biochem. J.*, 271, 487-491 (1990).
7. P. M. Jordan and S. C. Woodcock, *Biochem. J.*, 280, 445-449 (1991).
8. M. Lander, A. R. Pitt, P. R. Alefounder, D. Bardy, C. Abell and A. R. Battersby, *Biochem. J.*, 275, 447-452 (1991).
9. G. Song, Y. Li, C. Cheng, Y. Zhao, A. Gao, R. Zhang, A. Joachimiak, N. Shaw, and Z. J. Liu, *FASEB J.*, 23, 396-404 (2009).
10. P. M. Shoolingin-Jordan, A. Al-Dbass, L. A. McNiell, M. Sarwar and D. Butler, *Biochem. Soc. Trans.*, 31, 731-735 (2003).
11. Y. P. Nieh, J. Raftery, S. Weisgerber, J. Habash, F. Schotte, T. Ursby, M. Wulff, A. Hadener, J. W. Campbell, Q. Hao and J. R. Helliwell, *J.Synchrotron Radiat.*, 6, 995-1006 (1999).
12. A. Fiser and A. Šali, *Meth. Enzymol.*, 374, 461-491 (2003).
13. A. K. Malde, L. Zuo, M. Breeze, M. Stroet, D. Poger, P. C. Nair, C. Oostenbrink and A. E. Mark, *J. Chem. Theory Comput.*, 7, 4026-4037 (2011).
14. D. Van Der Spoel, E. Lindahl, B. Hess, G. Groenhof, A. E. Mark and H. J. C. Berendsen, *J. Comput. Chem.*, 26, 1701-1718 (2005).
15. L. D. Schuler, X. Daura and W. F. Van Gunsteren, *J. Comput. Chem.*, 22, 1205-1218 (2001).
16. G. Bussi, D. Donadio and M. Parrinello, *J. Chem. Phys.*, 126, 014101-014107 (2007).
17. M. Parrinello and A. Rahman, *J. Appl. Phys.*, 52, 7182-7190 (1981).
18. B. Hess, H. Bekker, H. J. C. Berendsen and J. G. E. M. Fraaije, *J. Comp. Chem.*, 18, 1463-1472 (1997).
19. T. Darden, D. York and L. Pedersen, *J. Chem. Phys.*, 98, 10089-10092 (1993).
20. U. Essmann, L. Perera, M. L. Berkowitz, T. Darden, H. Lee and L. G. Pedersen, *J. Chem. Phys.*, 103, 8577-8592 (1995).
21. W. Humphrey, A. Dalke, and K. Schulten, *J. Mol. Graph.*, 14, 33-38 (1996).
22. D. M. F. Van Aalten, A. Amadei, A. B. M. Linssen, V. G. H. Eijsink, G. Vriend, and H. J. C. Berendsen, *Proteins: Struct., Funct., and Bioinf.*, 22, 45-54 (1995).
23. B. F. Anderson, H. M. Baker, G. E. Norris, S. V. Rumball and E. N. Baker, *Nature*, 344, 784-787 (1990).

PLASTICITY OF BH3 DOMAIN-BINDING HYDROPHOBIC GROOVES IN THE ANTI-APOPTOTIC MCL-1 AND A1 PROTEINS

VIVEK MODI, DILRAJ LAMA[1] AND RAMASUBBU SANKARARAMAKRISHNAN[2]

Department of Biological Sciences & Bioengineering
Indian Institute of Technology Kanpur, Kanpur – 208016, India
[1]Current Address: Bioinformatics Institute
30 Biopolis Street, #07-01, Matrix, Singapore 138671
[2]E-mail: rsankar@iitk.ac.in

Mcl-1 and A1 constitute a subclass within the group of anti-apoptotic Bcl-2 proteins and are shown to be overexpressed in several human cancers. Diverse BH3 domains of different pro-apoptotic Bcl-2 proteins bind to the hydrophobic groove of both these helical bundle proteins with different affinities. Development of any inhibitors for Mcl-1 and A1 requires the knowledge of the behavior of the binding region. In this study, we have carried out molecular dynamics simulations of apo-Mcl-1 and holo-A1 systems for a period of 100 ns and compared the results with our earlier studies on Bcl-X_L, another anti-apoptotic protein from a different subclass. During the course of the simulation, the BH3-containing helix H2 is destabilized in Mcl-1, a behavior observed in Bcl-X_L also. The unwinding is attributed to the presence of glycine residues in the segment containing H2. Additionally, this region also contains eight basic residues including three doublets. We have hypothesized that the dibasic motifs could be the cleavage sites for the enzymes that will generate smaller isoforms of Mcl-1 and the unwinding of H2 can aid in the exposure of these cleavage sites. As far as A1 is concerned, the BH3-containing helix H2 is mostly stable except the last two helical turns. However, it adopts a completely different orientation within the first 10 ns. This change in orientation realigns the loop region that links H2 and the next helix so that most of the exposed hydrophobic residues can be shielded. The charged residues in both the proteins that are projected towards the hydrophobic groove in the experimentally determined structures are exposed to the solvent during the simulations. This is due to the destabilization of helical regions in which they are present. The unwinding of helix H2 in Mcl-1 and the change in orientation of the same helix in A1 point to the flexible nature of hydrophobic groove and this could be important for the binding of diverse BH3 peptides.

1. Introduction

Mitochondrial outer member permeabilization is a key event in the intrinsic pathway of apoptosis [1,2]. Interactions between anti- and pro-apoptotic class of Bcl-2 family of proteins is an important step in regulating this event which eventually determines the fate of the cell [3,4]. The pro-survival proteins, Bcl-2,

$Bcl-X_L$, Bcl-w, Mcl-1 and A1, have four conserved sequence motifs (BH1 to BH4 domains) and the pro-apoptotic proteins contain either three BH domains (BH1 to BH3) as observed in Bak and Bax or just a single BH3 domain as found in Bcl-2 proteins like Bim, Puma, Noxa, Bmf and Bid [5]. The diverse anti-apoptotic and multi-domain pro-apoptotic Bcl-2 proteins adopt a similar helical fold with a pronounced hydrophobic groove [6,7]. BH3 motifs of pro-apoptotic proteins and the BH3 peptides derived from these motifs bind to the hydrophobic grooves of anti-apoptotic proteins [8-10] and exert biological function. Anti-apoptotic proteins such as $Bcl-X_L$ and Mcl-1 are recognized by the BH3 domains of several pro-apoptotic proteins. Similarly, the BH3 domain of pro-apoptotic proteins like Bim and Puma can bind to a range of anti-apoptotic proteins [11-13]. Other pro-apoptotic proteins like Noxa and Bad bind selectively to specific anti-apoptotic partners [14]. Even when a specific anti/pro-apoptotic protein has multiple partners, they exhibit varying affinities. Over-expression of pro-survival Bcl-2 members in several human cancers has been reported and hence, developing therapeutic agents that target specific anti-apoptotic Bcl-2 members is an active area of research in cancer biology [15-18]. In this context, understanding the molecular determinants of pro- and anti-apoptotic protein-protein interactions is an important step and both experimental and computational approaches are being used for this purpose.

Structures of several anti-apoptotic proteins in complex with pro-apoptotic BH3 peptides have been determined [8-10,19,20]. Examination of structures does not clearly reveal the reasons for differential affinities of BH3 peptides. Molecular dynamics (MD) simulations on isolated BH3 peptides and $Bcl-X_L$ complex structures have been carried out to investigate the factors important for high affinity of pro-survival proteins for selected BH3 peptides [21-23]. MD technique was also employed to look into the features of hydrophobic groove of $Bcl-X_L$ which is the binding site for several BH3 peptides [24].

Mcl-1 and A1 form a subclass within the anti-apoptotic Bcl-2 group. The binding pattern of these proteins is similar and shows distinct differences with other anti-apoptotic proteins like Bcl-2, $Bcl-X_L$ and Bcl-w [25]. For example, the pro-apoptotic Bmf and Bad peptides exhibit weak binding affinities to Mcl-1 and A1 whereas Noxa BH3 peptide binds strongly to the same proteins. For Bcl-2, $Bcl-X_L$ and Bcl-w, the binding profiles for these peptides are exactly the opposite. The other main difference is the presence of charged residues in the hydrophobic grooves of Mcl-1 and A1. While A1 has Glu side-chains pointing towards the groove [8,26], Mcl-1's hydrophobic groove is influenced by basic residues [9,27]. In comparison, $Bcl-X_L$'s binding groove is uncharged and essentially dominated by the hydrophobic residues [20]. The nature of the hydrophobic groove of an anti-apoptotic protein should be such that it should be able to bind certain BH3 peptides with strong affinities and at the same time other BH3 peptides should be selectively excluded. The selective binding and differential affinities of BH3 peptides can be understood at molecular level only if we have a full understanding of the nature of the hydrophobic groove to which the BH3 peptides bind. Earlier simulation studies from our laboratory on apo- and holo-forms of Bcl-X_L helped to understand the hydrophobic groove of this protein [24]. They revealed that the BH3-containing helix in Bcl-X_L which is part of the groove is destabilized and indicated plasticity of the binding groove that can help to bind diverse BH3 peptides. The unwinding of this helix also explained its possible importance in homo- or hetero-dimerization of Bcl-2 proteins. Interactions among the hydrophobic residues in the groove and ability of the loop connecting the BH3-containing helix and the subsequent helix to shield the bulky hydrophobic residues are the major factors that explain how a large chunk of hydrophobic patch can exist in the solvent-exposed state. In this paper, we have performed MD simulations of free forms of two anti-apoptotic proteins Mcl-1 and A1 for a period of 100 ns. Two important questions are being addressed in this study. Does the BH3-containing helix in Mcl-1 and A1 undergo destabilization similar to that observed in Bcl-X_L? How does the presence of charged residues in the hydrophobic grooves of these two proteins affect the behavior of the BH3-binding region?

2. Molecular Dynamics Simulations of Mcl-1 and A1

The starting structures of Mcl-1 and A1 are those from mouse and the experimentally determined structures were downloaded from the Protein Data Bank (PDB) [28]. The corresponding PDB IDs are 1WSX [27] and 2VOH [8] respectively for Mcl-1 and A1. While Mcl-1 is an apo structure determined using NMR, the A1 is an X-ray structure in complex with Bak BH3 peptide (Resolution: 1.9 Å). Hence, we removed the Bak peptide from the structure and the holo-A1 was used in the simulation. The extra five N-terminal residues as a result of cloning were removed from both Mcl-1 and A1. The side-chain of Glu 141 in A1 is not resolved in the X-ray structure. It was built using the Biopolymer module available in Insight II software (Acclerys, San Diego, CA). The apo-Mcl-1 and holo-A1 structures were simulated using the protocol developed in our laboratory to simulate Bcl-X_L protein [21,24]. Briefly, MD simulations were carried out using GROMACS (ver. 4.0) suite of software [29,30] with the force field GROMOS96 43a1 [31,32]. The structures were initially placed in the center of a cubic box and the box size was determined with the criteria that the minimum distance between the protein and the box edge should be at least 15 Å. SPC water model [33] was used and all ionizable residues were assumed to have default protonation states. The solvated structures were first minimized followed by 2 ns of equilibration. The first 1 ns equilibration was carried out using NVT (constant number of atoms, volume and temperature) ensemble. After that, NPT (constant number of atoms, pressure and temperature) ensemble was used. A twin-range cut-off of 10 and 18 Å was used to evaluate the non-bonded interactions. The systems were simulated at 300 K with 1 bar pressure. Berendsen's algorithm [34] was used to maintain the reference temperature and pressure of the systems. All other simulation details are same as those reported in the earlier simulation studies of Bcl-X_L [21,24]. Both the systems were simulated for a period of 100 ns after equilibration. Details of the simulations are summarized in Table 1. Analysis of the simulations and important conclusions are presented below. The residue numbers of Mcl-1 and A1 presented in this paper correspond to the amino acid sequences from UniProt[35] accession codes P97287 and Q07440 respectively.

Table 1: MD simulations of Mcl-1 and A1.

Details	Mcl-1	A1
Number of residues	157	150
Number of atoms	50,733	45,898
Box size ($Å^3$)	80.1 X 80.1 X 80.1	77.0 X 77.0 X 77.0
Percentage helicity[a]	80.0 (Experimental)[c]	78.5 (Experimental)[c]
	66.9 (± 1.52)[d]	76.1 (± 1.15)[d]
RMSD (Å)[b]	4.1 (± 0.1)[d]	3.3 (± 0.1)[d]

[a]The percentage helicity of the structure was calculated as described in [21]

[b]Root Mean Square Deviation (RMSD) of each MD simulated structure was calculated by considering the Cα atoms of helical regions from the starting structure. The helical regions were defined as described in [21].

[c]The PDB IDs of Mcl-1 and A1 experimental structures are 1WSX and 2VOH respectively.

[d]Averages (± standard deviation) were calculated for the last 10 ns of the production runs.

3. Results and Discussion

3.1 *BH3-containing helix in Mcl-1 and A1*

The helical regions of the three experimentally determined structures of Mcl-1 (1WSX), A1 (2VOH) and Bcl-X_L (1PQ0) are superposed and are shown in

Figure 1. It is clear that all three structures adopt the same helical fold. However, there are differences in the packing of the helices (Figure 1a). The region forming the hydrophobic groove (Figure 1b) indicates that the conformations of the loop connecting the BH3-containing helix H2 and the subsequent helix H3 in all three structures are different. This shows the flexibility of this region and its potential to shield the exposed hydrophobic groove by adopting different conformations.

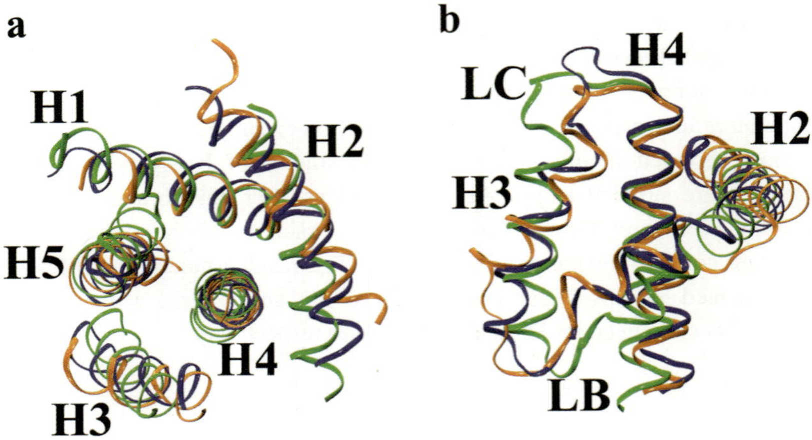

Figure 1: Superposition of the three experimentally determined structures of anti-apoptotic Bcl-2 proteins, namely, Mcl-1 (1WSX – blue), A1 (2VOH – green) and Bcl-X_L (1PQ0 – orange). Cα atoms of common helical regions obtained from structural alignment were used for superposition [7]. (a) Ribbon representation is shown viewed down the helix axis of central hydrophobic helix H4. The C-terminal helix and the loop regions are not shown for clarity. (b) The polypeptide region that contributes to the formation of the hydrophobic groove is shown with a view parallel to the central hydrophobic helix H4. Here also, the C-terminal helix is not displayed for clarity. The criteria for defining helical regions are as given in [21]. Individual helical segments and loop regions are labeled. All molecular plots are generated using VMD software [36].

Our previous simulation studies on Bcl-X_L apo/holo forms and Bcl-X_L structures in complex with BH3 peptides revealed that the helix H2 is destabilized and such conformational change may have functional significance [21,24]. It is not clear whether such behavior is common to all Bcl-2 proteins. Hence in the present study, we analyzed the stability of all helical segments in both Mcl-1 and A1 after 100 ns production run and compared them with the Bcl-X_L simulation results. Both the starting structure and the structure saved at the end of the production run were superposed (Figure 2a and 2b) and

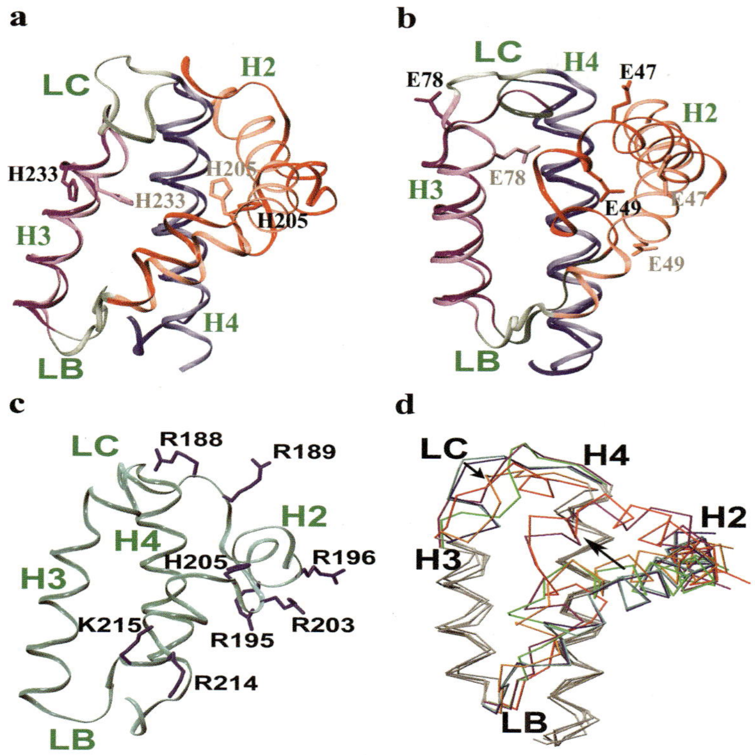

Figure 2: Superposition of starting structure and MD simulated structure(s) saved at the end of 100 ns MD simulations are shown for (a, c) Mcl-1 and (b, d) A1 viewed parallel to the central hydrophobic helix H4. Cα atoms of helices H3 and H4 were used for superposition. (a) and (b): Helices H2, H3, H4 and the loops are shown in red, purple, blue and gray colors respectively. Starting and simulated structures are distinguished by light and dark shades of the colors respectively. The basic and acidic residues that can respectively influence the hydrophobic grooves of Mcl-1 and A1 are shown. They are labeled either in gray (starting structures) or in black (simulated structures) fonts. (c) Structure of Mcl-1 saved at the end of 100 ns MD simulations. Helices H2, H3, H4 and loops LB and LC are shown in ribbon representation in light blue. The side-chains of basic residues in H2 and loop LB are displayed in stick representation in dark blue. (d) Cα-trace of A1 structures saved at various stages of MD simulations superposed on the starting structure using the Cα coordinates of H3 and H4. Starting structure and the structures saved at the end of 1 and 2 ns equilibration are displayed in blue, cyan and orange color respectively. Structures shown in green and purple colors represent the average structures calculated for the time period of 1 to 5 ns and 6 to 10 ns respectively. The structure saved at the end of 100 ns production run is shown in red color. Arrows depict the direction of movement of loops during the course of the simulation.

the stable helical segments H3 and H4 were considered for this purpose. While the helix H2 unwinds in Mcl-1 and is broken into two small segments, it remains mostly stable in A1. At least three factors leading to the unwinding of helix H2 in Bcl-X_L were identified in earlier simulations. Presence of a glycine in the middle of H2, presence of like-charged residue pairs (acidic or basic) separated by three to four residues and absence of strong inter-helical interactions involving H2 were suggested to be important for unwinding of helix H2. We were curious to know whether the same factors are applicable in the destabilization of helix H2 in Mcl-1 also. Moreover, we also investigated why helix H2 in A1 exhibits a stable helical character.

3.2 *Factors responsible for the helix H2 destabilization in Mcl-1*

In the case of Mcl-1, the mouse homolog used in this study has four glycines in the helical region. Multiple sequence alignment of Mcl-1 sequences from different species shows that at least three out of four glycines are conserved (Figure 3a). Glycine is known to be a helix breaker and introduce conformational flexibility in a polypeptide chain [37]. Hence, the presence a four glycines in the same helical segment could have destabilized the helical region (Figure 2a). The conformation thus adopted by the polypeptide chain in this region also seems to help burying a large number of hydrophobic residues (Figure 4a). Additionally, we also observe that the 32-residue segment [containing helix H2 and the loop region connecting H2 and H3 (LB)] is rich in basic residues. The eight basic residues in this segment include three doublets [(R188, R189), (R195, R196) and (R214, K215)]. Apart from contributing to the destabilization of helix H2, such large number of basic residues including three dibasic motifs in this region may have other functional significance. It is tempting to suggest that these three dibasic motifs may serve as a cleavage site for producing different isoforms of Mcl-1 protein which may have pro- or anti-apoptotic activity. A few isoforms of Mcl-1 have already been recognized [38-41] including a shorter form which is shown to have pro-apoptotic role. Dibasic motifs act as cleavage site for the enzymes to produce peptide hormones from a longer precursor protein [42-44].

All the basic residues in the 32-residue segment are completely solvent exposed (Figure 2c) and unwinding of the helix and exposure of these residues will help the enzymes to easily identify these sites and cleave the polypeptide chain. The only other dibasic motif (K257, H258) lies near the C-terminal end of the central hydrophobic helix H4.

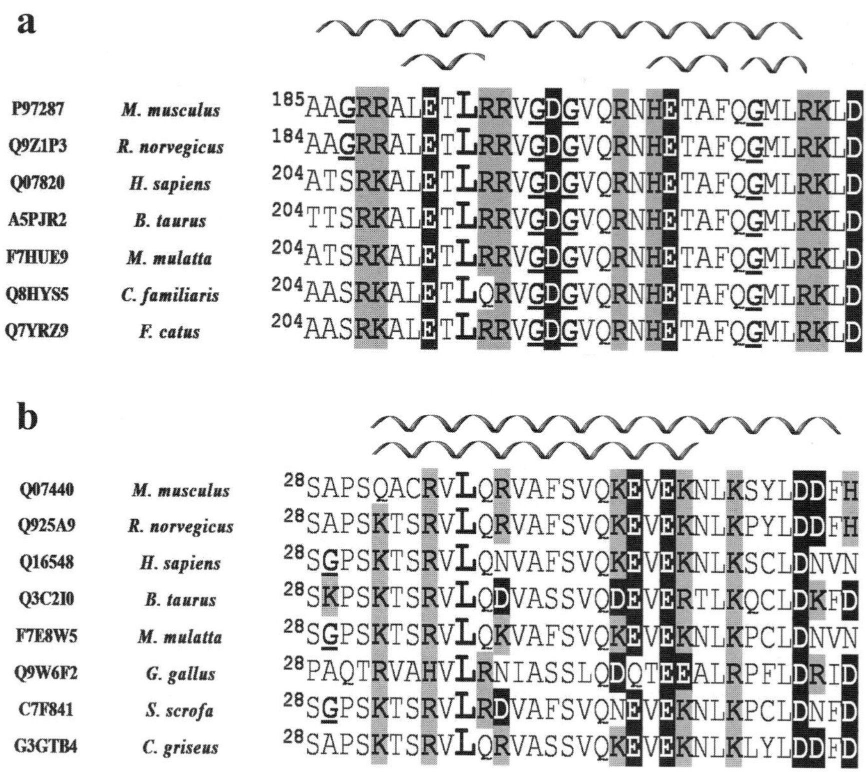

Figure 3: Multiple sequence alignment of the 32-residue segment encompassing the BH3-containing helical segment H2 is shown for (a) Mcl-1 and (b) A1. The UniProt [35] IDs and the respective species names are given on the left side of each sequence. Acidic and basic residues are shown in black and gray background respectively. Glycine residues are underlined. The highly conserved Leu is shown in bold font. The secondary structure corresponding to the polypeptide regions at the beginning and end of the simulations are shown respectively in the first and second rows at the top of each alignment.

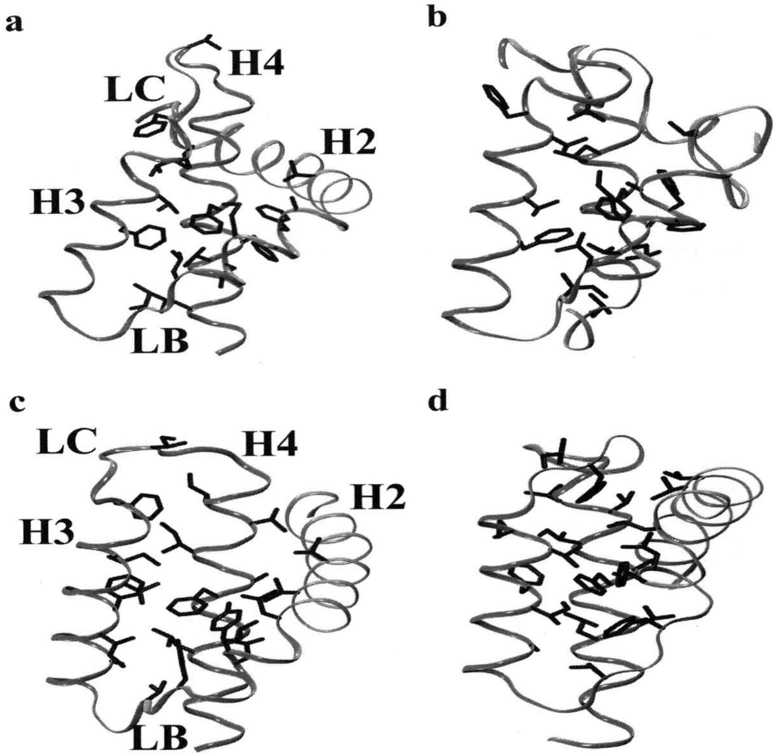

Figure 4: The hydrophobic residues from the hydrophobic groove are displayed in the starting structures (a and c) and the structures saved at the end of 100 ns MD simulations (b and d) for Mcl-1 (a and b) and A1 (c and d). The backbone of helices H2, H3 and H4 and the loops LB and LC are shown as gray ribbons. The side-chains of hydrophobic residues are depicted in stick representation.

3.3 *Why is helix H2 stable in A1?*

Unlike Mcl-1 and Bcl-X_L, the helix H2 of the anti-apoptotic protein A1 retained its helical character at the end of 100 ns production run (Figure 2b). Only in a small portion, unwinding of the helical region consisting of two helical turns near the C-terminus is observed. Multiple sequence alignment of A1 sequences indicates that there are no glycine residues in the structurally equivalent 32-residue segment comprising of helix H2 and loop LB (Figure 3b). In the nearly conserved "LXXXGD" motif in BH3 domains of Bcl-2 proteins, the residues Gly and Asp are substituted by Ala and Phe respectively in A1. This is highly unusual for a BH3 motif and further gives credence to the suggestion that the presence of one or more Gly residues is an important reason for the destabilization of helix H2 and this phenomenon was observed in Mcl-1 (see above) and Bcl-X_L [21,24]. Superposition of the polypeptide segment consisting of helices H2, H3, H4 and loop LB shows an intriguing phenomenon (Figure 2b and 2d). While the helical nature of H2 is preserved in A1, the orientation of the helix H2 is dramatically different at the end of the 100 ns simulation. The reason for change in orientation was further explored. The favorable interactions between the residues at the C-terminal end of helix H2 and the conserved "NWGR" motif in loop LC could be one driving force for the change in orientation. In an analysis of 16 non-redundant Bcl-2 protein structures, it has been observed that A1 has broader hydrophobic groove among all anti-apoptotic proteins [7]. Hence, the exposure of such a large hydrophobic patch in the holo-form of A1 may not be energetically favorable. The unwinding region near the C-terminal portion of helix H2 along with loop LB has resulted the polypeptide segment to spread the entire length of hydrophobic groove and thus helping it to shield the hydrophobic residues (Figure 2d, 4b and 4d). The change in the orientation of helix H2 can help to stretch the polypeptide segment to cover the hydrophobic region.

In summary, the BH3-containing helix H2 unwinds in Mcl-1 and this property is similar to that observed in Bcl-X_L in our earlier simulations [21,24]. In the case of A1, the helical character is retained for most of helix H2. The last two helical turn is lost near the C-terminal portion of H2. The stability of H2 in A1 can be attributed to the absence of Gly residue(s). However, a significant change in the orientation of helix H2 is observed presumably to have favorable interactions with the conserved "NWGR" motif from loop LC and to shield as many hydrophobic residues as possible from solvent exposure with the unwound helical part and loop LB.

3.4 *Charged residues in the hydrophobic groove of Mcl-1 and A1*

Both Mcl-1 and A1 have charged residues pointing towards the hydrophobic groove and the positive (Mcl-1) or negative (A1) charge(s) introduced in the BH3-motif binding region distinguishes these proteins from those observed in Bcl-2, Bcl-X_L and Bcl-w. It has been suggested that the presence of charges can potentially influence the BH3-domain binding properties of Mcl-1 and A1. We followed the interactions of the charged residues throughout the simulations. The two histidines (His 205 and His 233) in Mcl-1 move away from each other and are more solvent exposed. The glutamates in A1 (Glu 47, Glu 49 and Glu 78) also show a similar behavior and are completely exposed to the solvent at the end of 100 ns production run. To begin with, these residues were within the helical regions of H2 and H3. However, as the simulation progressed, the unwinding of helix H2 and shortening of H3 near its C-terminal segment made these residues to adopt an orientation in which they can interact easily with the aqueous medium. While this observation could indicate the binding region's plasticity, we must also remember that the default protonation states were considered for these ionizable residues. Depending upon whether a residue is protonated or not, its behavior within a protein may be affected. Chimenti et al. [45] recently examined 25 protein structures and showed that ionization of buried Lys residues in the hydrophobic environment either unfolded the protein or triggered localized changes. At least in A1, the protonation state of Glu 78 has been investigated [8] and the results suggest that Glu 78 is likely to be protonated when A1 is bound to BH3 ligand. Hence, an exhaustive study looking at various combinations of protonation states of ionizable residues within the hydrophobic groove is required that will help us to conclude their influence on BH3 peptide selectivity and binding.

4. Conclusions

The 100 ns MD simulations carried out on apo-Mcl-1 and holo-A1 proteins clearly indicate that the BH3-containing helix H2 in these proteins does not remain close to the experimentally determined structures. It is either completely destabilized as observed in Mcl-1 or it dramatically changes its orientation with respect to the other helices as seen in A1. The unwinding of helix H2 is observed in another Bcl-2 protein, namely, Bcl-X_L. The loss of helical character is due to the presence of four glycine residues in Mcl-1. In contrast to Mcl-1, the absence of any glycine residue helps to maintain the helical nature in H2 of A1. However, a change in the orientation of this helix is adopted to maximize favorable interactions with the conserved "NWGR" motif. Unwinding of H2 or a change in its orientation can help achieve two things. (i) This behavior introduces plasticity in the hydrophobic grooves of these proteins. (ii) It also helps to shield the hydrophobic residues in the groove in both proteins. Additionally, the presence of three dibasic motifs in this region of Mcl-1 could act as a cleavage site for the enzymes that will help to produce smaller isoforms. The simulations reveal that the charged residues present in the hydrophobic grooves of both Mcl-1 and A1 become solvent exposed at the end of the simulations. Lack of knowledge of exact protonation states of these charged residues is a major limitation. However, to understand more on the role of charged residues, MD simulations of Mcl-1 and A1 proteins in complex with the BH3 peptides have to be performed and they are currently in progress in our laboratory.

Acknowledgements

RS holds USV Chair Professorship in IIT-Kanpur. We thank all members of our laboratory for useful discussions.

References

1. N. J. Waterhouse, J. E. Ricci and D. R. Green, *Biochimie* **84**, 113 (2002).
2. D. R. Green and G. Kroemer, *Science* **305**, 626 (2004).
3. J. E. Chipuk and D. R. Green, *Trends Cell Biol.* **18**, 157 (2008).
4. R. J. Youle and A. Strasser, *Nature Rev. Mol. Cell Biol.* **9**, 47 (2008).
5. A. Kelekar and C. B. Thompson, *Trends Cell Biol.* **8**, 324 (1998).
6. A. M. Petros, E. T. Olejniczak and S. W. Fesik, *Biochim. Biophys. Acta* **1644**, 83 (2004).
7. D. Lama and R. Sankararamakrishnan, *Biochemistry* **49**, 2574 (2010).
8. C. Smits, P. E. Czabotar, M. G. Hinds and C. L. Day, *Structure* **16**, 818 (2008).
9. C. L. Day, C. Smits, C. F. Fan, E. F. Lee, W. D. Fairlie and M. G. Hinds, *J. Mol. Biol.* **380**, 958 (2008).
10. A. M. Petros, D. G. Nettesheim, Y. Wang, E. T. Olejniczak, R. P. Meadows, J. Mack, K. Swift, E. D. Matayoshi, H. Zhang, C. B. Thompson and S. F. Fesik, *Protein Sci.* **9**, 2528 (2000).
11. P. Gomez-Bougle, R. Bataille and M. Amiot, *Eur. J. Immunol.* **35**, 971 (2005).
12. D. Merino, A. Strasser and P. Bouillet, *Oncogene* **31**, 3392 (2012).
13. J. E. Chipuk, J. C. Fisher, C. P. Dillon, R. W. Kriwacki, T. Kuwana and D. R. Green, *Proc. Natl. Acad. Sci. U. S. A.* **105**, 20327 (2008).
14. L. Chen, S. N. Willis, A. Wei, B. J. Smith, J. I. Fletcher, M. G. Hinds, P. M. Colman, C. L. Day, J. M. Adams and D. C. S. Huang, *Mol. Cell* **17**, 393 (2005).
15. J. M. Adams and S. Cory, *Oncogene* **26**, 1324 (2007).
16. K. H. Kang and C. P. Reynolds, *Clinical Cancer Res.* **15**, 1126 (2009).
17. K. W. Yip and J. C. Reed, *Oncogene* **27**, 6398 (2008).
18. G. Lessene, P. E. Czabotar and P. M. Colman, *Nature Rev. Drug Discovery* **7**, 989 (2008).

19. M. Sattler, H. Liang, D. G. Nettesheim, R. P. Meadows, J. E. Harlan, M. Eberstadt, H. S. Yoon, S. B. Shuker, B. S. Chang, A. J. Minn, C. B. Thompson and S. F. Fesik, *Science* **275**, 983 (1997).
20. X. Liu, S. Dai, Y. Zhu, P. Marrack and J. W. Kappler, *Immunity* **19**, 341 (2003).
21. D. Lama and R. Sankararamakrishnan, *Proteins* **73**, 492 (2008).
22. D. Lama and R. Sankararamakrishnan, *J. Comput. Aided Mol. Des.* **25**, 413 (2011).
23. V. Modi, D. Lama and R. Sankararamakrishnan, *J. Biomol. Struct. Dyn.*, in press. (2012).
24. D. Lama, V. Modi and R. Sankararamakrishnan, *Manuscript submitted* (2012).
25. M. Vogler, D. Dinsdale, M. J. S. Dyer and G. M. Cohen, *Cell Death Diff.* **16**, 360 (2009).
26. M. D. Herman, T. Nyman, M. Welin, L. Lehtio, S. Flodin, L. Tresaugues, T. Kotenyova, A. Flores and P. Nordlund, *FEBS Lett.* **582**, 3590 (2008).
27. C. L. Day, L. Chen, S. J. Richardson, P. J. Harrison, D. C. S. Huang and M. G. Hinds, *J. Biol. Chem.* **280**, 4738 (2005).
28. H. M. Berman, J. Westbrook, Z. Feng, G. Gilliand, T. N. Bhat, H. Weissig, I. N. Shindyalov and P. E. Bourne, *Nucleic Acids Res.* **28**, 235 (2000).
29. B. Hess, C. Kutzner, D. van der Spoel and E. Lindahl, *J. Chem. Theory Comput.* **4**, 435 (2008).
30. D. van der Spoel, E. Lindahl, B. Hess, G. Groenhof, A. E. Mark and H. J. C. Berendsen, *J. Comput. Chem.* **26**, 1701 (2005).
31. X. Daura, A. E. Mark and W. F. van Gunsteren, *J. Comput. Chem.* **19**, 535 (1998).
32. W. F. van Gunsteren, S. R. Billeter, A. A. Eising, P. H. Hunenberger, P. Kruger, A. E. Mark, W. R. P. Scott and I. G. Tironi, *Vdf Hochschulverlag AG an der ETH Zurich, Switzerland*, 1 (1996).
33. H. J. C. Berendsen, J. P. M. Postma, W. F. van Gunsteren and J. Hermans. Interaction models for water in relation to protein hydration. In *Intermolecular Forces*; B. Pullman, Ed.; Reidel: Dordrecht, The Netherlands, 1981; pp 331.
34. H. J. C. Berendsen, J. P. M. Postma, W. F. van Gunsteren, A. DiNola and J. R. Haak, *J. Chem. Phys.* **81**, 3684 (1984).
35. The UniProt Consortium, *Nucleic Acids Res.* **40**, D71 (2012).
36. W. Humphrey, A. Dalke and K. Schulten, *J. Mol. Graph.* **14**, 33 (1996).

37. C. A. Rohl, A. Chakrabarthy and R. L. Baldwin, *Protein Sci.* **5**, 2623 (1996).
38. K. M. Kozopas, T. Yang, H. L. Buchan, P. Zhou and R. W. Craig, *Proc. Natl. Acad. Sci. U. S. A.* **90**, 3516 (1993).
39. C. D. Bingle, R. W. Craig, B. M. Swales, V. Singleton, P. Zhou and M. K. B. Whyte, *J. Biol. Chem.* **275**, 22136 (2000).
40. J. H. Kim, S. H. Sim, H. J. Ha, J. J. Ko, K. Lee and J. Bae, *FEBS Lett.* **583**, 2758 (2009).
41. C. R. Huang and H. F. Yang-Yen, *FEBS Lett.* **584**, 3233 (2010).
42. C. Tomasetto, S. M. Karam, S. Ribieras, R. Masson, O. Lefebvre, A. Staub, G. Alexander, M. P. Chenard and M. C. Rio, *Gastroenterolgy* **119**, 395 (2000).
43. Y. Rouille, G. Westermark, S. K. Martin and D. F. Steiner, *Proc. Natl. Acad. Sci. U. S. A.* **91**, 3242 (1994).
44. V. Chesneau, A. R. Pierotti, N. Barre, C. Creminon, C. Tougard and P. Cohen, *J. Biol. Chem.* **269**, 2056 (1994).
45. M. S. Chimenti, V. S. Khangulov, A. C. Robinson, A. Heroux, A. Majumdar, J. L. Schlessman and B. Garcia-Moreno, *Structure* **20**, 1071 (2012).

EFFECTIVENESS OF FRACTAL DIMENSION BASED MEASURES TO INVESTIGATE PROTEIN STRUCTURES

ANIRBAN BANERJI

Bioinformatics Centre, University of Pune
Pune-411007, Maharashtra, India

INDIRA GHOSH

School of Computational & Integrative Sciences
Jawaharlal Nehru University, New Delhi, 110067, India

Proteins are not classical (Debye) solids. They are deformable polymers and they belong to class of objects known as 'complex systems'. Fractal dimension based measures have been found to be an extremely useful tool to quantify various aspects of distributions of protein biophysical and biochemical properties. The present work shows that it is possible to separately quantify the self-similar symmetries in mass, hydrophobicity and polarizability distributions and in spatial correlation patterns amongst peptide–dipole units, charged residues, hydrophobic residues, residues with the π-electron cloud, etc. Current review presents multitudes of results to underscore the effectiveness of fractal measures for protein interior and exterior properties.

1. Introduction

1.1. *Why one needs fractal description of proteins?*

Proteins are enigmatic characters. They have been described as "complex mesoscopic systems"[1]. Although the structure of a protein can be accurately studied as that of a composite macromolecule, it has many a properties in common with that of macroscopic microparticles[2]. It has been established recently that proteins exist in a state of self-organized-criticality[3]. The standard 'compact object description' of proteins (characterized by small amplitude vibrations and by a Debye density of low frequency modes), cannot account for such "non-idealistic" behavior[4]. Moreover, inhomogeneous distributions of mass and hydrophobicity in protein interior[5] merely serve to a straightforward and linear scheme for description of protein interior. Similar difficulties prevail in the realm of protein exterior too, because neither protein shape nor protein surface conforms to simplistic Euclidean standards. Clearly, to model proteins, one needs to adopt an approach that has the capabilities to describe the inherently inhomogeneous and nonlinear behaviors of protein structural parameters as an innate capacity of it, rather than treating them as deviations from some standard forms. Quantification of self-similarity in protein properties can serve as a potent tool to achieve the aforementioned challenging task, because self-similarity can decipher the hidden symmetry that connects global patterns of macroscopic properties in proteins (like distribution of hydrophobicity, polarizability, etc.) with the local (atomic) interactions that produce them.

Fractal dimension (FD) can quantify self-similarity reliably. Thus, calculating FD to quantify the self-similarity prevailing in the distributions of protein biophysical and biochemical properties seems reasonable. In this article, we have discussed the concept of fractals, presented a short list of various methods that models protein properties with fractals reliably; and finally presented a small collage of protein FD results.

1.2. *Definition of fractal dimension*

While the Euclidean perception of symmetry revolves (essentially) around translation, rotation and reflection; self-similarity demonstrates itself through the symmetry of scale-invariance. This implies that when any part of a self-similar object is magnified, it is seen to bear a resemblance to the whole of it. For example, a magnified view of one part of the coastline may not precisely reproduce the full picture, but it will definitely possess the same qualitative appearance where the statistical parameters describing the object won't change. For regular structures, self-similarity (in general) is geometric (the so called 'geometric self-similarity'); for random or irregular objects however, one observes self-similarity to be statistical in nature. Statistical self-similarity implies that statistics of the spatial correlations in the patterns associated with the description of the system are similar across a range of length scales.

The dimension of an object measures its complexity. The intuitive concept of dimension, however, cannot satisfactorily characterize many "real-world" objects, which embody self-similarity[6]. The 'Hausdorff dimension'[7] can measure the self-similarity. A self-similar object is called a fractal object if the Hausdorff dimension describing it is found to be greater than the topological dimension that could have described the object. This definition of fractal objects, however, is not complete. For a formal introduction to fractals one can refer to ref. 8.

2. Fractal dimension associated with protein structures

2.1. *Studying protein interior with fractals*

2.1.1 *Backbone connectivity study*

In 1982, Havlin and Ben-Avraham[9] studied the fractal properties of polymer chains. Denoting the average end-to-end length of an unbranched polymer by L, the number of monomer segments (of length z) by N, and fractal dimension as FD, they found the scaling relationship to follow:

$$N = (L/z)^{1/v_F} = (L/z)^{FD}$$

where the exponent turns out to be the inverse Flory constant v_F in polymer theory. Theoretical considerations provide limits: $1 < FD < 2$, which corresponds to a linear structure given by $L = zN$, and a structure represented by unrestricted random walk, when FD approaches the topological dimension[10] magnitude 2, where $L = zN^{1/2}$. This approach of protein structure studies constitutes the FD-scheme of backbone connectivity analysis, where an appropriate description of protein length assumes immense importance. A detailed discussion on the various methods to define and describe protein length is included in the ref.11.

2.1.2 *Residue and atom distribution study*

This approach, does not use the connectivity profile of main-chain atoms, but concentrates upon the spatial distribution of amino acid residues and atoms in proteins. With this procedure (commonly called as the "mass fractal" calculation), the mass distributions within concentric spheres are studied[11]. Protein's mass scales as $M \propto R^{FD}$ (where M is the total mass of all the protein atoms and R is the characteristic length scale). Thus, from studying the linear (scale invariant) portion of the slope of log M versus log R plot, one can determine the FD. Though the generic name for calculation of this genre of FD is mass fractals, one can calculate other pertinent properties associated with protein atoms with this methodology too. More magnitude of mass-FD implies that within the protein interior the property under consideration has more compactness in its distribution. The mass-FD magnitude varies as 2.00 < mass-FD < 3.00. Proteins have an intrinsic self-similarity with respect to their compactness (atom packing). Mass-FD simply quantifies the symmetry of scale-invariance associated with any protein property that is dependent upon atom packing. It can be extended to the level of residues too, but not to the realm of organization of residues in secondary structures (a-helices are not composed of smaller helices). Moreover, one should take caution not to confuse between radius of gyration and mass-fractal dimension, while the former treats the proteins as classical solids, the later treats proteins as 'complex systems'.

2.2. *Studying protein exterior with fractals*

Protein surfaces are rough and they are characterized by constant vibrations (Figure 1). More roughness of protein surface often implies more probability of van der Waal contacts between the surface patch around the functional site and that of the interacting molecule. A rough binding site may indicate potentially stronger interactions between the protein and binding molecule[12,13]. Interestingly, protein surfaces, although irregular with varying degree of roughness, embodies a hidden symmetry in the form of self-similarity. Protein surfaces demonstrate statistical self-similarity (and not geometric self-similarity, which is inherently deterministic).

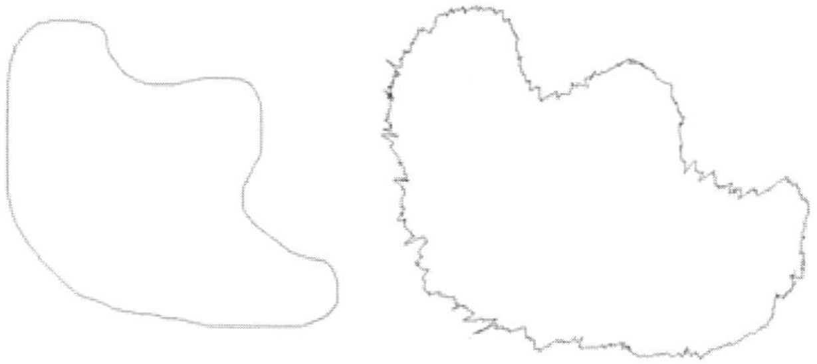

Figure 1. a) Idealized view of protein surface profile; b) Real view of protein surface profile.

One may intuitively suggest that surface roughness can be measured from the pattern of mean departure of the surface profile from the mean reference frame; whereby the surface roughness, in the continuous form, can be quantified as

$$R_{mean} = (1/L)\int_0^L |z|\,dx \quad \text{with RMSD} \quad R_{dev} = (1/L)\int_0^L z^2 dx,$$

where R_{mean} and R_{dev} represent the mean and RMSD of the profile from the mean reference line; $z(x)$ height information of the profile, L being the length of the profile. However, proteins are not classical rigid objects and they undergo (large) fluctuations to ensure proper functioning. Therefore the "mean reference frame" for any protein surface, even if calculated, will be a time-dependent and context-dependent nonlinear function of protein stability. This very problem of absence of an idealistic "reference frame" suggests clearly that geometrical constructs with their roots in Euclidean framework will not be appropriate to describe an object as complex as protein surface. Also a non-trivial probability of missing out on valuable biophysical information will always be present if geometric models for protein surface rely on idealistic and non-biological shapes like spheres, cones, cubes etc. Hence, only a geometric construct that can quantitatively describe the invariant patterns of a non-idealistic surface that is undergoing a spectrum of time-dependent, temperature-dependent, context-dependent fluctuations – *treating it as it is* – can be a suitable candidate to describe protein surfaces reliably. The FD is such a construct. In a landmark paper in 1985, Lewis and Rees[12] had shown that FD can quantify the symmetry in roughness profile of protein surfaces reliably.

If we model each atom by a sphere with its van der Waals radius R, the (irregular yet consistently defined) surface of the union of these spheres is referred to as the van der Waals surface (A_{VdW}) of the protein under consideration. A measure of the roughness of protein surface is measured on the van der Waals surface of it, with the help of a probe particle with radius (R_p). The degree of irregularity of this surface can then be described by FD ($2 < FD < 3$). As the surfaces become more irregular, magnitude of FD increases. Denoting the lateral area of protein surface patch under consideration by A_0, the basic relation that describes the entire process can be written as[14]:

$$\log(A_{VdW} / A_0) \propto \log(R_p^{2-FD}) \tag{1}$$

This shows why FD is a measure of irregularity, although magnitude of its own does not depend upon the chosen standard. The more readily implementable version of the equation (1) can be written as[12]:

$$FD = 2 - \frac{d(\log A_{VdW})}{d(\log R_p)} \tag{2}$$

While most of the papers on protein surface fractal use this definition, actual implementation of this scheme can be a bit problematic, two novel ways to achieve the same are presented below.

2.2.1 *Quantifying the extent of self-similarity of protein surface roughness by calculating the FD*

Since, to calculate the FD of a patch of protein surface (or the whole of it) one requires having magnitude of surface areas for a range of probe spheres, the original Lewis and Rees formula (equation 2) should be broken up in two steps; whereby, in the first step we calculate:

$$D_i = 2 - \frac{[(\log A_{VdW})_i / (\log A_{VdW})_{i-1}]}{[(\log R_p)_i / (\log R_p)_{i-1}]} \tag{3}$$

Then, the magnitude of *FD* is obtained through averaging, viz.,

$$FD = (1/n)\sum_i^n D_i$$

In the expressions above, $(A_{VdW})_i$ represents the van der Waals surface area as measured by probe sphere with radius $(R_p)_i$; and FD is calculated as the mean magnitude of several sets of D_i's; n being the number of such sets. For example when R_p is ranged between 1.4 Å to 4.4 Å, with an interval of 0.3 Å, n will be 11. When plotted in a log-log scale, it can be observed that for a particular range of varying R_p (plotted in the abscissa), the profile of A_{VdW} (plotted in the ordinate) is either suffering very little change or isn't changing at all. This unvarying profile of A_{VdW} (corresponding to aforementioned particular range of varying R_p), therefore, will demonstrate the scale invariance. Self-similarity of surface roughness corresponding to this specific R_p, can therefore be denoted as the FD of that specific patch of protein surface (Figure 2).

2.2.2 *Quantifying the extent of self-similarity of protein surface roughness by calculating the smoothed atomic fractal FD*

While studying the local variations in surface roughness at specific sites on protein surfaces, one must consider the fact that individual atomic areas are subject to statistical error. To account for this problem, viz., to smoothen the local roughness variations, Pettit and Bowie[13] introduced a variant of the aforementioned method by which FD can be calculated as:

$$f_i = 2 - \frac{d[\log(\sum_j A_{cs^j})]}{d(\log R_p)} \tag{4}$$

where f_i is defined as the particular magnitude of smoothed atomic fractal dimension (SAFD) of ith atom, A_{cs}^j represents the contact area due to *jth* atom, and the summation is carried out over all neighbor atoms j within 5Å of atom i. SAFD differs from the classical Lewis and Rees[12] and Aqvist and Tapia[15] method of FD calculation because it does not rely on summing up the atomic areas for atoms within base of the cones that are projected out from the protein center of mass.

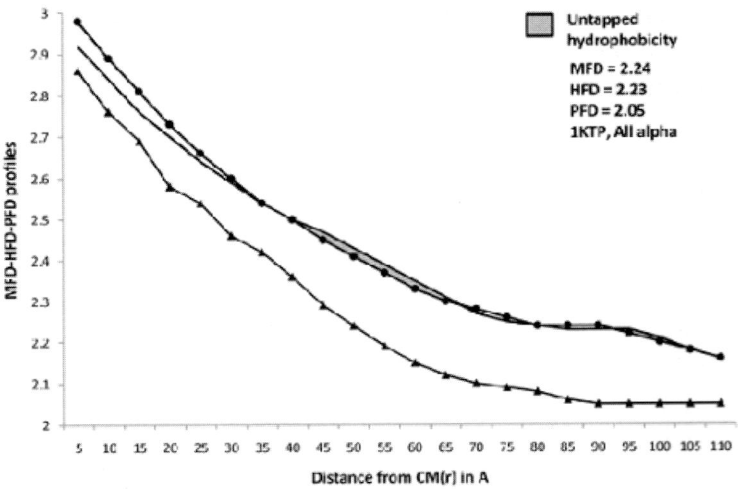

Figure 2. Profiles of mass-fractal (MFD), hydrophobicity-fractal (HFD) and polarizability-fractal (PFD) and Untapped Hydrophobicity in all-α protein is shown. Magnitudes of Mass, Hydrophobicity and Polarizability are plotted simultaneously for the (randomly chosen) all-α protein '1ktp'. Linear portions in their profiles (parallel to X-axis) denote the MFD, HFD and PFD magnitudes. — • — : MFD line. —— : HFD line. —▲— : PFD line . MFD can be calculated for an individual protein by plotting M contained inside concentric spheres of radius R from the center of mass (CM) of the protein, on a log-log scale. The invariant portion of slope of such a graph implies a scale-invariant (self-similar) distribution of the property under consideration for the protein; and MFD for that protein for that property is assigned this magnitude. The HFD was calculated in similar manner, by merely substituting atomic mass with (residue-specific) atomic hydrophobicity magnitudes. Calculation of PFD, however, involved the calculation of centroid of every amino acid (rather than the C_{α} atom of the same); because the centroid is a better representative of the location of an amino acid within any protein[11].

3. Results obtained with fractal-based investigations

3.1. *FD results obtained from protein interior properties*

Way back in 1982, Stapleton and his co-workers[16] had shown that changes in protein structure with solvent conditions could conveniently be monitored by it. Their studies had reported for the first time that the geometry of the carbon backbone determined from X-ray diffraction data and the vibrational dynamics of proteins as measured by Raman scattering, are both fractals. Their work proved that *FD* determined from the structure, correlates (almost) perfectly with that determined from the dynamics.

The recent trend of FD-based studies indicates a strong inclination towards discovering protein structural invariants. Enright and Leitner in 2005[17] formalized this intuitive notion by systematically calculating the mass-FD of a set of 200 proteins and found the magnitude to be 2.489 ∓ 0.172. Their result was consistent with those obtained from dispersion relations and the anomalous subdiffusion, computed for the same set of proteins[18].

Another interesting set of results[19] by Banerji and Ghosh revealed that barring the proteins belonging to all-α structural class, proteins belonging to all the major structural classes have an amount of unused hydrophobicity left in them (Figure 3). Amount of unused hydrophobicity was observed to be greater in thermophilic proteins, than that in their (structurally aligned) mesophilic counterparts. The set of all-β proteins (thermophilic, mesophilic alike) were identified with maximum amount of unused hydrophobicity, while all-α proteins have been found to possess minimum polarizability.

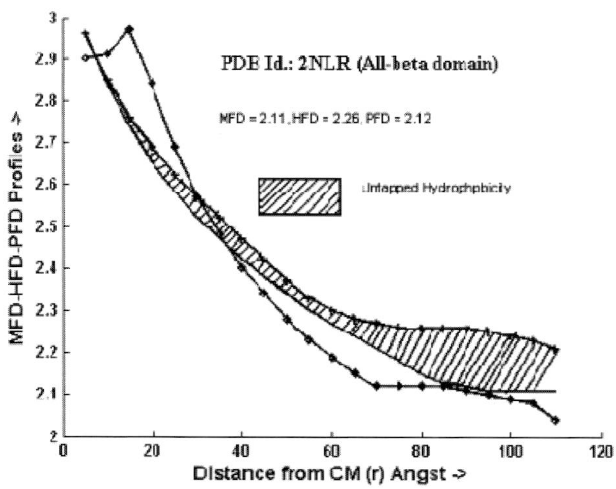

Figure 3. Profiles of (---) mass-fractal (MFD), (+) hydrophobicity-fractal (HFD) and (■) polarizability-fractal (PFD) for a (randomly chosen) protein (Hydrolase, All-beta) PDB id.:2NLR) are plotted simultaneously. Stretches in these profiles that are linear and parallel to abscissa (viz. radial distance from centre of mass of the protein), denote the extent to which these profiles are scale-invariant, with variations in radial distance. Ordinate magnitudes corresponding to these stretches provide the value of MFD, HFD and PFD for the protein. [HFD-MFD] parameter quantifies the magnitude of unused hydrophobicity left in the protein.

3.2. *FD results obtained from protein exterior properties*

In a landmark work in 1999, Pettit and Bowie[13] quantified the surface roughness of small molecular binding sites with FD. This was important because though many binding sites on proteins are found to be on the surface pockets and various pocket-finding algorithms were proposed to identify functional sites in protein structures[20], many functional sites are not concave and many of the existing techniques fell short of identify them. With an extensive analysis of high-resolution crystal structures (<2.5 angst) another work[21] could relate changes in side-chain conformation to FD magnitudes, before confirming the original assertion of[13] that "larger FD" magnitudes are observed "not only due to the larger number of pockets but also due to the increased roughness of pockets themselves".

Table 1. Evolution of FD-approaches to study proteins.

Year	Events
1967	Mandelbrot talks about "fractional dimensions", in his paper: "How Long Is the Coast of Britain? Statistical Self-Similarity and Fractional Dimension." Science, 156: 636-638.
1975	The word "fractal" was coined by Mendelbrot, in his book: "Les Objets fractals: forme, hasard et dimension, survol du langage fractal" (in French), Flammarion.
1980	Fractals are linked to interiors of proteins. Stapleton et al., Phys.Rev.Lett.45:1456-1459.
1982	FD of polymer chains calculated. Havlin-Ben-Avraham, J.Phys.A.15:L311-L316.
1984	Fractals are recognized as reliable tools to study surface roughness. Pfeifer P., Appl.Surf.Science., 18:146-164.
1985	Protein surface roughness is found to have self-similarity. Lewis-Rees, Science. 230:1163-1165.
1987	Protein surface roughness, as quantified with fractal dimension, was tried as a geometric marker to study protein-protein interactions. Aqvist-Tapia, J.Mol.Graph., 5:30-34.
1999	Prospects of fractal markers to study Protein-Ligand and Protein-Small molecule interactions were discussed. Pettit-Bowie, J.Mol.Biol.,285:1377-1382
2005	Mass fractal dimension calculated and linked to compactness. Enright-Leitner, Phys.Rev.E.,71:011912
2007	Protein surface roughness was found to be the 3rd best amongst a large set of physico-chemical properties, in the sphere of functional site identification. Pettit et al., J.Mol.Biol., 369(3):863-879.
2009	Fractal modeling of protein interior detected the presence of unused hydrophobicity. Banerji-Ghosh, PLoS One, 4(10):e7361.
2012	Proteins as Sponges: A Statistical Journey along Protein Structure Organization Principles. Luisa Di Paola et al., J. Chem. Inf. Model.52, 474–482

Within two years of Lewis and Rees publication, Aqvist and Tapia[15] attempted to test the efficiency of FD as a (reliable) marker to study protein-protein interaction. They defined residue fractal indices; and then, by suitably color coding the surface they proposed a method to visualize the roughness profile of protein surfaces. A recent work[22], employing SAFD methodology[13], has investigated interfaces of obligate and non-obligate protein-protein complexes. Table I given below enlists the prominent works in the evolution of FD-approaches to study protein structures.

FD-based protein study is still in its nascent state and a lot of unknown information[23] is yet to be unearthed by employing this method effectively. Purpose of the present review was to underline the necessity of using it, and also, to demonstrate the future potential of usage in the field of protein's interior and exterior characterization.

Acknowledgments

A. Banerji would like to thank COE-DBT, Govt. of India, for his scholarship.

References

1. Karplus M, *J. Phys. Chem.* **B 104**, 11 (2000).
2. Goetze T, Brickmann, J, *Biophys. J.* **61**,109 (1992).
3. Phillips JC, *Proc Natl Acad Sci USA* **106**, 3107 (2009).
4. De Leeuw M, Reuveni S, Klafter J, Granek R, *PLoS ONE* **4**, 7296 (2009).
5. Banerji A, Ghosh I , *Eur. Biophys. J.* **38,**577 (2009).
6. Mandelbrot BB, The Fractal Geometry of Nature. (W. H. Freeman and Company, San Francisco). (1982).
7. Hausdorff F, *Mathematische Annalen* **79**, 157 (1919).
8. Falconer K, Fractal Geometry: Mathematical Foundations and Applications. (John Wiley & Sons. New York) (1990).
9. Havlin S, Ben-Avraham D, *J. Phys. A.* **15**, L311 (1982).
10. Tanford C Physical Chemistry of Macromolecules. (Wiley, New York) (1961).
11. Banerji A, Ghosh I, *Cell. Mol. Life. Sc.* **68**, 2 711 (2011).
12. Lewis M, Rees DC, *Science.* **230**, 1163 (1985).
13. Pettit FK, Bowie JU, *J Mol Biol* **285**, 1377 (1999).
14. Buuren ARV, Characterization of Oil/Water Interfaces: A Molecular Dynamics Study; PhD thesis, University of Groningen. (1996).
15. Aqvist J, Tapia O, *J Mol Graph* **5**, 30 (1987).
16. Stapleton HJ, Allen JP, Flynn CP, Stinson DG, Kurtz SR , *Phys. Rev. Lett.* **45**, 1456 (1980).
17. Enright MB, Leitner DM , *Phys. Rev. E* 71 011912 (2005).
18. Yu X, Leitner DM, *J. Chem. Phys.* **119**, 12673 (2003).
19. Banerji A, Ghosh I, *PLoS ONE* **4**, 7361 (2009).
20. Laskowski R, Luscombe N, Swindells M, Thornton *J. Protein Sci.* **5**, 2438 (1996).
21. Chakravarty S, Wang L, Sanchez R, *Nucl Acid Res* **33**, 244 (2005).
22. Bera I, Ray S, *Bioinformation* **4**, 210 (2009).
23. Luisa Di Paola, Paola Paci, Daniele Santoni, De Ruvo, and Alessandro Giuliani, *J. Chem. Inf. Model.* **52**, 474 (2012).

THE INHERENT STRUCTURE LANDSCAPE OF MET-ENKEPHALIN DETERMINED BY THE MOLS TECHNIQUE

N. BALAJI, L. RAMYA AND N. GAUTHAM

C.A.S. in Crystallography and Biophysics, University of Madras
Maraimalai Campus (Guindy)
Chennai 600025, India

We have developed a method, called the MOLS method (for mutually orthogonal Latin squares) to exhaustively explore the conformational space of small peptides. Here we used this method to delineate the inherent structure (IS) landscape of the pentapeptide Met-Enkephalin at and above 'room temperature'. Inherent structures have been defined as structures located at the minima of the energy landscape. From the IS landscape we extracted the probability density and calculated the density of states. To compare the results, we have also sampled the conformations for the peptide through molecular dynamics simulations and periodic rapid quenching. The probability density distributions followed a similar pattern in both methods.

1. Introduction

The energy landscape of a protein or a peptide is a map of the potential energy of the molecule in its N-dimensional conformational space.[1,2] Here N is the number of conformational degrees of freedom, such as the variable torsion angles, that the molecule may possess. The energy landscapes of proteins have been studied extensively.[3-6] The concept of energy landscapes has helped to understand the transition states of the molecule of interest,[7] and the pathway, along with branch points, of transition from one low energy conformation to another.[1] Energy landscapes also indicate the high energy saddle points and the basins of attraction of the local minima, also called stationary points.[8-11]

Calculation of the complete and detailed energy landscape of even a small protein is a task that requires enormous computer resources.[1] Such a project would require the computation of the potential energy associated with every possible point in the N-dimensional conformational space. The number of such points grows as an exponential function of the number of dimensions (*i.e.* the number of conformational variables).[12]

However, it has been argued and demonstrated[13] that much information regarding structural and thermodynamic properties of the molecule may be obtained from the 'inherent structure' landscape, which is a subset of the energy landscape. Inherent structures (IS) are the structures located at the local minima of the energy landscape. Though only partial, such a view of the landscape can yield information about the folding and fluctuations in the structures relevant to their functional activity.[2,8] The delineation of the IS landscape can also lead to information about the thermodynamics of the molecule, such as the density of states.[14-16] The IS landscape has been obtained using different methods. Rao and Karplus[17] have computed the IS of the GS10 peptide using molecular dynamics quenching. Though this is the only technique that has been used so far to determine the IS landscape of biologically relevant molecules, other techniques have been used in material science. For example, Mauro *et al.*[18] used the self-consistent Monte Carlo method to compute the probability distribution of the discrete structures of selenium.

In this paper we describe the application of the MOLS technique to map the IS landscape of Met-enkephalin. This technique has been developed in our laboratory to rapidly and exhaustively explore the conformational space of a molecule and identify all its low energy structures.[19,20] For comparison purposes we also plot the IS landscape of the molecule obtained using molecular dynamics quenching, the method used earlier for the GS10 peptide.[17]

2. Methods

2.1 The MOLS method

The MOLS method, developed in our laboratory to exhaustively explore the conformational space of oligopeptides in reasonable computational time, has been extensively described previously by Vengadesan and Gautham[19,20] (http://www.unom.ac.in/Gautham/ mols.pdf). Recently[21] we have shown that for equivalent computational time, it outperforms sampling by Molecular Dynamics quenching, as well as Monte Carlo sampling, in the extent of the conformational space accessed. Here we used the MOLS method to generate a set of 1500 low-energy structures. We identify this set of low energy structures as the complete set of inherent structures of the peptide and use them to construct its IS landscape. The conformational energy of the molecule was calculated using the CHARMM22 potential.[22]

One of the aims of analyzing the IS landscape is to obtain the 'folding' temperature, i.e. the temperature at which the structures of the peptide or protein are equally distributed between a low energy population and another high energy one.[13] Such calculations have been carried out using molecular dynamics quenching by Rao and Karplus,[17] for the molecule GS10 peptide. Since the main purpose of the present study was to see how the MOLS method, which gives a more exhaustive sample of the conformational space,[19] would perform this task, and if it would yield more detailed results, we restricted ourselves to obtaining the IS landscape at four different temperatures. These temperatures were chosen to be at and above the room temperature. Experimental structures of the peptide at room temperature assume both well-folded[23] and extended[24] conformations. Thus the IS landscape at the room temperature and above it should show a shift from one conformational population to another, reflected as a shift between two populations centered on two different energy values.

The method was therefore used to generate four sets of 1500 structures for the molecule Met-enkephalin, each at a different temperature, set as T = 290K, 300K, 310K and 320K. These four sets were used to represent the IS landscapes of the peptide at the four different temperatures.

2.2 *Molecular Dynamics (MD) Simulation*

In order to compare the IS landscape obtained by the above method with one obtained using the molecular dynamics procedure reported earlier, we adapted the technique described by Nakagawa and Peyrard[13] and used it to generate IS landscapes of Met-enkephalin at the same four temperatures as above, i.e., T = 290K, 300K, 310K and 320K. We used the NAMD program package[25] to perform a 7.5 nanosecond molecular dynamics run on Met-enkephalin. The conformational space trajectory was sampled at intervals of 5 picoseconds, and the structure at every such interval was directly quenched by energy minimization. Thus these 1500 minimum-energy structures represent the local minima in the energy landscape as obtained at T_R = 290 K. The simulations were repeated at three other temperatures T = $1.03T_R$, $1.07T_R$ and $1.10T_R$ (i.e. T = 300 K, 310 K and 320 K). These four sets of 1500 structures each formed the basis for description of the inherent structure landscape of the molecule.

All computations were carried out on a cluster computer of Opteron processors at 1.8 GHz. Different numbers of processors were used at different stages of the calculations. Normalizing to a single dual core processor, it took 47 minutes to generate one inherent structure at a given specific temperature using the MOLS method, and 5 minutes to generate and to quench a structure using molecular dynamics simulation.

3. Results and Discussion

The 1500 structures generated at the four different temperatures, T = $1.0T_R$, $1.03T_R$, $1.07T_R$ and $1.10T_R$ were used to plot the probability distribution functions (pdf) for the molecule. Figure 1 displays the pdf in IS landscape generated using the MOLS method. Figure 2 shows the pdf derived using MD quenching. These figures were obtained as follows. The energies of the structures in each sample were normalized so that the energies are all positive, and brought on to the same scale. The structures were binned by energy, with the position and width of each energy bin being decided by $v(\mathbf{r}) = e_\alpha + \Delta v(\mathbf{r})$. Here $v(\mathbf{r})$ is the maximum energy value in the bin, e_α is the lowest energy in each bin, called the 'basin of attraction', and $\Delta v(\mathbf{r})$ is the width of the bin, here chosen as 0.5 k_BT.

Figure 1 shows a bimodal distribution of structures at room temperature with the peaks at 22 and 115 k_BT. As the temperature is raised the distribution becomes more and more unimodal, until the low energy population almost disappears at 1.1 T_R. This behavior is repeated in the structures generated by molecular dynamics sampling as shown in Figure 2. The differences between the two sampling techniques are in the relative heights of the peaks. Thus both techniques show that at room temperature the molecule exists with significant

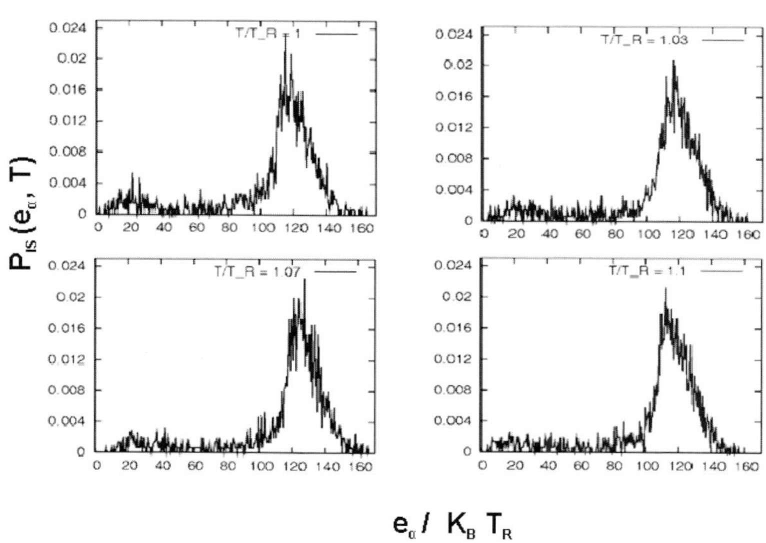

Figure 1: Probability distribution of the inherent state energies at T_R and above it in three steps derived using MOLS method.

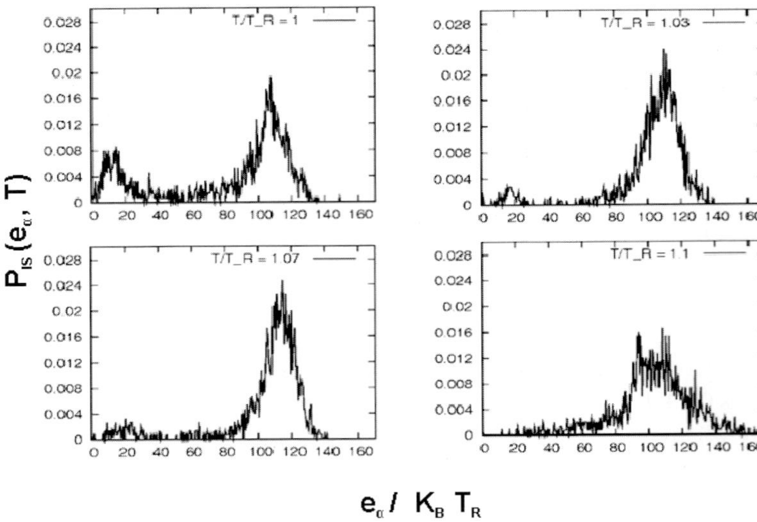

Figure 2: Probability distribution of the inherent state energies at T_R and above it in three steps derived using molecular dynamics method.

probability in a low energy form, and with greater probability, in a higher energy conformation. The low energy conformation may be identified with the folded form of the molecule (Figure 3a), while the higher energy form is identified with an extended structure (Figure 3b). Both types of structures have been noticed in experiments at room temperature.[23,24] The global energy minimum identified computationally in vacuum is a folded structure.[26]

As the temperature is raised to above the room temperature, the probabilities of the structure being folded, i.e. the low energy conformation almost disappears, while the numbers of structures at higher energies, i.e. the more extended conformations, increase. This is true in both the MOLS generated structures as well those generated by MD quenching.

Apart from the energies e_α, the density of states $\Omega_{IS}(e_\alpha)$ may be used to characterize the inherent structure landscape of a peptide. Owing to the very large number of possible states, it is not easy to describe the density of states of even a small peptide by exhaustive or systematic search. Thus we used the probability distributions described above to obtain the density of states of Met-Enkephalin at room temperature. Figure 4 shows the density of states $\Omega_{IS}(e_\alpha)$ obtained in this manner. We used the formulation of Nakagawa and Peyrard.[13] The probability $P_{IS}(e_\alpha,T)$ of occurrence of a structure (or state) in a particular basin of attraction labeled as $e_\alpha$at the temperature T is given by

$$P_{IS}(e_\alpha,T)=p_0(T)\Omega_{IS}(e_\alpha)e^{-\beta e_\alpha} \tag{1}$$

where $p_0(T)$ is the probability of occurrence of a structure in the ground state basin of attraction($e_{\alpha 0}=0$). β is 1/kT, where k is the Boltzmann's constant and the T is the absolute temperature. Thus the density of states $\Omega_{IS}(e_\alpha)$ is given as

$$\Omega_{IS}(e_\alpha)=e^{\beta e_\alpha}\frac{P_{IS}(e_\alpha,T)}{p_0(T)} \tag{2}$$

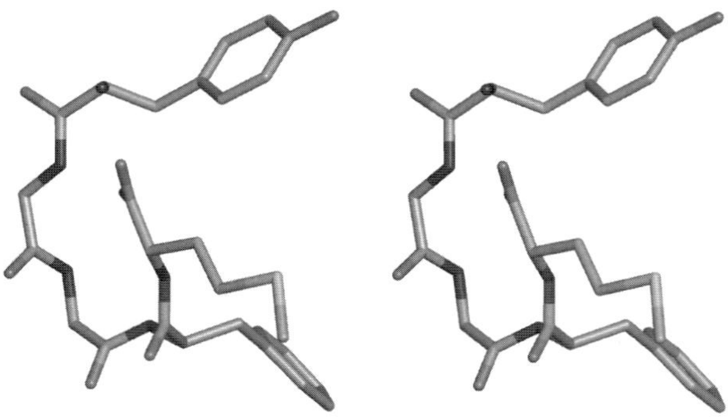

a)

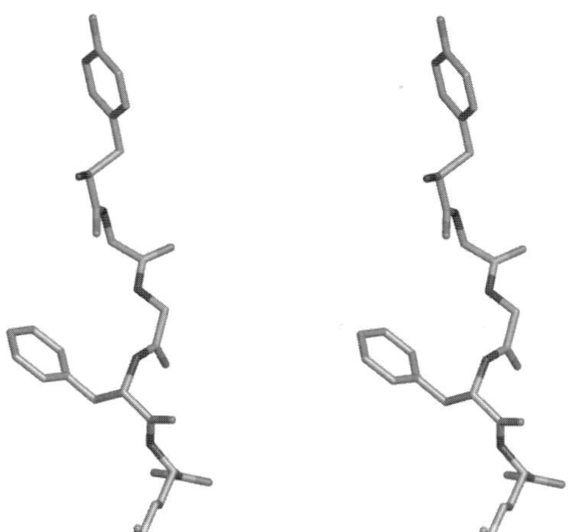

b)

Figure 3: a) The folded conformation (energy 22 k_BT), and b) the extended conformation (energy 115 k_BT) for Met-enkephalin. These examples have been obtained from the MOLS generated structures at T_R.

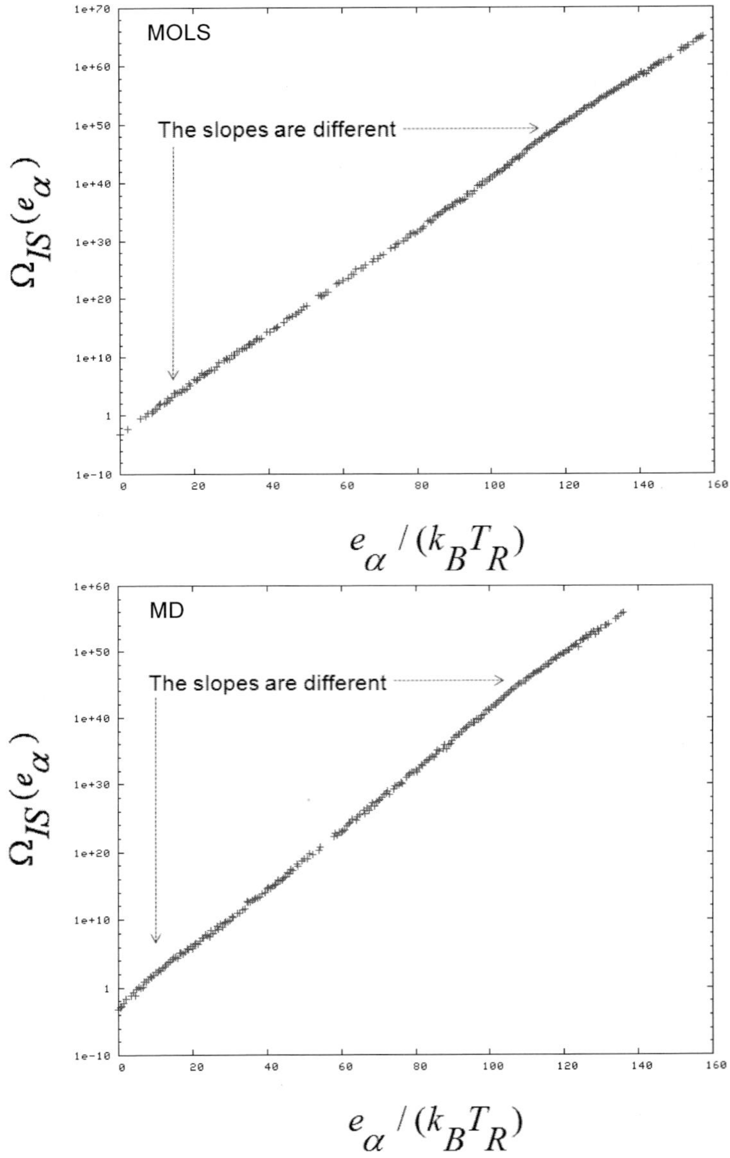

Figure 4: Density of states Ω_{IS} (e_α) according to Eq. 2 for the MOLS and the molecular dynamics (MD) generated energies shown at top and bottom respectively.

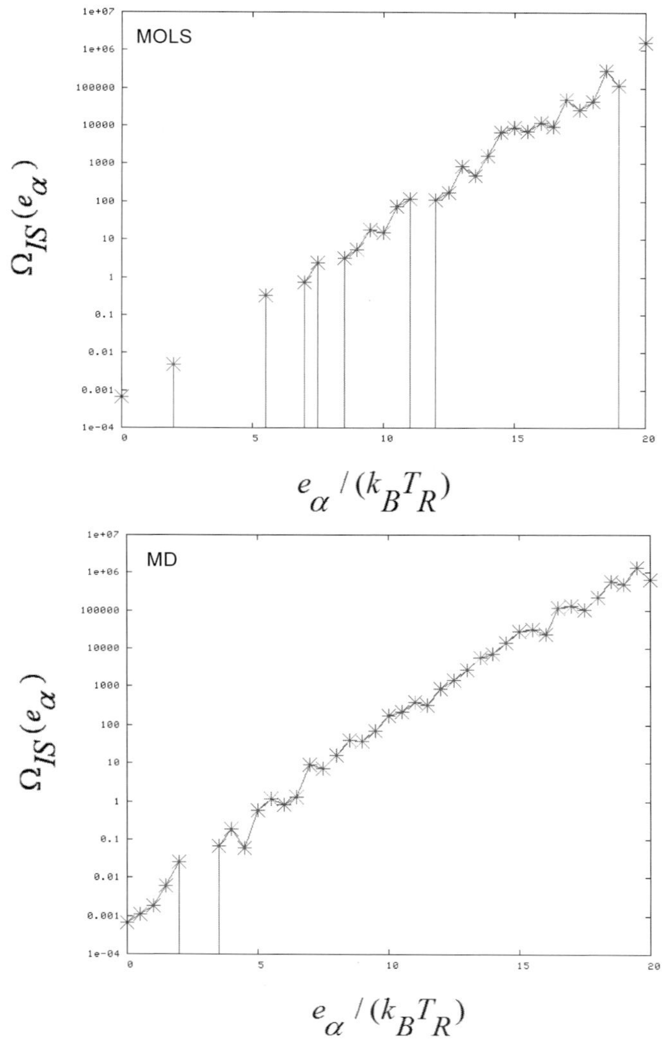

Figure 5: Density of states Ω_{IS} (e_α) of the inherent structures for the energies close to the ground state. Which is obtained from $P_{IS}(e_\alpha,T)$ for the MOLS and the Molecular dynamics generated energies shown at top and bottom respectively.

$P_{IS}(e_\alpha,T)$ and $p_0(T)$ are obtained from Figure 1 and Figure 2 for the two sets of calculations, i.e. using MOLS and molecular dynamics for each set respectively. (Apart from the potential energy represented by e_α, there is also contribution from the vibrational free energy within the basin of attraction of the inherent structure. Following Nakagawa and Peyrard,[13] we assumed that this energy does not significantly vary from one basin to another, and would therefore only add a constant factor to the above calculations). Figure 4 shows the density of states derived from the MOLS and molecular dynamics method at the room temperature T_R. Overall the two sets of results agree with each other. The distribution increases nearly linearly with temperature. However, in both the MOLS generated distribution and the MD generated one, there is a slight butdistinct deviation from linearity in two regions, corresponding, respectively, to the low energy folded conformations and the higher energy extended conformations. The slopes of the lines are thus different below and above the energy value of 60 k_BT.

The figures also indicate clearly that the density of states is low for the low energy conformations as compared to the high energy conformations. This implies that the low energy basins of attractions have much fewer associated substates than the ones at high energy. In other words, there are much fewer low energy conformations accessible to the molecule, while several different conformations could lead to the same high value of energy. This result is in line with the expectations of the 'folding funnel',[6,27-30] in which the folding process starts from extended high energy structures and proceeds towards folded lower and lower energy states which are more and more sparsely populated.

Figure 5 displays a magnified view of $\Omega_{IS}(e_\alpha)$ in the lower energy regions. The MOLS results yield distinct, well separated sets of substates at the ground state energy and the nearby energy states. The MD results show significantly different populations of structures. The energy substates at the ground state are identified with different possible 'native' conformations.[31,32] Thus for Met-enkephalin, the MOLS results show a few well separated basins of attraction which could form the near-native states of the molecule.

Acknowledgments

We thank the UGC India, for providing Research Fellowships in Science for Meritorious Students to N. Balaji and L. Ramya, and for providing support under the CAS programme. We also thank the DST India for support under FIST program.

References

1. D. J. Wales, *Energy Landscapes,* (Cambridge University Press, Cambridge 2003).
2. F. H. Stillinger and T. A. Weber, *Science,* **225**, 983 (1984).

3. J. D. Bryngelson and P. G. Wolynes, *Proc. Natl. Acad. Sci. U S A,* **84**, 7524 (1987).
4. J. D. Bryngelson and P. G. Wolynes, *J. Phys. Chem.* **93**, 6902 (1989).
5. J. D. Bryngelson, J.N. Onuchic, N. D. Socci and P. G. Wolynes, *Proteins,* **21**, 167 (1995).
6. J. N. Onuchic, Z. Luthey-Schulten and P. G. Wolynes, *Annu. Rev. Phys. Chem.* **48,** 545 (1997).
7. M. C. Prentiss, D. J. Wales, and P. G. Wolynes, *PLoS. Comput. Biol.* **6**, 1 (2010).
8. F. H. Stillinger and T. A. Weber, *Phys. Rev. A.* **25**, 978 (1982).
9. F. Sciortino, W. Kob and P. Tartaglia, *Phys. Rev. Lett.* **83**, 3214 (1999).
10. O. M. Becker and M. Karplus, *J. Chem. Phys.* **106**, 1495(1997).
11. M. T. Oakley, D. J. Wales and R. L. Johnston, *J. Phys. Chem. B.* **115**, 11525 (2011).
12. V. Kanagasabai, J. Arunachalam, P. Arun Prasad and N. Gautham, *Proteins: Struct. Funct. Bioinf.* **67**, 908 (2007).
13. N. Nakagawa and M. Peyrard, *Proc. Natl. Acad. Sci. U S A*, **103**, 5279 (2006).
14. J. Kim and T. Keyes, *J. Phys. Chem. B.* **111**, 2647 (2007).
15. A. K. Sangha and T. Keyes, *J. Phys. Chem. B.* **114**, 16908 (2010).
16. F. Rao, *J. Phys. Chem.* **1**, 1580 (2010).
17. F. Rao and M. Karplus, *Proc. Natl. Acad. Sci. U S A.* **107**, 9152 (2010).
18. J. C. Mauro, R. J. Loucks, J. Balakrishnan and S. Raghavan, *J. Chem. Phys.* **126**, 194103 (2007).
19. K. Vengadesan and N. Gautham, *Biophys. J.* **84**, 2897 (2003).
20. K. Vengadesan and N. Gautham, *Biopolymers*, **74**, 476 (2004).
21. L. Ramya and N. Gautham, *Biopolymers,* **97**, 165 (2012).
22. Jr., A. D. MacKerell, D. Bashford, M. Bellott, Jr., R. L. Dunbrack, J. D. Evanseck, M. J. Field, S. Fischer, J. Gao, H. Guo, S. Ha, D. Joseph-Mc Carthy, L. Kuchnir, K. Kuczera, F. T. K. Lau, C. Mattos, S. Michnick, T. Ngo, D. T. Nguyen, B. Prodhom, III., W. E. Reiher, B. Roux, M. Schlenkrich, J. C. Smith, R. Stote, J. Straub, M. Watanabe, J. Wiorkiewicz-Kuczera, D. Yin and M. Karplus, *J. Phys. Chem. B.* **102,** 3586 (1998).
23. I. Marcotte, F. Separovic, M. Auger and S. M. Gagne, *Biophys. J.* **86**, 1587 (2004).
24. J. F. Griffin, D. A. Langs, G. D. Smith, T. L. Blundell, I. J. Tickle and S. Bedarkar, *Proc. Natl. Acad. Sci. U S A.* **83**, 3272 (1986).
25. J. C. Phillips, R. Braun, W. Wang, J. Gumbart, E. Tajkhorshid, E. Villa, C. Chipot, R. D. Skeel, L. Kale and K. Schulten, *J. Comput. Chem.* **26**, 1781 (2005).
26. Y. Isogai, G. Nemethy and H. A. Scheraga, *Proc. Natl. Acad. Sci. U. S. A.* **74**, 414 (1977).
27. K. A. Dill and H. S. Chan, *Nat. Struct. Biol.* 4, 10 (1997).

28. D. J. Wales, J. P. K. Doye, M. A. Miller, P. N. Mortenson and T. R. Walsh, *Adv. Chem. Phys.* 115, 1 (2000).
29. S. S. Plotkin and J. N. Onuchic, *Q. Rev. Biophys.* 35, 111 (2002a).
30. S. S. Plotkin and J. N. Onuchic, *Q. Rev. Biophys.* 35, 205 (2002b).
31. A. J. Bray and M. A. Moore, *J. Phys. C: Solid State Phys.* 14, 1313 (1981).
32. H. Frauenfelder, F. Parak and R. D. Young, *Annu. Rev. Biophys. Biophys. Chem.* 17, 451 (1988).